METAL SURFACE TREATMENT

METAL SURFACE TREATMENT

Chemical and Electrochemical Surface Conversions

Edited by M.H. Gutcho

NOYES DATA CORPORATION
Park Ridge, New Jersey, U.S.A.
1982

Library of Congress Catalog Card Number: 82-3404
ISBN: 0-8155-0900-6
ISSN: 0198-6880
Printed in the United States

Published in the United States of America by
Noyes Data Corporation
Mill Road, Park Ridge, New Jersey 07656

10 9 8 7 6 5 4 3 2 1

Library of Congress Cataloging in Publication Data
Main entry under title:

Metal surface treatment.

(Chemical technology review, ISSN 0198-6880 ;
no. 208)
"Based on U.S. patents, issued from January
1977 to November 1981, that deal with metal
surface treatments"--Foreword.
Includes index.
1. Metals--Finishing--Patents. 2. Surface
preparation--Patents. 3. Protective coatings--
Patents. I. Gutcho, Marcia H. (Marcia Halpern),
1924- . II. Series.
TS213.M39 671.7'3 82-3404
ISBN 0-8155-0900-6 AACR2

FOREWORD

The detailed, descriptive information in this book is based on U.S. patents, issued from January 1977 to November 1981, that deal with metal surface treatments. These processes were developed for the purpose of increasing corrosion resistance, to provide a more receptive surface for application of various coatings, and to apply decorative finishes.

This book is a data-based publication, providing information retrieved and made available from the U.S. patent literature. It thus serves a double purpose in that it supplies detailed technical information and can be used as a guide to the patent literature in this field. By indicating all the information that is significant, and eliminating legal jargon and juristic phraseology, this book presents an advanced commercially oriented review of recent developments in the field of metal surface conversion coatings.

The U.S. patent literature is the largest and most comprehensive collection of technical information in the world. There is more practical, commercial, timely process information assembled here than is available from any other source. The technical information obtained from a patent is extremely reliable and comprehensive; sufficient information must be included to avoid rejection for "insufficient disclosure." These patents include practically all of those issued on the subject in the United States during the period under review; there has been no bias in the selection of patents for inclusion.

The patent literature covers a substantial amount of information not available in the journal literature. The patent literature is a prime source of basic commercially useful information. This information is overlooked by those who rely primarily on the periodical journal literature. It is realized that there is a lag between a patent application on a new process development and the granting of a patent, but it is felt that this may roughly parallel or even anticipate the lag in putting that development into commercial practice.

Many of these patents are being utilized commercially. Whether used or not, they offer opportunities for technological transfer. Also, a major purpose of this book is to describe the number of technical possibilities available, which may open up profitable areas of research and development. The information contained in this book will allow you to establish a sound background before launching into research in this field.

Advanced composition and production methods developed by Noyes Data are employed to bring these durably bound books to you in a minimum of time. Special techniques are used to close the gap between "manuscript" and "completed book." Industrial technology is progressing so rapidly that time-honored, conventional typesetting, binding and shipping methods are no longer suitable. We have bypassed the delays in the conventional book publishing cycle and provide the user with an effective and convenient means of reviewing up-to-date information in depth.

The table of contents is organized in such a way as to serve as a subject index. Other indexes by company, inventor and patent number help in providing easy access to the information contained in this book.

16 Reasons Why the U.S. Patent Office Literature Is Important to You

1. The U.S. patent literature is the largest and most comprehensive collection of technical information in the world. There is more practical commercial process information assembled here than is available from any other source. Most important technological advances are described in the patent literature.

2. The technical information obtained from the patent literature is extremely comprehensive; sufficient information must be included to avoid rejection for "insufficient disclosure."

3. The patent literature is a prime source of basic commercially utilizable information. This information is overlooked by those who rely primarily on the periodical journal literature.

4. An important feature of the patent literature is that it can serve to avoid duplication of research and development.

5. Patents, unlike periodical literature, are bound by definition to contain new information, data and ideas.

6. It can serve as a source of new ideas in a different but related field, and may be outside the patent protection offered the original invention.

7. Since claims are narrowly defined, much valuable information is included that may be outside the legal protection afforded by the claims.

8. Patents discuss the difficulties associated with previous research, development or production techniques, and offer a specific method of overcoming problems. This gives clues to current process information that has not been published in periodicals or books.

9. Can aid in process design by providing a selection of alternate techniques. A powerful research and engineering tool.

10. Obtain licenses–many U.S. chemical patents have not been developed commercially.

11. Patents provide an excellent starting point for the next investigator.

12. Frequently, innovations derived from research are first disclosed in the patent literature, prior to coverage in the periodical literature.

13. Patents offer a most valuable method of keeping abreast of latest technologies, serving an individual's own "current awareness" program.

14. Identifying potential new competitors.

15. It is a creative source of ideas for those with imagination.

16. Scrutiny of the patent literature has important profit-making potential.

CONTENTS AND SUBJECT INDEX

INTRODUCTION

In the field of metal surface treatment, there is a recognized need for treating the surface of both ferrous and nonferrous metals for improved corrosion resistance and paint adherence.

Protective coatings (surface conversion coatings) have been applied by chemical or electrochemical means, whereby the chemically active metal surface, subject to oxidation, is converted to one which is less active and more resistant to corrosion. The conversion coating serves as a substrate for the subsequent bonding of metal to paint or other organic finishes. The corrosion resistance of the painted article can be measured by humidity, salt-spray and other standardized corrosion tests.

One of the most widely used methods of providing such surface protection is phosphatizing. The phosphatizing of ferrous metal surfaces is a common practice.

In the phosphatizing process, the metal surface is rendered inactive by the formation or deposition of adherent, compact, insoluble, crystalline phosphate salts on the metal surface. The usual phosphate coatings, which are applied from aqueous phosphating solutions, are iron, zinc, zinc-calcium and manganese phosphates. Recent advances in phosphatizing described in this book cover these and other metal phosphate coating solutions, parameters of the phosphating bath, accelerators, phosphatizing techniques and apparatus, prephosphating treatments as well as means of controlling sludge formation common in acid phosphating baths. It has been common practice to seal the surface of phosphated metal components with a chromic acid rinse prior to painting. Since hexavalent chromium is toxic and environmentally undesirable, postphosphatizing substitutes for chromic acid, which are nontoxic and yet provide the necessary corrosion protection, are of particular importance. Of interest also are phosphatizing methods which have been developed where nonaqueous solvents are used as the liquid medium.

Chromate conversion coatings have been employed to impart brightness and improve the corrosion resistance of bare metal and as a substrate to provide improved paint adherence on aluminum, zinc, steel, magnesium and zinc-plated surfaces. These coatings are primarily obtained by immersion techniques, but they can also be obtained using electrochemical processes. A variety of chromate treatments are disclosed, using hexavalent chromium and trivalent chromium, as well as resinous coatings containing chromate.

In an ordinary environment, some metals like aluminum form a natural protective film. However, under more severe corrosive conditions this natural oxide coating is too thin to offer protection. Anodic oxidation of aluminum produces a hard corrosion-resistant oxide layer which increases the wear resistance of the surface of aluminum. The processes in this book relating to anodic oxidation involve additives for the electrolyte bath, and techniques for the sealing of anodized aluminum. Included also for the surface treatment of aluminum are processes for improved weathering, coatings for short-term protection, and a large variety of new types of coating compositions designed to improve paint receptivity and corrosion resistance.

A large number of processes in this book relate to ferrous metals, especially steel. These include treatments to form oxide and carbide surface layers of improved corrosion resistance, processes for handling rust, for improving subsequent coating, and for special situations in processing steel sheets and the manufacture of steel products.

Other chapters are devoted to the surface treatment of metal equipment and parts, to metal coloring, to zinc and zinc-plated metal, and to additional types of metal surface treating processes including composite coatings and organic coatings for imparting desirable characteristics.

PHOSPHATIZING

ZINC PHOSPHATE COATINGS

Controlled Ratio of Zinc to Nitrate Ions

Conventional zinc phosphate coating solutions generally contain zinc ions, alkali metal ions such as sodium and one or more of metal ions such as nickel, calcium, manganese, copper, iron ions and the like as cations, and phosphate ions and one or more of nitrate, nitrite, chlorate ions and the like as anions. Amounts of those ions may be varied depending mainly upon the desired characteristics of phosphate coatings and the conditions for metal surface treatment and of equipment therefor and/or the conditions for zinc phosphate coating processes.

The work of *S. Oka, T. Sobata, T. Kishimoto and K. Nobe; U.S. Patent 4,053,328; October 11, 1977; assigned to Nippon Paint Co., Ltd., Japan* relates to a zinc phosphate coating solution which is capable of being maintained at a relatively high pH value in forming zinc phosphate coatings on metal surfaces. Such zinc phosphate coatings possess sufficient coating properties as a paint base. Furthermore, the coating solution can serve in reducing the consumption of nitrites which are employed as accelerators for phosphating processes. The employment of this coating solution can also function to transform sludges which are formed as by-products in phosphate coating processes into a form easily removable and less sticky and at the same time reduce amounts of such sludges formed.

The zinc phosphate coating solution contains phosphate and zinc ions in particular amounts and having particular pH ranges and ratios of phosphate ion to nitrate ion. It is necessary that the phosphate ions be present in an amount of at least 0.5 wt %, preferably from 0.5 to 10 wt %, with respect to the coating solution and the molar weight ratio thereof to nitrate ion ranges from 1:0.7-1:1.3. The zinc ions are present preferably in an amount of at least 0.03 wt % based on the solution, preferably from 0.03 to 0.80 wt %, and in the molar weight ratio thereof to phosphate ion of from 0.116:1-0.005:1.

The zinc phosphate coating solutions also contain nitrate, nitrite and alkali metal ions, and/or ammonium ions. The presence of nickel ions is preferred. In this case, it is desired that the solution contain nickel ions in an amount of at least 0.01 wt %, preferably from 0.01 to 1.35 wt %, with respect to the solution and the molar weight ratio of nickel ion to zinc ion may vary from 1.89:1-0.014:1.

The zinc phosphate coating solution of the process is particularly applicable to an apparatus of the type designed to drain off little or no coating solution out of the phosphate coating system as that disclosed and claimed in U.S. Patent 3,906,895.

Figure 1.1 is a schematic plan view of the phosphating and subsequent water rinse stations of the apparatus applicable to the process.

Figure 1.1: Phosphating and Water Rinse Apparatus

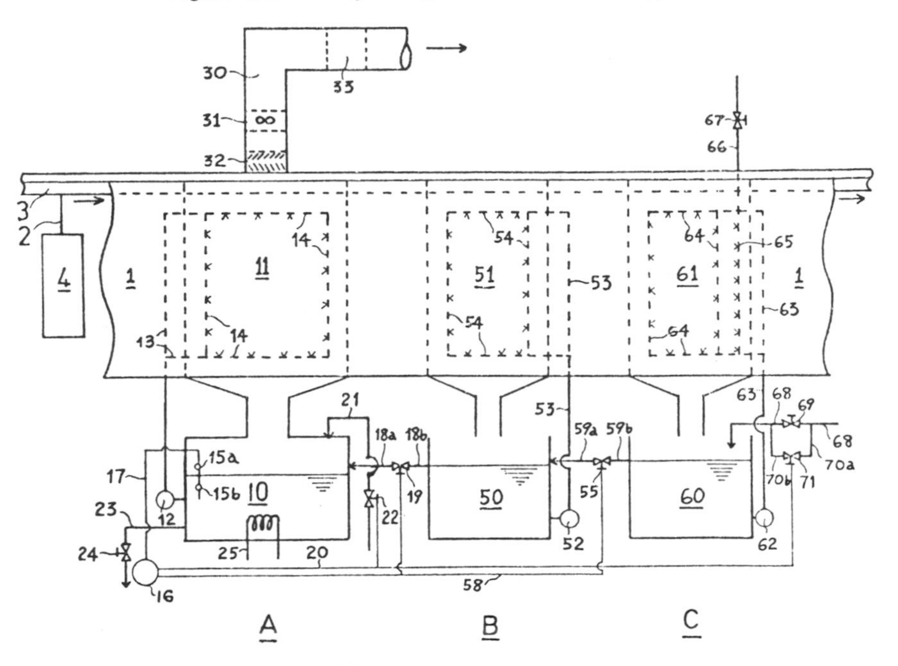

Source: U.S. Patent 4,053,328

A phosphate coating station is at station **A**; a first water rinse station is at station **B**; and a second water rinse station is at station **C**. The stations for the phosphating system as well as the subsequent water rinsing may be arranged in a treating chamber or tunnel **1** through which the metal object is treated during the passage through the treating chamber while being suspended from a plurality of work racks or hangers **2** provided in an endless horizontal conveyor **3** extending along the top and length of the treating chamber.

The phosphating station comprises a tank **10** and a spray chamber **11**, the tank incorporating a pump **12** therein which can pump solution from the tank up through a supply line **13** to a plurality of spray headers **14**, having a plurality of nozzles (not shown) so as to discharge the zinc phosphate coating solution against the surfaces of the metal object in all possible directions.

The tank **10** is provided with a pair of liquid level detectors comprising a maximum liquid level detector **15a** and a minimum liquid level detector **15b** for detecting the level of solution present in the tank **10**.

When the solution level in the tank goes outside the predetermined levels, the involved liquid level detector **15a** or **15b** operates to transmit a signal to a liquid level control **16**. The liquid level control which is connected by a circuit **17** to the liquid level detectors is designed such that it can control the amount of solution in the tank at a suitable level in response to the signal transmitted by either of the liquid level detectors **15a** or **15b** by adjusting the supply of fresh water and/or a water rinse solution from water rinse station at **B** and, at the same time, a water rinse solution to the station at **B** from station **C**.

The metal object **4** is then transferred to a next step at first water rinse station **B**. The first rinse station at **B** is composed of a tank **50** and a spray chamber **51**. The overflow of the water rinse solution from the tank **50** is supplied to the tank **10** through a discharge line **18b**. A solenoid valve **19** is provided in the discharge line and actuated through a circuit **58** from the liquid level control **16**, thereby adjusting the overflow of the rinse solution from the tank. The rinse solution may also be supplied to the tank **10** by spraying the rinse solution against the metal object and draining it thereinto. The spraying of the rinse solution may be made in combination with the overflowing of the rinse solution. In the tank **50** is a supply line **59a** through which the overflow of the rinse solution is supplied thereto from a tank at station **C**. The second water rinse station **C** has substantially the same structure as the first water rinse station **B**.

In the spray chamber **61** is provided another set of spray headers **65** having a plurality of nozzles (not shown) which are connected by a supply line **66** to a suitable source of fresh water, thereby fully rinsing with water the surfaces of the metal object advancing through the treating chamber. In the tank **60** is provided a supply line **68** through which fresh water is supplied thereto from a suitable source. The supply line **68** is provided with a valve **69** and diverted before the valve **69** through a branch line **70a** which is connected by a solenoid valve **71** to a branch line **70b** which in turn is combined again with the supply line **68** after the valve **69**. The solenoid valve **71** is actuated through a circuit **20** by the liquid level control **16**, thereby controlling an amount of fresh water to be supplied to the tank **60**.

A discharge line **59b** is provided in the tank **60** so as to have the excess of water rinse solution overflow to the tank **50** at station **B**. The overflow is controlled by a solenoid valve **55** which is actuated through the circuit **58** by the control **16**. The water rinse solution in the tank at station **C** may also be supplied to the tank **50** by spraying the rinse solution against the metal surfaces and draining the sprayed rinse solution into the tank at **B**. The supply of the rinse solution by spraying may be made in combination with the overflow thereof and can be controlled in substantially the same manner as with the overflow thereof. The metal object thus rinsed may then be advanced to a drying station where the metal object is dried in conventional manner for further processing.

Example: With a six-station apparatus having the phosphating system described, a cold rolled steel plate was subjected to zinc phosphate coating. The steel plate was first sprayed with a solution containing 1% of a weakly alkaline degreasing agent "Ridoline No. 1006N-1" (Nippon Paint Co., Ltd.) having a temperature of

50° to 60°C for 2 min and then transferred to water rinse stations. After being rinsed with water at room temperature twice, each for 1 min, the steel plate was then subjected to zinc phosphate coating at the phosphating station in which the plate was sprayed with a zinc phosphate coating solution having a pH value of 3.4 and the following composition at 50° to 55°C for 2 min.

Composition	Percent by Weight
PO_4^{3-}	1.10
NO_3^-	0.70
Zn^{2+}	0.05
Na^+	0.45
NO_2^-	0.01-0.015

This solution had the phosphate ion to nitrate ion ratio of 1:0.967. A sodium nitrite aqueous solution and an aqueous concentrate containing ions such as phosphate, nitrate and zinc ions necessary to maintain the desired composition of the zinc phosphate coating solution was replenished to the coating bath. The steel plate thus phosphate-coated was then rinsed twice with water at room temperature each for 1 min and dried.

It was found that the zinc phosphate coating solution was maintained within the range of pH 3.4±0.1.

The phosphate coating formed on the surface of the steel plate was found to be a good, uniform and fine coating layer. During the phosphate coating process, the amount of sodium nitrite consumed was 1.5 g/m² of the treating steel plate area. In the bottom of the tank in which the phosphate coating was effected, 20 g/m² of sludges (water content, 90%) were sedimented, and these sludges did not adhere to the tank and were in a form that was removed with ease.

It is to be noted that in this example, nitrate and sodium ions present in the solution showed little tendency to increase, and the phosphating solution used for 3 mo in which 800 m² of the steel plate per day were employed still maintained a good phosphating capacity. The steel plate thus treated had still a good phosphate coating thereon.

Comparative Example: With an apparatus having a conventional six-station construction, the procedure of the above Example was repeated using a conventional zinc phosphate coating solution having a pH value of 3.0 and the following composition.

Composition	Percent by Weight
PO_4^{3-}	1.50
NO_3^-	0.60
Zn^{2+}	0.16
Na^+	0.43
NO_2^-	0.013-0.016

The phosphate ion to nitrate ion molar weight ratio of this solution was 1:0.61. Although the steel plate thus treated possessed a good phosphate coating film thereon, the amount of sodium nitrite consumed was 3.0 g/m², the amount of sludges formed was 60 g/m² (water content, 90%), and such sludges tended to be sticky and difficult to remove.

Zinc Phosphate in Presence of Chlorate

It is well known that a satisfactory fine-grain zinc phosphate coating can be obtained on a ferrous substrate by treating the substrate with an acidic solution of zinc phosphate which also contains chlorate ions or both chlorate and nitrate ions as depolarizers for the principal process of coating formation. When employed on a continuous basis, this process suffers from the accumulation in the working solution of iron as ferrous ion which is liberated in the process reaction but which is only slowly oxidized to ferric ion by chlorate or chlorate/nitrate ions.

Two disadvantages of this accumulation of ferrous iron are recognized as follows: (1) the quality of the coating may be adversely affected, and (2) during any interruption of continuous working, for example, due to inadvertent or overnight stoppage, the slow oxidation of ferrous ion and the precipitation of the thus formed ferric ion as ferric phosphate causes an uncompensated rise in the acidity of the working solution with a resultant adverse effect on its coating performance when operation is resumed.

B.A. Cooke and M. Brock; U.S. Patent 4,071,379; January 31, 1978; assigned to Imperial Chemical Industries Limited, England provide a continuous process of producing a phosphate coating on a ferrous metal substrate which comprises treating the ferrous metal substrate with an acidic solution of zinc phosphate in the presence of chlorate ions, optionally in the absence of nitrate ions also, and adding to the solution as coating proceeds a proportion of a rapid-acting secondary oxidant for ferrous ion which is sufficient to maintain the concentration of ferrous ion at less than 112 ppm of the solution, there being present in the working solution when in the steady state a proportion of the secondary oxidant of from 0 to 0.6 mmol/ℓ of the solution.

Preferably, the concentration of ferrous ion is maintained at less than 56 ppm of the solution. Preferably, the acidic solution contains 0.5 to 5.0 g/ℓ of zinc as Zn, 3 to 50 g/ℓ of phosphate as PO_4, 0.5 to 5.0 g/ℓ of chlorate as ClO_3, and 0 to 15 g/ℓ of nitrate as NO_3. Preferably, the total acid content of the solution is not greater than 30 points and the ratio of free acid to total acid is in the range 0.02 to 0.1. (The number of points is the milliliters N/10 sodium hydroxide required to titrate a 10 ml sample of the solution using phenolphthalein as indicator for total acid and methyl orange for free acid.) Preferably the temperature of the solution does not exceed 65°C.

By the term "secondary oxidant" is meant an oxidant the function of which in the process is solely to oxidize the ferrous ion without taking part to any significant extent in the primary coating formation process. Any rapid acting oxidant will fulfill the function of the secondary oxidant. By a rapid acting oxidant is meant an oxidant which, when added to an acidic zinc phosphate solution containing ferrous ion, will within 10 min at the normal operating temperature of the solution reduce the concentration of ferrous ion by at least one-half of the extent theoretically possible.

Suitable rapid acting secondary oxidants include alkali metal nitrites or ammonium nitrite, hydrogen peroxide, compounds containing combined hydrogen peroxide which liberate hydrogen peroxide under acidic conditions, sodium hypochlorite peroxydiacid salts such as perphosphates and perborates. Particularly suitable oxidants are sodium nitrite and hydrogen peroxide.

It is conventional practice to replenish a working solution of acidic zinc phosphate with an acidic replenishment concentrate containing zinc phosphate and to add an alkaline solution ("toner") having as one of its functions the neutralization of the excess acidity which is present in the replenishment concentrate over that which is required in the bath. Commonly, the alkaline toner comprises sodium nitrite which functions both as an alkali (relative to the phosphating solution) and as a supply of oxidant to the process. The acidic concentrate is added at a rate sufficient to maintain the zinc and/or phosphate content of the solution at that desired and the sodium nitrite is fed independently at a rate which is sufficient to maintain an effective level of nitrite ion in the solution, commonly on the order of at least 2 mmol/ℓ.

Because the rapid-acting oxidant required supplies only part of the total oxidant requirement of the process, the quantity required to be fed is smaller than in those typical processes in which sodium nitrite is the sole additive apart from the zinc phosphate concentrate. But a substantial quantity of an alkaline substance, e.g., sodium hydroxide, is generally required to ensure the neutralization of excess free acidity in the zinc phosphate-containing concentrate. It, therefore, follows that it will be necessary to feed, in addition to the zinc phosphate-containing concentrate, both an alkaline toner and the required amount of rapid-acting oxidant.

Example: This example describes the coating of steel articles on a plant scale by the spray application of a working solution which comprises zinc, phosphate, chlorate, nitrate and sodium ions.

A phosphating tank of 5,400 ℓ capacity was charged with an initial ("start-up") phosphating solution prepared by mixing 102 pbw of a replenishment concentrate (A) which was compounded from the following ingredients: 122 pbw zinc oxide, 102 pbw 50% nitric acid, 338 pbw 81% phosphoric acid, and 79 pbw sodium chlorate. These ingredients were dissolved in water to give a total weight of 1,000 pbw, and 50 pbw of an intimately mixed solid starter powder consisting of: 145 pbw sodium dihydrogen phosphate, 67 pbw sodium chlorate, 213 pbw sodium nitrate, and 76 pbw sodium chloride. The mixture was dissolved further in water to a total weight of 5,000 pbw. The initial solution had a total acid pointage of 10.5 and a free acid pointage of 0.5.

Steel articles were sprayed with the solution prepared as described above at a temperature of 110° to 115°F to give a coating weight on the steel of 1.3 g/m^2. The replenishment concentrate (A) described above and a toner concentrate (B), which comprises: 44 pbw sodium hydroxide and 44 pbw sodium nitrite, these ingredients being dissolved in water to give a total weight of 1,000 pbw, were fed concurrently so that they were added to the working solution in equal volumes. Additions were initiated by an automatic controller so as to hold the conductivity of the solution constant as described in U.S. Patent 4,089,710, elsewhere in this book.

The chemical analysis of the solution was maintained substantially constant at: zinc as Zn, 2.00 g/ℓ; phosphate as PO_4, 7.04 g/ℓ; chlorate as ClO_3, 2.10 g/ℓ; and nitrate as NO_3, 3.95 g/ℓ. The concentration of nitrite ion in the solution under these conditions was substantially zero and that of ferrous ion was less than 20 ppm.

The process was continued for 12 hr a day over 20 working days and a total of 1.5×10^5 m^2 of steel was coated. It was found by scanning electron microscopy that the deposited phosphate coating completely covered the steel and was of fine grain. A coating of paint, applied subsequently by electrodeposition, gave excellent performance when subjected to accelerated tests for corrosion resistance and mechanical properties.

Low Temperature Zinc Phosphate Coating

W.C. Jones; U.S. Patent 4,140,551; February 20, 1979; assigned to Heatbath Corporation provides a microcrystalline zinc phosphate coating composition which is suitable for producing phosphate coatings on metal surfaces at low temperatures.

The microcrystalline zinc phosphate aqueous coating composition, useful at temperatures between about 70° and 120°F, comprises calcium ion, zinc ion, phosphate ion, nitrate ion and nitrite ion, wherein the sum of the total calcium and zinc concentration is at least 0.2 M, the total calcium to zinc molar ratio is from 2.8:1-5.8:1, respectively, the total phosphate to nitrate molar ratio is from 0.18:1-2.8:1, respectively, the concentration of nitrite, as NO_2^-, is from 0.13 to 0.33 g/ℓ of coating composition, and the ratio of total acid to free acid is from 8:1-40:1.

The coating weight will vary with the processing time and may be as little as 25 mg/ft^2 of surface area but is generally from 50 to 700 and typically 300 mg/ft^2. Processing times may vary from 1 to 30 min. In spray applications, processing times of 30 sec to 15 min are preferred.

Because of the high concentration of calcium utilized herein, the calcium is very susceptible to precipitation as the fluoride. Therefore, it is important that the formulations be essentially free of fluoride either in its simple or complex form.

Example 1: Two concentrates are prepared in accordance with the following formulations.

	Grams per Liter
Concentrate A	
Zinc oxide (2.0 M)	162.8
Phosphoric acid, 85% (6.0 M)	691.8
Nickel nitrate hexahydrate (0.07 M)	20
Water	599.4
Concentrate B	
Calcium nitrate* (4.55 M)	928
Water	612

*Assayed at 80.4 wt %.

A working solution is prepared by combining Concentrate A and Concentrate B and additional water in such a manner as to produce a 5% solution of Concentrate A and a 7.7% solution of Concentrate B by volume. The free acid of this solution is 10.3 points and the total acid is 67.5 points. Pointage is the number of milliliters of 0.1 N sodium hydroxide required to neutralize a 10 ml sample to the methyl orange endpoint for a measure of the free acid and to the phenolphthalein endpoint for the measure of the total acid. An addition of sodium hydroxide solution is made to the working solution to reduce the free acid to

5.0 points and the total acid to 63.6 points. The mol ratio of calcium to zinc is 3.5:1 and the mol ratio of phosphate to nitrate is 0.43:1. Sodium nitrite is added to the solution in sufficient quantity so as to provide a 0.5 g/ℓ concentration. The solution is heated to 100°F and a clean, weighed 2" x 3" 1010 steel panel is immersed for 15 min. The panel is then rinsed, dried, and reweighed.

The coating is then removed in a solution of 5% chromic acid at 165°F for 5 min. The panel is again rinsed, dried, and weighed. The stock loss is considered to be the difference in weight between the original and stripped weight and amounts to 717.6 mg/ft². The coating weight is measured by the difference between the coated weight and the stripped weight and amounts to 370.8 mg/ft². This is equivalent to a conversion ratio of 0.52 (coating weight divided by stock loss). This may be considered to be the efficiency of conversion of the bath.

Comparative Example: For comparison purposes a conventional high temperature calcium modified zinc phosphate concentrate is prepared in accordance with the following formulations.

	Grams per Liter
Concentrate C	
Zinc oxide (1.6 M)	130
Phosphoric acid, 75% (3.6 M)	467
Nitric acid, 42° Be′ (1.5 M)	135
Calcium Nitrate* (1.4 M)	290
Water	498
Concentrate D	
Calcium nitrate* (4.55 M)	928
Water	612

A working solution is prepared by combining Concentrate C and Concentrate D and additional water in such a manner as to provide a 4% solution of Concentrate C and a 2% solution of Concentrate D by volume. This solution shows a free acid of 6.3 points and a total acid of 34.5 points. The mol ratio of calcium to zinc is 1.4:1 and the mol ratio of phosphate to nitrate is 0.4:1. An addition of sodium hydroxide is used to reduce the free acid to 5.5 points and the total acid to 33.3 points. The solution is heated to 170°F and a 2" x 3" weighed steel panel is processed as described above. The stock loss is 513.6 mg/ft² and the coating weight is 285.6 mg/ft², giving a conversion ratio of 0.56.

It is seen, therefore, that the low temperature coating composition of the process (Example 1) produces a greater stock loss and a greater coating weight while keeping the efficiency of conversion relatively the same. The quality of the coatings produced in both formulations is fully comparable with respect to appearance, microcrystallinity, and uniformity of coverage.

Example 2: In order to assess the energy saving characteristic of the 100°F process of this method, the volume of steam condensate is collected for a period of 2 hr when the solution, as described in Example 1, is maintained at a fixed temperature in a 40 gal stainless steel tank having top surface dimensions of 15" by 24". At 100°F the condensate collection rate is 850 ml/hr. At 160°F the rate is 3,100 ml/hr, and at 180°F the rate is 4,950 ml/hr. Thus, the heat losses at 160°F are 3.6 times as great as at 100°F, and the loss at 180°F is 5.8 times as great as at 100°F.

Controlled $Zn:PO_4$ Ratio plus Fluoborate

It has been known for some time to prepare metal surfaces for subsequent painting by applying a phosphate coating to attain improved corrosion protection and better paint adhesion. The primarily used base coats for electric immersion coating are zinc phosphate coatings. The resulting coatings are, however, not satisfactory for subsequent painting because of their thickness and coarse crystallinity.

A disadvantage of the coatings with the known solutions on the base of mono-zinc-phosphate for subsequent electric immersion coating consists particularly in the fact that a considerable part of the phosphate coating is separated during the painting process and is absorbed by the paint-film with detrimental results.

German Patent P 22 32 067 avoids these disadvantages, in that it provides treatment solutions in which the zinc portion in relation to the phosphate ions is considerably lower than in the customary solutions on a base of mono-zinc-phosphate. The treatment results in improved thin and even phosphate coatings on metal surfaces, particularly iron and steel, with good adhesive strength and durability, and are particularly well suited for subsequent electric immersion coating.

G. Müller and W. Rausch; U.S. Patent 4,265,677; May 5, 1981; assigned to Oxy Metal Industries Corporation found that the considerable advantages obtained by the solution and process of the German Patent P 22 32 067, wherein the weight ratio of $Zn:PO_4$ is 1:12-110, can be further improved if the treatment solutions also contain fluoborate, preferably in amounts of 0.3 to 2.0 g/ℓ.

Example 1: Degreased sheets of steel were treated for 2 min by spraying with a phosphatizing solution at 58°C which contain in grams per liter the following: Zn, 0.69; Ni, 0.38; Fe(III), 0.018; PO_4, 11.4; NO_3, 1.6; NO_2, 0.07; ClO_3, 1.49; and Na, 2.8.

The value for free acid was at 0.8, for total acid 14.5. The sheets were then rinsed with water and fully deionized water, and were subsequently dried. The coating weight obtained was 1.8 g/m^2.

Thereafter, a modified epoxy-resin paint was cathodically deposited on the pretreated sheet. The electroimmersion bath was at room temperature; separation voltage and time were 180 V for 2 min. Thereafter, the paint was baked for 25 min at a temperature of 190°C. The paint coating obtained thereby was 15 μm, uniform and glossy.

The corrosion protection of the painted and scribed sheets was tested in the ASTM salt-spray test (1,000 hr). The under-migration found after the test was 1 to 2 mm.

Example 2: The abovedescribed process was repeated in every detail. However, a phosphatizing solution was used, which in addition to the components shown in Example 1 also contained 0.8 g/ℓ of BF_4^-. The salt-spray test established under-migration of less than 1 mm.

Example 3: The treatment process according to Example 2 was varied in that instead of a water rinse, the sheets were rinsed once with chromium(III)-acetate solution [150 mg/ℓ Cr(III)] and once with chromic acid/chromium(III)-acetate solution [150 mg/ℓ Cr(VI), 40 mg/ℓ Cr(III)]. The salt-spray test after painting gave the same results as under Example 2.

Comparison of the results shows that the corrosion protection, particularly protection against under-migration of the phosphatizing solution modified with fluoborate is considerably better than that which is obtained in phosphatizing solutions without fluoborate. It can also be seen that even without rinsing with Cr(III)-acetate/chromic acid solution, corrosion protection that practically equals that with the mentioned solutions is obtained.

Solution Containing Cyclic Trimetaphosphate

Phosphate coatings are conversion coatings for iron based metals. The coatings serve as a base for organic coatings to aid in cold forming, to improve wear resistance or to impart color and to provide corrosion resistance to the base metal. The coatings are for the most part phosphates of metals in the phosphating solution (the primary metal) and of iron from the base metal. Formation of a phosphate coating is by contact of the base metal with a phosphating composition for a time and at a temperature necessary to provide a coating of the desired thickness.

Though phosphate coatings have been used for many years, one limitation in the use of the conventional phosphating solutions is a sensitivity to contamination by excessive iron phosphates in solution. A fresh phosphating solution is typically free of iron phosphates. Thus, a coating produced therefrom would contain a minimum amount of iron phosphate derived from solubilization of iron by phosphoric acid. As the phosphating solution is used to phosphate additional surface area, the concentration of the dissolved iron in solution increases resulting in a concomitant increase of iron phosphate in the phosphate coating. Some iron phosphate in the coating is beneficial but excessive amounts detract from the quality of the coating.

W.A. Vittands and W.M. McGowan; U.S. Patent 4,168,983; September 25, 1979 provide a composition for producing phosphate coatings on ferrous metal surfaces which is capable of tolerating increased concentrations of dissolved iron, (typically twice as much) without adverse effects on the phosphate coating or the solution performance.

The phosphate coating composition will produce a dense and smooth phosphate coating at operative temperatures as low as 140° to 170°F, and exhibits improved corrosion resistance.

With the exception of the addition of a cyclic trimetaphosphate to the phosphating composition, the compositions of this process are those conventionally used. The primary metal of the phosphating composition is preferably zinc but manganese may be used alone or in admixture with zinc or combinations of metals may be used such as zinc-calcium and zinc-calcium-manganese.

The phosphating solutions are characterized by the addition of a cyclic trimetaphosphate conforming to the following formula:

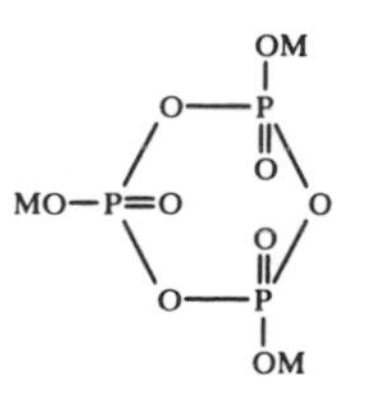

where M is a metal, preferably an alkali or alkaline earth metal though other metallic cations may be used provided they do not adversely affect the properties of the phosphate coating. The concentration of the trimetaphosphate is preferably maintained low, 0.001 mol/ℓ providing some benefit and increasing amounts providing increased benefits up to a maximum of 0.15 mol/ℓ. A preferred range varies between 0.01 and 0.1 mol/ℓ.

Example: A concentrated aqueous composition is prepared as follows: 380 g phosphoric acid (75%), 142 g nitric acid (67%), 160 g zinc oxide, 3.3 g sodium trimetaphosphate are combined and diluted to 1 ℓ with water.

To make an operating bath, 7.5 parts of the above concentrate are diluted with 92.5 parts of water and to simulate a used commercial formulation, 0.7 wt % iron in the form of steel wool is added.

A 26 ga. No. 87 steel test panel was prepared for phosphating by a sequence of steps comprising:

(a) Immerse in hot alkaline cleaner for 10 min at about 180°F (cleaner S-9, Lea Manufacturing),

(b) Hot water rinse (about 170°F),

(c) Pickle in 10 wt % hydrochloric acid by immersion for 10 min at room temperature,

(d) Cold water rinse,

(e) Immerse in conditioner of oxalic acid for 1 min at room temperature, and

(f) Cold water rinse.

Following preparation of the panel, it was immersed in the above composition maintained at a temperature of 170°F for 20 min, removed and rinsed. The phosphate coating so formed has a weight of about 2,250 mg/ft^2.

The coated part was tested for corrosion resistance by a salt spray following ASTM B-117 procedures. The test was discontinued after 24 hr without failure establishing that salt spray resistance exceeded 24 hr. Failure is defined for purposes herein as rust both on the sharp edges of the part and readily visible over the smooth surfaces. It should be understood that the test involves some subjectivity and there exists the possibility for experimental error.

The above procedure was repeated, but the phosphating formulation used was free of the trimetaphosphate. The coating weight was about 3,200 mg/ft^2. The salt spray test was repeated and the part failed between 4 and 6 hr of exposure.

METAL PHOSPHATE IS ZINC OR OTHER PHOSPHATE

Increasing Deposition Rate by Heating Metal

P. Paulus and J. Hancart; U.S. Patent 4,231,812; November 4, 1980; assigned to Centre de Recherches Metallurgiques-Centrum voor Research in de Metallurgie, Belgium observed that the deposition rate of the phosphate increases considerably when, instead of introducing the cold strip into the bath, it is preliminarily heated to a temperature higher than 250°C. This quite unexpected fact enables the metal strip to be coated with a relatively thick film in a very short time. The method comprises the following steps:

(a) The strip is heated to a temperature higher than 250°C, preferably higher than 300°C and preferably under a protective atmosphere if the heating temperature is higher than 500°C;

(b) The strip is quenched in a bath having a temperature higher than or equal to 80°C, preferably higher than or equal to 90°C, and containing one or more phosphates of the type $Me(H_2PO_4)_n$ where Me may be Zn, Ni, Mn, alkali metal, at concentrations of 1 to 20 g/ℓ, preferably of 5 to 15 g/ℓ. This bath may also possibly contain an inorganic acid, preferably of the phosphoric type, a reaction accelerator of the molybdate type, and a nonfoaming detergent agent. The pH of the solution is advantageously in the range from 3 to 6.

During treatment of strip, sufficient quantities of phosphating salts are added to prevent depletion of the bath. The metal strip to be coated may be steel strip.

The coating of the strip may be implemented during a thermal treatment in which the strip is introduced into a hot aqueous solution in order to rapidly cool it after recrystallization treatment and where it is possibly then introduced into a further hot aqueous solution for its final cooling after thermal overaging treatment, one or other of these solutions being used as a phosphating bath.

Operating modes which enable the thickness and the properties of the phosphate film to be regulated have been developed.

According to a first operating mode, in the case in which a film having a low thickness and a high phosphate content is required, a single immersion for a period of 0.5 to 40 sec is required.

In the case in which the phosphating operation is combined with heat treatment comprising quenching after recrystallization heating, but without tempering (for example, for a steel having a high tensile strength), the phosphating is carried out at the time of the quenching.

From these considerations a specific treatment for strip for deep-drawing may be established, this treatment comprising the following steps:

(a) Heating to a temperature higher than the recrystallization temperature of the strip and holding at this temperature for a sufficient period, this being carried out under a protective atmosphere;

(b) Quenching in a phosphating bath as disclosed above, the bath being at a temperature higher than or equal to 80°C with a holding time of 0.5 to 40 sec at a temperature lower than 150°C;

(c) Very careful rinsing of the strip after it has left the bath; and

(d) Tempering or overaging consisting in heating the sheet to a temperature of 300° to 500°C.

According to a first variant, the final cooling is carried out in a well-known customary manner, such as, for example, by pressure gas jets.

According to a second variant, in the case in which a film which is highly protective against corrosion is required, the strip is again quenched in a passivating solution containing, for example, a mixture of Cr^{3+} and Cr^{6+} after the overaging treatment.

One Step Film Forming Phosphatization

The work of *J.B. Fabregas and F. Gruber; U.S. Patent 4,006,041; February 1, 1977* relates to a method for forming in a single application, a corrosion resistant reticulated coating on metals, such as iron, zinc, aluminum, cadmium, steel, copper and alloys, which provides, in addition, an adherent surface for paint. The method includes the step of subjecting the metal to be treated to a treatment solution obtained by dissolving in phosphoric acid or a derivative thereof, metal salts selected from the group of zinc, manganese, iron and lead to provide a solution including primary phosphates of the metals.

Further steps include adding thereto an organic reducing agent, partially oxidizing the resultant composition through the addition of chromic acid or its salts, to produce in solution trivalent chromium ions, and applying the solution, preferably but not necessarily in heated condition, to the metal to be treated. The process consists generally of the following stages.

Stage 1: Primary phosphates of zinc, manganese, lead or iron are incorporated into a solution of phosphoric acid or derivatives, such as borophosphoric acid, the acid esters of phosphoric, etc. Phosphates may be formed by dissolving salts of the noted metals in the acid, by way of example oxides or carbonates.

Stage 2: An organic reducing agent is added to the solution of Stage 1. The preferred reducing agents are those containing groups of hydroxyls, amines, aldehydes and ketones. Among them are the following: carbohydrates (saccharoses); cellulose and its derivatives; starches; polyalcohols; primary, secondary and tertiary alcohols; aliphatics; aromatics or heterocyclics; esters; primary, secondary and tertiary amines; and in general substances containing hydroxylated groups or free amines, whether aliphatic, aromatic or heterocyclic form suitable reducing agents.

The reducing agents may be added directly to the solution of Stage 1 or may be themselves dissolved in one of the noted phosphoric acids or derivatives, and then added.

Stage 3: The resultant composition is partially oxidized under control by the action of chromic acid and/or its salts. The quantity of chromic acid required to be added will depend upon the quantity of reducing agent added in the second phase.

In Stage 3 a reaction is initiated which partially reduces the hexavalent chrome to trivalent chrome, forming a phosphate of soluble chrome (chromi-chromates-phosphoric acid) and obtaining a transparent green solution signifying the presence of the trivalent chromium which will have been formed as a result of the redox reaction. It is usually preferable to have a certain amount of hexavalent chromium present since the passivation properties will be better. The quantity may be varied according to the metal to be treated, and preferably should be maintained between 0 and 5 wt % expressed in chromic anhydride (CrO_3), optimum values being between 0.05 to 0.8 wt % in the solution to be used.

Stage 4: The resultant solution may be applied to a metallic surface, e.g., iron, zinc, aluminum, etc. This initiates the reaction with the metal and a consequent deposit of metallic phosphates, together with the formation of a transparent film which adheres well to the metallic surface, is highly resistant to corrosion and completely insoluble in water. The film forming process is accelerated when the solution is heated.

The chromium hexavalent ions and trivalent ions present in the composition have passivating characteristics.

Reticulation and insolubilization can also be improved by the addition of tannic acid, acrylic acid, or their derivatives and also by polycarboxylic or hydrocarboxylic acids.

Example 1: Zinc oxide is added to commercial phosphoric acid (85%), preferably when hot, in order to form a primary soluble phosphate of zinc. Reducing substances as described above, such as ethylene glycol, are then added to the solution, which is mixed when cold. These reducing groups are oxidized next with chromic acid or its salt. This reaction, which will occur when cold, can be accelerated by raising the temperature. The product then acquires a green color, proving at the same time that at least a portion of the hexavalent chrome has been reduced to trivalent chrome. Substances with film-forming properties, such as acrylic dispersions, can then be added, and the metal treated.

Example 2: The following specific formulations may be mentioned.

Composition A	Parts by Weight
Phosphoric acid, 75%	30
Zinc oxide	3
Ethylene glycol	4
10% water dispersion of chromic acid	63

Composition B	Parts by Weight
Phosphoric acid, 85%	50
Zinc oxide	4
Iron oxide	2
Starch	3
Water	3
10% solution of sodium dichromate in water	38

To the compositions of A and B, solutions, dispersions or emulsions of acrylic polymers are added (containing approximately 30 to 60% of solids) in the following proportions:

	Parts by Weight
Composition A or B	1
Acrylic polymer	0.25-20

The most usual proportions are from 1:1-1:6 parts, respectively. These mixtures are stable, without forming any precipitations or separations. The total content of water is normally between 25 and 75%.

Primary Coating of Fine Multinuclear Crystalline Phosphate Structure

One stage in the pretreatment of automobile bodies is that of chemically treating the bodies with a phosphate solution prior to application of the first paint coat. This has hitherto been done by immersing the bodies in a tank containing the solution, or in some cases by spraying the phosphate solution on the bodies.

Problems can arise after the phosphating treatment, when paint is applied by electro-deposition, which lead to a phenomena known as "scab corrosion" in the finished vehicle. This is body corrosion which results from an ineffective bonding between the crystalline phosphate coating and the metal surface, and is usually associated with too coarse a crystalline growth on the body with excessive and inconsistent weight deposition.

It has been found that when using electro-deposition paint priming processes, scab corrosion can be substantially reduced if a smaller or finer phosphate crystalline structure is obtained than that normally produced by conventional immersion techniques.

In a process developed by *A.J. Rowe; U.S. Patent 4,196,023; April 1, 1980; assigned to Carrier Drysys Limited, England* an automobile body is prepared for painting by a surface treatment which results in a primary coating comprising a very fine multinuclear crystalline phosphate structure having a uniform surface density of between 10^5 and 10^6 crystals/cm^2 and a substantially consistent weight between 1.5 and 2.0 g/m^2.

The body is conveyed roof uppermost past high pressure sprays of phosphate solution which produce a maximum droplet size of between 100 and 250 μ and have an impact velocity on the body surfaces of at least 7.5 m/sec. These sprays initiate the growth of the fine crystalline structure from a multitude of crystal nuclei. The sprays are applied for a predetermined time, preferably in the region of 14 to 16 sec.

The growth of the crystalline structure is then completed by immediately immersing the body in a tank containing the phosphate solution generally to a level about that of the window openings. A flow of the solution is generated from one end of the tank to the other to cause the solution to move relative to the body during its movement through the tank. This achieves constant replenishment of solution and avoids contamination.

The unimmersed upper portion of the body is sprayed with the solution while the body moves along the tank and at the same time phosphate solution is flooded through the window openings so that a head of solution builds up inside the body to cause a flow of solution from the interior of the body to the outside. The process ensures that a fine, even crystalline coating is present on all exposed body panels.

Generally, iron and zinc phosphates are used in aqueous solution at a pH of from 2 to 4 and at a concentration of between 10 and 38 points. The concentration of zinc phosphate bath is generally expressed on a "point" scale. A point is equivalent to 1 ml of 0.1 N NaOH and the concentration of the phosphate solution is expressed as the number of milliliters of 0.1 N NaOH required to neutralize the total acid in a 10 ml sample with phenolphthalein as indicator. A typical concentration is between 10 and 38 points.

A typical analysis of a 27 point solution would be zinc 0.24%, phosphate 1.0%, nickel 0.2%, nitrate 0.2%, and fluoride 0.2%. A preferred phosphate solution is prepared from Granodine 38 TC at a pointage of between 13 and 15. Apparatus for carrying out the process is presented in the patent.

METAL PHOSPHATES OTHER THAN ZINC

Nickel Phosphate Coating

Y. Matsushima, S. Tanaka and A. Niizuma; U.S. Patent 4,063,968; December 20, 1977; assigned to Oxy Metal Industries Corporation found that the formation of a crystalline film consisting of strong, fine and adherent nickel phosphate,

$$Ni_3(PO_4)_2 \cdot 7H_2O$$

on the surface of iron or steel can be obtained by surface-preconditioning and phosphate conversion with an aqueous nickel phosphate solution which contains a hydroxycarboxylic acid.

Preconditioning may be accomplished by scotch abrasion using Sumitomo 3M Company's abrasives. As an alternate, an alkali phosphate solution which contains suspended nickel phosphate crystals can be prepared by dissolving preferably 1 to 3 g/ℓ of sodium dihydrogen phosphate, disodium hydrogen phosphate, sodium pyrophosphate or sodium phosphate in the water and then suspending 0.5 to 20 g/ℓ of crystalline fine powders of nickel phosphate in this aqueous solution.

As the preferred treating condition, the pH value is 10 to 12, the concentration of crystalline powders of nickel phosphate is 3 to 5 g/ℓ, the period of immersion is 1 min or the period of spraying is 15 to 30 sec.

After the foregoing preconditioning, the surface is conversion-coated with nickel phosphate. The nickel phosphate conversion solution contains preferably 3 to 5 g/ℓ of hydroxycarboxylic acid such as salicylic, gallic, lactic, tartaric, citric, malic, glyceric, glycolic, mandelic and tropic acids, preferably 5 to 20 g/ℓ of phosphate (as P_2O_5), preferably 2 to 6 g/ℓ of nickel, and preferably 1 to 10 g/ℓ of nitric acid (as NO_3).

If desired, known accelerators such as chlorates and fluorides may be used. The acceptable ranges of total acid value of the phosphating solution is 10 to 50 points and the range of free acid value is 1.0 to 5 points. (The total acid is the number of milliliters of 0.1 N NaOH solution required to neutralize 10 ml of the sample solution using phenolphthalein as the indicator, and the free acid is determined in the same manner using bromophenol blue.) As the treating conditions of the methods of immersion and spraying for iron or steel, the temperature of the bath is maintained at 50° to 70°C during the time from 30 sec to 15 min. After this treatment, the material is washed and then dried.

The external appearance of the phosphate film is of bluish-green color and the weight of film is 3 to 10 g/m^2.

Nickel phosphate film formed according to this process has excellent adhesion and corrosion resisting properties as an undercoat for the application of organic paints and also has a good adhesive property as an undercoat for ceramic coating, e.g., porcelain enamel.

Example: A steel panel of thickness 0.8 mm was cleaned by the usual alkali degreasing solution. It was then treated by first immersing for 1 min into the solution which had suspended therein 5 g/ℓ of fine powder of nickel phosphate in 1 g/ℓ of sodium phosphate solution having pH 12, at room temperature and subsequently by immersing into a phosphate conversion solution heated to a temperature of 60°C. The phosphate solution contained 5 g/ℓ of Ni, 10.6 g/ℓ of PO_4, 5.3 g/ℓ of NO_3 and 5 g/ℓ of salicylic acid and has 20 points total acid value and 1.3 points of free acid value. The surface was then rinsed with water and dried by warm air. The weight of nickel phosphate film of this example was 8 to 10 g/m^2.

Subsequently, the thus treated steel plate was coated with alkyd-melamine resin to a thickness of about 20 μ and baked in hot air for 30 min at a temperature of 130°C. 100% adhesion of the paint was obtained when the surface was cross-hatched at 1 mm intervals and tape-pulled. When an identical pane was treated in the same manner and subjected to salt spray for 200 hr (according to JIS Z 2371), results were as good as those obtained using a zinc phosphate comparative.

Tin Phosphate Coating

The process of *Y. Matsushima, N. Oda and H. Terada; U.S. Patent 4,220,486; September 2, 1980; assigned to Nihon Parkerizing Co., Ltd., Japan* is a conversion coating solution which has a pH value in the range of 5.5 and 6.5 and contains therein 0.1 to 50 g/ℓ as phosphate ion of acid salts of alkali phosphates, 0.01 to 0.5 g/ℓ of stannous ion, and 2 to 12 fold by weight as much fluorine ion as the stannous ion. Further, 0.2 to 5 g/ℓ of one or more members of pyrazole, hydroxylamine and hydrazine compounds are added to the above treating solution. With the existence of stannous ion and fluorine ion in the form of complex ions, both substances are able to exist stably and the treating solution becomes also stable.

By treating metallic surfaces with the above conversion coating solution, the coating film containing insoluble tin phosphate can be continuously formed in a uniform state. The bonding between a substrate and the conversion coating film itself is quite strong with the result of improvement in anticorrosiveness and the adhesiveness and the anticorrosiveness and gloss of the subsequently applied overcoating become good.

The metallic products to be treated are first cleaned by a weak alkaline degreasing agent. Then they are sprayed with the conversion coating solution or they are dipped into the conversion coating solution. After that, the metallic products are rinsed with water and further dried by hot blast at a temperature of 50° to 220°C. The temperature of the conversion coating solution may be the normal temperature, however, if the treatment is carried out by heating the coating solution at a temperature of 45° to 70°C so as to promote the conversion reaction, coating films of excellent anticorrosiveness can be obtained. The preferable time of contact between the metallic products to be treated and the conversion coating solution may be in the range of 10 to 60 sec.

Example 1: D1S cans having exposed steel surfaces were cleaned by using a 1% hot water solution of a weak alkaline degreasing agent (Fine Cleaner 4361, Nihon Parkerizing Co., Ltd.). After that, the conversion coating treatment was carried out by spraying the following conversion coating solution 1 on the cans for 30 sec. After this spraying, the cans were washed with city water and further sprayed with deionized water of above 500,000 Ω cm in specific resistance for 10 sec. Then, the cans were dried for 3 min in a hot blast furnace at 200°C. These treated cans were immersed into the city water at 60°C for 30 min, thereby performing the test of anticorrosiveness. As shown in the table on the next page, the results were better than those of the following Comparative Example 1.

Conversion Coating Solution 1

NaH_2PO_4	15 g/ℓ
$SnSO_4$	0.2 g/ℓ
NH_4F	1 g/ℓ
pH*	5.6
Temperature of coating solution	60°C

*Adjusted by NaOH aqueous solution

Comparative Example 1: The components of comparative conversion coating solution 1 were as Example 1 above, but without $SnSO_4$.

Example 2: With using the following conversion coating solution 2, the conversion coating treatment was carried out in like manner as the foregoing Example 1 and the tests of anticorrosiveness were carried out likewise.

Conversion Coating Solution 2

Na_2HPO_4	20 g/ℓ
NaF	1.6 g/ℓ
3-methyl-5-hydroxypyrazole	0.5 g/ℓ
$(NH_2OH)_2 \cdot H_2SO_4$	0.5 g/ℓ
$SnSO_4$	0.4 g/ℓ
pH*	6.0
Temperature of coating solution	65°C

*Adjusted by 75% H_3PO_4 aqueous solution

Comparative Example 2:

Comparative Conversion Coating Solution 2

H_3PO_4, 75%	15 g/ℓ

(continued)

Comparative Conversion Coating Solution 2

NaF	2.0 g/ℓ
$SnSO_4$	0.06 g/ℓ
$(NH_2OH)_2 \cdot H_2SO_4$	1.8 g/ℓ
pH*	5.0
Temperature of coating solution	55°C

*Adjusted by NaOH aqueous solution

Results of Anticorrosiveness Tests

	. . Example (%) . .		Comparative . . Example (%) . .	
	1	2	1	2
Red rust formation	~30	⩽5	100	>80

Iron Phosphate Coating at Ambient Temperature

J.A. Saus and L.P. McCartney; U.S. Patent 4,298,405; November 3, 1981; assigned to Intex Products, Inc. describe a process by which iron phosphate coatings of excellent quality and weight can be produced on a ferrous metal surface at ambient temperatures of from 50° to 100°F by treating the metal surface with an aqueous phosphating solution of specified composition and pH and including a nitrite-containing ambient temperature activator system.

The phosphating solution contains, as essential ingredients, orthophosphate ions as the source of phosphate and an ambient temperature activator system containing nitrite ions at a concentration of at least 0.08 g/ℓ. The ambient temperature activator system may utilize nitrite ions as the sole activator, or the nitrite ions may be used in combination with one or more auxiliary activators or adjuvants selected from the following: nitrates, molybdates, chlorates, thiosulfates, thiosulfites, borates, perborates, peroxides, bisulfites, organic nitroaromatic compounds, organic nitroaliphatic compounds, hydroxylammonium salts, organic molybdate salts, and organic molybdate complexes. The phosphating solution may also contain inert fillers and conventional additives or processing aids.

Ambient temperature phosphating compositions may be used in either spray washers or dip tanks, and either as an iron phosphating composition alone, if so desired, or as a combination metal cleaner/iron phosphate. The compositions may thus be used in either a three-stage system where simultaneous cleaning and phosphating occur in the first stage, or in a five-stage system where precleaning is accomplished in the first stage, followed by a water rinse, and with a separate phosphating operation being carried out in the third stage.

When used as a combination metal cleaner/iron phosphate, the composition preferably includes a cleaning agent selected from the group consisting of low foaming surfactants, organic solvents, and mixtures of low foaming surfactants and organic solvents. It is also envisioned that thickened phosphating compositions may be produced for brush application on a large, irregular metal surface that would not lend itself to traditional dip-tank or spray washer application.

Regardless of the particular method of application employed, or whether the phosphating operation is carried out in a three-stage, five-stage or other type of system, the phosphating composition, and any separate cleaning baths which might be used, are applied to the metal at ambient temperature (which typically is within the range of 50° to 100°F) without the necessity of any heating.

Concentrates for producing ambient temperature phosphating solutions in accordance with this process consist essentially of the following compounds in the amounts specified.

	Broad	Preferred
	. . Percent by Weight. .	
Alkali metal or ammonium dihydrogen orthophosphate	30-95	70-90
Alkali metal or ammonium nitrite activator	0.5-15	2-10
Auxiliary activators selected from the group specified herein	qs15	qs10
Cleaning agent	qs15	qs10

The concentrate may also contain inert fillers and other nonreactive compounds having no significant effect on the fundamental reactions which take place during the phosphating operation. For example, the concentrate may contain up to 40 wt % of an inert filler, such as sodium sulfate. The concentrate may also contain conventional pH buffers, defoamers, and other processing aids conventionally used in the iron phosphating field.

The following examples are intended to illustrate various specific compositions and how they may be used for producing phosphate coatings at ambient temperature.

Example 1: *General Purpose Cleaning/Phosphating Bath* – A concentrate having the following composition is prepared as follows.

Ingredient	Percent by Weight
Ammonium dihydrogen phosphate	82
Potassium nitrite (activator)	8
Nonionic surfactants (cleaning agent)	10

This concentrate is diluted in water at a rate of 25 g/ℓ. The resulting solution, having a pH of about 3.8, is used in the first stage of a conventional three-stage washer. A ferrous metal article is directed into the washer, and the solution is sprayed at a pressure of 15 psi and a temperature of 60°F onto the metal article for a period of 1 min so as to clean and simultaneously form an iron phosphate coating on the metal article. The metal article is then directed through an ambient temperature water rinse, followed by a final ambient temperature chromic acid rinse. The iron phosphate coating formed on the metal article has a weight of about 30 mg/ft^2 and has a generally bluish color.

Example 2: *Oxidizing Cleaning/Phosphating Bath* –

Ingredient	Percent by Weight
Potassium dihydrogen phosphate	72
Sodium nitrite (activator)	10
Sodium nitrate (auxiliary activator)	8
Nonionic surfactants (cleaning agent)	10

This concentrate, when diluted in water and applied as in Example 1, gives a heavier phosphate coating than in the previous example for similar treatment time. The exact coating weight will depend on the time in the phosphating stage and the nature of the metal being processed. The appearance of the coating is usually a shade of blue or gray.

Example 3: *Oxidizing Phosphating Bath –*

Ingredient	Percent by Weight
Sodium dihydrogen phosphate	87
Sodium nitrite (activator)	5
Sodium metanitrobenzene sulfonate (auxiliary activator)	3
Sodium nitrate (auxiliary activator)	5

This concentrate produces a coating similar to that of the previous example and is particularly useful in systems where the metal surface has been suitably cleaned in a previous stage, such as in a five-stage system.

Calcium Phosphate Coating

A method is disclosed by *F.P. Heller; U.S. Patent 4,108,690; August 22, 1978; assigned to Amchem Products, Inc.* for producing a noncrystalline, lightweight tightly adherent coating of calcium phosphate on ferrous metal surfaces.

In this method, the ferrous metal surface is treated with a coating solution containing calcium phosphate together with an oxidizing agent at a pH which closely approaches but does not exceed the saturation point of the calcium phosphate in solution. The calcium phosphate content of the coating bath is preselected from the range of from 0.01 to 1.0 mol/ℓ as measured by the Ca^{++} cation. Next, a bath temperature of from 50° to 160°F is selected. The pH of the bath is raised to a pH approaching but not exceeding the saturation point of calcium phosphate at the selected temperature and the bath is brought to the selected temperature.

Coating of the ferrous metal with the calcium phosphate solution is accomplished by conventional dip or spray methods. The resultant coating, though amorphous and with very low coating weight, provides the same superior bonding characteristics associated with heavier and crystalline prior phosphate coating materials.

A coating solution is provided first as a concentrate of the following ratio: calcium carbonate (98.5% pure), 9.28%; phosphoric acid (as a 75% solution), 33.9%; and water to make 100.00%.

From this concentration, which is 1.0 M in calcium (as Ca^{++}), a coating bath is made by diluting each liter of the concentrate with sufficient water to make a 0.025 M (as measured by Ca^{++}) bath solution.

In the operation of the process, ferrous metal articles such as steel panels are prepared for the coating process by a cleaning and degreasing step following known methods. Following the cleaning step, the panels are rinsed in water. The cleaned panels are then spray coated with the calcium phosphate solution as prepared above.

Following the spray coating of the calcium phosphate the panels are preferably rinsed with water and dried. A final "after" rinse is performed to enhance corrosion resistance. The after-rinse, which utilizes hexavalent chromates or other suitable materials for this purpose is well known, as disclosed, for example, in U.S. Patents 3,063,877 and 3,450,579, and forms no part of this process. Following the after-rinse, the coated panels are dried and are then ready to receive paint or lacquer.

Example: In a comparison test example, five sets of steel panels of approximate size of 4" x 12" were cleaned, rinsed and then coated according to the process by spraying them with a coating solution of calcium phosphate containing 0.025 mol/ℓ calculated as Ca^{++} and various amounts of sodium nitrite at various pHs. The temperature of the coating solution was 100°F, and the spray time was 60 sec.

The physical properties of the test panels after preparation and before testing are summarized in Table 1 and represented as a single number, but it should be appreciated that this number is generally an average of two or three test runs.

Table 1: Physical Properties of Test Panels After Preparation but Before Testing Represented as an Average Value

Panel	Total Acid (points)	$NaNO_2$ (ppm)	pH	Coating Wt. (mg/ft^2)	Iron Loss (mg/ft^2)	Coating Eff.
A	8.8	217	3.30	30.3	93.9	0.32
B	8.8	229	3.65	34.5	72.9	0.47
C	8.4	236	3.80	27.3	66.6	0.41
D	8.7	279	4.02	27.0	40.2	0.67
E	8.6	248	4.20	26.7	39.6	0.67

In Table 1, column 1 indicates total acid points which is a measurement of the number of milliliters of 0.1 M sodium hydroxide required to neutralize a 10 cc bath sample to a phenolphthalein end point. Total acid points are used to indicate the phosphoric acid concentration. Column 2 shows the amounts of sodium nitrite accelerator utilized measured in parts per million. Column 3 indicates the various pH units utilized. Column 4 indicates the weight of coating deposited on the panel as measured in milligrams per square foot. Column 5 indicates iron loss from the panel as a result of the method of this process measured in milligrams per square foot. Column 6 gives the coating efficiency which is a ratio of coating weight/iron loss.

These five panels were utilized in tests as outlined in Table 2. Table 2 contains data on corrosion resistance of the panels as measured in a salt water spray test and a water immersion test as compared to zinc phosphate coated panels and iron phosphate coated steel panels as standards.

The five sets of test panels whose properties and specifications are listed in Table 1 together with the two sets of standard panels were subjected to the following tests under the following described test conditions. The panels were coated with various commonly used test paint primers plus top coat and some of the thus coated panels were scribed through the coating layers to bare metal. All the panels were then subjected to a salt spray or water immersion test and the results are summarized in Table 2.

Table 2

Paint system	1	1	1	1	1 + TC	2	2
Test used	SS	SS	SS	SS	SS	SS	WS
Length (hr)	96	168	240	336	336	240	240
Scribed panels	No	No	No	No	Yes	Yes	Yes
Corrosion Rating							
Panel A	10	9.7	8.3	7.2	10	10	10
Panel B	10	10	10	8.0	9.7	10	10
Panel C	10	10	9.0	9.0	9.9	10	10
Panel D	10	9.8	9.2	8.7	9.9	9.9	10
Panel E	10	9.8	9.2	8.7	9.9	9.9	10
Iron phos. standard	10	8.7	7.7	7.7	9.0	9.0	10
Zinc phos. standard	10	8.0	8.0	8.0	8.3	10	10

In Table 2, reading across, beginning with the top row, paint systems 1 and 2 refer to paint systems employed by the auto industry as standards to test the efficacy of proposed phosphate or phosphate type coatings. System 1, employed by General Motors, consists of a PPG water-based paint applied electrophoretically as a primer. System 1 plus TC utilizes the aforesaid prime coat plus an E.I. Du Pont de Nemours spray surfacer and spray topcoat for a total of three coats of organic finish. System 2, used at Ford Motor Company, utilizes a solvent based first and second primer and an internally developed Ford Motor Company topcoat.

In the test used row, SS refers to a standard salt water spray test as described in detail in the American Society of Testing Materials Bulletin No. ASTM-B 117. WS refers to a standard water immersion test, also an American Society of Testing Materials test, described in their Bulletin No. ASTM-D 870. Length refers to the duration of the exposure, measured in hours.

Scribed indicates whether the panel was scribed or not and the purpose of this test is to evaluate the extent of the corrosion emanating outwardly into the painted area from the exposed metal of the scribe mark. Panels A through E and the two control panels have already been described.

The rating of the panels is done visually on a scale averaging from 1 to 10 with 10 representing the best results.

As can readily be seen from Table 2, the calcium phosphate coated panels of this process give results overall which are equal to and sometimes superior to prior phosphate coated panels under the same test conditions.

Two-Stage Spray Process

The work of *H. Gotta and F.-H. Schroll; U.S. Patent 4,003,761; January 18, 1977; assigned to Gerhard Collardin GmbH, Germany* relates to a process for applying a phosphate coating to a ferric surface which comprises spraying an aqueous acidic solution at a pH of 4.3 to 6.5 containing an orthophosphate salt of a cation selected from the group consisting of alkali metals and ammonium, in the presence of oxidizing agent or reducing agent accelerators onto the surface. The improvement in the process consists of adding to the aqueous acidic solution from 0.05 to 1 g/ℓ of a short-chain alkylolamine having from 2 to 4 carbon atoms in each alkylol and from 0.01 to 1.5 g/ℓ of at least one nonionic surface-active wetting agent.

The acid phosphate solutions used contain orthophosphates in a concentration of about 1.0 to 20.0 g/ℓ in the form of the alkali metal and/or ammonium phosphates, such as sodium, potassium, and/or ammonium orthophosphate.

The oxidizing agent or reducing agent accelerators are such compounds as alkali metal nitrites, alkali metal perborates, alkali metal bromates, hydroxylamine salts, as well as alkali metal or ammonium molybdates, and organic nitro compounds. The accelerators are used in amounts of 0.05 to 5 g/ℓ, preferably 0.1 to 3 g/ℓ.

Suitable short-chain alkylolamines are those having from 2 to 4 carbon atoms in each alkylol group, particularly monoethylolamine, diethylolamine, triethylolamine and the corresponding propylolamines.

The nonionic surface-active wetting agents are, in particular, the water-soluble reaction products of ethylene oxide alone or with propylene oxide, with organic compounds having an active hydrogen atom and a hydrophobic moiety of at least 8 carbon atoms, such as alkylphenols having from 8 to 20 carbon atoms in the alkyl, higher fatty alcohols having from 8 to 20 carbon atoms, higher fatty acid amides having from 8 to 20 carbon atoms, etc. The turbidity point of the wetting agents used is generally between 20° and 70°C.

The duration of the treatment of the iron and steel surfaces in the spraying process is 0.5 to 5 min, preferably, 2 to 4 min. The process can be carried out at temperatures between 40° and 95°C, preferably 50° to 70°C.

It was also found that the good corrosion protection achieved with the above-described procedure can be further improved if the solutions also contain aliphatic monocarboxylic acids with 6 to 10 carbon atoms or aromatic monocarboxylic acids in the form of benzoic acid or alkylated benzoic acid in amounts of 0.05 to 0.5 g/ℓ in each case. The aliphatic monocarboxylic acids which can be used are particularly alkanoic acids having 6 to 10 carbon atoms, such as capronic acid, caprylic acid, as well as capric acid.

A special variation of the process consists in that the phosphatization is effected in two stages, with the concentration of orthophosphate being increased in the second stage by about 50 to 100%, compared to the first stage. Furthermore, it was found that it is generally of advantage in this two-stage process if the tenside concentration of nonionic wetting agents, which is between 0.1 and 1.5 g/ℓ, is reduced in the second stage by about 20 to 30%, compared to the first stage.

Example: *(A)* – Deep-drawn quality steel sheets were treated in the spraying process at a temperature of 65°C and a spraying pressure of 1.5 kg/cm^2 for 3 min with an acid solution of the following composition.

	Grams per Liter
Primary sodium orthophosphate	9
Hydroxylamine sulfate	0.4
Nonionic wetting agent*	0.6
Diethylolamine	0.4

*Addition product of 10 mols of ethylene oxide to nonylphenol

(B) – Another series of deep-drawn quality sheets was treated by the same procedure with an acid solution of the composition indicated under (A), which contained, however, in addition 0.2 g/ℓ of caprylic acid.

(C) – In a third series (comparison test) the treatment was effected with an acid solution according to (A) which did not contain the diethylolamine. The pH values in the solutions (A), (B) and (C) were adjusted with sodium hydroxide solution to 5.4 in each case.

The sheets which were processed according to (A) to (C) were further coated with a gray prime coat applied by electrodipping, as is customary in the automobile industry. The coat thickness was about 18 μ.

The coated sample sheets were subject to the salt-spray test according to SS DIN 50,021 with crosscut. After an exposure of over 240 hr to the salt-spray, the evaluation according to the degree of blistering on the surface (DIN 53,209) and the subsurface rusting in millimeters, starting from the crosscut, are indicated in the following table where the value indicated under (C) represents the reference example without the additions according to the process.

Treatment of Solution	(A)	(B)	(C)
Degree of blistering DIN 53,209	m0/g0	m0/g0	m1/g3
Rust in the crosscut	2-3 mm	1.5-2 mm	6-7 mm

Acid Metal Phosphates plus Ligand Forming Organic Polymer

The process of *M. Kronstein; U.S. Patent 4,233,088; November 11, 1980; assigned to International Lead Zinc Research Organization, Inc.* is concerned with the modification of steel surfaces and metal coated surfaces, in particular zinc-coated (including galvanized) steel, with the objective of making them suitable to serve as substrates for protective and decorative organic coating systems.

The process for inhibiting corrosion and providing a foundation for subsequent application of organic coating systems to metal surfaces, such as steel surfaces and zinc-, lead-, copper- and tin-coated surfaces, comprises the development of a protective phosphatizing reaction coating based on a metal other than the metal which is to be protected either by an immersion treatment or by a spray treatment with a phosphatizing bath which contains the metal phosphate or metal acid phosphate matter for such a treatment in a status nascendi.

Such a state is obtained by the use of an aqueous medium containing phosphate ions derived from an alkali metal phosphate, an alkali metal acid phosphate, phosphoric acid or combinations of those. Further, introducing into the aqueous medium a metal oxide based on a metal other than that which is to be treated, preferably, an oxide of the metal group of molybdenum, vanadium, tungsten, titanium, lead, manganese and copper. In accordance with this, the metal oxide forms with the phosphate ions of the aqueous medium the desired freshly prepared metal phosphate or metal acid phosphate to develop on the treated metal surface the required protective reaction coating.

The phosphatizing bath can be modified further by introducing a ligand-forming organic polymer which is capable of entering the reaction coating formation.

The polymer can further be influenced also by the addition into the aqueous medium of a small amount of an acetylenic alcohol or a dialdehyde. Also, a dispersing agent, such as formamide or an alkyl-substituted formamide, can be employed to increase the reactivity of the metal oxide component.

Molybdenum, vanadium, tungsten, lead, titanium, manganese or copper phosphates or acid phosphates can be obtained for the use in the phosphatizing processes either by introducing anhydrides or oxides of the metals directly into an aqueous medium containing alkali metal phosphate (such as sodium phosphate) or alkali metal acid phosphate (preferably under addition of some phosphoric acid) or by introducing the metal anhydrides or oxides into phosphoric acid using approximately stoichiometric amounts (e.g., using about 1 mol molybdic anhydride introduced to about 2 mols of available phosphate ions) or into its aqueous solutions, and proceeding from there in the formation of the phosphatizing solution.

Alternatively, a coating of increased density and coherence on the zinc-coated (galvanized) steel surfaces can be obtained by introducing organic polymer groupings as "ligands" into the phosphatizing solution and hence thereafter into the reaction products formed on the phosphatized galvanized steel surfaces. The interreaction between the ligand-forming organic polymer and the reaction products is even increased when the ligand-forming organic polymer is introduced into the phosphatizing solution while the new metal phosphate or metal acid phosphate is being formed therein.

Example: *Introducing an organic polymer component into the phosphatizing solution* – Coatings of a higher coating weight and of desirable properties are obtained by incorporating organic polymer groupings into the phosphatizing solution and hence into the reaction products of the coating. Thus, instead of using the molybdenum (or other above-listed metals) acid phosphate (or phosphate) formations in an inorganic form as a protective coating to the galvanized steel surface, an organic polymer component can be added to the phosphatizing solution and forms a joint compound or ligand with the metal acid phosphate (or phosphate), whereby the organic polymer grouping also becomes a part of the reaction coating.

The organic polymer component can be a water-dispersible polyvinyl alcohol, methyl cellulose or any other water-dispersible organic polymer capable of forming a ligand with the metal acid phosphate (or phosphate) and capable of becoming a part of the reaction coating on galvanized steel. The degree of the final cure of the developed treatment with its polymer component can further be increased by the addition of certain accelerator materials, such as dialdehydes, e.g., glyoxal.

Therefore, even when the applied reaction product is being removed from the phosphatized galvanized steel surface by a conventional hydrochloric acid/formaldehyde stripping solution and the stripping solution has been further diluted with water, a subsequent ether extract of the organic polymer matter from the water solution can still identify the introduced organic polymer groupings in the infrared spectrum. This established that in the use of the organic polymer component according to this example, the resulting phosphatized coating represents a different product from that where the phosphatization has been applied without such organic polymer component.

Such a phosphatizing treatment can be obtained following the procedure below. As an initially separate dispersion, 30 g of a water-dispersible polyvinyl alcohol (Elvanol 90-50 of the Du Pont de Nemours Company) was added to 480 g water. It was heated at around 60° to 80°C with mixing or magnetic stirring until clear.

In a second step, 275 g sodium acid phosphate (monobasic) were dissolved in 960 g water with addition of 29 g phosphoric acid (85%) and heating at 60° to 80°C until a clear solution was obtained.

In a third step, 30 g molybdenum trioxide (commercial Type M, Climax Molybdenum Company) was first dispersed in 90 g water, or it was dispersed first in 5 g dimethyl formamide and 90 g water was added to the dispersion, or the 30 g molybdenum trioxide was refluxed in 5 g dimethyl formamide until the initial greenish dispersion had turned yellowish and 90 g water was added to the dispersion. Then the third step solution was combined with the second step solution with heating and stirring at 60° to 80°C until a clear solution was formed. The same procedure can be carried out using here vanadium pentoxide, tungsten trioxide, titanium dioxide, cuprous oxide, lead monoxide or manganese dioxide as the metal oxide material.

After the first step dispersion of the organic polymer matter had been diluted with one and a half gallons of water, the combination of the two other solutions was added and the resulting solution was stirred by a rotating pump and heated at 60° to 80°C. Hereby within the solution the water-dispersed organic polymer ligand can participate in the formation of an interreaction between the molybdenum trioxide with the phosphoric ions of the solution of the second step and can so become a component of the complex reaction coating. Eventually 5 g of an antifoaming agent (Nopco NXZ, Diamond Shamrock Co.) was added.

The so prepared solution can be used directly for the phosphatizing of galvanized steel. Moreover, it can be used also for the phosphatizing of plain or automotive steel or of joints between plain and galvanized steel. In addition, it can be used for the phosphatizing of one-sided galvanized steel, i.e., steel which has been galvanized or zinc-coated on one side only, but whose other side still is a plain steel surface.

The preferred pH of the phosphatizing solution is between 2.8 and 3.4. The coating which contains the organic polymer ligand is a denser and more coherent coating than the inorganic coating. The ligand also provides a higher uniformity of the coating. The ligand-containing coating is less readily damaged by mechanical scratching and hence the steel surface underneath is better protected from progressive corrosion.

Corresponding dispersions were obtained from vanadium pentoxide (V_2O_5), but the resulting clear brown phosphatizing solution required some filtration, because the vanadium pentoxide is less reactive than molybdenum trioxide. Also, tungsten trioxide (WO_3) or the other above-listed metal oxides were introduced into such phosphatizing solutions in a corresponding manner.

The application of the phosphatizing solution can be performed by immersing the metal surfaces into the heated solution or by spraying the heated solution under pressure upon the metal surfaces.

Acidic Solution Including Molybdate, Tungstate and Vanadate Ions

D. Oppen and K. Lampatzer; U.S. Patent 4,264,378; April 28, 1981; assigned to Oxy Metal Industries Corporation have developed a process for the preparation of metal surfaces of iron, zinc, or aluminum or their alloys for the subsequent application of organic coatings. This is achieved by applying a phosphate coating by means of wetting with a phosphatizing liquid, containing at least one metal cation of valence two or greater, and subsequent drying in situ of the liquid film.

The metal surface is wetted with a phosphatizing liquid, which possesses a pH value of from 1.5 to 3.0, is free from chromium and, apart from metal phosphate, contains at least one ion selected from the group consisting of soluble molybdate, tungstate, vanadate, niobate and tantalate ions. The molecular ratio of metal phosphate, calculated as $Me^{n+}(H_2PO_4)_n$, wherein n is an integer of two or more, to molybdate, tungstate, vanadate, niobate and/or tantalate ion, calculated as MoO_3, WO_3, V_2O_5, Nb_2O_5 and Ta_2O_5, lies within the range of 1:0.01-0.4.

Some preferred cationic components which simply form well-adhering tertiary phosphates are mentioned in the examples below. To improve anchoring simple or complex-bound fluoride ions are incorporated in the phosphatizing liquid. Reducing substances and finely-divided silica or organic polymers may also be included.

The liquids used in accordance with the process preferably contain the components in such a quantity that they show an evaporation residue of from 5 to 150 g/ℓ. Preferably, wetting is effected with a quantity of liquid film of between 2.5 and 25 ml/m^2 of working part surface.

Particularly good application results are achieved if the film of the phosphatizing liquid is measured in such a way that, after drying in situ, a coating weight of from 0.03 to 0.6 g/m^2 is obtained. The drying in situ, which follows the wetting of the metal surface, can be effected in principle, at room temperature. Better results are attained at higher temperatures, preferably between 50° and 100°C.

Examples 1 through 5: In each example the aluminum strip was cleaned with a pH 1.3 solution containing, per liter: 5 g H_2SO_4 (96%), 0.5 g ethoxylated alkyl phenol, and 0.05 g hydrofluoric acid (100%). The strips were then wetted with the phosphatizing liquids by means of a roller coating machine and dried at 80°C. The details are tabulated below.

The samples, thus pretreated, were coated with a vinyl lacquer and with an epoxy/phenolic resin paint and tested for adhesion in the bending test as well as for corrosion resistance in the pasteurizing test. In these cases, technological values were found, which, in comparison with the use of solutions based on CR(III)/SiO_2, showed at least equivalent, partly even better, results for the procedure according to the process.

	Example				
	1	2	3	4	5
PO_4, g/ℓ	30	20	40	40	20
Al, g/ℓ	2.7	–	–	–	–
Zn, g/ℓ	–	6.5	–	–	6.5

(continued)

	Example				
	1	2	3	4	5
Mg, g/ℓ	–	–	2.6	2.6	–
Mn, g/ℓ	–	–	–	5.5	–
Co, g/ℓ	–	–	–	–	–
Ni, g/ℓ	–	–	5.5	–	5.5
Molybdate, g/ℓ	–	–	5.0	–	–
Tungstate, g/ℓ	3.3	–	–	0.5	–
Vanadate, g/ℓ	–	1.06	–	–	5.3
Fluoride					
Type	–	H_2TiF_6	HBF_4	H_2ZrF_6	H_2SiF_6
Quantity, g/ℓ	–	1.6	8.8	2.4	14.4
Reducing Agent					
Type	Glucose	Ascorbic Acid	Hydrazine	Sodium Hypophosphite	Glucose
Quantity, g/ℓ	5.0	5.0	1.0	3.0	5.3
SiO_2, g/ℓ	6.0	1.2	–	2.4	12.0
Polyacrylate, g/ℓ	–	–	–	–	–
Liquid quantity, ml/m^2	8.0	4.0	8.0	2.0	8.0
Evaporation residue, g/ℓ	44.5	31.0	61.0	56.0	66.0
Coating weight, mg/m^2	356	124	488	112	528

CONTROLLING SLUDGE AND SCALE FORMATION ON EQUIPMENT

Addition of Low Molecular Weight Polyacrylamide

Phosphate coatings may be applied to metal substrates, notably ferrous substrates, by reaction of the substrate with an aqueous acidic solution of certain metal phosphates, e.g., phosphates of iron, manganese and zinc. There are certain by-products of the reaction with the substrate some of which are precipitated from the phosphating solution as coating proceeds. These by-products will usually include an insoluble phosphate salt of the substrate metal, for example, ferric phosphate in the case of a ferrous substrate.

The precipitate is a hindrance to efficient coating since it may form a crust on the walls of the coating bath and its associated equipment, in particular heat-transfer surfaces. Also, a layer of precipitate accumulates as a sludge in the bottom of the coating bath or of the reservoir of working coating solution which may be difficult to remove when its removal is desirable. A further possibility is that the phosphated work pieces may become contaminated. The formation of an insulating crust on the heat transfer surfaces located in the coating bath and, in the case of a spray process, on the spray nozzles, necessitates frequent scraping of these and other parts of the equipment in order to maintain the efficiency of the process; for example, good heat transfer and temperature control.

M. Brock and B.A. Cooke; U.S. Patent 4,052,232; October 4, 1977; assigned to Imperial Chemical Industries Limited, England have found that the precipitate which is produced in phosphating processes of the type described above can be modified in its physical form, so that it is less likely to cake into a rigid mass and has a reduced tendency to form a crust on, for example, heat transfer surfaces and spray nozzles, by the addition to the bath of certain water soluble polymers of molecular weight not greater than 5 x 10^5, the polymer comprising moieties of monomers selected from acrylic acid, methacrylic acid, acrylamide and methacrylamide.

Preferably, the effective water-soluble polymer has a molecular weight of at least 150. More preferably, the molecular weight of the water-soluble polymer is in the range of 1,000 to 50,000. Particularly suitable polymers are polyacrylic acid and polyacrylamide.

A suitable polymer is a polyacrylamide of molecular weight of 10 to 20,000. Commercially available polyacrylamides are Versicol W 11 and Versicol E 5. Preferably, the phosphating solution contains at least 1 ppm of the water-soluble polymer and preferably at least 5 ppm. A suitable concentration is in the region of 50 ppm.

Example 1: The heat transfer conditions existing in an industrial phosphating bath were simulated in the laboratory by the following procedure. 4 ℓ of an aqueous phosphating solution were prepared which contained 1.24% Zn, 1.0% PO_4, and 2.4% NO_3, and which had a total acid pointage of 38 points (number of ml N/10 NaOH required to titrate a 10 ml sample of the solution using phenolphthalein as indicator). The phosphating solution was stirred slowly to maintain its homogeneity and its temperature maintained at 71°C by a tubular mild steel heating jacket containing a silicone oil which was heated to about 160°C by an electrically heated element. The silicone oil was stirred rapidly to ensure an even temperature over the exterior of the heating jacket.

An initial small addition of sodium nitrite toner was made to the solution (to provide a titer of 2 ml N/10 $KMnO_4$ against 50 ml of the solution in the presence of 50% H_2SO_4). A mild steel panel was passed through the phosphating solution every 15 min, the total time of immersion of each panel being 5 min. The solution was regularly replenished with a concentrate containing 11.8% Zn, 34.5% PO_4 and 13.9% NO_3 to maintain a pointage of 38 to 42, and after a total of 126 panels had been treated in the solution, the zinc content of the solution had been replaced twice.

Example 2: In parallel experiments using the above procedure in which the solution contained no other additive, and in which the solution contained 50 ppm of Versicol W 11 (polyacrylamide having a reduced viscosity of 0.22 compared with water at 25°C which indicates a molecular weight of 10,000 to 20,000) the following observations were made:

(a) The heating jacket was coated with an adherent crust about ⅛" thick and the sludge (when removed by decantation and placed in a measuring cylinder to a sludge height of 20 cm) was very fine and was virtually impenetrable by a glass rod; and

(b) The heating jacket had a thin loose crust which was readily dislodged and the sludge (tested as above) was relatively mobile and could be easily penetrated by the glass rod.

Example 3: In a conventional spray phosphating process employing a phosphating solution which contained ingredients similar to those used in Example 1, it was found that the presence of 50 ppm of Versicol W 11 significantly improved the capability of the heat exchanger and prevented the formation of the solid crust which occurred in the absence of the additive. The crust became detached in large pieces which blocked the orifice of the spray jets whereas in the presence of the additive the coating formed on the heat exchanger was relatively soft and became detached only in small pieces which did not block the orifices of the spray jets.

Addition of Water-Soluble Lignosulfonate

M. Brock and B.A. Cooke; U.S. Patent 4,147,567; April 3, 1979; assigned to Imperial Chemical Industries Limited, England provide an improved process of applying a phosphate coating to a metal substrate by treating the substrate with an acidic metal phosphate solution in the presence of a water-soluble lignosulfonate, which is soluble in the aqueous coating solution. It may be added in any suitable form, for example, as a sodium, potassium, ammonium or other metal salt. One particularly convenient material is that comprising sodium lignosulfonate which is a by-product in the manufacture of paper. A suitable material is "Wanin S" (Steetley Chemicals).

Preferably, the phosphating solution contains at least 1 ppm of the water-soluble lignosulfonate and preferably at least 5 ppm. A suitable concentration is in the region of 50 ppm. The process is applicable to all conventional phosphating processes, for example, to spray and dip processes. Preferably, it is applied to ferrous substrates, but may be applied to zinc, aluminum or mixed metal surfaces.

Example 1: See Example 1 of U.S. Patent 4,052,232 above.

Example 2: In parallel experiments using the above procedure in which the solution contained: (a) no other additive and (b) 50 ppm of a lignosulfonate, Wanin S. Wanin S comprises a mixture of 60% sodium lignosulfonate, 26% sugars and other impurities. The following observations were made.

(1) The heating jacket was coated with an adherent crust about ⅛" thick and the sludge (when removed by decantation and placed in a measuring cylinder to a sludge height of 20 cm) was very fine and was virtually impenetrable by a glass rod; and

(2) The heating jacket had a thin loose crust which was readily dislodged and the sludge (tested as above) was relatively mobile and could be easily penetrated by the glass rod.

Composition Containing Citric or Tartaric Acid

Zinc phosphate coating has conventionally been applied on surfaces of iron and steel as a base coating for subsequent painting to improve the durability of the top coated paints. However, conventional zinc phosphate coating solutions contain, in general, an oxidant such as nitrate ion, nitrite ion, chlorate ion, bromate ion and the like as a promoter or accelerator. Such oxidants are decomposed during the chemical conversion coating generating undesirable by-product gases. Metals dissolved from the surface of the base metal during the chemical conversion coating contribute to sludge formation. The presence of an oxidant accelerates the rate of dissolution and also the rate of sludge formation.

Y. Ayano, K. Yashiro and A. Niizuma; U.S. Patent 4,153,479; May 8, 1979; assigned to Oxy Metal Industries Corporation found that the rate of sludge formation can be reduced without sacrificing quality of the zinc phosphate coating by including at least one member selected from the group consisting of tartaric acid, citric acid and soluble salts thereof in an aqueous solution containing from 0.01 to 0.2 wt % of zinc ion, from 0.3 to 5 wt % of phosphate ion and free from any oxidant and adjusting the pH to 3.5 to 4.7. The weight ratio of tartaric or citric acid in relation to zinc ion should be between 0.1 to 20. In a preferred modification, it is desirable to further include from 0.01 to 0.2 wt % of nickel ion in the solution.

Example 1: A zinc phosphate coating solution was prepared to contain: 0.95 wt % phosphate ion, 0.12 wt % zinc ion and 0.08 wt % tartaric acid.

The pH of the solution was then adjusted to 3.5 by adding sodium carbonate. A cold rolled steel sheet having a size of 7 x 15 cm was sprayed with a weakly alkaline degreasing agent and then with the abovementioned phosphating solution at 55°C for 2 min to provide a zinc phosphate coating.

The treated sheet was washed with cold water and then dried by means of hot air. The resulting coating had a microcrystalline structure and appearance satisfactory as an undercoating for painting. The amount of metal dissolved from the base metal during treatment amounted to 0.7 g/m^2, and the coating weight was 1.6 g/m^2.

Comparative Example 1: A phosphating solution was prepared as in Example 1 except tartaric acid was omitted resulting in a precipitate of zinc phosphate at a pH of 3.5. A steel sheet was treated with that phosphating solution in the manner of Example 1. A blue iron phosphate coating rather than a zinc phosphate coating was formed.

Example 2: The phosphating was carried out on a steel sheet cleaned in the same manner as in Example 1 with the zinc phosphate coating solution composed of 0.95 wt % phosphate ion, 0.08 wt % zinc ion, 0.08 wt % tartaric acid, and 0.04 wt % nickel ion at a pH of 3.6 for 2 min at 55°C.

The resulting coating had very uniform microcrystalline structure. The amount of base metal dissolved was 0.8 g/m^2 and the coating weight was 1.5 g/m^2.

Addition of Dialkyl Triamine Pentakismethylene Phosphonic Acid

A chronic problem associated with the formation of zinc phosphate coatings on metal is the build-up of a hard rocklike scale on heating elements, headers, nozzles and the tank used to hold the phosphating solution. If spray nozzles are being used to apply the phosphating solution the scale build-up decreases the amount of spray, changes the spray patterns and given nonuniform coatings. Excessive build-up on heating coils acts as an insulating medium and leads to poor heat transfer and eventual shut-down of the phosphating operation. Cleaning of the heating coils is a costly time consuming operation and damage to the coils occurs often.

T.C. Atkiss and W. E. Keen, Jr.; U.S. Patent 4,057,440; November 8, 1977; assigned to Pennwalt Corporation rendered the scale soft and dispersible by adding a scale-reducing amount of dialkyl triamine pentakismethylene phosphonic acid or its alkaline salt in the zinc phosphating solution.

The phosphonic acid sludge reducer is incorporated in the liquid concentrate used to prepare the zinc phosphating bath in a sludge-reducing amount. This is the amount required to obtain at least a 90 wt % reduction in hard sludge formation. This quantity will vary to some extent depending on the concentration of other ions such as zinc, nitrate, nitrite and phosphate. The minimum concentration of the sludge-reducer is about 0.7 g/ℓ. The concentration range of the dialkyl triamine pentakismethylene phosphonic acid will be about 0.7 to 15 g/ℓ in the concentrate. A concentration of as little as 0.0675 g/ℓ of dialkyl triamine pentakismethylene phosphonic acid salt gave a sludge reduction of 96.5 wt %. It was found that the sludge reduction was uniformly high at 96 to 96.5 wt % when the concentration of the sludge reducer ranged from 0.05 to 3 g/ℓ in the phosphating bath.

Example 1: A zinc phosphating bath was prepared in a stainless steel tank from liquid concentrate having the following average composition: 33.3 wt % phosphoric acid (75%), 26.8 wt % nitric acid (38°Bé), 15.7 wt % zinc oxide, and 24.2 wt % water. The concentrate was diluted with 6.4 times its weight of water to give a phosphating bath containing about 1.69 wt % zinc and 3.17 wt % phosphate as PO_4 ion. Titration of the bath with one-tenth normal sodium hydroxide showed that the bath had a total acidity of 35 points, a free acid of 7.5 points and an acid ratio of 4.7:1. The nitrite content titrated at 1.2 points based on titration of a 25 ml sample with 0.05 N potassium permanganate.

A stainless steel heating coil was weighed and immersed in the above zinc phosphating bath to obtain a representative sludge deposit in zinc phosphating baths. Steam was applied to the coil until the zinc phosphating solution reached a temperature of 180°F. Steel wool was then added to break in the bath. The stainless steel heating coil remained immersed in the bath at 180°F for 24 hr. After the 24 hr period the steam coil was removed, thoroughly spray rinsed with water, dried and weighed. This procedure was repeated for several measurements. The average weight of the residual hard sludge deposited on the steam coil for the 24 hr period was found to be 77,317 mg/ft^2.

Example 2: Varying amounts of diethyl triamine pentakismethylene phosphonic acid sodium salt sludge reducer were added to separate portions of the liquid concentrate used in Example 1 as shown below.

Liquid Zinc Concentrates

Sludge reducer, % by wt	0.01	0.05	0.1	1.0
Sludge reducer, g/ℓ	0.15	0.765	1.53	15.3

Portions of these liquid concentrates were then diluted with water to prepare zinc phosphating solutions. Following the procedure used in Example 1, the stainless steel heating oil was cleaned, weighed and then immersed in the phosphating solutions having varying concentrations of the sludge reducer as shown in the table below. After 24 hr immersion at 180°F, the coil was removed and the reduction in the amount of scale formation was determined by weighing the coil and comparing it with the scale formation without any sludge-reducer additive.

Zinc Phosphating Solutions

Sludge reducer, g/ℓ	0.0135	0.0675	0.135	1.35
Weight of sludge, g	1.39	0.09	0.09	0.13
Sludge coating, mg/ft^2	33,900	2,190	2,190	3,170
Sludge reduction, %	56	96.5	96.5	96.0

ACCELERATORS

Gaseous Nitrogen Peroxide

One of the main tasks which researchers in the phosphate coating industry have been and are engaged in is the provision of means for accelerating the formation of such coatings. A number of methods have been advanced for achieving this and the most extensively used methods involve the use of substances classed as accelerators. These accelerators are added to the coating solution in carefully worked out quantities and function as depolarizers and metal oxidizers. The exact reaction mechanisms by which these accelerators influence the formation of the phosphate coating are extremely complicated.

K.J. Woods; U.S. Patent 4,086,103; April 25, 1978 provides a method for accelerating the formation of a zinc or manganese phosphate coating on a metal substrate, which comprises dissolving gaseous nitrogen peroxide in an acidic zinc or manganese phosphate coating solution and contacting the metal substrate with the solution thus produced.

The nitrogen peroxide gas may be introduced into the phosphate coating solution by any convenient means such as, for instance, by bubbling or aspirating the gas through the solution. The gas may be provided directly from storage cylinders or may be produced immediately before use by any one of a number of well-known methods. Such methods include heating nitric acid above 256°C; heating nitrate salts such as lead nitrate above their decomposition temperatures; heating ammonia with air over a platinum gauze at 760° to 900°C and combining the nitric oxide obtained with oxygen; passing air through an electric arc and reacting the product thus obtained with oxygen; and combining nitric oxide from storage cylinders with oxygen.

A system for generating and introducing nitrogen peroxide into the phosphate coating solution is described in the process.

Example 1: A zinc phosphating solution was prepared with the following composition: zinc oxide, 4.0 g; phosphoric acid (75%), 8.8 g; nitric acid (70%), 5.0 g; and water to 500 ml.

The solution was heated to 35°C and clean steel panels were introduced into it for 2 minutes. No visible coating was formed.

The vessel holding the solution was partially evacuated and connected to a commercial gas cylinder of nitrogen peroxide in such a way that sufficient nitrogen peroxide was introduced to produce a strong blue color on starch-iodide test paper. The nitrogen peroxide cylinder was then disconnected and clean steel test panels were introduced into the solution for 2 minutes. A fine tight zinc phosphate coating was formed.

Example 2: A manganese phosphating solution was prepared with the following composition: manganese carbonate, 5.7 g; phosphoric acid (75%), 17.6 g; and water to 500 ml.

The solution was heated to 55°C and clean steel panels were introduced into it for 2 minutes. No visible coating was formed.

Nitrogen peroxide was introduced into the solution from a gas cylinder using the procedure described in Example 1. Clean steel test panels were again introduced into the solution for 2 minutes when lightweight adherent coating formed.

Tungstate Ion

The process of *A. Askienazy, V. Ken and J.-C. Souchet; U.S. Patent 4,089,708; May 16, 1978; assigned to Compagnie Francaise de Produits Industriels, France* relates to a treatment of iron or steel surfaces to give them enhanced anticorrosive and paint-retaining characteristics. This comprises the application, by spraying, dipping or otherwise, of a composition in the form of a solution of phosphates which includes an accelerating agent represented by a quantity of tungstate ion and which has a pH value in the range of 5.8 to 6.5, and preferably 6.0 to 6.4.

The content of tungsten ion, which is preferably introduced in the form of hydrated ammonium tungstate $(NH_4)_{10}W_{12}O_{41}\cdot 7H_2O$ or $Na_2WO_4\cdot 2H_2O$, is 0.01 to 10 g/ℓ and preferably 0.1 to 0.3 g/ℓ.

The duration of the treatment, when it is effected by spraying, is generally 1 to 3 minutes.

In the case of application of the solution by dipping the subject having the surface to be treated in the solution, the duration is 2 to 5 minutes.

Example 1: By dissolving the necessary quantities of the different constituents referred to below, a bath of phosphating solution is prepared with the following composition: monosodium phosphate, 9 g/ℓ; disodium phosphate, 0.35 g/ℓ; ammonium tungstate, 0.1 g/ℓ; and condensate of 10 molecules of ethylene oxide on the nonylphenol (serving as the surfactive agent), 1 g/ℓ. This bath has a pH of 6.

Applying this solution by spraying for 3 minutes at a temperature of 60°C and at a pressure of 1.5 bars to steel test panels of 7 x 14 cm and 0.8 mm thickness, cut from a steel sheet of the type used in the manufacture of automobile bodies, a deposit of blue grey phosphate of a weight of 0.80 g/m^2 and good adherence is obtained.

Example 2: To show the superiority of phosphating solutions in accordance with the process over those already known, comparative tests were made between the solution described in Example 1 and two commercial solutions which contain ammonium molybdate as an accelerating agent and have the following compositions:

	Solution A	Solution B
Monosodium phosphate, g/ℓ	8.80	8.80
Phosphoric acid in 75% aqueous solution, g/ℓ	0.14	0.25
Sodium molybdate ($Na_2MoO_4\cdot 2H_2O$), g/ℓ	0.08	0.12
Surfactive agent described in Ex. 1, g/ℓ	0.93	0.92

The pH value of these two compositions is respectively 5 and 4.4.

Steel test panels of the type identified in Example 1 were treated by spraying with each of these three solutions for a period of 3 minutes at 60°C and under a pressure of 1.5 bars. In the case of the solution in accordance with this process, one obtained a coating of phosphate having a weight in g/m^2 of 0.80 as indicated above, while this weight was 0.55 g/m^2 in the case of solution A and 0.63 g/m^2 in the case of solution B.

The phosphated test panels rinsed and dried, were coated with a primer of automobile quality (epoxy primer PF 26-516, Duco Company) by the electrodeposition technique, the final coating having a thickness of about 22 microns.

After a fresh baking the so-treated test panels were cross-scratched to bare metal and then submitted to a corrosion test of the salt-spray-type, or ASTM B-117-62, for 200 hours.

After this test the panels were rinsed, dried and then scraped with a pointed tool on both sides of the scratch to remove the nonadherent paint, the results being expressed as millimeters of paint failure at the scratch. Under these conditions it was found that the paint on the test panels which had been phosphated in accordance with this process was peeled away to a width of 1 mm, while the width of failure was 5 mm in the case of test panels treated by the two prior art solutions.

Chlorate or Bromate with Hydroxylamine Sulfate

A.J. Hamilton; U.S. Patent 4,149,909; April 17, 1979; assigned to Amchem Products, Inc. discloses that a high-quality iron phosphate conversion coating of moderate coating weight is developed on ferrous metal surfaces by a spray-applied sodium acid phosphate solution employing a combination accelerator comprising hydroxylamine sulfate and an oxidizing agent such as a chlorate or a bromate. A preferred accelerator combination is sodium chlorate and hydroxylamine sulfate in a ratio of about 4 parts by wt of sodium chlorate to 1 part by wt of hydroxylamine sulfate. The coatings are formed by spray application of a coating solution having a sodium acid phosphate concentration of about 4% by wt and a total chlorate/hydroxylamine accelerator concentration of about 0.4% by wt.

The solution is applied at a bath temperature of between 90° and 130°F in a conventional five-stage spray treatment employing an alkaline cleaner and a partially reduced chrome final rinse. The treated metal is preferably dried at a temperature of about 250° to 350°F for a period of about 5 to 10 minutes. The conversion coatings produced at low temperatures have unusually good salt-spray resistance. Coatings of equivalent coating weight and salt-spray resistance cannot be obtained by the use of either type of accelerator alone.

Example: A concentrated phosphating composition containing a chlorate accelerator is prepared by mixing the following:

	Pounds	Percent by Weight
Phosphoric acid (25% by wt)	2.942	26.54
Sodium carbonate	0.883	7.97
Chromic nitrate (color additive)	0.007	0.06
Sodium chlorate	1.242	11.21
Water	6.009	54.22

The sodium carbonate is slurried in one-half the formula amount of water and carefully blended with phosphoric acid by slow addition to control the effervescence. The remaining water, the chromic nitrate and sodium chlorate are mixed with stirring to produce a clear solution having a specific gravity of 1.289.

The concentrate is utilized to prepare a phosphating bath at a concentration of 4%.

To the above bath there is then added 12.5 g of hydroxylamine sulfate and 1 ml of caustic soda to adjust pH. Panels are treated with results as shown below.

Panel	Cleaner Temp (°F)	Phosphate Bath Temp (°F)	Alkalinity	Total Acid	pH	Hydroxylamine Sulfate (% of formula)	Coating Weight (mg/ft²)	Coating Appearance
1	–	–	1.2	12.3	5.29	3.0	–	–
2	176	114	1.4	12.2	5.40	3.0	24.3	*
3	168	116	–	–	–	3.0	–	*
4	164	115	–	–	–	3.0	–	*
5**	152	114	–	–	–	3.0	24.3	*
6***	–	–	1.2	14.0	5.15	6.0	–	*
7†	–	–	1.6	13.6	5.25	6.0	–	*
8	167	115	1.8	13.5	5.32	6.0	–	*
9	160	115	–	–	–	6.0	–	*
10	158	115	–	–	–	6.0	–	*
11	–	–	1.4	–	5.25	6.0	–	*
12	160	110	1.1	14.5	5.04	6.0	–	*
13	165	118	–	–	–	6.0	–	*

*Very good heavier coating.

**After panel #5 was coated an additional 12.5 g of hydroxylamine sulfate was added to the bath.

***After each of panels 6 and 7 there were added 3.5 and 2 ml respectively of caustic solution to reduce total acid.

†After each of panels 10 and 11 there were added 1 and 1.5 ml respectively of phosphoric acid to reduce the alkalinity.

The coatings obtained with hydroxylamine sulfate additions to the phosphating bath demonstrated exceptional salt-spray resistance when treated in standard salt-spray tests for 96 and 168 hours.

Nitrite Accelerator Supplied by Electrolytic Reduction of Nitrate

The work of *S. Oka, R. Murakami, and A. Sueyoshi; U.S. Patent 4,180,417; December 25, 1979; assigned to Nippon Paint Co., Ltd, Japan* relates to a method for formation of a phosphate coating film having good coating properties on the surface of a metallic substrate by treatment with a phosphating solution while maintaining appropriate concentrations of useful nitrite ions in the phosphating solution without accumulation of unfavorable ions.

It was found that the application of a direct current to electrodes available as the cathode and the anode and dipped in a phosphating solution containing nitrate ions reduces electrolytically the nitrate ions to nitrite ions and that the appropriate control of the conditions in the electrolytic reduction makes it possible to maintain a constant concentration of nitrite ions in the phosphating solution. When the reduction is continued for a long time, however, precipitates adhere to the electrode as the cathode, reducing the conversion efficiency of nitrate ions to nitrite. The ratio of useful ions is also decreased.

When the direction of the current flow is alternately changed at a fixed interval of time and the electrode is switched over from plus to minus or from minus to plus, the precipitates accumulated on the electrode are redissolved to recover normal surface of electrode, and thereby, the nitrite ions are constantly produced and also other useful components such as zinc ions can be kept in a constant concentration.

Both electrodes can be made of stainless steel or carbon, or one of the electrodes can be made of stainless steel or carbon and the other electrode can be made of platinum, platinum-plated titanium, oxidized noble metal, lead dioxide, or triiron tetraoxide.

Preferable phosphating solutions to be used possess a pH value of from about 1.0 to 4.0. The acidic phosphate coating solution may include an acidic zinc phosphate coating solution, acidic zinc calcium phosphate coating solution and acidic zinc manganese phosphate coating solution.

A favorable initial concentration of nitrite ions is within a range of 0.002 to 0.1% by wt, and this level may be preferably maintained during the phosphating. The initial concentration of nitrate ions in the phosphating solution is usually 0.2% by wt or higher, and the concentration of nitrate ions during the phosphating is preferred to be kept within a range of 0.2 to 5% by wt.

Example: A phosphating solution comprising the following ion components (pH, 3.0) is employed:

Ion	Concentration (%)
Zn^{2+}	0.10
Na^{+}	0.44
PO_4^{3-}	1.20
NO_3^{-}	0.60
NO_2^{-}	0.0

The phosphating process is carried out by using a spray-type installation as usually used in zinc phosphate coating (the device for electrolysis being used in connection with the tank in the phosphate coating step of the phosphating process). The phosphating process using such a spray-tube installation comprises a degreasing step **1**, two water-rinsing steps **2** and **3**, phosphate coating step **4**, three water-rinsing steps **5**, **6**, and **7**, and drying step **8**, as is shown in Figure 1.2a. That is, after degreasings and water-rinsing, an iron plate to be treated is subjected to the phosphating coating and thereafter washed with water and dried to give the desired plate which has an excellent phosphate film coating.

The tank in the phosphate coating step **4** has a volume of 300 liters. The phosphating solution is circulated into the device for electrolysis **15** by a pump **14**. As is shown in Figure 1.2b, in the device for electrolysis **15**, electrodes **16** and **17** are mutually arranged parallel with each other, both being made of stainless steel, (18-Cr stainless steel), and these electrodes **16** and **17** are connected to an electric source of direct current **19** via switch **18**. The total effective area of these electrodes is 0.4 m^2 and the distance between the electrodes is 2 cm. The phosphating solution (300 liters) is charged into a tank for phosphate coating and is circulated into the device for electrolysis via pump, wherein the temperature in the tank for phosphate coating is kept at 50° to 55°C.

Figure 1.2: Nitrite Accelerator Obtained by Electrolytic Reduction of Nitrate in Phosphating Solution

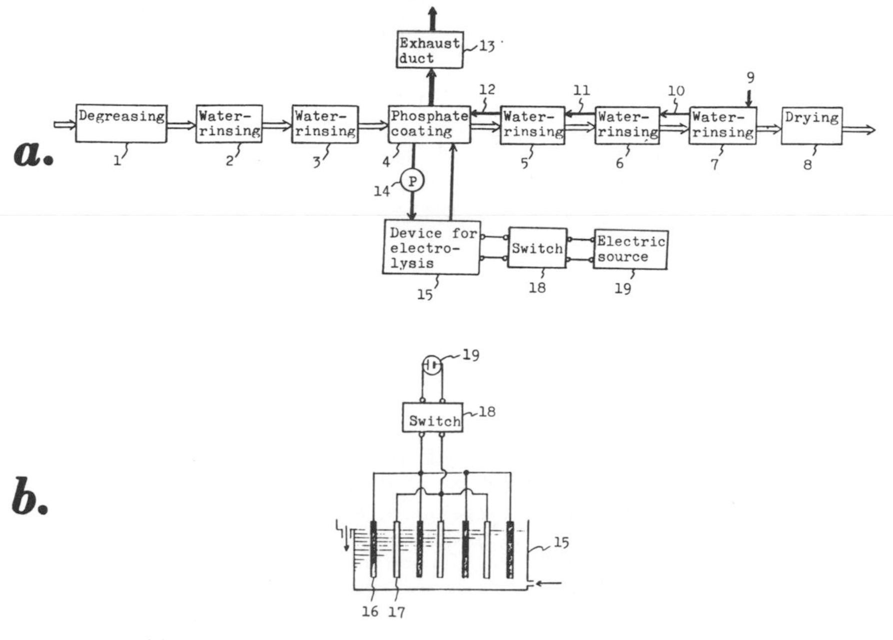

(a) Flow sheet of process
(b) Device for electrolysis of phosphating solution

U.S. Patent 4,180,417

The electrolysis is then effected under the following conditions: electric current density, 1.5 A/dm^2; value of total electric current, 30 A; voltage, 5 V; interval of changing of the direction of the current flow, 5 minutes; and total period of time of hydrolysis, 95 minutes. As a result, nitrite ions are produced in the phosphating solution to make a concentration of 0.008%. It is thus confirmed that nitrate ions are reduced into nitrite ions with an electric current efficiency of 60%.

Using the nitrate and nitrite ions-containing phosphating solution thus obtained, phosphating of an iron plate is carried out for 2 minutes. The iron plate thus phosphated is then washed with water three times and dried as shown in the figure. For the water-rinsing step, fresh water **9** is supplied to the water-rinsing step **7**, the overflow **10** from this step is supplied to the water-rinsing step **6**, the overflow **11** from this step is supplied to the water-rinsing step **5**, and the overflow **12** from this step is supplied to the phosphate coating step **4**, respectively. By the exhaust duct **13**, evaporation of water in an amount corresponding to the overflow supplied to the phosphate coating step **4** is effected.

According to the above installation, the dragout from the phosphate coating step **4** can be recovered and returned to the tank in the phosphate coating step **4** without exhaustion of the ions in the phosphating solution to the outside of the system. Further, the amount of fresh water **9** to be used at the water-rinsing step **7** can be reduced.

The zinc phosphate film formed on the article is uniform and fine and has good properties.

S. Oka, R. Murakami, and A. Sueyoshi; U.S. Patent 4,113,519; September 12, 1978; assigned to Nippon Paint Co., Ltd., Japan also relate to a phosphating solution comprising nitrate ions subjected to electrolytic reduction for conversion of the nitrate ions into nitrite ions so as to attain a desired level of nitrite ions in the phosphating solution.

A favorable initial concentration of nitrite ions is within a range of 0.002 to 0.1% by wt, and this level may be preferably maintained during the phosphating. The initial concentration of nitrate ions in the phosphating solution is usually 0.2% by wt or higher, and the concentration of nitrate ions during the phosphating is preferred to be kept within a range of 0.2 to 5% by wt.

The electrolytic reduction may be carried out by passing a direct current between at least one electrode as the cathode and at least one electrode as the anode, which are dipped in the phosphating solution, whereby the conversion of nitrate ions into nitrite ions takes place at the cathode. The electric current density at the cathode is usually from 0.01 to 15 A/dm^2, preferably from 0.1 to 8 A/dm^2, particularly from 0.5 to 3 A/dm^2. The electric current density at the anode may vary within a wide range and is usually not more than 30 A/dm^2.

As the cathode, there is advantageously employed an electrode having a relatively large hydrogen overvoltage, which generates in the electrolysis little or substantially no hydrogen gas. Examples of such electrodes are those made of mercury, zinc, copper, lead, tin, titanium, etc. Among them, the use of a zinc electrode is particularly preferred. As the anode, there may be used an electrode made of a material hardly soluble or insoluble in the phosphating solution. Examples of such electrodes are those made of platinum, platinum-plated titanium, oxidized noble metal such as oxides of noble metals (e.g., Ru, Ir) coated on Ti or Ta, lead dioxide, stainless steel, triiron tetroxide (magnetite), carbon, etc. A zinc electrode, which can be dissolved in the phosphating solution on the electrolysis, may be also used as the anode.

While the use of a hardly soluble or insoluble electrode as exemplified above requires the supplementation of zinc ions in an amount corresponding to the consumption, it may be advantageous in not requiring the frequent exchange of the electrode and the occasional control of the pH of the phosphating solution. The use of a zinc electrode is advantageous in attaining automatically the supplementation of zinc ions into the phosphating solution. The frequent exchange of the electrode and the occasional control of the pH will be necessary, however.

Preferable phosphating solutions to be used possess a pH value of from about 1.0 to 4.0. The acidic phosphate coating solutions which are applicable to the process may include an acidic zinc phosphate coating solution, acidic zinc calcium phosphate coating solution and acidic zinc manganese phosphate coating solution. It has the following composition: zinc ions, from 0.05 to 0.5% by wt; nickel ions, from 0 to 0.2% by wt; sodium ions, from 0 to 0.5% by wt; phosphate ions, from 0.2 to 2.0% by wt; nitrate ions, from 0.2 to 2.0% by wt; and nitrite ions, from 0.005 to 0.5% by wt. The acidic zinc calcium phosphate coating solution may contain calcium ions in an amount of from 0.01 to 2.0% by wt in addition to the composition of the acidic zinc phosphate coating solution. The acidic zinc manganese phosphate coating solution may contain manganese ions in an amount of from 0.01 to 0.5% by wt in addition to the composition of acidic zinc phosphate coating solution.

The electrolysis device can be used in connection with a tank in the phosphate coating step of the phosphating process, or the device for electrolysis can be used per se as the tank in the phosphate coating step of the phosphating process.

Example 1: A phosphating solution comprising the following ion components (pH, 3.0) is employed:

Ion	Concentration (%)
Zn^{2+}	0.10
Na^{+}	0.44
PO_4^{3-}	1.20
NO_3^{-}	0.60
NO_2^{-}	0.0

The phosphating solution (300 liters) is charged in a device as shown in Figure 1.3.

Figure 1.3: Electrolysis Device

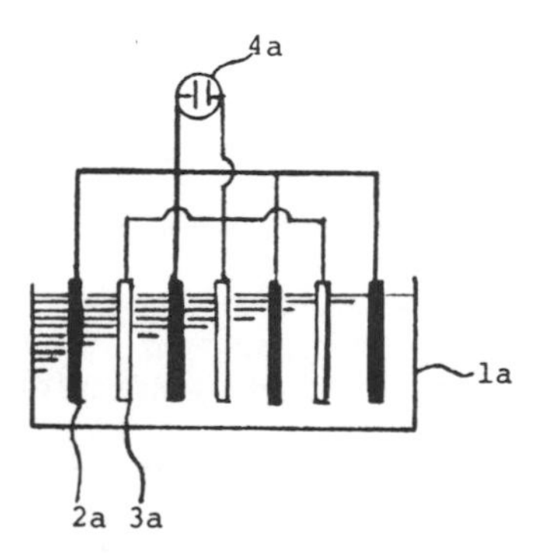

Source: U.S. Patent 4,113,519

In the device **1a**, four electrodes as the anode **2a** and three electrodes as the cathode **3a**, both being made of zinc, are alternately arranged in a row (no diaphragm being present between the electrodes), and these electrodes **2a** and **3a** are connected to an electric source of direct current **4a**. The area of each of these electrodes **2a** and **3a** is 527 cm^2. The total effective area of anode is 3 x 527 cm^2, and that of cathode is also 3 x 527 cm^2. Then, electrolysis is effected by sending to the solution a direct electric current for 95 minutes under the following conditions: electric current density for anode and cathode, 2.5 A/dm^2; value of total electric current, 39.5 A; voltage, 4.0 V (the inner temperature of the device **1a** being kept at 50° to 55°C). As a result, nitrite ions are produced in the phosphating solution to make a concentration of 0.008%. It is thus confirmed that nitrate ions are reduced into nitrite ions with an electric current efficiency of 45%.

Using the nitrate and nitrite ion-containing phosphating solution thus obtained, phosphating of an iron plate is carried out in an installation (spray-type) wherein the device for electrolysis is per se used as a tank in the phosphate coating step. The installation comprises the steps shown in U.S. Patent 4,180,417 above. The iron plate to be phosphated proceeds in the above order of the steps and is treated with the phosphating solution in the phosphate coating step for 2 minutes.

According to the above installation, the dragout from the phosphate coating step can be recovered and returned to the tank in the phosphate coating step without exhaustion of the ions in the phosphating solution to the outside of the system. Further, the amount of fresh water to be used at the water-rinsing step can be reduced.

The zinc phosphate film formed on the article by this example is uniform and fine and has good properties.

Example 2: A phosphating solution comprising the following ion components (pH, 3.0) is employed:

Ion	Concentration (%)
Zn^{2+}	0.10
Na^{+}	0.44
PO_4^{3-}	1.20
NO_3^{-}	0.60
NO_2^{-}	0.0

The same installation as used in Example 1 is employed, but the device for electrolysis is set up outside the phosphate coating tank.

As the cathode, a zinc electrode is used, and as the anode, a stainless steel (NTK 430 18-Cr stainless steel) electrode is employed. The total effective area of the anode is 0.16 m^2, and that of the cathode is also 0.16 m^2.

The phosphating solution (300 liters) is charged into the phosphate coating tank and made to circulate to the device for electrolysis by the aid of the pump. The inner temperature of the phosphate coating tank is kept to 50° to 55°C. Then, electrolysis is effected by sending a direct electric current for 95 minutes under the following conditions: electric current density, 2.5 A/dm^2; value of total electric current, 39.5 A; voltage between electrodes, 6 V. As the result, nitrite ions are produced in the phosphating solution to make a concentration of 0.008 %. It is thus confirmed that nitrate ions are reduced into nitrite ions with an electric current efficiency of 45%.

Using the nitrate and nitrite ion-containing phosphating solution thus obtained, phosphating of an iron plate is carried out with a treating time of 2 minutes, whereby a uniform and fine zinc phosphate film having excellent properties is formed.

REPLENISHMENT OF BATH CONSTITUENTS IN CONTINUOUS PROCESSES

Maintaining Constant Conductivity of Phosphating Solutions

In order to maintain or to achieve that optimum concentration of essential constituents which is necessary in a working solution for achieving a consistent and satisfactory phosphate coating, it is necessary to add to the solution one or more replenishment concentrates which make up for the depletion of each constituent.

It has been recognized previously that some form of continuous control of the concentration of the important constituents of the working solution is essential for satisfactory operation.

B.A. Cooke; U.S. Patent 4,089,710; May 16, 1978; assigned to Imperial Chemical Industries Limited, England developed a process whereby the composition of the acidic phosphating solution is brought to that composition which is characteristic of the steady state at the desired optimum, a continuous metal surface or a series of metal surfaces is passed through the acidic phosphating solution, and thereafter additions are made to the acidic phosphating solution of a material (a) comprising zinc and phosphate ions and of material (b) comprising alkali metal ions so as to maintain constant its electrolytic conductivity at a given temperature, the addition rates of (a) and (b) made in response to any change in conductivity being in a definite ratio of addition rates.

There is provided an improved and consistent method of controlling the composition of an acidic zinc phosphate solution when used in a continuous phosphating process.

This process is applicable to a phosphating process in which the phosphating solution has reached the steady state and in which the steady state can be maintained by addition of essential replenishment ingredients in a definite ratio of addition rates.

By the term "steady state" of a phosphating solution in a given process is meant that the composition of the solution does not vary systematically with time of operation, the criterion of systematic variation being established over periods on the order of several hours.

In a preferred process the acidic phosphating solution comprises as essential ingredients zinc, phosphate, chlorate and optionally nitrate ions, and in such a case, for example, material (a) comprises zinc, phosphate, nitrate and chlorate ions and material (b) comprises sodium ions. However, other suitable depolarizing oxidants may be used in the process, for example, nitrite, perchlorate, persulfate, perborate and hydrogen peroxide. Another suitable alkali metal ion for use in material (b) is potassium ion.

Example: This example describes the coating of steel panels with zinc phosphate according to the method using a phosphating solution which comprised zinc, phosphate, chlorate, nitrate and sodium ions. The optimum composition of the solution at the steady state was determined by analysis of prior phosphating baths of this type which were known to be in the steady state and which give satisfactory coatings at the steady state.

Replenishment materials (a) and (b) were as follows:

(a)	Zinc/Phosphate/Nitrate/Chlorate	Parts by Weight
	Zinc oxide	122
	59% Nitric acid	102
	81% Phosphoric acid	338
	Sodium chlorate	79
	Water, qs to 1,000 parts	

(b)	Sodium/Oxidant (Toner)	Parts by Weight
	Sodium hydroxide	84
	Sodium nitrite	25
	Water, qs to 1,000 parts	

An initial acidic phosphating solution was prepared by mixing 102 parts of the solution of replenishment material (a) with 50 parts of an intimately mixed solid starter powder (consisting of 145 parts sodium dihydrogen phosphate, 67 parts sodium chlorate, 213 parts sodium nitrate and 76 parts sodium chloride) the mixture being dissolved in further water to a total weight of 5,000 parts. This initial solution (also containing a small proportion of sodium carbonate) had a total acid pointage of 10.5 and a free acid pointage of 0.5 (Pointage = ml of 0.1 N sodium hydroxide required to titrate a 10 ml sample of the solution using methyl orange as indicator for free acid and phenolphthalein as indicator for total acid). The conductivity of the solution was 2.32×10^{-2} ohm^{-1} cm^{-1} at 50°C.

Rolled mild steel panels measuring 30.5 cm x 22.9 cm x 0.9 mm thick were treated by spray application with the above solution at a temperature of 50°C and at a rate of 4 panels per hour. The rate of metal treatment was thus 0.112 m^2/ℓ of bath per hour and at this rate of treatment after 12 hours total running there had been a complete turnover of the zinc content of the bath.

Coating was continued for a total time of 24 hours but in four separate periods of 6 hours each.

The replenishment of the phosphating solution was effected by simultaneous additions of the above solutions (a) and (b) in a constant ratio of feed rates, 0.43 g of (b) being added for every 1 g of (a), in response to changes in the electrical conductivity of the phosphating solution. The electrical conductivity was measured by conventional means, there being provided means for preventing insulation of the conductivity sensor by precipitated materials. 50 parts by volume portions of the bath were rejected at one-half hour intervals and the original volume restored in order to simulate the carry-over in an operational plant. No additions were made to the bath other than those mentioned. At no time did the concentration of ferrous ion in the phosphating solution exceed 56 ppm and the concentration of nitrite ion did not exceed 0.3 mmol/ℓ.

A high standard of coating was maintained throughout the experiment, the coating weight being approximately 1.9 g/m^2. The final free acid pointage was 0.5, the final total acid pointage 10.4 and the conductivity 2.23×10^{-2} ohm^{-1} cm^{-1}. The analysis of the bath remained substantially as it was at the beginning of the experiment when it was as follows: 2 g/ℓ of zinc as Zn; 7.7 g/ℓ of phosphate as PO_4; 2.3 g/ℓ of chlorate as ClO_3; 4.3 g/ℓ of nitrate as NO_3; 3.2 g/ℓ of sodium as Na; and 0.93 g/ℓ of chloride as Cl. The phosphated panels were subsequently satisfactorily painted by electrodeposition or by spraying and the finished panels were consistent in appearance and corrosion resistance.

Voltammetric Sensing

B.A. Cooke; U.S. Patent 4,182,638; January 8, 1980; assigned to Imperial Chemical Industries Limited, England found that in a coating process wherein a metal substrate is coated by reaction of the metal with a coating solution, the effective concentration of an electroactive constituent of the coating solution can be sensed and its concentration adjusted to a desired level by suitable means (for example, by addition of a suitable ingredient) in response to the magnitude of electrical current which flows between an indicating electrode comprised of a noble metal and a counterelectrode immersed in the coating solution. Conditions of potential appropriate to the electroactive constituent are established at the indicating electrode with reference to the neighboring solution or to a suitable reference electrode immersed therein. Such sensing of the current flow at an indicating electrode is herein referred to as voltammetry.

By " an electroactive constituent" is meant a constituent taking part in an electrode reaction at a definite electrode potential or within a definite electrode potential range, giving rise to a current whose magnitude is a function of the concentration of that constituent. Examples of electroactive constituents relevant to metal pretreatment are nitrite, copper ions, peroxide, zinc ions and protons.

By "the effective concentration" is meant the concentration of the constituent which governs its function in the coating process.

Particularly suitable processes include:

(a) A process of coating metal substrates by reaction of the substrate with an aqueous acidic metal phosphate solution. Preferably the metal phosphate is zinc phosphate. In such a process the substrate is coated with a metal phosphate.

(b) A process of coating a ferrous metal substrate with copper by reaction of the substrate with a solution containing copper sulfate, for example, an aqueous solution comprising copper sulfate, sulfuric acid and a substance which inhibits acid attack on the ferrous metal substrate.

Example 1: In this example is described the application of a copper metal coating to steel wire using a coppering solution comprising copper sulfate. The effective concentration of copper ion was controlled in response to the sensing of the voltammetric current at a platinum electrode at –0.30 volt relative to the saturated calomel electrode.

6,800 liters of a coppering solution comprising sodium chloride, copper sulfate, sulfuric acid and an inhibitor, were employed at 60°C in a coating bath to treat pickled steel wire at a rate of 470 m^2 of wire surface/hour. The effective concentration of copper ion in the initial coppering solution was such that a 1.00 ml sample of the solution was equivalent to 4.3 ml of 0.02 M ethylenediaminetetraacetic acid using alizarin yellow indicator. A replenishment material containing copper sulfate was added to the coating bath in response to measurement of the voltammetric current due to copper metal ion at –0.30 volt.

The process was continued over a period of 30 hours and during this time, at suitable intervals of time the voltammetric current due to the copper ion constituent was periodically sensed in a sample cell consisting of a chamber to contain a 120 ml sample of the coating solution (withdrawn at regular intervals from the coating bath), a platinum indicating electrode (1.5 mm length and 0.5 mm diameter platinum wire), a standard calomel reference electrode fitted with a saturated potassium chloride salt bridge and a platinized titanium counterelectrode (of 50 cm^2 exposed area).

The following voltage/time sequence was observed in the sample cell over a total period of 5 minutes, and the sequence was repeated throughout the time of operation of the process.

Time (seconds)	Voltage (at indicating electrode)
0-30	+2.5
30-90	0*
90-120	-0.3**
120-300	0

*Sample received in cell
**After 28 seconds sense voltammetric current

Throughout the process the effective concentration of copper ion was maintained by automatic replenishment and 1.00 ml of the solution was consistently equivalent to 4.3±0.2 ml of 0.02 M EDTA.

The coppered steel wire obtained over the 30-hour period of the process was of consistently high quality.

Example 2: This example illustrates the application of voltammetry at a platinum indicating electrode to the control of hydrogen peroxide concentration in a solution suitable for treating a ferrous substrate with a coating comprising zinc phosphate.

A zinc phosphating solution of the following composition in gram-mol per liter was prepared and held at 31°C: zinc, 0.069; phosphate, 0.138; nitrate, 0.087; and sodium 0.083.

3% hydrogen peroxide solution was added to the above solution such that an acidified 50 ml sample of solution was equivalent to 4.0 ml 0.1 N potassium permanganate solution. The voltammetric current due to peroxide was sensed at regular intervals of 15 minutes in the same manner as described in Example 1, but with the following voltage/time sequence.

Time (seconds)	Voltage (at indicating electrode relative to saturated calomel reference electrode)
0-30	+2.3
30-35	0
35-55	-0.45
55-135	0
135-165	+0.95 (after 20 sec)*
165-900	

*i.e., at 155 sec, sense voltammetric current

The voltammetric current was determined under the above conditions with hydrogen peroxide content corresponding to 4.0 ml 0.1 N $KMnO_4$ on an acidified 50 ml sample. A quantity of ferrous sulfate solution was then added so as to depress the $KMnO_4$ titration to 1.8 ml. A fall of 56% was observed in the voltammetric current. The voltammetric sensing circuit was then linked to an automatic dosing device supplying 3% hydrogen peroxide solution to the phosphating solution under test. This device replenished the hydrogen peroxide automatically until the original voltammetric current was restored. When this replenishment was complete, an acidified 50 ml sample of the phosphating solution was found to be equivalent to 4.2 ml 0.1 N $KMnO_4$ solution.

Replenishment Feed to Maintain Orthophosphate:Zinc Ratio

B.A. Cooke; U.S. Patent 4,233,087; November 11, 1980; assigned to Imperial Chemical Industries Limited, England has developed a continuous process of applying a phosphate coating to a ferrous or zinciferous metal substrate by treating the substrate with an acidic phosphating solution of zinc phosphate in the presence of hydrogen peroxide or of a hydrogen peroxide-liberating substance.

The phosphating solution comprises:

(a) 0.005 to 0.5 g atom of zinc (Zn) per liter of solution;

(b) 0.0002 to 0.02 g mol of hydrogen peroxide per liter of solution; and

(c) orthophosphate (PO_4) such that the molar ratio PO_4/Zn in solution is in the range of 0.5 to 3.7.

As phosphating proceeds, the solution is replenished with hydrogen peroxide or with a hydrogen peroxide-liberating substance and with at least two other replenishment feeds, replenishment feed (1) and replenishment feed (2), to maintain the solution at a desired composition.

The replenishment feed (1) comprises sufficient zinc (Zn) to maintain the desired concentration (a) in the solution, and sufficient orthophosphate (PO_4) together with another anion N^{n-} to maintain the molar ratio PO_4/Zn in the solution within the range defined in (c). The replenishment feed (1) has a free acidity of F gram equivalents/kilogram of feed. Replenishment feed (2) comprises an alkaline material selected from the group consisting of the hydroxides, carbonates and bicarbonates of alkali and alkaline earth metals, the alkaline material being soluble in water at the concentration of the replenishment feed (2) and has a total alkalinity of A gram equivalents/kilogram of replenishment feed (2).

The ratio of the quantities of replenishment feeds (2) and (1) respectively which are added to the phosphating solution within a significant period of time is (XF/A) where the value of X is in the range of 0.5 to 1.5. The anion N^{n-} is selected such that the acid H_nN has a pK_a value in the nth dissociation step not greater than 3. Preferably the anion N^{n-} is selected from NO_3^-, SO_4^{2-} and Cl^-.

Example 1: This example illustrates the use of a phosphating bath in coating steel panels wherein (a) the molar ratio of phosphate to zinc in the bath is initially within the limits which are specified in the process and a satisfactory phosphate coating is produced on a metal panel; but wherein (b) as the phosphating of further panels proceeds and the bath is replenished conventionally to maintain a constant level of zinc, this ratio moves outside the specified limits and a negligible phosphate coating is ultimately produced. Example 1(c) illustrates a continuous coating process according to the process wherein there is present NO_3^- as the anion N^{n-}.

(a) An acidic phosphating solution of zinc phosphate containing hydrogen peroxide as accelerator was prepared which comprised:

0.143 g mol orthophosphate (as PO_4) per liter of solution;

0.051 g atom zinc (as Zn) per liter of solution;

0.037 g atom sodium (as Na) per liter of solution; and

0.002 g mol hydrogen peroxide (as H_2O_2) per liter of solution.

The molar ratio PO_4/Zn in the solution was 2.8 which is within the limits defined in the process. The ratio of free acid to the total acid of this phosphating solution at 30°C was less than 0.05.

The above solution was sprayed onto a degreased rolled steel panel for 90 seconds at 30°C to produce a coating of zinc phosphate which when painted showed excellent resistance to corrosion. The coating weight was 1.6 g/m^2.

(b) As a succession of degreased steel panels was coated by spraying as described in (a), the phosphating solution was replenished to maintain the zinc content constant by adding appropriate quantities of (1) an acidic concentrate of zinc phosphate (containing 9.4% Zn and 38.5% PO_4) and (2) a solution of sodium hydroxide (which was necessary to control the ratio of free to total acidity in the bath at 30°C below a value of 0.05). Further hydrogen peroxide was also added in order to maintain the concentration defined in Example 1(a).

The quality of the phosphate coating diminished with the number of panels processed and it was noted that the phosphating solution reached a steady state only when it had the composition:

0.219 g mol phosphate (as PO_4) per liter of solution;

0.051 g atom zinc (as Zn) per liter of solution; and

0.113 g atom sodium (as Na) per liter of solution.

This solution gave a coating weight of only 0.06 g/m^2 on the degreased steel panels when sprayed for 90 seconds at 30°C. The molar ratio PO_4/Zn was then 4.3 which is outside the limits of the process.

(c) An acidic phosphating solution of zinc phosphate containing hydrogen peroxide as accelerator and additional nitrate ions (where $NO_3^- = N^{n-}$ according to the process; pK_a of HNO_3 is less than 1) was prepared which comprised:

0.143 g mol phosphate (as PO_4) per liter of solution;

0.051 g atom zinc (as Zn) per liter of solution;

0.113 g atom sodium (as Na) per liter of solution;

0.076 g mol nitrate (as NO_3) per liter of solution; and

0.002 mol hydrogen peroxide (as H_2O_2) per liter of solution.

The molar ratio PO_4/Zn was 2.8 as in Example 1(a). When sprayed onto degreased rolled steel panels for 90 seconds at 30°C, a coating of zinc phosphate was produced which when painted showed excellent resistance to corrosion. The coating weight was 1.6 g/m^2.

The composition of the above coating solution could be maintained substantially as given above (and hence the molar ratio PO_4/Zn and the Zn content were both maintained substantially constant) by addition to the coating solution of further hydrogen peroxide and two replenishment feeds (1) and (2), (1) comprising:

Percent		Parts by Weight
19	Nitric acid	11.5
	Zinc oxide	12.2
81	Phosphoric acid	32.8
	Water	43.5

which had a free acidity (F) of 0.79 gram equivalent/kilogram of feed (1); and (2) comprising 5.5 parts by weight of sodium hydroxide and 94.5 parts by weight of water which had a free alkalinity (A) of 1.38 gram equivalents/kilogram of feed (2) in a delivery ratio (by wt) (2)/(1) of 0.57 kilogram of (2)/1 kilogram of (1).

Thus according to the requirement of the process

$$X = 1$$

since

$$F/A = 0.79/1.38 = 0.57$$

Throughout the operation, in which a succession of panels was sprayed with the bath solution at 30°C and with a contact time of 90 seconds per panel, highly satisfactory coatings were produced of uniform coating weight close to 1.6 g/m^2.

PHOSPHATING TECHNIQUES AND APPARATUS

Dip Phosphating

R. Murakami, H. Shimizu and S. Sato; U.S. Patent 4,287,004; September 1, 1981; assigned to Nippon Paint Co., Ltd., Japan provide a dip phosphating process which comprises introducing a substrate to be phosphated into a phosphating bath, dipping the substrate in the bath until the phosphating is accomplished at the surface of the substrate and taking out the phosphated substrate from the bath, characterized in that the introduction of the substrate into the bath is effected while undulating the phosphating solution in the bath at the entrance section of the bath.

The process will be illustrated more in detail with reference to Figures 1.4a and 1.4b.

In the figures, **1** is a phosphating bath, **2** is a settling bath provided between the entrance section and the central section of the bath **1**, and **3** is an overflow tank provided in front of the entrance section of the bath **1**.

Figure 1.4: Dip Phosphating: Undulating Phosphating Solution at Bath Entrance

a.

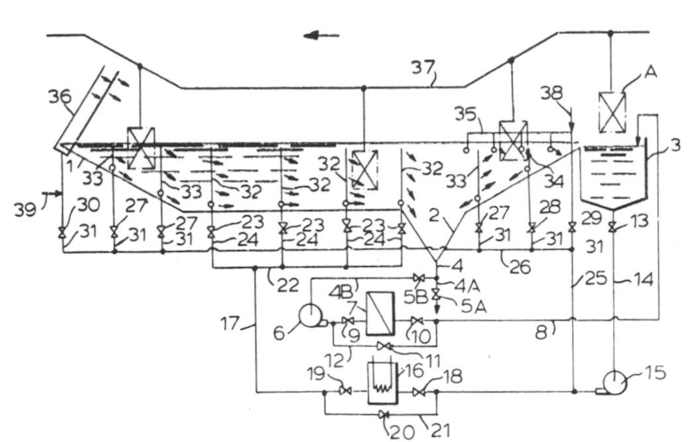

b.

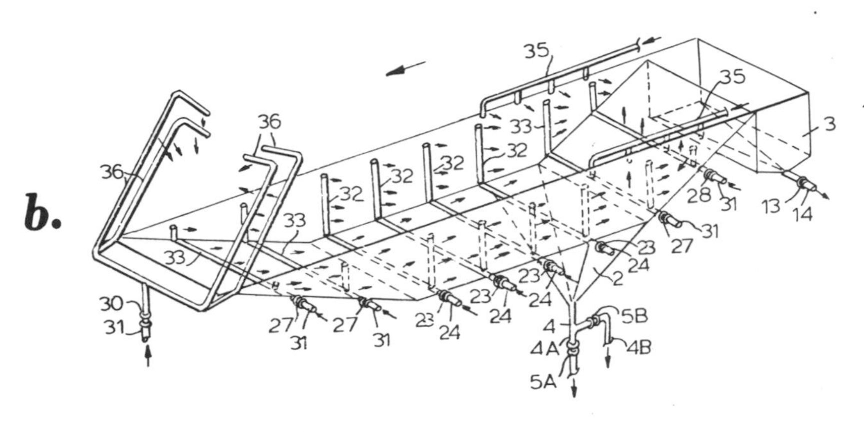

(a) Flow sheet of apparatus
(b) Details of spouting mechanism

Source: U.S. Patent 4,287,004

At the bottom of the settling bath **2**, there is provided a discharge pipe **4**, which is branched into a pipe **4A** having a valve **5A** and a pipe **4B** having a valve **5B**. The pipe **4B** is connected to a pump **6**, which is connected to a filter **7** by a pipe **8**. At the front and rear parts of the filter **7**, there are provided valves **9** and **10**. To the pipe **8** is connected a pipe **12** having a valve **11** without passing through the filter **7**.

At the bottom of the overflow tank **3**, there is provided an outlet pipe **14** with a valve **13**, the pipe **14** being connected to a pump **15**. To the pump **15**, a heat exchanger **16** is connected by a pipe **17**, and valves **18** and **19** are provided on the front and rear parts of the heat exchanger **16**. To the pipe **17** is connected a pipe **21** having a valve **20** as a detour.

The pipe **17** is connected to a main pipe **22**. To the main pipe **22**, there are connected branch pipes **24** each having a valve **23**. A pipe **25** branched from the pipe **17** is connected to the main pipe **26**, to which branch pipes **31** having respectively valves **27**, **28**, **29** and **30** are provided. As apparent from Figure 1.4b, the branch pipes **24** and **31** are arranged at a suitable distance on the bottom surface of the phosphating bath **1** and inside the lateral wall, and are connected to risers **32**, **33** and **34** which have nozzle holes for spouting. The other part of branch pipe **31** is arranged at the entrance section of the bath **1** and connected to a riser **35** which has the nozzle hole for spouting. A further part of branch pipe **31** is disposed on the lateral side at the exit section of the bath **1**, and is connected with a riser **36** having the nozzle hole for spouting. **37** is a device for conveyance of a substrate **A** to be phosphated. **38** and **39** are pipes for supplying water (which may be the water used in the washing step after the phosphating step).

The nozzle holes may be each opened into an appropriate direction so as to make the desired flows of the phosphating solution including the undulation at the entrance section. The arrow marks indicate an example of the directions of the nozzle holes which are practically applicable.

The phosphating treatment by the use of the apparatus may be carried out as follows.

First, the required amounts of a phosphating solution are supplied to the phosphating bath **1** and the overflow tank **3**. Then, the valves **13**, **18**, **19** and **23** are opened, the pump **15** actuated to spout the phosphating solution in the tank **3** into the bath **1** from the riser **32** through the heat exchanger **16**, whereby the phosphating solution in the bath **1** flows into the tank **3**. The heat exchanger **16** is a means of maintaining the temperature of the phosphating solution to the required temperature; when heating is not necessary, the valves **18** and **19** may be closed with opening of the valve **20**.

When the phosphating solution comes to the required temperature, the valves **27**, **28**, **29** and **30** are opened, and the phosphating solution is spouted from the risers **33**, **34**, **35** and **36**. Then, by means of the conveying device **37**, the substrate **A** is brought into the bath **1**, and phosphating treatment is started.

By the actions of the phosphating solution and/or water spouted through the riser **35**, the phosphating solution at the entrance section of the bath **1** is undulated. By this, there can be formed a uniform phosphating film without stepped unevenness of phosphating on the surface of the substrate A. Also, by spouting the phosphating solution upward from the riser **34** on the bottom surface at the entrance section of the bath **1**, the phosphating solution is sufficiently allowed to contact the portion which is otherwise not readily phosphated by the development of an air pocket, such as the recess portion on the bottom surface of the substrate A.

Also, by spouting the phosphating solution from the risers **32** and **33**, the phosphating solution in the bath **1** is stirred to run into the tank **3**. By conveying the substrate A through such phosphating solution, a uniform film of phosphating can be formed.

The substrate A after phosphating treatment is carried to the step of water washing by means of the conveying device **37**. In the bath **1**, the sludge by-produced in the phosphating treatment is floating, and the sludge so formed is stuck to the substrate **A**. Before the sludge is firmly deposited, the phosphating solution or water is spouted from the riser **36** to wash and remove the sludge. By the phosphating treatment, a large quantity of the sludge comprising mainly phosphate compounds is by-produced in the bath **1**. The sludge precipitated at the bottom of the bath **1** moves readily by its own gravity and under the spouting action of the phosphating solution from the risers **32** and **33** to fall into the settling bath **2**. When the sludge in the settling bath **2** becomes a high concentration, the valves **5B, 9** and **10** are opened to operate the pump **6**, by which the above slurry is sent to the filter **7** for separation between the solid and the liquid.

A test plate of commercially available cold-rolled steel (100 x 300 x 0.8 mm) was subjected to dip degreasing with a weak alkali degreasing agent (2%), (Ridoline SD 200, Nippon Paint Co., Ltd.) at a temperature of 60°C for 2 minutes. The plate was thereafter washed with water and subjected to dipping treatment with a metal surface conditioning agent (0.1%) (Fixodin 5N-5, Nippon Paint Co., Ltd.) at room temperature for 30 seconds. Then, after the dipping treatment with a phosphating solution (Zn, 0.15%; Ni, 0.04%; PO_4, 1.4%; NO_3, 0.5%; ClO_3, 0.1%; Cl, 0.1%; NO_2, 0.0065%) under the condition of a total acidity of 17 points, a free acid degree of 0.9 point, a toner value of 1.5 points and a temperature of 50°C for 2 minutes, the plate was washed with water and dried.

The conveyance speed of a substrate (i.e., an automobile body) in the automobile coating line is usually 4 to 5 m/min, and dipping of the substrate in a phosphating solution is made with inclination at an angle of 20 to 25 degrees to the solution surface. Therefore, in practicing the above operation, phosphating treatment was effected with adjustment thereto, and the finished appearance of the phosphating film was judged by visual inspection.

Comparative Example 1: By the use of a conventional zinc phosphating equipment of dip type, phosphating treatment was carried out.

Example 1: A zinc phosphating equipment of dip type different from a conventional one in the following respect was used: At the entrance section of the phosphating bath, the nozzle holes of the riser **35** were provided slightly below the surface of the phosphating solution so that the phosphating solution was stirred by undulation under the spouting of the phosphating solution.

Example 2: In addition to the construction of Example 1, the riser **35** was provided to the lateral side at the exit section of the phosphating bath so that the phosphating solution was spouted to the plate which had been just phosphated.

Example 3: In addition to the construction of Example 2, the risers **32, 33** and **34** were arranged at the bottom part and the lateral side so that the phosphating solution was stirred by spouting.

The results are shown in the following table.

	...... Appearance of Phosphating Film		
Example	Stepped Uneveness	Film Quality	Deposited Sludge
1	No	Fine	Deposited
2	No	Fine	None

(continued)

Example	Stepped Uneveness	Film Quality	Deposited Sludge
	......Appearance of Phosphating Film......		
3	No	Fine	None
Comparative	Present	Rough	Deposited

Closed-System Phosphating

R. Murakami, Y. Anegawa, M. Ishida and K. Masada; U.S. Patent 4,181,539; January 1, 1980; assigned to Nippon Paint Co., Ltd., Japan provide a process for iron phosphating of an iron substrate by treatment of the iron substrate with a phosphating solution in an installation of closed system, characterized in that the phosphating solution is an aqueous solution comprising at least one alkali metal or ammonium phosphate and at least one aromatic nitro compound as a phosphating accelerator and having a pH of about 3 to 6.5.

The alkali metal or ammonium phosphate is used usually in an amount of about 1 to 15 g (calculated in terms of phosphate ion) per liter, preferably in an amount of about 2 to 12 g/ℓ. The usual amount of the aromatic nitro compound is from about 0.05 to 5 g/ℓ, and the preferred amount is from about 0.2 to 2 g/ℓ. The temperature at the phosphating stage is usually from about 40° to 70°C, preferably from about 40° to 55°C. The time for phosphating treatment may be from a period of about 30 seconds to 10 minutes.

Example: In this example, iron phosphating of cold-rolled steel plates is effected by the use of an installation of closed system (spray-type) as shown in Figure 1.5. Each plate is first degreased at the degreasing step **1**. Degreasing is carried out by spraying an aqueous solution containing a weak alkaline degreasing agent known as Lidolin 75N-4 (Nippon Paint Co., Ltd.) in a concentration of 1.5% on the plate at 55°C for 1 minute. The degreased plate is rinsed with water at the rinsing steps **2** and **3** and then introduced into the phosphating step **4**.

In the phosphating step, a phosphating solution is sprayed onto the plate at a temperature of 50° to 55°C for 1 minute, during which the treatment area is 30 m^2/hr. As the phosphating solution, there is used an aqueous solution (pH 5.6) comprising sodium ion, phosphate ion and m-nitrobenzenesulfonate ion respectively in concentrations of 0.12%, 0.43%, and 0.05% and having a total acidity of 4.0 at the initial stage. In order to maintain the pH and the total acidity of the phosphate solution at the initial values during the treatment, an aqueous solution comprising sodium ion, phosphate ion and m-nitrobenzenesulfonate ion respectively in concentrations of 43 g/ℓ, 252 g/ℓ and 4.6 g/ℓ is occasionally supplied to the phosphating solution. The ion concentrations in the phosphating solution after the phosphating for 100 hr and 300 hr are as shown in the table below.

The thus-phosphated plate is rinsed with water at the rinsing steps **5**, **6** and **7** in order and finally dried at the drying step **8**. At the rinsing step **7**, fresh water **9** is sprayed on the plate, under which a tank is located as a reservoir. The overflow **10** from this tank is sent to a tank as a reservoir at the rinsing step **6**. The overflow **11** from this tank is then sent to a tank as a reservoir at the rinsing step **5**. The overflow **12** from this tank is further sent to the tank at the phosphating step **4** where water in an amount substantially equal to that of the overflow **12** is vaporized and exhausted through the duct **13**. The exhausted vapor is condensed by cooling, and the resulting water is used as fresh water in any rinsing step, usually as the fresh water **9**.

Figure 1.5: Flow Sheet for Phosphating Iron Substrate in a Closed System

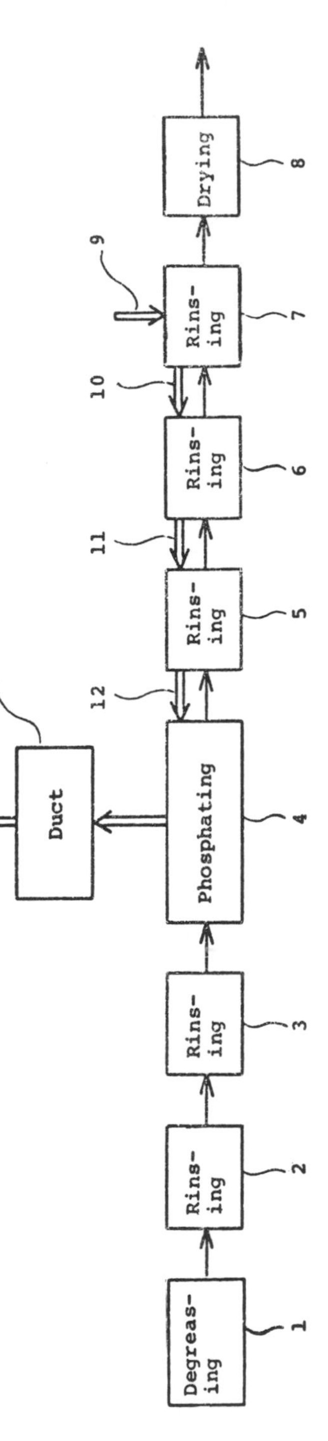

Source: U.S. Patent 4,181,539

The appearance of the plates as phosphated according to the above process is shown in the table below, from which it is understood that the concentrations of various ions (except iron ions) in the phosphating solution are substantially unchanged even after the treatment for 300 hours. A slight increase of the iron ion concentration is seen, but it is not so serious as to impart an unfavorable influence to the resulting phosphate film.

	Initial	...Phosphating After... 100 hr	300 hr
Ion Concentration, %			
Na^+	0.12	0.12	0.13
PO_4^{3-}	0.43	0.43	0.43
m-Nitrobenzenesulfonate*	0.05	0.048	0.051
Iron ion, ppm			
Fe^{2+}, Fe^{3+}	0	7	9
Total acidity	4.0	4.0	4.0
pH	5.6	5.6	5.6
Appearance	.. Uniform, reddish gold, excellent...		

*The phosphating solution (10 ml) is taken out and, after addition of 5 ml HCl and 0.5 g Zn powder, heated in a water bath for 30 minutes to cause reduction. The mixture is filtered by filter paper, and the filtrate titrated with a 1/40 N $NaNO_2$ solution using a potassium iodide-starch paper as an indicator.

In the installation used, the drag-out or take-out of the phosphating solution from the phosphating step can be recovered and returned to the phosphating step without removal from the installation. This is quite advantageous in causing no environmental pollution problem. Further, since the evaporated water at the phosphating step can be condensed and reused as fresh water in the rinsing steps, the amount of water to be supplied to the installation is much decreased. This is meritorious from the economical viewpoint.

Dip plus Spray Treatment

R. Murakami, H. Shimizu, T. Yoshii, M. Ishida, and H. Yonekura; U.S. Patent 4,292,096; September 29, 1981; assigned to Nippon Paint Co., Ltd., Japan found that the performances of a phosphate film such as adhesion and corrosion resistance are much improved when the film is formed by the use of a phosphating solution having a certain composition according to a certain specific treatment procedure. Advantageously, such film is quite suitable as a base for cationic electrocoating.

According to this process, a metal surface is first dipped in an acidic phosphating solution comprising a zinc compound in a concentration of 0.5 to 1.5 g/ℓ as zinc ion and a phosphate in a concentration of 5 to 30 g/ℓ as phosphate ion with at least either one of a nitrite in a concentration of 0.01 to 0.2 g/ℓ as nitrite ion and an aromatic nitro compound of 0.05 to 2 g/ℓ in water at a temperature of 40° to 70°C for not less than 15 seconds, and then sprayed with the same phosphating solution as above at the same temperature as above for 2 to 60 seconds. By such treatment, a uniform and fine phosphate film of low film weight (e.g., 1.5 to 3 g/m^2), especially having good adhesion and corrosion resistance suitable as a base for cationic electrocoating, is formed on the metal surface.

The apparatus for phosphating used in the examples is shown in Figure 1.6a and Figure 1.6b wherein Figure 1.6a represents a schematic view of the vertical section of such apparatus and Figure 1.6b is an enlarged perspective view of the frame-type hanger and the test plates as shown in Figure 1.6a, each of the hanger and the plates being separated. In these figures, a frame-type hanger **3** has a hook **1** on the upper surface and plural holes **2** on the peripheral surfaces. The two open sides of the hanger are fixed with test plates **4, 5**. The hanger **3** thus furnished with the test plates **4, 5** is suspended in the tank **6** which contains a phosphating solution. In case of spray treatment, the phosphating solution is sprayed to the test plates from the risers **7** located within the tank. In the case of dip treatment, the hanger with the test plates is dipped in the phosphating solution.

Figure 1.6: Apparatus for Dip plus Spray Technique

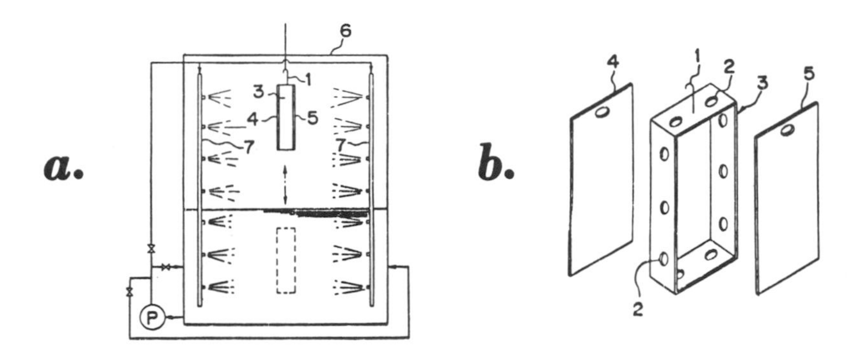

(a) Vertical section of phosphating apparatus
(b) Frame-type hanger and test plates—each hanger and test plate separated

Source: U.S. Patent 4,292,096

Examples 1 through 3: Commercialized cold-rolled steel plates (70 x 150 x 0.8 mm) were treated with an alkali degreasing agent (Ridoline SD 200, 2% by wt) at 60°C by spraying for 1 minute, followed by dipping for 2 minutes. The plates were then washed with water and dipped in a surface conditioning agent (Fixodine, 0.1% by wt) for 15 seconds. Thereafter, they were subjected to dip treatment in a phosphating solution containing 0.8 g/ℓ of Zn ion, 0.5 g/ℓ of Ni ion, 14 g/ℓ of PO_4 ion, 3 g/ℓ of NO_3 ion, 0.5 g/ℓ of ClO_3 ion and 0.08 g/ℓ of NO_2 ion under the conditions of a total acidity of 17 points, a free acidity of 0.9 point and a toner value of 1.5 points at a temperature of 52°C for 30 to 90 seconds, followed by spray treatment under the same conditions as above for 10 to 60 seconds. Thereafter, the plates were washed with tap water and deionized water in order and dried.

On the plates phosphated as above, the appearance of the film, the film amount and the film crystal were inspected with the inner surface (the surface facing the inside of the hanger) and the outer surface (the surface opposite to the inner surface). The results are shown in the table below wherein the photographs indicating the film crystals were taken on a scanning-type electron microscope (JSM-T20, Nippon Denshi Co.) at an angle of 45° and a magnification of 1:1500.

The above-phosphated plate was coated with a cationic electrocoating composition (Power Top U-30 Black, Nippon Paint Co., Ltd.) under application of an electric voltage of 250 V for 3 minutes to make a coating film having a thickness of 20 microns and was baked at a temperature of 180°C for 30 minutes.

The resulting electrocoated plate was subjected to 5% salt spray test, JIS (Japanese Industrial Standard) Z-2371, for 1,000 hours. The results are shown in the table below.

The above-electrocoated plate was coated with an intermediate coating composition (Orga To 778 Gray, Nippon Paint Co., Ltd.) to make a coating film having a thickness of 30 microns, followed by baking.

The resultant plate was then coated with a top coating composition (Orga To 226 Margaret White, Nippon Paint Co., Ltd.) to make a coating film having a thickness of 40 microns, followed by baking.

The thus-obtained 3-coated and 3-baked plate was dipped in deionized water at 50°C for 10 days and cut into sections at intervals of 2 mm on both sides to make 100 squares in total. To the surface an adhesive tape was stuck, after which it was peeled off and the number of the squares remaining on the plate were counted to inspect adhesion.

Another 3-coated and 3-baked plate was installed at an inclination of 15 degrees to a horizontal plane. Onto this plate, a steel arrow of 1.00 g in weight and 140 mm in total length having a conical head (material quality: JIS G-4404; hardness, HV, more than 700) was dropped perpendicularly from a distance of 150 cm above the plate to make 25 flaws. Then, the plate was subjected to corrosion test (hereinafter referred to as "spot rust test") of 4 cycles, each cycle comprising salt-spray test (JIS Z-2371) for 24 hours, wet test (temperature, 40°C; relative humidity, 85%) for 120 hours and allowing to stand in a room for 24 hours. Survey was made on the mean values of the maximum sizes of filiform corrosion and blisters on the surface after the test. The results are shown in the table below.

Comparative Examples 1 through 3: Phosphating treatment and subsequent electrocoating and normal coatings were carried out in the same manner as adopted in Examples 1 through 3. However, in Comparative Example 1, phosphating treatment was made only by spraying for 2 minutes. In Comparative Example 2, phosphating treatment was made by spraying for 15 seconds and dipping for 2 minutes. In comparative Example 3, phosphating treatment was made by spraying for 30 seconds and dipping for 2 minutes. Performances of the produced films and coatings were tested in the same manner as in Examples 1 through 3. The results are shown in the table.

From the data in the table it can be seen that according to the spray process and the spray dip process given in Comparative Examples 1 through 3, a uniform, satisfactory film is formed on the outer surface of the plate, but an uneven film containing yellow rust and/or blue colored iron phosphate is formed on the inner surface. Further, even the film formed on the outer surface is inferior in water-resistant adhesion, resistance to salt spraying and performance against spot rust after cationic electrocoating. According to the process, normal, fine and good films are formed on both the inner and outer surfaces, and water-resistant adhesion, resistance to salt spray and performance against spot rust after cationic electrocoating are excellent.

Example No.	Phosphating Procedure	Film Appearance Outer Surface	Film Appearance Inner Surface	. .Weight of Film. . Outer Surface (g/m^2)	. .Weight of Film. . Inner Surface (g/m^2)
1	Dipping 30 sec. Spraying 60 sec.	Even, fine, excellent film		2.1	2.0
2	Dipping 60 sec Spraying 30 sec.	Even, fine excellent film		2.3	2.2
3	Dipping 90 sec. Spraying 10 sec.	Even, fine excellent film		2.3	2.3
1*	Spraying 2 min.	**	***	2.5	0.3
2*	Spraying 15 sec. Dipping 2 min.	**	†	2.4	1.4
3*	Spraying 30 sec. Dipping 2 min.	**	††	2.5	1.0

Example No.	. . . .Salt Spraying. . . . Outer Surface (mm)	Inner Surface (mm)	. . . . . . Adhesion . . . . Outer Surface	Inner Surface	. .Spot Rust Test . . Outer Surface (mm)	Inner Surface (mm)
1	<1	<1	100/100	100/100	0.95	1.1
2	<1	<1	100/100	100/100	0.94	0.91
3	<1	<1	100/100	100/100	0.90	0.90
1*	4.0	†††	0/100	30/100	2.81	5.03
2*	2.5	2	51/100	73/100	1.93	2.04
3*	3	4	0/100	20/100	2.12	3.27

*Comparative
**Even, excellent film
***Iron phosphate film with yellow rust
†Uneven, zinc phosphate film
††Blue colored iron phosphate film
†††Tape width

Foam Phosphatizing

G.W. Chunat and J.E. Maloney; U.S. Patent 4,060,433; November 29, 1977; assigned to Economics Laboratory, Inc. found that adequate control over the phosphatizing of a nonhorizontal or overhead metallic surface (e.g., a ferrous metal surface) can be obtained with a phosphatizing composition and a spraying technique which produces an adherent foam.

The method involves providing a mixture containing: (a) a phosphatizing composition containing at least 0.2% by wt of phosphate anions and having a pH ranging from about 3.0 to about 5.5; (b) at least about 0.25 part by wt, per 100 parts by wt of the phosphatizing composition of a high-foaming surfactant; and (c) a volume of gaseous fluid (e.g., air) which is relatively larger than the combined volume of the phosphatizing composition and the high-foaming surfactant. This mixture is used to generate a foam spray which is directed onto a nonhorizontal or overhead metallic surface to obtain an adherent foam deposit on this surface (the foam spray has adherent properties with respect to the surface). The foam deposit which results on the metallic surface is allowed to remain in place until a phosphate coating weight of at least 10 mg/ft^2 (about

105 mg/m^2) is obtained and until the adherent foam deposit has been in place for a single-pass, actual dwell time which inherently corresponds to a single-pass, vertical, ASTM D 609-61 panel dwell time of about 10 to 180 seconds.

The high-foaming surfactant should have an initial Ross-Miles foam height in excess of 10 cm, determined at 0.1% by wt concentration in water in a column of 0 hardness water maintained at 50°C.

The following illustrates typical ranges of ingredients for a foamable phosphatizing composition (i.e., a phosphatizing composition combined with a high-foaming surfactant).

Component	Parts by Weight
Phosphatizing agent, e.g., phosphoric acid	0.75-4
Accelerator system*	0.1-1.5
High-foaming surfactant	0.25-10

*Typically a compound selected from the group consisting of a molybdenum oxide, a molybdate salt, a nitrated aromatic sulfonic acid, a nitrated aromatic sulfonate, or mixtures thereof.

The abovedescribed phosphatizing composition can be made into a liquid concentrate by adding up to 100 parts by wt of water or a suitable organic carrier, e.g., a hydrocarbon liquid solvent having a flash point in excess of 85°F and an initial boiling point in excess of 285°F. It is preferred that the phosphoric acid be neutralized. Liquid concentrates with excellent compatibility with water are obtained when the phosphoric acid is neutralized with about 1 to 3 equivalents of an amine for each 3 equivalents of phosphoric acid. The pH of the liquid concentrate as well as the diluted concentrate or a dissolved solid phosphatizing composition should be within the range of about 3.0 to 5.5. The dibasic phosphate salts (e.g., salts of the formula MH_2PO_4, wherein M is ammonium or monovalent metal) are generally ideal from a pH standpoint. When amines are used for neutralization of the phosphoric acid, the preferred amines are of the alkanol-substituted type, e.g., mono-, di-, or triethanolamine.

The high-foaming surfactants particularly preferred for foamable phosphatizing compositions generally belong to one of the following classes of materials: alkyl sulfonates (including straight-chain alkyl sulfonates and aralkyl sulfonates); higher alkyl sulfates (particularly the straight chain type); alkyl ether sulfates; corresponding acids of these sulfonates and sulfates; alkylphenoxypolyethoxy ethanols (particularly the nonylphenoxy and isooctyl types containing several oxyethylene units); alkyl, ethyl cycloimidinium, 1-hydroxy, 3-ethyl alcoholate, 2-methyl carboxylate, alkyl amine oxides; polypeptides; and fatty acid diethanol amides.

Examples 1 through 5: Five liquid foamable phosphatizing compositions were used in a realistic field test of this process. The foam applicator was the 10 gallon standard Foam and Clean unit (Economics Laboratory, Inc.). The Foam and Clean unit was pressurized to 60 psi, the air valve was set in the full open position, and the valve for the foamable phosphatizing composition in the tank was set in the one-third open position. No hot water was used in applying the foam. The surface of the three-dimensional metal part to be phosphatized was oily. The part was fully coated with the foam, and the foam was allowed to dwell on the part until it began to slide off under the influence of gravity. The part was then totally rinsed with water heated to 180°F. The part was allowed to air dry and was then painted.

Each of the five foamable phosphatizing compositions contained the following active ingredients and diluents:

(a) a phosphatizing concentrate including phosphoric acid, hydrofluosilicic acid (for buffering), molybdenum trioxide (antibronzing agent), and xylene sulfonic acid (coupler), all neutralized with monoethanol amine;

(b) sodium m-nitrobenzene sulfonate (the accelerator);

(c) a high-foaming surfactant system (described subsequently);

(d) except in the case of Example 4, a hydrocarbon solvent containing over 90% aromatics with a flash point of 150°F, COC, (Cleveland Open Cup) known as Aromatic 150, (Exxon); and

(e) water, 6 to 9.1 parts by volume for each 0.9 to 4 parts by volume of the combination of components (a) through (d).

Laboratory phosphatizing was also carried out on vertically hung ASTM D-609 test panels, Q-Panel, Type S, (Q-Panel Company) with both single and double passes of the foam spray, thereby providing a contact (dwell) time of 60 seconds (for the single pass) and 90 seconds (for the double pass). The foam on the vertical ASTM D-609 panels slid off more or less completely at the end of the dwell time. Typical coating weights obtained for these dwell times were 40 to 45 mg/ft^2.

The combination of the above-listed components (a) through (e) provided for the five liquid foamable phosphatizing compositions described in the following table.

Components	1	2	3	4	5
	. . . . Example	(% by wt).			
Phosphoric acid	3.90	1.30	1.82	1.82	3.90
Hydrofluosilicic acid	0.90	0.30	0.42	0.42	0.90
Monoethanolamine	2.40	0.80	1.12	1.12	2.40
m-Nitrobenzene sulfonate	0.15	0.05	0.07	0.07	0.15
Molybdenum trioxide	0.03	0.01	0.014	0.014	0.03
Water	83.87	89.29	90.006	95.006	74.87
Aromatic solvent*	4.75	4.75	4.75	–	4.75
Xylene sulfonic acid	0.45	0.15	0.21	0.21	0.45
High-foaming surfactant system:					
n-alkyl benzene sulfonate	0.30	0.10	0.14	0.14	0.30
sodium lauryl ether sulfate	3.00	3.00	1.20	1.20	12.00
nonyl phenolethylene oxide adduct**	0.25	0.25	0.25	–	0.25

*Aromatic 150
**9-10 ethyleneoxy units

POSTPHOSPHATIZING TREATMENTS

Zinc Sealant Rinse Composition

It has been common practice to seal the surface of phosphatized metal components with a chromic acid rinse prior to painting. Hexavalent chromium is highly toxic and environmental considerations have resulted in a search for a less toxic substitute which provides corrosion protection for the metal components being treated.

The process of *G.D. Howell and D.A. Lange; U.S. Patent 4,220,485; September 2, 1980; assigned to Calgon Corporation* relates to a composition and method for sealing phosphatized metal components with a non-chromic-acid-based-material. The composition consists essentially of phosphoric acid, a zinc compound(s), a heavy metal accelerator and/or crystal refiner, a phosphonate corrosion inhibitor and sufficient water to dilute the composition to its desired strength.

The components of the composition of the process are present in the following amounts:

Component	Broad	Preferred
	.... (wt %)	
Phosphoric acid	5 to 80	22.5 to 60
Zinc compound(s)	1 to 16	3 to 12
Heavy metal accelerator and/or crystal refiner	0.1 to 10	2 to 7
Phosphonate	1 to 80	10 to 30

75% by wt phosphoric acid is the preferred material. Zinc oxide is the preferred form of zinc.

Heavy metal accelerators useful in the compositions include compounds of such metals as vanadium, titanium, zirconium, tungsten and molybdenum. The compounds utilized most frequently are the molybdates. In combination with or in place of accelerators, an optional crystal refiner, such as acid-soluble salts of nickel, cobalt, magnesium and calcium may be utilized.

The most preferred phosphonate compounds are aminotris(methylene phosphonic acid) and hydroxyethylidene-1,1-diphosphonic acid (HEDP) and water-soluble salts thereof.

The zinc sealant rinse composition may be applied by conventional immersion or spray processes. Typical processes which may be used include a three-stage process which comprises a cleaning and phosphating step, a water-rinse step and the zinc sealant rinse step. Better coatings may be obtained by using a five-stage process which comprises an alkaline cleaning step, a water-rinse step, a phosphatizing step, an additional water-rinse step and the zinc sealant rinse step. The zinc sealant rinse step is carried out at temperatures of from 55° to 180°F and contact times of from 10 seconds to 2 minutes.

Example: Metal panels were evaluated in salt-spray tests using a 5% salt spray at 95°F for 120 hours in accordance with the procedures set forth in ASTM Procedure B-117-64 and the panels were evaluated by ASTM Procedure D-1654-61 for corrosion creepage from a scratch as well as the degree of body blisters on the test area. The ratings are based on a scale of 1 to 10, with 10 being the best possible rating and 1 being the worst. A representative composition of the process was compared to prior compositions and the results are set forth in the table below.

In the following table A is a composition consisting of 5% by wt sodium molybdate, 50% by wt of 75% phosphoric acid, 20% by wt of Dequest 2000 [aminotris(methylene phosphonic acid)] and 25% by wt water; and B is a composition consisting of 5% by wt zinc oxide, 5% by wt sodium molybdate, 50% by wt of 75% phosphoric acid, 20% by wt of Dequest 2000, and 20% by wt water.

Panel Description	Water Rinse	0.05% Chromic Acid	A at 3 oz/gal	B at 3 oz/gal	B at 5 oz/gal	B at 7 oz/gal
Blister creepage	½" to ¾"	⅛" to ¼"	to ¼"	to 1/16"	to 1/32"	to 1/64"
ASTM D-1654-61						
Schedule #1 rating	1	5	4	7	8	9
Body blisters	None	None	None	None	None	None
ASTM D-1654-61						
Schedule #2 rating	10	10	10	10	10	10
Body pinhole rusting	None	None	None	None	None	None
Moisture penetration of paint	Nil	Nil	Nil	Nil	Nil	Nil
Paint thickness, mil	1.3	1.3	1.3	1.3	1.3	1.3
Paint coverage	Good	Good	Good	Good	Good	Good

The results set forth in the above table demonstrate the improvements obtained when using the compositions of the process.

Melamine-Formaldehyde Resin and Vegetable Tannin

The work of *L. Kulick and J.K. Howell, Jr.; U.S. Patent 4,039,353; August 2, 1977; assigned to Oxy Metal Industries Corp.* relates to a method of posttreating a phosphate or chromate conversion coated metal surface without the use of chromium chemicals, the improvement comprising contacting the surface with an aqueous solution comprising at least 0.1 g/ℓ of a melamine-formaldehyde resin and at least 0.01 g/ℓ of a vegetable tannin wherein the weight ratio of resin:tannin is at least 1:1.

The melamine-formaldehyde resin is selected from the group consisting of the methylolated melamines and the C_{1-4} alkyl derivatives thereof.

The tannin is selected from the group of tannin extracts consisting of tannic acid, chestnut, quebracho, wattle and cutch.

Example: An aqueous alkali metal phosphate solution was prepared in a five-gallon laboratory spray tank from sodium dihydrogen phosphate and sodium chlorate. The phosphate solution, which contained 10 g/ℓ phosphate ions and 5 g/ℓ chlorate ions, was heated to 160°F. A number of 4" x 12" CRS (cold-rolled steel) panels were conventionally spray-cleaned for one minute at 160°F, hot-water-rinsed for one-half minute, spray-coated in the alkali metal phosphate solution for 1 minute, cold-water-rinsed for one-half minute and posttreated.

The posttreatment solution was made up to contain 4 g/ℓ of resin (Tanak MRX from American Cyanamid Co.) plus tannin (nonbisulfited quebracho from Arthur C. Trask Corp.). The pH was adjusted to 8.5 and the weight ratio between the resin and the tannin in the solution was varied while maintaining the total concentration at 4 g/ℓ. Panels were separately tested with the Dulux 704-6731 and the Duracron 200 paint. The salt-spray results after 168 hours are given in Table 1. These results show that a weight ratio of resin to tannin of at least 1:1 is desired in order to obtain salt-spray results approaching those obtained with the conventional dilute chromium rinse. A ratio of 3.75:1 or above produces results as good as or better than those obtained with a dilute chromium rinse. While the paint employed affects the over-all spray results, the posttreatment improves the corrosion resistance for both paint systems.

Salt-spray corrosion resistance – Salt-spray corrosion resistance was measured in accordance with the procedure of ASTM B-117-61. The panels were rated in terms of the amount of paint loss from a scribe in 1/16" increments (N for no loss of paint at any point). The principal numbers represent the general range of the creepage from the scribe along its length whereas the superscripts represent spot or nonrepresentative creepage at the point of maximum creepage along the length of the scribe. Thus, $2\text{-}7^{10s}$ means representative creepage varied from 2/16" to 7/16" with a maximum of 10/16" at one or two spots. Where corrosion was extensive, the results were expressed as percent peel over the entire panel surface, e.g., 60% P.

Table 1: Resin and Tannin = 4 g/ℓ

Wt. Ratio	.... Salt Spray (168 hr)	
Resin/Tannin	Dulux 704-6731	Duracron 200 Paint
1/30	7-9	70% P
1/15	4-7	60% P
1/7.5	4-5	5-8
1/3.75	5-8	85% P
1/1	1-4	3.3^{4s}
2.75/1	$0\text{-}1^{s}$	0-1
7.5/1	N	$0\text{-}1^{s}$
15/1	N	0-1
30/1	0-1	$0\text{-}1^{2s}$
Controls		
Tap water	9-10	3-4
Deionized water	8-10	2-3
Chromium	$0\text{-}1^{s}$	0-1

The procedure of the example was repeated except that resin Cymel 7273-7 (American Cyanamid Co.) was employed and the weight ratio of the resin to the tannin was adjusted to 7.5:1. Results are shown in Table 2. These results show that at the weight ratio and pH values employed a resin concentration in excess of 0.45 g/ℓ is desirable in order to obtain salt-spray results comparable to those obtained with the dilute chromium rinse.

Table 2: Resin/Tannin = 7.5

Cymel 7273-7 Concentration, g/ℓ	pH	Salt Spray (168 hr)
0.45	9.2	3-5
0.90	9.7	1-1
1.35	9.9	$0\text{-}1^{s}$
1.8	9.9	1-2
1.8	4.0	1-2
Controls		
Tap water		7-10
Deionized water		9-10
Chromium		$0\text{-}1^{s}$

Inhibiting Rinse Containing Sodium Nitrite and Citric Acid

In accordance with the process of *J.V. Otrhalek, R.M. Ajluni, and G.S. Gomes; U.S. Patent 4,182,637; January 8, 1980; assigned to BASF Wyandotte Corporation* a chrome-free final rinse is provided in a process for treating surfaces of ferrous metal.

Ferrous metal surfaces are rendered corrosion-resistant by a first step in which the surface is phosphated to provide a conversion coating, followed by a rinse step in which the conversion coating is sealed to enhance its corrosion-resistant properties.

The inhibiting rinse for sealing phosphated ferrous metal surfaces comprises a solution consisting essentially of water, from about 0.40 to 2.2 g/ℓ of nitrite ions, and about 0.05 to 0.5 g/ℓ of citrate ion, with the pH of the solution being in the range of about 4 to 5. Best results are obtained when the inhibiting rinse contains about 1.6 g/ℓ of sodium nitrite and about 0.2 g/ℓ of citric acid.

Example 1: This example illustrates the preparation of corrosion-resistant ferrous surfaces in accordance with the process.

Low carbon steel panels were cleaned by immersion for about 5 minutes in a caustic solution maintained at about 170° to 180°F. The caustic bath consists of by wt, 2% caustic soda; 2% soda ash; 0.9% sodium metasilicate; 0.8% sodium tripolyphosphate; 0.3% sodium lignin sulfonate; 0.1% sodium gluconate; 0.6% sodium linear alkylate sulfonate and 93.3% water.

After cleaning, the panels were then rinsed with water to remove any residuals of the caustic solution.

The panels were then zinc phosphated by immersing them in a zinc phosphating solution maintained at about 150° to 165°F for about 10 minutes. The zinc phosphating solution used herein consists essentially of, by weight, 0.8% zinc ions, 2.4% phosphate ions, 0.07% nickel ions, 0.6% nitrate ions and 0.3% ferrous ions. The panels were then removed from the solution, rinsed with water at a temperature of about 130°F to promote dry-off, and dried with a forced air drier at 300°F.

The panels were then immersed for 30 seconds at about 60°C in an aqueous solution containing 1.6 g/ℓ of sodium nitrite and 0.1 g/ℓ of citric acid. The panels were then removed from the rinse solution and dried.

Example 2: The procedure of Example 1 was repeated, except that the citric acid was omitted from the rinse solution to provide a comparison result.

Example 3: The procedure of Example 1 was repeated, except that the sodium nitrite was omitted from the rinse solution to provide another comparison result.

Example 4: The procedure of Example 1 was repeated, except that the rinse was pure water in order to provide a control example.

The panels prepared according to Examples 1 through 4 were then exposed for 2 hours in a 5% salt fog according to ASTM-B-117. The amount of rust forming is shown in the table below in which an average result for 3 panels is given for each example.

Inhibiting Rinse	Percent Rust after 2 Hours in Salt Spray
Example 1	<10
Example 2	40
Example 3	40
Example 4	100

From the table above, it is seen that the use of either citric acid or sodium nitrite in the rinse water improves the resistance to corrosion.

However, it is also seen that the combination of citric acid and sodium nitrite is considerably better than the use of either ingredient alone.

Passivating with Fluorophosphate Salts After Phosphatizing

B. Parant, L. Cot, W. Granier and J.-H. Durand; U.S. Patent 4,153,478; May 8, 1979; assigned to The Diversey Corporation are concerned with a process for the treatment of metal surfaces by means of an aqueous solution of a fluorophosphate salt.

It is to be understood that by treatment of metal surfaces is meant passivation and the preparation of metal surfaces for painting. The metals which can be treated are steels, aluminum and its alloys and zinc and its alloys.

It is known that for the treatment of metal surfaces, and specifically passivation and paint bonding, oxychromium(VI) compounds are used which, however, have the disadvantages of being toxic and having carcinogenic action. The process provides metal treating compositions which are less toxic than those containing chromium(VI) compounds with respect to use for passivation of metal surface. The treatment may also be used to prepare metal surfaces for painting.

The fluorophosphate salts generally most suitable for use in the working solutions, and in solid premix concentrates, are those of the formulas:

$$(1) \qquad M^a_2PO_3F \cdot nH_2O$$

wherein M^a represents Na, K, Rb, Cs and NH_4 and n is 1 when M^a is NH_4 and n is 0 when M^a is Na, K, Rb and Cs;

$$(2) \qquad LiM^aPO_3F \cdot nH_2O$$

wherein M^a represents Na, K, Rb, Cs and NH_4 and n is 0 no matter which listed element M^a represents, n can be 1 when M^a is K and n is 3 when M^a is Na;

$$(3) \qquad NaM^aPO_3F \cdot nH_2O$$

wherein M^a represents K, Rb and NH_4, n is 1 when M^a is NH_4 and n is 0 when M^a is K or Rb;

$$(4) \qquad M^bPO_3F \cdot nH_2O$$

wherein M^b represents Cd, Mn, Ni and Zn, n is 8/3 when M^b is Cd, n is 4 when M^b is Mn and n is 6 when M^b is Zn or Ni;

$$(5) \qquad M^a_2M^b(PO_3F)_2 \cdot nH_2O$$

wherein M^a represents K or NH_4, M^b is Ni or Zn and n is 6 when M^a is NH_4 and n is 2 when M^a is K; and

$$(6) \qquad M^c_2(PO_3F)_3 \cdot nH_2O$$

wherein M^c is trivalent Cr, Fe and Al and n is 0 to 24.

Sodium fluorophosphate (Na_2PO_3F) and potassium fluorophosphate (K_2PO_3F) are considered the salts of choice for use in the process because they are very effective and low cost. The best salt is believed to be the potassium salt, especially when used in combination with potassium hydroxide in use solutions.

The use solutions will contain an effective amount, up to its maximum solubility in water, of at least one such compound in water. A concentration of about 0.25 to 100 g, and preferably about 2 to 10 g, of one or more of the compounds per liter of use solution is suitable for treating metal surfaces. From about 0.04 to 4.0 g/ℓ of use solution of an alkali metal hydroxide such as sodium or potassium hydroxide may be included in the use solution to give a desired alkaline pH. For paint adherence treatment of metal surfaces, better results are obtained with potassium hydroxide than sodium hydroxide.

The use solutions may also include one or more suitable surfactants, which may be amphoteric, cationic, anionic or nonionic. Some suitable surfactants which may be included in the use solutions are octylphenoxy poly(ethyleneoxy)ethanol, polyoxyethylene sorbitol oleate, diethanolamine fatty acid amide, sodium lauryl sulfate, fluorinated anionic surfactant (Florochemical FC-95) and sorbitan monooleate. Including a surfactant, such as in the range of 0.1 to 5% by wt, in a use solution generally will aid in wetting the metal surface to be treated and in subsequent rinsing of the treated surface.

Example 1: An SPCI steel (French designation) sheet was degreased and exposed to an amorphous phosphatizing bath under the following conditions. During processing in a spraying tunnel the part was first treated with a phosphatization product containing 90% by wt of sodium dihydrogen phosphate, 5% by wt of a nonionic surfactant, 4% by wt of butyl glycol and 1% by wt of sodium molybdate and used in a concentration of 10 g/ℓ at a temperature of 70°C for 1½ min.

After rinsing the phosphatized sheet, passivation according to the process was carried out in a hot aqueous solution (50° to 70°C) containing 6 g/ℓ of K_2PO_3F (the pH having been adjusted to 12 by adding KOH) for about 20 to 60 sec.

The thus-treated sheet was compared with an identical sheet for which passivation was carried out with chromic anhydride at the same concentration (6 g/ℓ). It was found that the sheet (unpainted) treated according to the process was able to resist a salt spray fog for 16 hours, whereas that treated with chromic anhydride only resisted for 7 hours.

Example 2: A truck chassis made of mild AG4 MC steel (French designation) parts and with various zinc-coated or galvanized parts was exposed to an amorphous phosphating solution at 60°C for 2 minutes.

Following rinsing, passivation was carried out in a solution of 0.8 g/ℓ of $NiPO_3F \cdot 6H_2O$, adjusted to pH 12 by adding KOH, for about 20 to 60 seconds.

The surfaces (unpainted) treated in this way had a resistance to salt-spray fog of 16 hours, as compared with 6 hours resistance for surfaces treated with chromic anhydride. Furthermore, paint adhesion tests performed according to French Standard NF T 30 038 yield a 100% rating with respect to glycerophthalic paint in the case of surfaces treated according to the process. Treatment with chromic anhydride gave an identical paint adhesion reading.

VARIOUS TREATMENTS

Cleaning and Activating Surfaces Prior to Phosphatizing

The method of *D.J. Guhde; U.S. Patent 4,152,176; May 1, 1979; assigned to R.O. Hull & Company, Inc.* comprises the steps of first preparing a mixture consisting essentially of water, sodium tripolyphosphate, disodium phosphate and a titanium-containing compound and adding this mixture to solid disodium phosphate with mixing whereby a solid titanium phosphate composition is obtained. Generally, the mixture prepared in the first step will contain from about 25 to 35 parts of water, about 12 to 25 parts of sodium tripolyphosphate, about 25 to 50 parts of disodium phosphate and about 0.02 to 10 parts of a titanium-containing compound.

In general, the method of preparing solid titanium phosphate compositions comprises the addition of sodium tripolyphosphate to water which is heated to a temperature of between about 65° to 95°C whereupon the titanium-containing compound is added while maintaining the mixture at the desired temperature. After thorough mixing in a blender (about 3 to 10 minutes), the disodium phosphate is added and blended into the mixture at the desired temperature for a period of from about 5 minutes to 1 hour or more.

This mixture may then be added to solid disodium phosphate either while hot or the mixture may be precooled. The amount of solid disodium phosphate should be sufficient to produce a dry powder when thoroughly blended with the above mixture and cooled.

Example: 35.2 parts (all parts by wt) of water and 12.6 parts of sodium tripolyphosphate is blended in a Cowles dissolver at 5,800 rpm at a temperature of from about 65° to 70°C for about 5 minutes whereupon 2.6 parts of titanium potassium fluoride is added to the mixture. This mixture is heated at 70°C for 5 minutes, and 49.6 parts of disodium phosphate is added. The contents of the blender are heated to and maintained at a temperature of about 75° to 80°C for approximately 15 minutes. The temperature of the mixture rises slowly to about 88°C. This mixture, which is a slurry, is added slowly to 100 parts of anhydrous disodium phosphate with good mixing and cooling. The solid product is crushed to produce a dry powder containing about 15% moisture.

In one variation of the above procedure, the slurry is cooled to room temperature prior to addition to the solid anhydrous disodium phosphate. When the slurry is precooled, it is somewhat easier to control the exothermic reaction obtained when the slurry is mixed with the anhydrous disodium phosphate.

The titanium-containing phosphate compositions prepared in accordance with the process are useful in cleaning and activating the surfaces of ferrous, zinc or aluminum metals and alloys thereof particularly for subsequent reaction with phosphate coating solutions. The solid titanium-containing phosphate compositions can be dissolved in water to form pretreatment solutions of different desirable concentrations. Thus, useful aqueous pretreatment solutions can be prepared containing a titanium ion concentration between about 0.004 and 0.05% and from about 0.01 to 2% of the sodium phosphates.

An example of a workable activating solution is prepared by dissolving about 1 g of the solid titanium phosphate composition in about 1 ℓ of water. The metal surfaces are cleaned and conditioned by treating the surfaces with the activating solution such as by immersion or in a spray line while maintaining the solution at a temperature of from about 35° to 50°C.

After the metal surface has been treated with the activating solution until clean, usually for a short period of time, the metal surface is subjected to a water rinse at about 50°C to remove any materials present which may not be desirable when the surfaces are subjected to phosphate coating compositions. When ferrous, zinc or aluminum metal surfaces are cleaned and activated by the method described above, improved quality phosphate coatings can be applied to the activated surfaces utilizing phosphating compositions and techniques well known in the art.

The solid titanium-containing phosphate conditioner may be added to aqueous solutions containing cleaning compounds normally used for cleaning metal surfaces such as sodium silicates, sodium phosphates, wetting agents. The concentration of the titanium-containing phosphate conditioner compound in activating solutions for treating metal surfaces can range between 5 to about 25%.

Preferably, a slurry containing about 4.5 to about 9 kg of the titanium-containing compositions in about 190 ℓ of water is metered into the last rinse solution just ahead of the phosphating stage. The conditioning treatment at this stage results in a phosphate coating having excellent characteristics.

Treatment of Rinse Water by Reverse Osmosis and Ion Exchange

The work of *R. Murakami and M. Zinnouti; U.S. Patent 4,130,446; Dec. 19, 1978; assigned to Nippon Paint Co., Ltd, Japan* relates to a process for phosphating iron and steel which includes a phosphate conversion coating stage and a multistep water rinsing stage. The method comprises subjecting a part of the rinsing water used therein to a reverse osmosis treatment, returning the concentrated liquid thereof to the conversion coating stage, and using the filtrate as replenishing water for the conversion coating and/or the degreasing and water rinsing stages after subjecting it to ion exchange treatment.

Example 1: The treating system illustrated in Figure 1.7 is adopted, in which the zinc phosphate treating liquid is used for the conversion coating **4**. The rinsing water **15** from the water rinsing **5** (0.6 ℓ/min) is passed through the filter **9** (sponge filter R-2410, Kanegafuchi Spinning Co., Ltd.), and then introduced into the reverse osmosis membrane device **10** under the pressure of 50 kg/cm^2 to obtain the membrane-filtered water **17** (0.53 ℓ/min).

Figure 1.7: Flow Sheet for Rinse-Water Treatment

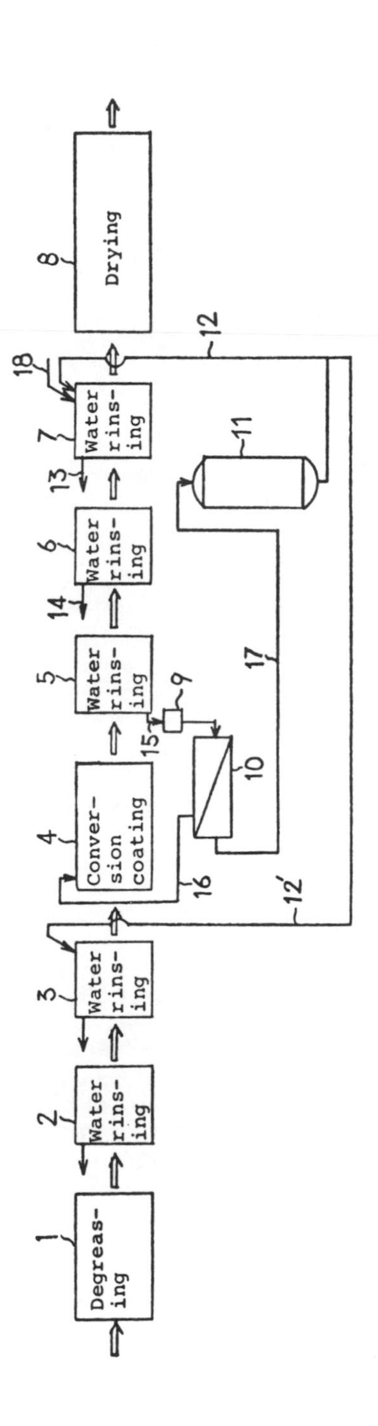

Source: U.S. Patent 4,130,446

This membrane-filtered water **17** is passed through the ion exchange resin column **11** to remove miscellaneous ions and used as the replenishing water **12** for the water rinsing **7**. On the other hand, the concentrated water **16** (0.967 ℓ/min) concentrated through the reverse osmosis membrane device **10** is returned to the conversion coating **4** for reutilization. The reverse osmosis membrane device and the membrane to be used are the BRO-type reverse osmosis membrane device (Paterson Candy Int.) and the reverse osmosis membrane T-2-15 (Paterson Candy Int.), and the ion exchange resins are, as H-type, Diaion SK 1B (Mitsubishi Kasei Kogyo Co., Ltd.) treated with 1 N HCl into H-type, and as OH-type, Diaion SA 10B (Mitsubishi Kasei Kogyo Co., Ltd.) treated with 1 N NaOH into OH-type.

The liquid compositions (ppm) of the above rinsing water **15**, concentrated water **16**, membrane-filtered water **17**, and replenishing water **12** are shown in Table 1.

Table 1

	Rinsing Water	Concentrated Water	Membrane-Filtered Water	Replenishing Water
Zn^{2+}	100	895	0.6	ND
Ni^{2+}	35	315	qs 0.1	ND
Na^{+}	400	1,600	245	qs 0.1
PO_4^{3-}	1,100	9,700	25	ND
NO_3^{-}	420	2,120	206	ND
NO_2^{-}	ND	ND	ND	ND

Note: ND is not detected.

Example 2: A dull steel plate is continuously subjected to conversion coating treatment, and the compositions (% by wt) of the treating liquid and the appearances of the conversion-coated films after lapse of 100 and 300 hours are examined.

As the degreasing solution there is used a 2% solution of Ridoline No. 75 (Nippon Paint Co., Ltd.), at the degreasing treatment temperature of 60°C and the treating time of 1 min.

The zinc phosphate treating liquid used is that originally having total acidity 16, acid ratio 20, and pH 3.0. The conversion coating treating temperature is 50° to 55°C and the treating time is 1 minute and 30 seconds. For replenishment of the zinc phosphate treating liquid an aqueous solution containing as the main components 2.4 mols/ℓ of zinc ion, 5.8 mols/ℓ of phosphate ion, and 0.25 mol/ℓ of nickel ion is used, and the replenishment is effected at the rate of 0.188 ℓ/hr. As the conversion coating reaction accelerator, an aqueous solution of 40% sodium nitrite is used, which is continuously added dropwise so as to make the concentration of nitrite ion in the treating liquid 0.008%.

According to the process, the conversion coating is continuously practiced while a part of the rinsing water is being taken out and treated, in the same manner as in Example 1. On the other hand, according to the conventional process, the conversion coating is continuously made without taking out or treatment of such rinsing water while the treating liquid is controlled by the closed-type system. In both this process and the conventional process, zinc phosphate conversion coating is made at the rate of the treating area of 30 m^2/hr, and the compositions of the treating liquid and the appearances of the conversion-coated films after lapse of 100 and 300 hours are examined, the results of which are shown in Table 2.

Table 2

	Initial	. . . Conventional Process. . .		. . . . . This Process.	
	Conc.	After 100 hr	After 300 hr	After 100 hr	After 300 hr
Treating liquid composition, wt %					
Zn^{2+}	0.100	0.079	0.038	0.098	0.099
Ni^{2+}	0.035	0.035	0.034	0.035	0.035
Na^+	0.30	0.45	0.70	0.32	0.32
PO_4^{3-}	1.1	1.1	1.0	1.1	1.1
NO_3^-	0.42	0.76	1.40	0.43	0.44
NO_2^-	0.008	0.008	0.008	0.008	0.008
Appearance of conversion coated film	*	*	**	*	*

*Uniform, fine, excellent.
**Yellow rust, coarse coating, not uniform, bad.

As apparent from Table 2, according to the conventional process, the nitrate ion and the sodium ion are accumulated at high concentrations in the treating liquid and the zinc ion concentration is decreased to invite unsatisfactory conversion coating. Against this, according to the process, excellent conversion coating treatment can be made with scarce variation in the ion concentrations of nitrate ion and sodium ion and without the lowering of the concentration of zinc ion even after the lapse of 300 hours.

pH Stable Liquid Concentrate

J.E. Maloney; U.S. Patent 4,017,335; April 12, 1977; assigned to Economics Laboratory, Inc. found that a substantially pH-stable liquid concentrate can be provided by neutralizing phosphoric acid with an organic amine having a pK_b within the range of about 3 to 10, preferably 3 to 5, thereby obtaining an organic ammonium phosphate salt (preferably an organic ammonium dihydrogen phosphate salt) solution having a pH within the range of about 3.0 to about 5.5.

The organic ammonium phosphate salt concentrates can be diluted with water in the weight ratio of from 1:5 to 1:250 (concentrate:water) and used in bath, spray, or steam phosphatizing (sometimes called phosphating). Preferably, the concentrates contain a buffer, various surfactants, and particularly for low temperature use (e.g., 20° to 55°C) an accelerator system. The concentrates are well suited to automatic dispensing of a phosphatizing spray or make-up solution (for a bath); they are efficient at low temperatures; and they have a good storage stability.

The table on the following page gives broad, preferred, and optimum proportions of the various components described previously. In footnotes to the table, it is pointed out that certain components are not necessary or desirable in concentrates for phosphate baths or spray compositions. As buffers, inorganic fluorinated acids are preferred, i.e., acids of the formula H_2MF_6, where M is an element of Group IV-A or IV-B of the Periodic Table.

Proportions Used in Concentrate in Percent by Weight

Component	Broad	Preferred	Optimum
Water*	qs	qs	qs
75% H_3PO_4 **	10-50	20-30	25
Buffer acid (H_2MF_6, 100%)	0.25-10.0	1.0-3.0	1.2
Amine ($HOCH_2CH_2NH_2$)	***	***	***
Surfactants			
Organic phosphate ester†	0.1-30.0††	2.0-15.0††	5.0-8.0††
Oxyalkylene-containing nonionic†	0.1-10.0	1.0-2.5	1.8
n-Alkylbenzene sulfonic acid (to be neutralized with amine)†††	0.1-10.0	1.0-3.0	2.0
Hydrotropic aromatic sulfonic acid (to be neutralized with amine)	30.0-1.0††	15.0-2.0††	8.0-5.0††
Accelerator system			
Molybdenum trioxide§	0.02-0.4	0.03-0.25	0.06-0.22
Sodium molybdate§	0.025-0.5	0.05-0.3	0.08-0.25
Organic nitro-compound	0.1-5.0	0.2-1.50	0.25-1.25

*Preferred, but optional, in all concentrates; can be omitted or replaced with organic solvent.

**Preferred ratio of H_3PO_4:buffer is 10-20:1, broad range can be 5-40:1.

***As needed to neutralize acids to pH of 3.0 to 4.0 in concentrate. Proportions will vary greatly with acid concentration and molecular and equivalent weights.

†Useful in spray-type phosphatizing concentrate, not required for bath type.

††See discussion in original patent regarding combinations with phosphate esters.

†††Useful in bath-type phosphatizing concentrate, not required in spray type.

§Molybdenum trioxide and sodium molybdate generally equivalent; both not needed. Sodium molybdate better suited for spray type, trioxide in bath type; used primarily to control bronzing.

Manufacture and Uses of Concentrate: Concentrates of this process should be diluted at least 1:5 but preferably not more than 1:250 for use in virtually any type of phosphatizing zone including zones provided by sprays (spray washers), baths, steam guns, pressure, etc. The preferred dilution range is from 1:20 to 1:50. Use solutions thus typically contain about 0.5 to 15% by wt of the concentrate, more preferably 2 to 5% by wt. The concentrates are preferably free of chromium-containing compounds (except for incidental amounts due to impurities or the like, e.g., amounts less than 0.1%).

Using the nomenclature of Ross et al, U.S. Patent 3,060,066, which is incorporated herein by reference, the use solutions produced from concentrates of this process are classifiable as noncoating, iron, and iron-on-iron. Thus, phosphatizing compositions of this process provide dihydrogen phosphate ions which can dissociate to form hydrogen ions and hydrogen phosphate ions. The hydrogen ions can attack the ferrous metal surface being treated to produce iron phosphate (e.g., ferrous or ferric hydrogen phosphate) crystals which adhere to the ferrous metal surface. Virtually any ferrous metal surface (iron, steel, etc.) can be treated. Good results are obtained at normal ambient temperatures and moderately elevated temperatures (e.g., 25° to 35°C) which are not overly energy consuming can also be used. (For energy conservation, operating temperatures below 50° or 55°C are preferred.)

The preferred manufacturing procedure is as follows:

(1) Charge H_3PO_4 and buffer (H_2MF_6) and all other acids (e.g., sulfonic acids) to mixer;

(2) Add sufficient amine, e.g., primary amine (RNH_2) for the reaction:

$$H_3PO_4 + H_2MF_6 + 3RNH_2 \rightarrow (RNH_3)H_2PO_4 + (RNH_3)_2MF_6$$

and, if applicable, for the reactions:

$$R^1SO_3H + RNH_2 \rightarrow (RNH_3)SO_3R^1$$

$$MoO_3 + H_2O + 2RNH_2 \rightarrow \text{organic ammonium molybdate.}$$

(3) Add surfactants, including coupling agents, if needed.

Example: *Spray-type phosphatizing concentrate* – The following components in the indicated amounts were blended to form monoethanolamine salts. The monoethanolamine salt of xylene sulfonic acid appears to provide good coupling effects.

	Amounts, Wt %
Water	40.55
Phosphoric acid, aqueous, 75%	23.10
Fluosilicic acid, aqueous, 30%	5.33
Xylene sulfonic acid	6.76
Antara LP-700*	6.22
Monoethanolamine	15.81
Sodium molybdate	0.09
m-Nitrobenzene sodium sulfonate	0.36
Ethoxylated alcohol benzyl ether, U.S. Patent 3,444,242**	1.78

*Organic phosphate ester surfactant
**$RO(CH_2CH_2O)_nCH_2C_6H_5$

The concentrate of this example was diluted to 3 wt % concentration with water and tested at various pHs and temperatures, using standard industrial Q panels. The phosphatizing time in all cases was 2 minutes. In the table below, results are given in coating weights (mg/ft^2).

Coating Weights for Phosphatizing Concentrate at 3% Concentration

	Temperature				
pH	80°F	90°F	105°F	120°F	160°F
5.6	17	–	–	–	–
4.3	43	–	–	88	133
3.5	38	–	–	–	–
3.0	40	43	58	60	–

A further run was made at 4.2% concentration and a temperature of only 75.2°F. A coating weight of 48.2 mg/ft^2 was obtained.

Improving Corrosion Resistance of Phosphating Oil by Addition of PVP and Zinc Powder

It is also known that the degree of corrosion resistance provided by phosphated metal surfaces can be enhanced by application of an oil composition which adheres to the phosphated surface and further prevents penetration of water, oxygen or other corrosive agents. Typically such an oil composition, commonly referred to as a phosphating oil, comprises an aqueous emulsion of an oil composition, which may contain stabilizers, emulsifying agents and extenders and other additives.

J. Hyner, J.M. Hage and S. Gradowski; U.S. Patent 4,216,032; August 5, 1980 provide an aqueous emulsion/or nonaqueous solution of oil used for treatment of phosphated metal or metallic surfaces to improve corrosion resistance. This comprises an aqueous emulsified or nonaqueous solution of oil containing an effective amount of polyvinylpyrrolidone and dispersed metallic zinc powder.

Polyvinylpyrrolidone polymers having an average molecular weight of between 10,000 to 360,000 are utilized. Amounts of polyvinylpyrrolidone ranging from about 1 to 60 g/ℓ have been found to be particularly effective.

It is believed that the zinc dust contributes to the improvement in corrosion resistance of the improved phosphating oil composition because it is able to accumulate or be deposited in micropores formed in the crystalline matrix of the phosphating coating on the phosphated metal surface. It has been found that an amount of zinc dust ranging from 1 to 100 g/ℓ is effective. Preferably, zinc dust ranging from about 0.5 to 25 microns in size is utilized.

Example: A 1 ℓ solution of the aqueous emulsified oil composition was prepared by diluting 20% by volume of a conventional phosphating oil, Rustarest 53253 (R.O. Hull and Company, Inc.) with water. To this solution was added 20 g of polyvinylpyrrolidone, having an average molecular weight of about 40,000 and 20 g of zinc dust. The composition was mixed thoroughly, so as to uniformly disperse the zinc powder and maintain its dispersion in the phosphating oil composition.

A steel fastener, having previously been phosphated by application of Irco Bond 51800 phosphating concentrate (R.O. Hull and Company, Inc.), was immersed in the composition of the process for a period of 30 seconds, while the composition was maintained at a temperature of 140°F. Following immersion of the steel fastener, it was allowed to drain and dry, and thereafter was subjected to the standard 72-hour neutral salt-spray test, ASTM B-117.

For purposes of comparison, an identical phosphated steel fastener was similarily treated with the same phosphated oil (Rustarest 53253), but without the polyvinylpyrrolidone and zinc dust additives of the process. Likewise, following drying, this fastener was subjected to the same salt-spray test and was found to endure 90 hours before failure. By comparison, the phosphatized steel fastener treated with the process composition endured 180 hours of exposure before failure in the salt-spray test. This represents a 100% improvement in corrosion resistance.

SOLVENT PHOSPHATIZING

METHYLENE CHLORIDE PHOSPHATIZING

Methylene Chloride, Methanol, Dimethylformamide, and Phosphoric Acid

E.A. Rowe, Jr. and W.H. Cawley; U.S. Patents 4,073,066; February 14, 1978; 4,008,101; February 15, 1977; and 4,070,521; January 24, 1978; all assigned to Diamond Shamrock Corp. found that a chlorinated hydrocarbon phosphatizing composition can produce a highly desirable coating when such composition is maintained in a "more wet" condition. An initial key ingredient for the composition is methylene chloride. A further critical ingredient, in addition to a phosphatizing proportion of phosphoric acid, is an amount of water exceeding such proportion of phosphoric acid. But such water is not present in sufficient amount to provide a liquid composition that does not retain liquid phase homogeneity. Moreover, it has been found possible to increase the coating weight of the resulting phosphate coating, by increasing the water content of the phosphatizing composition well beyond a content of just minute amounts.

A further and most significant theory is the achievement of phosphatized coatings of extremely reduced water sensitivity. Because of this, phosphate coatings are achieved wherein the coatings can be successfully topcoated with water-based compositions. Such compositions can include aqueous chrome rinses. They can additionally include such coatings as water reduced paints and electrocoat primers.

With the ingredients that are in the phosphatizing composition, including a solubilizing solvent capable of solubilizing the phosphoric acid in the methylene chloride, it has further been found that a vapor zone can be achieved in connection with the phosphatizing solution in which zone there is obtained enhanced rinsing. For example, with the solubilizing solvent methanol, an especially desirable vapor zone can be obtained.

Broadly, the process is directed to a methylene chloride and water-containing liquid composition having a continuous and homogenous liquid phase. The composition is suitable for phosphatizing metal with a water-resistant coating, while the liquid phase contains water in minor amount. More particularly, the composition comprises methylene chloride, solubilizing solvent capable of solubilizing phosphoric acid in methylene chloride, a phosphatizing proportion of phosphoric acid, and water in an amount exceeding the proportion of phosphoric acid while being sufficient for the composition to provide a phosphatized coating of substantial water insolubility, and while retaining liquid phase homogeneity.

Preparation of Test Panels: Bare steel test panels, typically 6" x 4" or 3" x 4" unless otherwise specified, and all being cold-rolled, low carbon steel panels are typically prepared for phosphatizing by degreasing for 15 seconds in a commercial, methylene chloride degreasing solution maintained at about 104°F. Panels are removed from the solution, permitted to dry in the vapor above the solution, and are thereafter ready for phosphatizing.

Phosphatizing of Test Panels and Coating Weight: In the examples, cleaned and degreased steel panels are phosphatized by typically immersing the panels into hot phosphatizing solution maintained at its boiling point for from 1 to 3 minutes each. Panels removed from the solution pass through the vapor zone above the phosphatizing solution until liquid drains from the panel; dry panels are then removed from the vapor zone.

Unless otherwise specified in the examples, the phosphatized coating weight for selected panels, expressed as weight per unit of surface area, is determined by first weighing the coated panel and then stripping the coating by immersing the coated panel in an aqueous solution of 5% chromic acid which is heated to 160° to 180°F during immersion. After panel immersion in the chromic acid solution for 5 minutes, the stripped panel is removed, rinsed first with water, then acetone and air dried. Upon reweighing, coating weight determinations are readily calculated. Coating weight data is presented in milligrams per square foot (mg/ft^2).

Example 1: To 288 parts of methylene chloride there is added, with vigorous agitation, 102.4 parts methanol, 1.3 parts orthophosphoric acid, and 15.8 parts N,N-dimethylformamide. These blended ingredients are thereafter boiled for 1 hour using a reflux condenser and the solution is permitted to cool. The water content of the resulting boiled solution, provided principally by the phosphoric acid, is found to be about 0.1 wt %. This water content is directly determined by gas chromatograph analysis of a sample wherein the column packing is Porapak Q (Waters Associates, Inc.). The resulting solution is then heated to 102° to 103°F and panels are phosphatized in the manner described hereinabove.

Some of the resulting coated panels, selected in sets of two with each panel in the set being coated under identical conditions, are then subjected to testing. One panel in the set is used for coating weight determination in the manner described hereinabove. The other panel in the set is subjected to the water-solubility test. For this test the panel is weighed and then immersed in distilled water for 10 minutes, the water being maintained at ambient temperature and with no agitation. Thereafter, the test panel is removed from the water, rinsed in acetone and air dried. Subsequently, on reweighing, the amount of water-solubility of the coating is shown by the weight loss. This loss, basis total original coating weight, is reported in the table below as the percentage or degree of coating loss.

Coating weights and water-solubility of coatings, are determined initially for test panels that have been phosphatized in the abovedescribed phosphatizing composition. Such data are determined thereafter for additional coated panels that have been phosphatized in compositions of differing water contents, all as shown in the table below. These baths of varying water content are prepared in stepwise fashion by starting with the abovedescribed bath, and then adding about 1 wt % water to the bath followed by boiling the resulting solution for 1 hour. This procedure is repeated with additional water increments of 1 wt %, as shown in the table below. The phosphatizing coating operation for each bath of varying water content has been described hereinabove. For each phosphatizing bath, water content determinations are made prior to phosphatizing by the abovedescribed method.

Water Content of Coating Bath, wt %	Panel Coating Weight, mg/ft^2	Solubility of Coating in Water, %
0.1	4	60
1.1	6	50
2.1	10	20
3.1	13	<5
4.1	24	<5

The tabulated results demonstrate the enhancement in the degree of water-insolubility for the phosphate coating as the water content in the phosphatizing bath increases. As determined by visual inspection, it is also noted that the degree of uniformity of the phosphate coating is increasing as the water content of the phosphatizing bath increases. For the particular system of this example, the desirable water content is deemed to be between about 2 wt % and about 5 wt %. Below about 2 wt %, the degree of water-solubility for the coated panels is regarded as being excessive. By continuing the stepwise water addition discussed hereinabove, this system is found to separate free water, i.e., lose liquid phase homogeneity, when the water content reaches 5.1 wt %.

Example 2: In the manner described hereinabove, a phosphatizing solution is prepared to contain, by weight, the following ingredients: 60 parts water, 1,188 parts methylene chloride, 253 parts methanol, 7.3 parts orthophosphoric acid, 47.2 parts N,N-dimethylformamide and 1.0 part dinitrotoluene. Hereinafter, for convenience, the resulting phosphatizing solution is referred to as the "process organic phosphatizing composition."

Steel panels were phosphatized in this process organic phosphatizing composition. Further, in the manner described hereinbefore, but for comparative purposes, panels were phosphatized in a well-known and extensively used commercial phosphatizing bath based on trichloroethylene, referred to as the "standard organic phosphatizing composition." This standard organic phosphatizing composition was prepared by blending together orthophosphoric acid with two products, Triclene-L and Triclene-R, to contain a commercially acceptable amount of phosphoric acid in the blend. The use of such a commercial phosphatizing bath has been described, for example, in U.S. Patent 3,356,540.

Additional comparative test panels used herein for evaluation are panels with an aqueous phosphatizing composition and prepared in accordance with specifications that are generally accepted as standards for performance in the automotive and household appliance industries. These comparative test panels are generally referred to herein as prepared from the "comparative aqueous phosphatizing composition." Such composition is a solution that can contain zinc acid phosphate, with the test panels being dipped in this aqueous solution typically for 1 minute. Thereafter, the test panels are rinsed and then immersed in a dilute solution of chromic acid. Such test panels are then dried and are thus provided with a chromic acid rinse coating.

All test panels are painted, before testing, with a commercial enamel topcoat. The enamel is a commercial white alkyl baking enamel; the enamel ostensibly contains a modified alkyd resin based upon a system of partially polymerized phthalic acid and glycerin, and has 50 wt % solids. After coating panels with the enamel, the coating is cured on all panels by baking in a convection oven for 20 minutes at a temperature of 320° to 325°F.

Panels are then selected and subjected to the various tests described hereinbefore for testing paint film retention and integrity. The tests used and the results obtained are listed in the table below. In the conical mandrel test, the numbers listed in the table are centimeters of paint removal after taping; the reverse impact test is conducted at 64 inch-pounds. For the reverse impact test and the conical mandrel test, where a range is presented in the table, such range results from the testing of a series of panels.

In the following table, the efficacy of the total coating obtained on the coated parts in the cross hatch and reverse impact tests is quantitatively evaluated on a numerical scale from 0 to 10. The parts are visually inspected and compared with one another and the system is used for convenience in the reviewing of results. In the rating system the following numbers are used to cover the following results:

(10) Complete retention of film, exceptionally good for the test used;

(8) Some initial coating degradation;

(6) Moderate loss of film integrity;

(4) Significant film loss, unacceptable degradation of film integrity;

(2) Some coating retention only; and

(0) Complete film loss.

Phosphatizing Composition	Cross Hatch	Conical Mandrel	Reverse Impact	Coin Adhesion
Process Organic	10	0-1.7	6-9	Good
Standard Organic	10	0.4-1.9	4-8	Good
Comparative Aqueous	10	1.9	4-9	Good

The tabulated results above show that the phosphate coating from the process organic phosphatizing composition can provide paint adhesion that will compare under a variety of tests as the equal of or superior to comparative systems based either on organic commercial baths or aqueous compositions.

Adjusting Content of Water and Solubilizing Solvent

In U.S. Patent 4,008,101 there are disclosed phosphatizing compositions which contain methylene chloride and water, while having a continuous and homogeneous liquid phase.

E.A. Rowe, Jr. and W.H. Cawley; U.S. Patent 4,056,409; November 1, 1977; assigned to Diamond Shamrock Corp. have found that topcoat adhesion can be significantly enhanced following the making of adjustments to the constituency of the phosphatizing composition. And importantly, other attractive aspects of the phosphatized coating are not deleteriously affected. Through augmenting the concentration in the phosphatizing composition of water and/or solubilizing solvent, e.g., methanol, or by adding to the composition, or by augmenting the concentration, of aprotic polar organic compound, phosphate coatings from the resulting composition can have significantly increased topcoat adhesion. Additions of such ingredients in combinations are also most effective.

The resulting adjusted solution will preferably contain above about 4 wt % thereof, up to and including water saturation at the operating temperature of the solution. And when this is accompanied by addition of solubilizing solvent, especially for the C_{1-4} alcohols, the resulting phosphatizing medium will typically contain from about 16 to about 20 wt % of such alcohol and will have a specific gravity at or near its boiling point of between about 1.12 and about 1.14. Such a specific gravity, as can be determined by hydrometer, will thus be obtained at a temperature within the range of from about 95° to about 105°F.

It is typical practice during the phosphatizing operation to maintain the phosphatizing solution at or near boiling condition. Under these conditions of temperature and specific gravity, commercially desirable coatings which can have augmented topcoated adhesion, will be efficiently achieved.

Example: A fresh phosphatizing solution that is recognized as a standard for a methylene chloride phosphatizing composition, is prepared from 100 parts of methylene chloride, 21.48 parts methanol, 0.91 part orthophosphoric acid, 4.6 parts N,N-dimethylformamide (DMF), 0.09 part dinitrotoluene, 0.04 part p-tert-amylphenol, 4.98 parts water, and 0.03 part p-benzoquinone. All parts are weight parts. These ingredients are simply mixed together and the resulting phosphatizing composition is identified hereinafter as the "standard" composition.

A phosphatizing composition, having an adjusted constituency, is then prepared in like manner. The ingredients in this composition are the same as for the standard composition, except for the following: water content is increased from 4.98 to 6.15; methanol content is increased from 21.48 to 25.64; and the DMF content is increased from 4.6 to 5.87. Hereinafter this composition is identified as the "first adjusted" composition.

In like manner, a "second adjusted" composition is prepared wherein, on the basis of the standard composition, the following constituency is changed: the water is adjusted from 4.98 to 6.82; the methanol from 21.48 to 28.03; and the DMF from 4.6 to 6.59.

Bare steel test panels, being 6" x 4" cold-rolled, low carbon steel panels are used for phosphatizing. The panels are phosphatized by immersing the panels in the selected solution, first heated to its boiling point and maintained there during phosphatizing. Immersion times are varied to provide for a low coating weight and for a high coating weight, as shown in the table below. Panels removed from the respective phosphatizing solution pass through the vapor zone above such solution until liquid drains from the panel; dry panels are then removed from the vapor zone.

For all compositions, that is, the standard composition, the first adjusted composition, and the second adjusted composition, the low coating weight is 30 mg/ft^2 and the high coating weight is 80 mg/ft^2. The coating weight determination is described in U.S. Patent 4,008,101.

As shown in the table below, selected panels are subjected to the reverse impact test described below in U.S. Patent 4,102,710. The results are tabulated as shown.

	400°F Cure Paint	. . Enamel: 225°F Cure . .	
	80 in-lb	80 in-lb	160 in-lb
Bath	. . . . 30 mg/ft^2 Coating Weight.		
Standard	4.5	6.5	1.0
First adjusted	5.5	10.0	9.0
Second adjusted	3.0	10.0	9.0
	. . . . 80 mg/ft^2 Coating Weight		
Standard	2.5	9.5	8.0
First adjusted	4.0	9.5	8.5
Second adjusted	7.0	10.0	9.0

As shown in the table, the topcoat adhesion at the 160 in-lb test for the enamel is most dramatically enhanced. And this, plus the other improvements in topcoat adhesion, are accomplished by adjustment of the composition constituency, that is, without bringing a new ingredient into the composition.

Use of Phenolic Stabilizer

E.A. Rowe, Jr. and W.H. Cawley; U.S. Patent 4,102,710; July 25, 1978; assigned to Diamond Shamrock Corp. found that topcoat adhesion can be increased for coatings applied over a solvent phosphatized metallic surface.

The composition capable of forming a phosphatized coating of substantial water-insolubility on a metal surface comprises methylene chloride, solubilizing solvent capable of solubilizing phosphoric acid in methylene chloride, a phosphatizing proportion of phosphoric acid, water in an amount exceeding the proportion of phosphoric acid, while being sufficient for the composition to provide a phosphatized coating of substantial water-insolubility, and while retaining liquid phase homogeneity, with the composition being particularly characterized by containing phenolic stabilizing substance in an amount above about 0.2 wt %, basis total composition weight. The methylene chloride phosphatizing compositions are disclosed in U.S. Patent 4,008,101.

The phenolic stabilizing substance can be phenol or substituted phenol having one or more ring substituents that are either alkyl or alkoxy. The phenolic stabilizing substance can also be a bisphenol or a bisphenol having connecting alkyl ring linkage, i.e., bisphenol A. Further, the bisphenols can have, on each ring, one or more of alkyl and alkoxy substituents. All of such alkyl substituents are preferably lower alkyl of 6 carbon atoms or less. Likewise, the alkoxy substituents will have 6 carbon atoms or less. Although these stabilizing substances are likely solid before typical use, many if not all are soluble in methylene chloride as well as in the solubilizing solvent.

Preferably for efficiency and economy, such substance is present in the composition in an amount between about 0.25 to 0.6 wt % basis total composition weight.

Substituted phenols that are useful as the stabilizing substance in addition to such mentioned hereinabove, include thymol, p-tert-butylphenol, p,p'-bisphenylmethane, butylated hydroxytoluene, p-tert-amylphenol, hydroquinone monomethyl ether, and their mixtures.

Example: A composition for sustaining phosphatizing by addition to a phosphatizing bath is prepared by blending together 93.28 parts methylene chloride, 5.99 parts methanol, 0.71 part water, 0.01 part p-tert-amylphenol (referred to hereinafter as "Pentaphen") and 0.01 part p-benzoquinone. Hereinafter, the resulting homogeneous, stable solution is referred to as the "sustaining solution."

For comparative purposes there is separately prepared, by blending together into a homogeneous solution, 62.64 parts methanol, 17.57 parts water, 19.24 parts N,N-dimethylformamide, 0.38 part dinitrotoluene, 0.12 part Pentaphen, and 0.044 part p-benzoquinone. This is referred to hereinafter as a "comparative" precursor composition. One part by volume of this resulting uniform solution is then blended with 3 parts by volume of the sustaining solution. To this resulting homogeneous blend there is then added sufficient orthophosphoric acid to provide about 0.55% by weight of the orthophosphoric acid in the resulting blend. Such blend is then termed the "comparative bath."

For improvement purposes other precursor compositions are prepared as described hereinabove, except that such compositions contain 1.37 and 2.11 parts of pentaphen. These are the improved precursor compositions. Phosphatizing baths are made with these improved precursor compositions in the manner described above to contain 0.2 and 0.4 wt % of Pentaphen, as shown in the table below. These are improved baths. These baths, and the comparative bath, are each then separately heated to reflux and maintained in such condition for about 7 days to prepare same for phosphatizing.

Bare steel test panels, being 6" x 4" cold-rolled, low carbon steel panels are used for phosphatizing. The panels are typically prepared for phosphatizing by degreasing for 15 seconds in a commercial, methylene chloride degreasing solution maintained at about 104°F. Panels are removed from the solution, permitted to dry in the vapor, and are thereafter ready for phosphatizing. The panels are phosphatized by immersing the panels in the selected bath, first heated to its boiling point and maintained there during phosphatizing. Immersion time is 1 minute. Panels removed from the respective phosphatizing baths pass through the vapor zone above such bath until liquid drains from the panel; dry panels are then removed from the vapor zone.

For all compositions, that is, the comparative bath as well as the improved baths, the coating weight is approximately 40 mg/ft^2 of phosphate coating on the panels as determined by visual observation of a blue-gold coating coloration.

Some of the test panels, selected at random, are spray painted with a commercial gray enamel topcoat. The spray equipment is a disposable Sprayon Jet Pad using a dichlorodifluoromethane propellant. The enamel is a baking enamel that ostensibly contains a modified alkyd resin based upon a system of partially polymerized phthalic acid and glycerine, and has 50 wt % solids. After coating, coated panels are cured by baking in a convection oven for 25 minutes at a temperature of 300°F. Coating thickness is about 1 to about 1.5 mils, which from experience, is regarded as only a minor variation in paint thickness for the reverse impact test when using the same paint system with each panel.

As shown in the table below, selected panels are subjected to the reverse impact test. In this test the weight of the metal ram is coordinated with the height of the drop to provide both 80 in-lb and 160 in-lb as shown in the table below. Paint removal is determined qualitatively by first taping the convex (reverse) surface of comparative panels before testing, then removing the tape following testing and rating them by visual observation of the tape. In the rating system the following numbers are used to cover the following results:

(10) Complete retention of film, exceptionally good for the test used;

(8) Some initial coating degradation;

(6) Moderate loss of film integrity;

(4) Significant film loss, unacceptable degradation of film integrity;

(2) Some coating retention only;

(0) Complete film loss.

Bath	Pentaphen* Content of Bath, pbw	....Reverse Impact....	
		80 in-lb	160 in-lb
Comparative Bath	0.12	5	0
Improved Baths:			
First	0.2	10	10
Second	0.4	10	10

*p-tert-amylphenol.

Nonaqueous Rinse Composition

The rinsing of phosphatized metal substrates is a usual procedure and such is typically done with a chromium-containing rinse composition. The metal substrate that has been phosphatized, as by immersion in a phosphatizing bath, while in wet condition and often after rinsing, is then typically immersed or flooded with a chromium-containing rinse composition.

W.H. Cawley and E.A. Rowe, Jr.; U.S. Patent 4,186,035; January 29, 1980; assigned to Diamond Shamrock Corp. describe a chromium-containing treating composition, which is especially useful for rinsing phosphatized metal surfaces. This contains hexavalent chromium-containing compound, methylene chloride and a substance that can be acetamide, N,N-dimethylformamide, acetone oxime and mixtures thereof. The composition is a homogeneous solution that is particularly useful where aqueous systems may create potential water pollution problems.

Preparation of Test Panels: Bare steel test panels, 6" x 4" or unless otherwise specified, and all being cold-rolled, low carbon steel panels are typically prepared for phosphatizing by degreasing for 15 seconds in a commercial degreasing solution maintained at its boiling point. Panels are removed from the solution, permitted to dry in the vapor above the solution and are thereafter ready for phosphatizing. Phosphatizing of test panels and coating weight is described above in U.S. Patent 4,073,066, page 77.

Example: To 100 parts of stabilized, commercially available technical grade methylene chloride there is added, with agitation, 21.5 parts of methanol, 5.0 parts deionized water, 4.6 parts N,N-dimethylformamide, 0.09 part dinitrotoluene, 0.04 part Pentaphen, 0.02 part p-benzoquinone, and 0.33 part of 85% strength orthophosphoric acid. The bath is heated to reflux and panels, prepared as described above, are coated in the above discussed manner. Coating weights for panels are determined in the previously described manner and are found to be 40 mg/ft^2.

Several different rinse compositions are then prepared, each to contain 594.9 parts of methylene chloride. The No. 1 composition additionally contains 9.44 parts of N,N-dimethylformamide and 0.725 part of chromic acid. The No. 2 composition contains 18.88 parts of N,N-dimethylformamide and 0.736 part of chromic acid. The No. 3 composition contains 0.44 part of acetone oxime and 0.6 part of chromic acid. The No. 4 composition contains 1.32 part of acetone oxime and 0.6 part of chromic acid. The No. 5 composition contains 14.16 parts of N,N-dimethylformamide, 0.18 part of acetamide, and 1.2 parts of chromic acid. The No. 6 composition contains 21.24 parts of N,N-dimethylformamide, 0.44 part of acetamide and 0.75 part of chromic acid.

In each case, chromium-free blends were first prepared, and the chromic acid then carefully added to these blends, with agitation. Moreover, with both the No. 5 and No. 6 compositions, the first preparation step involved the dissolving of the acetamide in the N,N-dimethylformamide.

Each bath is used by taking a phosphatized panel, prepared in the abovedescribed manner, and immersing in one of the baths for 15 seconds. The panel is then removed from the bath and permitted to air-dry in the ambient atmosphere for 5 minutes before being painted.

An additional, comparative, test panel is used herein for evaluation purposes. It is a panel phosphatized with an aqueous phosphatizing composition and prepared in accordance with specifications that are generally accepted as standards for performance in the automotive and household appliance industries. This comparative test panel, for convenience, is generally referred to herein as the "comparative panel." Such aqueous composition is a solution that can contain iron phosphate, with the test panel being dipped in this aqueous solution typically for 1 minute. Thereafter, the test panel is rinsed and then immersed in a dilute solution of chromic acid. Such test panel is then dried and is thus provided with a chromic acid rinse coating.

All test panels are painted, before testing, with a commercial enamel topcoat applied with a draw bar. The enamel is a commercial white alkyd baking enamel; the enamel ostensibly contains a modified alkyd resin based upon a system of partially polymerized phthalic acid and glycerin and has 50 wt % solids. After coating panels with the enamel, the coating is cured on all panels by baking in a convection oven for 35 minutes at a temperature of 300°F.

All resulting panels are then scribed with an X configuration and subjected to salt spray testing. The duration of the test, and the test results are reported in the table below.

Chrome Rinse Solution No.	Salt Spray Paint Loss* 118 Hours
Comparative panel	5.5/32
No. 1	1/32
No. 2	1/32
No. 3	3/32
No. 4	3/32
No. 5	1/32
No. 6	1.5/32

*Average of two panels.

Corrosion resistance of coated panels is measured by means of the standard salt spray (fog) test for paints and varnishes, ASTM B-117-64. The figures presented in the examples, e.g., 1/32, indicate the inches, expressed in thirty-seconds of an inch, of coating failure away from scribe lines.

OTHER SOLVENTS

Use of Hydrocarbon and Other Solvents

The process of *E.A. Rowe, Jr. and W.H. Cawley; U.S. Patent 4,029,523; June 14, 1977; assigned to Diamond Shamrock Corp.* is directed to an organic phosphatizing composition having a continuous and homogeneous liquid phase. The composition is suitable for phosphatizing metal with a water-resistant coating, while the liquid phase contains water in minor amount.

The composition comprises an organic solvent providing liquid phase homogeneity with a solubilizing liquid, while being a nonsolvent for a phosphatizing proportion of phosphoric acid in the composition, with the organic solvent being unreactive with phosphoric acid in the composition. The composition further comprises a solubilizing liquid capable of solubilizing phosphoric acid in the composition while retaining liquid phase composition homogeneity, such solubilizing liquid being unreactive with phosphoric acid in the composition.

Further, the composition comprises a phosphatizing proportion of phosphoric acid and water in an amount exceeding such proportion of phosphoric acid while being sufficient for the composition to provide a phosphatized coating of substantial water-insolubility on a ferrous metal substrate in phosphatizing contact with the composition and while retaining liquid phase homogeneity.

For efficient operation, the organic solvent is liquid at normal pressure and temperature and has a boiling point at normal pressure above about 35°C. Solvents that are contemplated for use are the chlorinated solvents such as 1,1,1-trichloroethane, fluorine-containing hydrocarbon solvents, e.g., trichlorofluoromethane, solvents containing only hydrogen and carbon, including aliphatic solvents such as n-heptane and aromatic liquids of which benzene is exemplary, as well as high boiling nitrogen-containing compounds which would include 2-allylpyridine, 2-bromopyridine, 2,3-dimethylpyridine, 2-ethylenepyridine and 1-tert-butylpiperidine, and further the aliphatic ketones, such as ethyl butyl ketone, having molecular weight above about 100 and below 200.

Most advantageously for efficiency of operation the solubilizing liquid is an alcohol having less than 6 carbon atoms.

Preferably, for most efficient coating operation, the phosphoric acid is present in an amount between about 0.2 to 0.8 wt %, basis the phosphatizing solution, although an amount below even 0.1 wt % can be serviceable.

The amount of the phosphatizing substance in the phosphatizing solution is exceeded by the amount of water present in such solution. Water must be present in at least an amount sufficent to provide a phosphatized coating on ferrous metal of substantial water-insolubility. This means that the coating will be, at most, about 20% water-soluble.

For many phosphatizing solutions of the process, on the one hand water-insoluble coatings are achieved, coupled with an acceptable coating weight, when the water content of the solution reaches about 1 to 2 wt %. On the other hand, phase separation for many solutions can occur when the water content reaches about 5 to 7 wt %, basis total solution weight. But, since the solubilizing liquid can affect the ability of a phosphatizing solution to solubilize water, then especially those solutions wherein the solubilizing liquid predominates, may be solutions able to contain substantial amounts of water, for example, 10 to 25 wt % of water might be reached without achieving saturation. But the water will always provide a minor weight amount of the phosphatizing solution.

A further substance that may be present in the phosphatizing solution is an aprotic organic substance. Although it is contemplated to use aprotic polar organic compounds for such substance, it is preferred for efficient coating operation to use dipolar aprotic organic compounds. These compounds act in the coating solution to retard the formation of an undesirable, grainy coating.

It is preferred, for extended retention of the aprotic organic compound in the phosphatizing solution during the phosphatizing operation, that such compound have a boiling point above the boiling point of the organic solvent in the solution. Preferably, for most extended presence in the coating solution, such compound boils at least about 20°C higher than the organic solvent. The aprotic organic compound is often a nitrogen-containing compound; these plus other useful compounds include N,N-dimethylformamide, dimethyl sulfoxide, acetonitrile, acetone, nitromethane, nitrobenzene, tetramethylene sulfone and their inert and homogeneous liquid mixtures where such exist.

Another substance generally found in the phosphatizing composition is the organic accelerator compound; compounds that can be used include urea, pyridine, thiourea, dimethyl sulfoxide, dimethylisobutyleneamine, ethylenediaminetetraacetic acid and dinitrotoluene. The use of stabilizers can include p-benzoquinone, p-tert-amylphenyl, thymol, hydroquinone and hydroquinone monomethyl ether.

For efficient operation, it is therefore preferred to formulate a replenishing liquid composition containing organic solvent, solubilizing liquid and water. Further, such replenishing liquid can be used for sustaining the phosphatizing composition, and may form a homogeneous and storage-stable blend before use. Thus, for convenience, this liquid is often referred to herein as the "sustaining solution." The sustaining solution can be prepared ahead for later use after storage and/or shipment.

In the makeup of the sustaining solution, the organic solvent will be the predominant ingredient; in the balance, the solubilizing liquid will supply the major amount, with water the minor amount. Generally, the solution will contain from about 70 wt % to greater than 95 wt % of organic solvent, with above about 2 wt %, but not more than about 25 wt % of solubilizing liquid. The water will most always be present in the sustaining solution in an amount of about 0.4 to 4 wt %. Preferably, for enhanced phosphatizing operation, the water-solubilizing liquid and organic solvent will be combined in the sustaining solution in the equivalent proportions of such substances in the phosphatizing medium vapor zone.

The phosphatizing composition will typically provide a desirable phosphate coating, i.e., one having a weight of 20 mg/ft^2 or more on ferrous metal, in fast operation. Although contact times for ferrous metal articles and the phosphatizing composition may be as short as 15 seconds for spray application, it will typically be on the order of about 45 seconds to 3 minutes for dip coating, and may even be longer. The coating weights, in milligrams per square foot, can be on the order as low as 10 to 20 to be acceptable, i.e., provide incipient corrosion protection with initial enhancement of topcoat adhesion, and generally on the order of as great as 100 to 150 although much greater weights, e.g., 300 or so, are contemplated.

Preferably, for best coating characteristics including augmented topcoat adhesion and corrosion protection, the coating will be present in an amount between about 20 to 100 mg/ft^2. Such coatings are readily and consistently produced with desirable coating uniformity. The coatings that are obtained on ferrous metal will have at least substantial water-insolubility, and hence are also termed herein to be water-resistant coatings.

Preparation of test panels is described above in U.S. Patent 4,186,035. Phosphatizing of test panels and coating weight is described in U.S. Patent 4,073,066.

Example: To 219.7 parts of benzene there is added, with vigorous agitation, 118.7 parts methanol, 3.64 parts orthophosphoric acid, and 23.6 parts N,N-dimethylformamide. These blended ingredients are thereafter boiled for 1 hour using a reflux condenser and the solution is permitted to cool. The water content of the resulting boiled solution is found to be about 0.1 wt %. This water content is directly determined by gas chromatograph analysis of a sample wherein the column packing is Porapak Q (Waters Associates, Inc.). The resulting solution is then heated to boiling and panels are phosphatized in the manner described previously. (In the above formulation, parts are by weight.)

Some of the resulting coated panels, selected in sets of two with each panel in the set being coated under identical conditions for the other panel in the set, are then subjected to testing. One panel in the set is used for coating weight determination in the manner described on page 78. The other panel in the set is subjected to the water-solubility test. For this test the panel is weighed and then immersed in distilled water for 10 minutes, the water being maintained at ambient temperature and with agitation. Thereafter, the test panel is removed from the water, rinsed in acetone and air dried. Subsequently, on reweighing, the amount of water-solubility of the coating is shown by the weight loss. The loss, basis total original coating weight, is reported in the table below as the percentage or degree of coating loss.

Coating weights and water-solubility of coatings are determined initially for test panels that have been phosphatized in the abovedescribed phosphatizing composition. Such data are determined thereafter for additional coated panels that have been phosphatized in compositions of differing water contents, all as shown in the table below. These baths of varying water content are prepared in stepwise fashion by starting with the abovedescribed bath and then adding about 1 wt % water to the bath followed by boiling the resulting solution for 1 hour. This procedure is repeated with additional water increments of 1 wt %, as shown in the table below. The phosphatizing coating operation for each bath of varying water content has been described hereinabove. For each phosphatizing bath, water content determinations are made prior to phosphatizing by the abovedescribed method.

Coating Bath Water Content, wt %	Panel Coating Weight, mg/ft^2	Degree of Solubility of Coating in Water, %
0.1	18	82
1.1	28	11
2.1	26	<5
3.1	27	<5
4.1	21	<5
5.1	35	<5

The tabulated results demonstrate the enhancement in the degree of water-insolubility for the phosphate coating as the water content in the phosphatizing bath increases. As determined by visual inspection, it is also noted that the degree of uniformity of the phosphate coating is increasing as the water content of the phosphatizing bath increases above about 1%. For the particular system of this example, the desirable water content is deemed to be from about 1.5 wt % to above 5 wt %. At 1.1 wt % and below, the degree of water-solubility for the coated panels is regarded as being undesirable, since it can be easily improved. By continuing the stepwise water addition discussed above, this system is found to separate free water, i.e., lose liquid phase homogeneity, when the water content reaches 6.1 wt %.

Nonflammable Composition Containing Halogenated Solvent

This process of *E.A. Rowe, Jr. and W.H. Cawley; U.S. Patent 4,118,253; Oct. 3, 1978; assigned to Diamond Shamrock Corp.* is directed to a nonflammable organic phosphatizing composition having a continuous and homogeneous liquid phase suitable for phosphatizing metal with a coating of at least substantial water-insolubility, with the liquid phase containing water in minor amount. The composition comprises halogenated organic solvent providing liquid phase homogeneity with an organic solubilizing liquid, while being a nonsolvent for a phosphatizing proportion of phosphoric acid in the composition. The halogenated organic solvent is unreactive with phosphoric acid in the composition and has halogens selected from the group consisting of chlorine, fluorine and mixtures thereof.

The composition further comprises solubilizing liquid capable of solubilizing phosphoric acid in the composition, while retaining liquid phase composition homogeneity. The solubilizing liquid is unreactive with phosphoric acid in the composition, and either has a boiling point greater than the boiling point of the halogenated organic solvent, or forms an azeotrope with the halogenated organic solvent. The azeotrope must boil at a point below the boiling point of the solubilizing liquid and contain above about 80 mol % of the halogenated organic solvent.

Further, the composition comprises a phosphatizing proportion of phosphoric acid, and contains water in an amount exceeding such proportion of phosphoric acid, while being sufficient for the composition to provide a phosphatized coating of substantial water-insolubility on a ferrous metal substrate in phosphatizing contact with the composition and while retaining liquid phase homogeneity.

The halogenated organic will generally provide the major amount of the phosphatizing solution and will typically provide between about 60 to about 90 wt % of such solution. It is most preferable, for efficient phosphatizing composition preparation, that the halogenated organic solvent and the solubilizing liquid form storage stable blends. That is, that they form blends that on extended storage are free from phase separation.

Most preferably for efficient operation, the halogenated organic solvent is liquid at normal pressure and temperature and has a boiling point at normal pressure above about 35°C. Solvents that are contemplated for use are the chlorinated solvents such as 1,1,1-trichloroethane and the fluorine-containing hydrocarbon solvents, e.g., trichlorofluoromethane. Other useful organic solvents in addition to those already mentioned and which can or have been used include chloroform, 1,1,3-trichlorotrifluoroethane, perchloroethylene, trichloroethylene, and 1,1,2,2-tetrachloro-1,2-difluoroethane, as well as the inert and homogeneous liquid mixtures of all the solvents mentioned herein, where such exist, as for example, azeotropic mixtures.

The solubilizing liquid, phosphatizing portion of phosphoric acid, the aprotic polar organic substance, accelerator and stabilizers have been discussed above in U.S. Patent 4,029,523. Also discussed above are the makeup of the solution and desirable coating weights. In the example, all parts are by weight.

Example: The panels were prepared and phosphatized as described above in U.S. Patent 4,029,523. To 434 parts of trichlorotrifluoroethane there is added, with vigorous agitation, 95 parts methanol, 2.7 parts orthophosphoric acid and 17 parts N,N-dimethylformamide. These blended ingredients are thereafter boiled for 1 hour using a reflux condenser and the solution is permitted to cool. The water content of the resulting boiled solution is found to be about 0.1 wt %. This water content is directly determined by gas chromatograph analysis of a sample wherein the column packing is Porapak Q. The resulting solution is then heated to boiling and panels are phosphatized in the manner described.

Some of the resulting coated panels, selected in sets of two with each panel in the set being coated under identical conditions for the other panel in the set, are then subjected to testing as described in U.S. Patent 4,029,523. Results are tabulated below.

Coating Bath Water Content, wt %	Panel Coating Weight, mg/ft^2	Degree of Solubility of Coating in Water, %
0.1	25	52
1.1	35	14
1.3	39	<5
1.4	37	<5

The results show the enhancement in the degree of water-insolubility of the phosphate coating as the water content in the phosphatizing bath increases; also, visual inspection confirms that the degree of uniformity of the phosphate coating is increasing as the water content of the phosphatizing bath increases. For this particular system, the range for the desirable water content is quite narrow, with further water addition to the bath being found to separate free water when the water content reaches only 1.6 wt %.

A solution of 78 parts trichlorotrifluoroethane, 17.07 parts methanol, 3.06 parts N,N-dimethylformamide, 1.38 parts water and 0.49 part of orthophosphoric acid was then tested for flammability. The test used a Setaflask Tester (Stanhope-Seta Ltd.). The test was run in accordance with ASTM D-3278-73. There was no flash at a temperature of 104°F, which was chosen because the water-trichlorotrifluoroethane azeotrope boils at 103.8°F.

Chlorinated Hydrocarbon, Chromic Anhydride, Phosphoric Acid, Solubilizer and Stabilizer

H. Wada, A. Ushio and M. Koganei; U.S. Patent 4,257,828; March 24, 1981; assigned to Nippon Paint Co., Ltd., Japan provide a nonaqueous composition for chemical treatment of the surface of a metallic substrate, which can form a nonsticky, uniform film firmly adhered to the surface. The coating resulting from such chemical treatment has a high corrosion resistance and a good adhesion with a paint coating film provided thereon.

The term "nonaqueous" as used above is not intended to absolutely exclude the presence of any amount of water but to exclude the presence of a substantial or considerable amount of water. In other words, the presence of water in such a small amount that it is insufficient to serve as a liquid medium by itself is permissible.

The nonaqueous composition comprises a chlorinated hydrocarbon solvent, chromic anhydride, phosphoric acid, a solubilizer and a stabilizer. The term "chlorinated hydrocarbon" covers any chlorinated and fluorinated hydrocarbon.

Among specific chlorinated (and fluorinated) hydrocarbons, particularly preferred are methylene chloride and 1,1,2-trichloro-1,2,2-trifluoroethane. The chlorinated hydrocarbons may be used alone or in combination. Since the chlorinated hydrocarbon solvent is generally noninflammable, the resulting nonaqueous composition is hardly inflammable and can be used safely in industry.

The chromic anhydride is alternatively called "chromium trioxide" (CrO_3). It may be used in an amount of 0.003 to 0.03 pbw to 100 pbw of the chlorinated hydrocarbon solvent.

The use of orthophosphoric acid is particularly favorable. Commercially available aqueous phosphoric acid wherein the phosphoric acid concentration is 75% or higher may be as such used effectively. The amount of phosphoric acid to be used may be from 0.003 to 0.12 pbw to 100 pbw of the chlorinated hydrocarbon solvent.

The solubilizer is required to be one which is per se soluble in the chlorinated hydrocarbon solvent and can solubilize chromic anhydride and phosphoric acid therein. Specific examples of the solubilizer are monovalent saturated alcohols having 1 to 7 carbon atoms such as sec-propanol, tert-butanol and tert-pentanol. These may be used alone or in combination. The amount of the solubilizer is usually from 0.1 to 2 pbw to 100 pbw of the chlorinated hydrocarbon solvent.

Particularly preferred stabilizers are the combination of zinc fluoride and p-benzoquinone, the combination of zinc oxide and 1-nitroso-2-naphthol, the combination of a trialkyl phosphite and methanol, the combination of eugenol and diisobutylamine, the combination of dimethoxyethane and isoprene, etc. The amount of the stabilizer may be from 0.001 to 5 pbw to 100 pbw of the chlorinated hydrocarbon solvent.

Examples 1 through 5: Nonaqueous compositions for chemical treatment were prepared according to the prescriptions as shown in Table 1.

Table 1

	Example				
	1	2	3	4	5
Materials	Parts by Weight				
Trichloroethylene	100	100	100	100	100
Chromic anhydride	0.015	0.001	0.015	0.015	0.015
75% orthophosphoric acid	0.06	0.06	0.2	0.06	0.06
tert-Pentanol	0.5	0.5	0.5	0.5	0.5
p-Benzoquinone	1	1	1	0	1
Zinc fluoride	0.005	0.005	0.005	0.005	0

A clean steel plate (150 x 70 x 0.8 mm) previously degreased with trichloroethylene vapor was immersed in the nonaqueous composition heated at 84°C for 1 second. Then, the plate was taken out and dried in the air. On the thus treated steel plate, an acrylic resin paint composition, Super Lac D-4TX-64 Enamel (Nippon Paint Co., Ltd.) was applied by spraying to make a dry film thickness of 30 μ and baked in a furnace of 160°C for 20 minutes to obtain a coated plate, of which the performance is shown in Table 2.

Table 2

	Example				
Test Item	1	2	3	4	5
Impact Strength*					
30 cm	0	0	0	0	0
50 cm	0	Δ	0	Δ	0
Resistance Against Salt Spray**					
240 hr	0	0	0	Δ	Δ
500 hr	0	Δ	Δ	x	x
Acid Resistance***	0	Δ	Δ	Δ	Δ

*Du Pont impact tester, ¼ inch R, 500 g, 30 cm, 50 cm. 0 = paint film not peeled off; Δ = paint film partially peeled off; and x = 50% or more of paint film peeled off.

**Salt spray tester, 5% aqueous sodium chloride solution, 35°C cross-cut on paint film, 240 hours, 500 hours. Swelling of paint film was observed within 1 mm from cross-cut line, 0; within 3 mm from cross-cut line, Δ; and beyond 3 mm from cross-cut line, x.

***Immersion into 5% acetic acid, 20°C, 72 hours, peeling-off cross-cut by tape. Paint film peeled off is 2% or less, 0; is 2 to 20%, Δ; and is 20% or more, x.

CHROMATE TREATMENTS

USING HEXAVALENT CHROMIUM

Cr(VI) plus Nonionic Surfactant or Glycine

Chromate coatings have conventionally been employed to improve the corrosion resistance of the bare metal surface and as a paint base coating for enhanced corrosion resistance on aluminum, zinc, galvanized steel plate, tin plate and the like and as a sealing, anticorrosive or coating on iron or steel pretreated with a phosphatizing solution and as an insulating or anticorrosive coating for electrolytic iron plate. As the chromating solution, there have been employed chromic acid together with aminoalcohols, polyvalent alcohols, fatty acids and the like as a reducing agent suitable for the acid. However, such coatings have been unsuccessful because they have unsatisfactory corrosion resistance and adhesion of topcoated paints.

M. Nishijima, N. Oda and H. Terada; U.S. Patent 4,059,452; November 22, 1977; assigned to Oxy Metal Industries Corporation found that improved corrosion resistance and adhesion of topcoated paints can be achieved by coating a metal surface with a chromating solution comprising (1) a hexavalent chromium compound, (2) one or more compounds selected from the group comprising: (a) nonionic surface active agents having the general formula of $RO(CH_2CH_2O)_nH$ wherein R represents a saturated or unsaturated aliphatic hydrocarbon group of 5 to 25 carbon atoms and n represents an integer from 2 to 30 and (b) glycine, followed by drying and baking the applied coating. Preferably, the solution also contains urea.

Example 1: Cold-rolled steel plate (SPCC-D Shin Nippon Seitetsu Co.) having a size of 0.8 x 70 x 150 mm was degreased with an alkali, cleaned and mechanically polished. Separate plates were coated with chromating solutions having compositions as indicated in Table 1 by means of a roller, dried and baked for 6 min in an electrically heated forced air oven at a temperature ranging from 170° to 200°C to form a coating having a coating weight ranging from 170 to 190 mg/m². The corrosion resistance of the coated specimen was tested by the saline spraying test according to JIS-Z-2871.

Identically treated steel specimens were coated with a modified epoxy resin to a film thickness of 25±5 μ and baked at 160°C for 20 min in a forced-air oven to produce a coating having a pencil hardness of 3H as measured according to JIS-G-3312. These panels were tested for corrosion resistance and adhesion with a topcoated paint. For comparison purposes, control panels were treated with a chromate solution containing a nonionic surfactant outside the scope of the process and painted as above. Table 2 shows the results obtained in Runs 1 through 4 and in the comparative test.

Table 1

Run No.	Chromating Solution: Compound	Grams per Liter
1	CrO_3	40
	$C_{12}H_{25}O(CH_2CH_2O)_7H$	5
	$C_{13}H_{27}O(CH_2CH_2O)_7H$	5
2	CrO_3	40
	$C_{18}H_{35}O(CH_2CH_2O)_9H$	10
3	CrO_3	40
	ZnO	16
	75% H_3PO_4	13
	$C_{12}H_{25}O(CH_2CH_2O)_7H$	5
	$C_{13}H_{27}O(CH_2CH_2O)_7H$	5
4	CrO_3	40
	Glycine	20
	$C_{12}H_{25}O(CH_2CH_2O)_7H$	0.25
	$C_{13}H_{27}O(CH_2CH_2O)_7H$	0.25
*	CrO_3	40
	C_9H_{19}–⬡–$O-(CH_2CH_2O)_{14}H$	10

*Control

Table 2

Run No.	Unpainted Surface, Saline Spraying Test, Percent Corroded: 1 hr	2 hr	5 hr	6 hr	30 hr	Painted Surface, Saline Spraying Test, Creepage (mm): 240 hr	500 hr	Painted Surface: Adherence (%)
1	0	5	5-10	–	–	–	–	–
2	0-5	5-10	5-10	–	–	–	–	–
3	0	0	0	0	8-10	0	0-0.5	98-100
4	0	0	0	0	5-10	0	0-0.5	98-100
*	40-60	50-90	100	100	100	2.0	3.0-5.0	90-97

* Control

Example 2: Cold-rolled steel plate (SPCC-D) of 0.8 x 70 x 150 mm was cleaned with an alkaline degreasing solution and mechanically polished. The polished plate was coated with chromating solutions having the compositions of Table 3 by means of a rubber roller and dried and baked for 6 min in an electrically heated forced-air oven at 170° to 200°C. The treated plates were then coated with a zinc-rich paint (SD Zinc Primer ZE No. 100, Kansai Paints Co.) to a thickness from 15 to 20 μ and allowed to air dry. Table 3 shows the results obtained.

Table 3

Run No.	...... Chromating Solution...... Compound	Grams per Liter	Salt Spray Corrosion Resistance After 500 hr Creepage (mm)
5	CrO_3	40	0
	$C_{12}H_{25}O(CH_2CH_2O)_7H$	5	
	$C_{13}H_{27}O(CH_2CH_2O)_7H$	5	
	NH_2CONH_2	20	
6	CrO_3	40	0
	Glycine	20	
	NH_2CONH_2	20	
	$C_{12}H_{25}O(CH_2CH_2O)_7H$	0.25	
	$C_{13}H_{27}O(CH_2CH_2O)_7H$	0.25	

The saline spraying test was carried out according to JIS-Z-2371.

Cr(VI), Urea Reducing Agent, Zinc Compound to Adjust pH

A composition has been formulated by *A.W. Kennedy; U.S. Patent 4,123,290; October 31, 1978; assigned to Diamond Shamrock Corporation* which will provide a coating yielding excellent coated substrate corrosion resistance.

This is achieved by incorporating into the composition, as at least a part of the reducing agent thereof, the substance urea. The urea can be used alone as the reducing agent or its benefits can be achieved when it is used in combination with other reducing substances, particularly carboxylic acids. It is also important for obtaining the full benefits of the composition that it be formulated to contain a pH adjusting agent which is a compound of zinc.

The pulverulent metal-containing aqueous coating composition contains hexavalent chromium, pH adjusting agent and a reducing agent for the hexavalent chromium. The corrosion resistance of resulting coated metal substrates will be enhanced by incorporating into the composition urea as at least a portion of the reducing agent. Further, the pH adjusting agent should be from the group of zinc oxides, basic zinc peroxides, zinc salts of weak acids and mixtures thereof.

Preparation of Test Panels: Unless otherwise specifically described, test panels are typically 4 x 8 cold-rolled, low-carbon steel panels. They are prepared for coating by first scrubbing with a cleaning pad which is a porous, fibrous pad of synthetic fiber impregnated with an abrasive. Thereafter, the scrubbed panels are immersed in a cleaning solution typically containing chlorinated hydrocarbon and maintained at about 180°F, or containing 1 to 5 oz/gal of water, of a mixture of 25 weight percent tripotassium phosphate and 75 weight percent potassium hydroxide. This alkaline bath is maintained at a temperature of about 150° to 180°F. Following the cleaning, the panels are rinsed with warm water and preferably dried.

Primer Topcoating and Application: When prepainted panels are primer topcoated, the primer used is a commercially available primer which is a zinc-rich weldable primer having a weight per gallon of about 15.2 lb, a solids volume of about 27%, and containing about 62 weight percent of nonvolatiles. The binder component is prepared from a high molecular weight epoxy resin. The primer has a typical viscosity of about 80 sec as measured on a No. 4 Ford cup. This primer is typically applied to prepainted panels by drawing the primer down over the panel with a drawbar to provide a smooth, uniform primer coat, generally of about 0.5 mil thickness. Resulting coated panels are usually cured for about 3½ to 4 min in an oven at about 500° to 550°F.

Corrosion resistance of coated parts is measured by means of the standard salt spray (fog) test for paints and varnishes ASTM B-117.

Example: There is formulated, with blending, a control prepaint coating composition for comparative use and containing 20 g/ℓ chromic acid, 3.33 g/ℓ succinic acid, 1.67 g/ℓ succinimide, and 1.5 g/ℓ xanthan gum hydrophilic colloid, a heteropolysaccharide prepared from the bacteria species *Xanthomonas campestris* and having a molecular weight above 200,000. Additionally, this control composition contains 1 ml formalin, 7 g/ℓ zinc oxide, 120 g/ℓ zinc dust having an average particle size of about 5 μ and having all particles finer than 16 μ, and 1 drop per liter of a wetter which is a nonionic, modified polyethoxide adduct having a viscosity in centipoises at 25°C of 180 and a density at 25°C of 8.7 lb/gal. After mixing all of these constituents, this control composition is then ready for coating test panels.

For further test purposes, there is subsequently blended together a separate composition which is representative of the process. With the two compositions, the only difference is that the process composition contains 0.8 g/ℓ urea in place of the 1.67 g/ℓ succinimide.

Panels, prepared as described hereinabove, are dipcoated, some into freshly prepared control coating composition and some into freshly prepared process prepaint coating composition. Panels are removed from these compositions and excess composition is drained from the panels. If the panels are not further coated, they are then baked for 3.5 min in an oven at an oven temperature of 500°F.

Some panels, selected for topcoating, are baked for 1 min in the oven at the 500°F temperature. These panels are then topcoated with the primer, and the primer cured, all in the manner as described hereinbefore. Some of these topcoated panels, as well as some not topcoated, are then selected for testing in the abovedescribed corrosion resistance (salt spray) test. All panels are scribed before testing. The results of the corrosion resistance testing are reported in the table and are for the scribe lines only.

In the manner as described above, panels are quantitatively evaluated for the scribe lines on a numerical scale. This is done by visual inspection, comparing panels with one another and employing a rating system using the following numbers that cover the following results:

(10) no red rust in the scribe;

(8) up to 30% red rust in the total scribe and initial run-off from the intersection of the scribe lines;

(6) up to 80% red rust in the scribe with run-off from up to 50% of the scribe lines;

(4) complete red rust of the scribe with moderate to heavy run-off from all portions of the scribe; and

(2) heavy run-off from all portions of the scribe lines.

Coating weights for panels, as chromium, and not as CrO_3, and as zinc, both being in weights in milligrams per square foot of coated substrate, are listed for panels coated with both the control and the process composition in the table below. Such weights are determined by a Porta-Spec x-ray fluorescence spectroscope manufactured by Pitchford Corporation. The lithium fluoride analyzing crystal is set at the required angle to determine chromium, and at the required angle to determine zinc. The instrument is initially standardized with coatings containing known amounts of these elements. The machine is adapted with a counter unit and the count for any particular coating is translated into milligrams per square foot by comparison with a preplotted curve.

Prepaint Coating	Prepaint Coating Weight*		Topcoating	240 Hours Salt Spray Corrosion for Scribe Only
	Cr	Zinc		
Control	31.5	365	No	1
Control	33.5	380	No	1
Process	26.5	470	No	10
Process	26.0	450	No	10
Control	–	–	Yes	3**
Process	–	–	Yes	10**

*In mg/ft^2 determined after testing; no determination for topcoated panels.
**Average of two panels.

Partial Substitution of Boric Acid for Chromic Acid in Coating Composition

T. Higashiyama and T. Nishikawa; U.S. Patent 4,266,975; May 12, 1981; assigned to Diamond Shamrock Corporation offer a substantially resin-free anticorrosive coating composition for metals comprising at least one boric acid compound and at least one water-soluble chromic acid compound and particulate metal and at least one high-boiling organic liquid and water and/or organic solvent which, when necessary, contains a pH modifier. Further, the composition may contain nonionic dispersing agent and/or viscosity modifier.

In the matrix of the bonding material formed by the composition, the leaching rate of the boric acid ions present therein is far slower than that of the hexavalent chromium, the former ions retain the activity of the metal particles well, and they maintain the conductivity of the coating layer well. Preferably, the composition comprises 10 to 40 weight percent of particulate metal; 1 to 12 weight percent of water-soluble acid compounds (boric acid component plus chromic acid constituent); 7 to 30 weight percent of at least one high-boiling organic liquid; and with the remainder being water or water mixed with a solvent, with the composition optionally containing ingredients such as pH modifier and viscosity modifier.

As the particulate metal ingredient, Zn, Al or their mixture or an alloy of Zn and Al is used. The preferred particulate form is flake, of which thicknesses are on the order of 0.1 to 0.5 μ, and the longest part has a length of 150 μ, generally 15 μ or less.

The high-boiling organic liquid compounds used in the composition correspond to oxohydroxy low molecular weight organic compounds, e.g., of a molecular weight of 300 or less, that is, polymers of glycol and their low molecular weight ethers.

The pH modifier is used to adjust the pH of the final mixture to from 3.0 to 6.0.

The process is explained specifically by way of working examples. The corrosion resistance tests and the evaluation of the test results employed in these examples are as follows.

(1) Saltwater (fog) spray test: The neutral saltwater spray test described in JIS Z-2371 was as follows. The degree of corrosion of the test samples was visually observed and evaluated in accordance with the following standards:

5 points – Absolutely no formation of red rust.
4 points – Formation of ten or less pinholes of red rust.
3 points – Rust spots are distributed and some flow of rust is observed.
2 points – The flow of rust is remarkable.
1 point – The entire surface is covered with red rust.

(2) CASS test. The test method of JIS D-0201-1971 was followed, except that a spray liquid of pH 3.5 was used. Standards for evaluation of the formation of rust are the same as above.

(3) Outdoor exposure test. The test pieces were exposed attached to exposure stands (surfaces facing the south inclined at 30°) in Yokohama, Japan. Standards for evaluation of formation of rust are the same as above.

The used test pieces were 15 x 15 cm, 0.8 mm thick soft steel plates.

Example 1: 60 parts of metallic zinc flakes (0.1 to 0.3 μ thick, about 15 μ long on average in the longest part) were dispersed in diethylene glycol containing 0.3 part of Nopco 1529 (alkylphenol polyethoxy adduct surfactant, Diamond Shamrock Corporation of U.S.A.) so as to make the total amount 100 parts. (This mixture is the first component.) Separately orthoboric acid and chromic acid anhydride are dissolved in deionized water so that the orthoboric acid content was 5.17% and the chromic acid content was 1.72%, and calcium oxide was added as the pH modifier so that the content thereof would be 1.72%. The boric acid concentration, basis boric plus chromic acid, is 75 weight percent. (This mixture is the second component.)

The first and second components were mixed in the weight ratio 42:58 by pouring the former into the latter while slowly stirring, and stirring was continued overnight at room temperature. In this mixture, the orthoboric acid concentration was 3%, the chromic acid anhydride concentration was 1% and the calcium oxide concentration was 1%. The mixture thus obtained was applied onto soft steel plates by means of a bar coater to form a uniform film thereon, the plates having been washed with alkali and sufficiently polished with a Scotch Bright very fine polishing cloth (3M Company of U.S.A.) and the plates were heated in an electrically heated hot air circulating furnace, the temperature of the soft steel plates being held at 300°C for 4 min after the plates reached that temperature, and they were then left standing to cool to room temperature. The amount of the applied composition was 1 μ in thickness and 250 mg/ft²(2.7 g/m²), by weight, per area.

Example 2: Coating films were formed under the same condition as Example 1 except that the final mixture contained 3.6% orthoboric acid and 0.4% chromic acid anhydride, thereby providing a boric acid concentration of 90%.

Example 3: Coating films were formed under the same conditions as in Example 1 except that the final mixture contained 2% orthoboric acid and 2% chromic acid anhydride, for 50% boric acid concentration.

Example 4: Coating films were formed under the same conditions as in Example 1 except that the final mixture contained 1% orthoboric acid and 3% chromic acid anhydride, for a boric acid ratio of 25%.

Example 5: Coating films were formed under the same conditions as in Example 1 except that the final mixture contained 0.4% orthoboric acid and 3.6% chromic acid anhydride, for a 10% boric acid concentration.

Comparative Example: Coating films were formed under the same conditions as in Example 1 except that no boric acid compound was used (that is, ingredients were blended so that the chromic acid anhydride content was 4%).

The tests described above were carried out with these test pieces on which the coating films were formed as described above. The results are shown in the following table.

Ex. No.	Boric Acid Concentration (%)	Saltwater Spray Test (144 hr)	CASS Test (20 hr)	Outdoor Exposure (6 mo) Flat Surface	Scribe
*	0	5	1	3	1
5	10	5	2	4	3
4	25	5	3	5	4
3	50	5	3	5	5
1	75	4	3	5	5
2	90	1	2	5	5

*Comparative.

As will be understood from the table, the composition of the process exhibits excellent anticorrosive effect against both saltwater and freshwater.

Zinc Flake Dispersion, Cr(VI), Organic Liquid

The work of *A.W. Kennedy; U.S. Patent 4,026,710; May 31, 1977; assigned to Diamond Shamrock Corporation* is directed to a process for making a zinc flake and hexavalent chromium-containing coating composition, which can provide an adherent and corrosion-resistant coating on a metal substrate. The medium of this coating composition is supplied by water plus organic substance. The process comprises combining finely divided zinc metal having nonflaked zinc particles with a medium containing organic substance, with the flaking medium organic substance being present in an amount sufficient to provide a weight ratio of zinc metal to such substance of between about 3:1 and 1:3. The substance is selected from the group consisting of diacetone alcohol, 2-ethoxyethanol and 2-butoxyethanol. The process also comprises mechanically flattening and polishing the zinc particles while contained in such medium, thereby preparing a dispersion of flaked metal particles in the medium.

Next the process calls for admixing the zinc flake dispersion with additional coating composition ingredients including hexavalent-chromium-containing substance and water, with there being sufficient zinc flake and organic substance in the coating composition whereby it contains: (1) between about 50 and 500 g/ℓ of the metal particles; and (2) above 15 but below about 50 volume percent of organic substance, basis the total volume of the composition liquid medium; and there being sufficient of the hexavalent-chromium-containing substance to supply the composition with a weight ratio of chromium, expressed as CrO_3, to the metal particles of between about 1:1 and 1:15.

In addition to the abovementioned ingredients, the flaking mill will generally be supplied with a very minor amount of weldment inhibitor, such as finely divided aluminium metal.

Example 1: 800 g of finely divided zinc dust having an average particle size of about 5 to 6 μ, together with 16 g of stearic acid and 700 ml of diacetone alcohol were charged to a 0.3 gal porcelain mill jar. Also in the jar was about 0.5 lb of ½-inch cylindrical porcelain grinding medium. At 90 rpm the mill was operated for about 18 hr. After milling and on opening the jar, no gassing or pressure release was noted.

The fluid decanted from the jar was seen to be very uniform on visual inspection. After filtration of this fluid, 554 g of the resulting wet cake along with 3 g of nonionic dispersing agent, 50 g of chromic acid and 19 g of zinc oxide were all blended together with sufficient deionized water to make 1 ℓ. The 554 g of the wet cake contain sufficient diacetone alcohol for this substituent to contribute 20 volume percent of alcohol to the liter of coating composition. Upon visual inspection, the resulting coating composition was judged to be suitable for preparing desirable, corrosion-resistant coatings on metal substrates.

Example 2: In an additional production run, a milled paste, containing desirable flaked zinc, is prepared to contain, after press filtration, 11 weight percent 2-butoxyethanol, 1 weight percent stearic acid, and 88% of zinc flake. Next, to 160 g of this paste, including the 11 weight percent 2-butoxyethanol, there is added 50 g of dipropylene glycol, 8 g of nonionic wetter, 4 g of the abovedescribed hydroxyethylcellulose (which is a cream to white colored powder having a specific gravity of 1.38 to 1.40 at 20°/20°C, an apparent density of 22 to 38 lb/cf, and with all particles passing through 80 U.S. mesh), 40 g of chromic acid, and 800 ml of water.

A 4 x 8 inch cold-rolled steel panel, is coated in this composition by dipping therein and then permitting the panel to drain after removal. The panel has a total coating weight, after curing, of 1,100 mg/ft^2. Curing is achieved by placing the panel in an oven for 20 min, the oven being at temperature of 570°F. The appearance of the faces of the cured panel, on visual inspection, is one of a smooth and uniform coating.

The panel is subjected to conical mandrel testing carried out by the procedure of of ASTM D-522. Under such testing, and where a rating of 10 is given for complete film retention of a coating subjected to the test, the tested panel receives an exceptionally good rating of 9.5.

Example 3: To an additional 320 g of the Example 2 paste containing the 2-butoxyethanol, there was added 30 g of aluminum flake, 210 g of dipropylene glycol, 9 g of nonionic wetter, 3 g of the abovedescribed hydroxyethylcellulose, 50 g chromic acid, 4 g of zinc oxide and 730 ml water. Small steel fastening clips were coated in a manner as described above. The clips had a total coating weight after curing of 1,100 mg/ft^2. Curing was achieved as described above.

A sample of these coated clips had a smooth and uniform appearance. In salt spray testing, according to the ASTM B-117-64 test, no red rust was observed in 500 hr of testing. Another sample of the coated clips was left exposed on an outdoor test rack, in the Northern Hemisphere, for a 5-month period from fall through midwinter. After this test, no red rust was visually observed on the coated clips of this sample.

No Rinse Composition

In accordance with the process of *W.D. Krippes; U.S. Patent 4,266,988; May 12, 1981; assigned to J.M. Eltzroth & Associates, Inc.* a ferrous or nonferrous metal-surfaced article is treated with an aqueous chromate depositing solution containing hexavalent chromium but no trivalent chromium, together with fluoboric acid, hydrofluoric acid, sulfuric acid, with or without hydrofluosilicic acid. The solution also contains as an additive zinc oxide and/or magnesium oxide, and/or magnesium hydroxide, and/or aluminum sulfate, and/or aluminum hydroxide, the ratio of the additive to the total acids being such as to give a pH within the range of 1.5 to 3.6 at 22°C and a chromate concentration of 0.05 to 10.0 g/ℓ, as Cr.

Compositions of the type described when employed in treating clean ferrous or nonferrous metal-surfaced articles provide enhanced adherency of the treated surface to organic film-forming polymers which dry to a water-resistant coating and do not require rinsing of the treated surface prior to the application of the organic film-forming polymers, thereby avoiding environmental contamination that would otherwise be caused by rinse waters.

The coating composition is normally prepared as a concentrate which is then diluted with water to the desired concentration for coating a particular type of metal, the concentration also depending upon the amount of the coating to be deposited upon the metal.

In carrying out the process the temperature of the chromate depositing solution for use on ferrous or nonferrous metal-surfaced articles is normally within the range of 70° to 210°F and usually 120° to 140°F.

The time of contact between the chromate depositing solution and the ferrous or nonferrous metal-surfaced article will normally be within the range of 1 to 3 sec. In the latter case the pH of the solution can also be somewhat higher, but would be within the range of 0.8 to 5.0.

The chromate depositing solution can have a solids content within the range from

0.2 g/ℓ to 75.0 g/ℓ, the remainder being water, and the chemical composition should be essentially the following:

Ingredients	Grams per Liter
Chromic acid (CrO_3), (expressed as Cr)	0.10-50.0
Hydrofluoric acid (H_2F_2) (expressed as F)	0.01-5.0
Fluoboric acid (HBF_4) (expressed as BF_4)	0.01-50.0
Sulfuric acid (H_2SO_4) (expressed as SO_4)	0.01-5.0
Hydrofluosilicic acid (H_2SiF_6) (expressed as SiF_6)	0.0-5.0
Additive*	0.01**

*From the group consisting of zinc oxide, magnesium oxide, magnesium hydroxide, aluminum sulfate, aluminum hydroxide and mixtures thereof.

**Saturation solubility at 22°C.

Especially good results are obtained by using zinc oxide as the additive in proportions of 1 to 6 g/ℓ. It will be understood, of course, that zinc oxide can combine with chromic acid to form zinc chromate ($ZnCrO_4$). Normally, however, it is preferable for the quantity of chromic acid to exceed the quantity of zinc oxide which would combine with the chromic acid to form zinc chromate. Thus, in a preferred concentrate formula the weight ratio of CrO_3 to ZnO is approximately 1.8:1 or a molar ratio of CrO_3 to ZnO of 1.4:1 whereas the molar ratio of CrO_3 to ZnO in zinc chromate is 1:1. The hydrofluosilicic acid can be omitted from the formula. It is not required in coating aluminum and only a small amount is desired in coating cold-rolled steel. The ratio of the additive, (e.g., zinc oxide) to total acids in the coating bath is preferably such as to give a pH within the range of 1.8 to 3.5 at predetermined concentrations used in the coating process.

Example 1: A concentrate was prepared by mixing together the following ingredients:

Ingredients	Percent
CrO_3	12.0
ZnO (French processed)	0.8
H_2F_2 (48% concentration)	2.6
H_2SiF_6 (26% concentration)	0.4
HBF_4 (48% concentration)	0.8
H_2SO_4 (78 to 80% concentration)	0.8
H_2O	82.6

This concentrate has a specific gravity of approximately 1.11.

Water is added to the foregoing concentrate in sufficient amount to give a running bath having a concentration of 0.5 to 1% with a pH of approximately 1.8.

The metal to be processed or coated can be, for example, cold-rolled steel, aluminized and galvanized iron and steel, aluminum, aluminum-zinc alloys, magnesium or magnesium-aluminum alloys. Typical examples of cold-rolled steel are SAE 1005 or 1010.

The concentration of 0.5 to 1.0% is given as Cr. The weight ratio of the amount of water added to the concentrate is approximately 15:1. This ratio may vary depending upon the desired concentration of the depositing solution but will usually be within the range of 3:1 to 50:1.

The metal to be coated is carefully cleaned with an alkaline cleaner at 160°F, hot water rinsed at 140° to 180°F and then coated in a coating bath containing a predetermined concentration of the foregoing composition and having a predetermined pH which is adjusted by adding more or less of the zinc oxide or other additive previously described to the concentrates. The coating weight on the metal will depend upon the particular metal and the pH of the coating bath. Thus, on aluminum, lowering the pH from 2.7 to 1.8 increases the coating weight with the concentration of chromate, as Cr, increasing from 7 to 14.4 mg/ft^2. Likewise on galvanized iron lowering the pH from 2.7 to 1.8 increases the coating weight from 2.8 to 12 mg/ft^2, as Cr. An optimum pH is 1.8 to 2.0.

On cold-rolled steel lowering the pH reduces the coating weight. Thus, at a pH of 3 the coating weight is approximately 34.5 mg/ft^2, as Cr, and at a pH of 2 the coating weight is approximately 20.0 mg/ft^2, as Cr. As the Cr concentration is increased from 0.05 to 10.0 g/ℓ, with constant pH, the total coating weight may increase from 3 mg/ft^2.

The foregoing coating weights are based on an application of 3 sec contact time using a roll coater, dip, spray or other type of coating, followed by a squeegee to remove excess coating composition. Removal of excess composition is quite important. The application of the coating composition to the metal is preferably with a time period range of 1 to 10 sec. After coating, most of the excess is removed by passing the metal in strip form through a squeegee and it is desirable to dehydrate the coated metal as much as possible before painting. Paints are preferably baked on the metal at temperatures up to 550°F. Any kind of synthetic resin coating composition can be applied which dries to a water-resistant film. Both corrosion resistance and adherence are enhanced.

Aluminum coated with a coating composition of the type described above is coated with a polyvinyl chloride primer and top coat baked on in the manner described above will withstand standard salt spray tests for at least 2,000 to 3,000 hr. Galvanized steel similarly coated will withstand standard salt spray tests at least 900 to 1,000 hr. Cold-rolled steel similarly coated will withstand standard salt spray tests at least 600 hr.

No Rinse Paint Precoat

S.C. Williamson; U.S. Patent 4,227,946; October 14, 1980; assigned to Oakite Products, Inc. provides an aqueous composition effective for providing the surfaces of metals, such as iron, steel, galvanized iron or steel, zinc, aluminum, copper and brass, a so-called no-rinse, paint precoat that enables any number of paints, such as alkyd, polyester, vinyl, epoxy, polyurethane, silicone ester, and alkyd melamine paints and others, to bond excellently to the selected metal substrate.

The aqueous compositions of the process contain dissolved in their water:

chromic acid (from solution of chromium trioxide in water) and chromium phosphate as resulting from chromic acid and phosphoric acid as derived from reacting chromic acid and hypophosphorous acid with the chromic acid present in a stoichiometrically equivalent amount from about 1 to 100% over that needed to oxidize the hypophosphorous acid to phosphoric acid,

from 0 to about 0.01% of a nonfoaming or low-foaming wetting agent, and

from 0 to about 7% of finely divided silica of from about 4 to 60 mμ particle size, (i.e., colloidal) and optimally from 4 to about 25 mμ (the particles in the 4 mμ range being actually dissolved); which compositions can be:

(1) a concentrate with its solids (dissolved and any colloidally dispersed) content, being from about 2% to an amount below that at which the composition will gel under the conditions at which it is stored; and

(2) a metal treating bath composition with its total solids content being from 2 to about 15%, and optimally from about 5 to 10%, of that of the concentrate.

In the concentrate the excess of chromium trioxide should be at least an amount sufficient to enable the concentrate to remain stable, (i.e., against gelling) at least for the interval of time from its preparation to that when it is to be diluted to provide a metal treating bath. It occurs that the extent of excess of chromic acid over the hypophosphorous acid in the compositions influences the time over which the product remains stable. For example, a concentrate having a minimum of about a 40% excess of chromic acid stoichiometrically over the hypophosphorous acid can be expected to remain stable indefinitely.

The storage life of the concentrate decreases as the excess chromic acid content decreases. For example, a concentrate including 0.0045% of the Monoflor 31 (anionic wetting agent product of ICI UK Ltd.) and having a 10% stoichiometrical excess over the hypophosphorous acid remained stable for about a month when stored at 50°C.

Such a concentrate, however, would remain stable for a longer period of time if stored at an ambient temperature of from 20° to 25°C and can be used before it reaches its gelling stage if it is diluted to a concentration suitable for use as a treating bath. A concentrate with as little as 1% excess chromic acid would gel in 2 to 3 days if stored at 47°C and in a somewhat longer period if stored at the lower ambient temperature, but could be used before it gels if diluted to the extent needed for use as a treating bath.

The treating baths, because of the excess chromic acid, are generally quite acidic, for example, those prepared from 5 to 10% by volume of concentrate show pH 2.0 to 2.2.

The concentrates are prepared by mere diluting of the 50% aqueous hypophosphorous acid with the water and admixing the chromic acid into the diluted hypophosphorous acid beneficially at a rate to avoid boiling (from the exothermicity). When any wetting agent and/or silica is to be added, it is desirable to admix whichever of them is to be added into the water before the aqueous hypophosphorous acid.

The metal treating baths of the process then are readily prepared by admixing the quantity of concentrate into the required amount of water to provide the selected or desired treating bath dilution.

Example 1: *Solution of both acids –*

7.11 pbw 50% hypophosphorous acid
10.80 pbw chromium trioxide
82.09 pbw water

Example 2: *Both acids and a wetting agent –*

7.11 pbw 50% hypophosphorous acid
10.80 pbw chromium trioxide
0.0045 pbw anionic wetting agent (Monoflor 31)
82.0855 pbw water

Example 3: *Both acids and colloidal silica –*

7.2 pbw 50% hypophosphorous acid
10.80 pbw chromic acid (as CrO_3)
10.0 pbw colloidal silica aqueous dispersion (34% SiO_2)
72.0 pbw water

Example 4: *Both acids, colloidal silica and wetting agent –*

7.2 pbw 50% hypophosphorous acid
10.80 pbw chromic acid (as CrO_3)
10.0 pbw colloidal silica aqueous dispersion (34% SiO_2)
0.0045 pbw anionic wetting agent (Monoflor 31)
71.9955 pbw water

Example 5: *Treating film on galvanized sheet steel –* A treating-film application bath, prepared by admixing 6.5 parts by volume of the concentrate of Example 2 with 93.5 parts by volume of water, was sprayed under conditions to provide a break-free continuous film over galvanized sheet steel drawn from a roll of it and passed through the spraying zone at a rate of 250 ft/min.

The thus-treating-film-coated galvanized sheet steel then was passed through a drying oven equipped to remove the water from the applied film. The thus-paint-precoated galvanized sheet steel then continued on to the paint spray zone. Where the metal to which there is to be applied a paint precoat with a bath embraced by this process, is to be painted with an alkyd or a polyester paint, it is beneficial to apply the precoat-providing film from one of the treating baths which contains silica.

The surface of the metals to which the paint precoat is applied, particularly the iron, steel, galvanized iron and steel, and also copper, manifest enhanced corrosion resistance.

Emulsion Coating Composition Containing Film-Forming Organic Resin

There is a continuing need for coatings used for paint bonding and corrosion resistance. This is especially true where the metal is steel, aluminum, magnesium, aluminum alloys and zinc-surfaced articles including galvanized iron or steel, where such coatings are required to protect the articles against deterioration.

Chromate conversion coatings have been used with varying degrees of success. The effective protective ingredient in these coatings seems to be chromium in the hexavalent state. It has long been recognized that it would be desirable to combine in a single coating composition an organic resinous film-forming component which is water-insoluble and chromium in a hexavalent state.

A process is provided by *R.C. Miller; U.S. Patents 4,138,276; February 6, 1979; 4,137,368; January 30, 1979; 4,088,621; May 9, 1978 and 4,067,837; January 10, 1978; all assigned to J.M. Eltzroth & Associates, Inc.* for preparing a hexavalent chromium-containing emulsion coating composition without causing coagulation of the emulsion and for producing an emulsion coating composition containing chromium in the hexavalent state and at least one water-insoluble particulate film-forming organic thermoplastic resin. This is achieved by mixing a body of an emulsion coating composition containing water in the continuous phase and in the discontinuous phase at least one water-insoluble particulate film-forming organic thermoplastic resin with gradually added dilute aqueous solution of an inorganic ionizable water-soluble hexavalent chromium compound having a concentration of 1 to 10% by weight, calculated as Cr, while cooling and maintaining a pH within the range of 2 to 10.5 and maintaining the body in a nonfoaming state.

The process is preferably carried out by causing the body of emulsion to rotate and gradually adding the dilute solution to a peripheral portion of the rotating body.

If it is desired to incorporate a water-insoluble pigment into the emulsion coating composition, the thermoplastic resin used in forming the emulsion is divided into two parts, the first part being emulsified in water and mixed with the aqueous solution of water-soluble hexavalent chromium compound in the manner previously described, and the second part being mixed with the pigment in sufficient proportions to coat the pigment, and thereafter combining the first and second parts.

Another modification involves the use of a particular type of pigment and a process of making them, the pigment consisting essentially of $CaSiF_6$ and 0 to 10% by weight, calculated as Cr, of hexavalent chromium added as an inorganic ionizable water-soluble hexavalent chromium compound.

To prepare a pigment of the type previously described an inorganic calcium compound and an inorganic silicofluoride are mixed in equimolecular proportions in sufficient water to form a thick slurry, the resultant mixture is neutralized and thereafter dried and ground.

These emulsion coating compositions can be used under controlled conditions for coating a metal substrate, more particularly, steel, aluminum, magnesium, and zinc-surfaced metals including galvanized iron or steel, wherein the applied coatings not only enhance corrosion resistance but also, when used as primer coatings, will adhere to subsequently applied finishing coatings.

The success of the process appears to depend upon a combination of factors, namely, the gradual addition of a dilute aqueous solution of an inorganic ionizable water-soluble hexavalent chromium compound having a concentration of 1 to 10% by weight, calculated as Cr, cooling so that the temperature does not exceed 90°F, maintaining a pH within 2 to 10.5 and maintaining the body of emulsion in a nonfoaming state. The use of these conditions reduces esterification and aldehyde formation and the amount of sediment formed in the process.

In the final analysis the amount of sediment formed is a measure of the success of the process. Under the worst conditions large amounts are formed whereas under the conditions employed herein it is possible to practice the process so that no more than 25 to 35 g of sediment is produced in a 55 gal batch.

Calcium silicofluoride is preferred when a pigment is employed in the practice of the process because when the coating composition containing it is applied as a coating to a substrate and baked, decomposition occurs to form calcium silicate which is very water-insoluble. Similarly, magnesium silicofluoride ($MgSiF_6$), strontium silicofluoride ($SrSiF_6$) and/or barium silicofluoride ($BaSiF_6$) can be used, although calcium silicofluoride is preferred.

In carrying out the process, a mixed pigment can be prepared by adding at least one chromium compound selected from the group consisting of chromic acid, strontium chromate, zinc chromate and lead chromate to a slurry of an inorganic calcium, magnesium, strontium or barium compound and the inorganic silicofluoride, the quantity of the chromium compound being within the range of 0.05 to 50% by weight of the total solids.

The preferred pH control materials are ammonium hydroxide, dimethylethanolamine, and diethanolamine. These substances are alkaline as contrasted with the aqueous solutions of the dichromates which are acidic. Tertiary amines such as dimethylethanolamine are preferred from the standpoint of enhancing the stability of the resultant emulsions with which the hexavalent chromium compounds have been incorporated.

These compositions fall into two main categories depending upon whether the resinous binder is a thermoplastic resin or a combination of thermoplastic and thermosetting resins. Where the resinous binder is solely a thermoplastic resin, the composition of the emulsion is preferably as follows:

Thermoplastic resin, (e.g., Rhoplex MV-1)	50-95% (vol)
Water	45-0% (vol)
Dilute inorganic chromate solution, (e.g., $Na_2Cr_2O_7$)	Enough to give a pH of 5.5-9.0
Pigment, (e.g., $SrCrO_4$ or $CaSiF_6$, or chromate containing $CaSiF_6$)	0-1% (wt)
Coalescing agents, (e.g., tributyl phosphate)	0.5-3% of the resin (vol)
Surfactant (Triton CF10)	Trace to 0.1% of resin
Thickening agent (Acrysol G-110)	Trace to 0.5% of resin
Titanium dioxide	0-5% (wt)
Dispersing agent, (e.g., Tamol 850)	0-0.5% pigment (wt)

In the case of a composition containing both thermoplastic and thermosetting resins a preferred composition contains:

Thermoplastic resin, (e.g., Rhoplex MV-1)	15-60% (vol)
Thermosetting resin, (e.g., Rhoplex AC-604)	10-60% (vol)
Water	10-4%
Dilute chromate solution, (e.g., $Na_2Cr_2O_7$)	Enough to give pH of 6.8 to 9.5
Pigment, (e.g., $SrCrO_4$, $CaSiF_6$ or chromate containing $CaSiF_6$)	0-1% (wt)
Coalescing agents, (e.g., tributyl phosphate)	0.5-3% of the resin (vol)
Surfactant (Triton CF10)	Trace to 0.1% of resin
Thickening agent (Acrysol G-110)	Trace to 0.5% of resin
Titanium dioxide	0-5% (wt)
Dispersing agent, (e.g., Tamol 850)	0-0.5% of pigment (wt)

Since all of the chromate-type primers have a decided color, whenever a top coating is to be white or pastel in color, there is produced a decided color detraction after the baking operation. To offset this effect rutile titanium dioxide is incorporated and sometimes there is added a blue pigment dispersible in water, (e.g., Hercules Imperial A-984, or X2925) to improve the final color effect for whites, blues and greens. For other shades some comparable color pigments akin to the desired color tone can be added. These tend to block out or hide the color effect of the active chromates contained in the prime coating.

Preferred coating compositions of the process consist essentially of the following:

	Ingredients	Weight Percent
(a)	Resin solids of at least one organic film-forming resin selected from the group consisting of thermoplastic resins, thermosetting resins, and mixtures thereof	10-60
(b)	Calcium, magnesium, strontium or barium silicofluoride	0.5-20
(c)	Hexavalent inorganic chromium compound as Cr	0-10
(d)	Pigments and pigment extenders other than (b)	0-100 of resin
(e)	Driers	up to 0.5 of total solids
(f)	Coalescent agents	0.05-10
(g)	Dispersants	0.05-6
(h)	Defoamers	0-0.05
(i)	Surfactants	0.01-3
(j)	Thickening agents (viscosity and required film thickness determinants)	trace to 10.0 by weight of resin
(k)	Water including water-soluble additives	5-65

The organic film-forming compositions herein described containing calcium, magnesium, strontium and/or barium silicofluorides are not only useful as primer coatings which accept other organic coatings such as paints, e.g., water-based or water-reducible paints, but also act as adhesive coatings for self-supporting organic sheets or films such as polyacrylic films, polyethylene, polypropylene, polyvinyl fluoride (Tedlar) and regenerated cellulose (cellophane). The self-supporting sheets can be applied to aluminum, steel, or zinc-surfaced sheets which have previously been coated with 0.1 to 0.3 mil thickness coating of the organic film-forming composition and baked for 30 to 60 sec at 450° to 500°F (metal temperature 350° to 400°F). The laminating sheet is applied while the coated metal is still hot enough to soften the laminating sheet and the laminated sheet is subjected to pressures of 300 to 400 psig between rollers. These products are useful as panel stock, gutters, industrial siding, and for many other purposes.

Aluminum coated with organic film-forming compositions of the process has resisted salt spray for at least 3,000 hr and as much as 7,200 hr without failure.

In the foregoing preferred coating compositions, where a hexavalent inorganic chromium compound is added the amount usually constitutes 0.5 to 10% by weight of the coating composition, calculated as Cr.

Example 1: An emulsion coating composition was prepared as follows: 4 gal of acrylic emulsion polymer containing 46±0.5% solids and having a pH of 9 to 11 (Rhoplex MV-1) was mixed with premix of 15 to 30 ml of tributyl phosphate and 3 to 10 ml of a defoaming agent (Nopco NXZ) and agitated thoroughly until no fisheyes or agglomerates appeared on a fineness of grind gauge. 1 to 2 gal of water was added and to the resultant emulsion while it was being rotated with a paddle-type agitator there was added dropwise at the periphery of the emulsion body a dilute solution of sodium dichromate in water containing 1 g of sodium dichromate per 10 ml of solution until a pH value of 6.6 to 6.8 was obtained. The resultant emulsion was then filtered.

A coating of the foregoing emulsion was applied using a draw bar to a thickness of 0.05 mil to 1.0 mil thickness on aluminum panels of No. 3003 alloy having a gauge thickness of 0.019 to 0.025. These panels had previously been alkaline cleaned. Some were chromate conversion coated. Others were rinsed with water and then acid rinsed with chromic acid-phosphoric acid mixtures and dried prior to the application of the coating.

The coatings were cured by preheating at 120° to 160°F for 20 to 40 sec followed by complete cure in 45 to 60 sec at 600°F. Panels of the cured coating were bent and the edge of the bend exposed to 60 inch-pounds direct impact. There was no pullaway using 3M 600 tape.

Example 2: A pigment was prepared by mixing together in water 1 lb of slaked lime $(CaOH)_2$, 1 ℓ of water and 15 ml of a surfactant (Triton CF-10) to form a paste. 26% fluosilicic acid was then slowly added with thorough mixing until a pH of 8.0 was obtained. The reaction was highly exothermic and the mixer was water cooled. Thereafter, chromic acid was added until the pH was 7.0. This gave a pigment containing hexavalent chromium. If the addition of chromium is not desired, another acid such as phosphoric is used to neutralize the composition to the desired pH of 7.0, immediately before using–this is a catalyst addition. The pH value can be altered.

In either case the resultant composition was then ground in a ball mill and about ½ lb of synthetic talc was used to reduce gloss.

1 to 2 lb of titanium dioxide were then added (DuPont 966 TiO_2). A premix was prepared by mixing together 500 ml of tributyl phosphate, 100 ml of a defoaming agent (Nopco NXZ), 1 gal of propylene glycol and 75 g of a pigment dispersant and stabilizer (Tamol 830) added to the foregoing composition. The resultant mixture was milled until no fisheyes, agglomerates or large particles appeared on a ground gauge and the particles had a size within the range of 0.5 to 5 μ.

1,000 ml of thermoplastic acrylic emulsion polymer (Rhoplex MV-1) was then added and milled with the pigment until all of the pigment particles were uniformly coated.

In a slow paddle wheel mixer 5 gal of the thermoplastic acrylic emulsion polymer was mixed with 2 gal of water and the pigmented mixture slowly added thereto. Thereafter a dilute solution of sodium dichromate containing 1 to 10% by weight hexavalent chromium, as Cr, was added dropwise to the peripheral portion of the rotating mixture until a pH of 7.6 to 7.8 had been obtained.

A diluted thickening agent was then added (Acrysol G-110) until a viscosity reading of 25 to 30 sec was obtained using a Zahn No. 2 cup. 2 gal of an acrylic thermosetting polymer (Rhoplex AC-604) were then added and the mixture was mixed slowly for at least 6 hr.

A coating of the foregoing emulsion was applied to both 3003 aluminum sheet and spangle hot dip galvanized sheet to a wet thickness of 0.2 mil using a draw bar. The coated hot dip galvanized sheet was preheated to 160°F for 30 sec and cured at 600°F for 75 sec. The coated aluminum sheet was preheated at 160°F for 30 sec and cured at 600°F for 60 sec. In each case the resultant coatings had a pencil hardness of 2H/H.

The hot dip galvanized coated sheet withstood the impact of 80 to 100 inch-pounds and the coated aluminum sheet withstood impact of 30 inch-pounds. The aluminum sheet after being bent withstood an edge impact of 60 inch-pounds. The impact value (lb/in^2) is that just before metal or crystal fracture occurs; thus, this value must be a variable.

Salt spray tests on both sheets were conducted to failure. The coated hot dip galvanized sheet ran 400 to 600 hr and the coated aluminum sheet ran over 1,000 hr. On embossed aluminum sheeting a coating of the foregoing emulsion applied as previously described gave very superior results on sharply rounded areas as tested by the modified Preece copper sulfate test.

Example 3: 50 g of strontium chromate, 50 g of water and 13 ml of dispersant (Tamol 850) were mixed with a premix of 2 ml tributyl phosphate, 1 ml surfactant (Triton CF10), and 1 ml defoaming agent (Nopco NXZ) and ball milled for 4 hr. 500 ml of thermoplastic acrylic emulsion polymer (Rhoplex MV-1) was added to the mixture and ball milled for 30 min.

The resultant mixture was then subjected to slow agitation and 1,000 ml of thermoplastic acrylic polymer (MV-1) and 500 ml of an acrylic thermosetting polymer (AC-604) were added. While the resultant emulsion was being agitated slowly a dilute aqueous solution of sodium dichromate (1 g in 10 ml of water) was slowly added at the periphery of the rotating body until a pH of 8.4 was obtained.

The resultant emulsion coating composition was then applied to properly cleaned and surface-prepared spangle hot dip galvanized panels to produce a wet coating having a thickness of 0.2 ml using a 0.2 mil draw bar. The coated panels were preheated at 160°F for 30 sec, cured at 600°F for 60 sec, 90 sec and 120 sec, respectively. They were then subjected to indirect impact tests at 120 inch-pounds and showed no pullaway but with some fracturing of galvanized crystals appearing. A pencil hardness test gave a 3 to 2H to all panels.

All of the panels were then placed in boiling water for 1 hr and tested for pencil hardness after a 15 min recovery period. The panel with the 60 sec cure softened badly and absorbed water. The panel with the 90 sec cure softened somewhat and the panel with the 120 sec cure recovered with very little change as compared with the panel before the boiling water test. The 120 sec cure panels when subjected to indirect impact testing at 120 inch-pounds gave no tape pullaway.

Coated panels prepared as previously described will adhere to finishing coats including, for example, polyester, (e.g., PPG-JJ-487 Duracron), acrylic, (e.g., PPG 11W30, Duracron 100, DuPont 876-5461, 876-559 and 876-5484), polyvinyl chloride, (e.g., Sherwin-Williams G-77WC198), modified silicone, (e.g., 64X423), modified polyester, (e.g., Dexstar 5X100A), modified epoxy, (e.g., Dexstar 9X165), and aqueous resin coatings, (e.g., Armorcote 11 White).

For Aluminum and Its Alloys: Zinc, Fluoride and Molybdate Ions

The process of *E.R. Reinhold; U.S. Patent 4,146,410; March 27, 1979; assigned to Amchem Products, Inc.* relates to a composition for applying a conversion coating on aluminum and aluminum alloy surfaces. In the method an aluminum-containing metal surface is contacted with a ferricyanide free aqueous acidic solution containing zinc, hexavalent chromium, fluoride, and molybdate ions. The coated metal surface shows enhanced resistance to corrosion and improved paint adhesion. The absence of ferricyanide in the composition promotes easier disposal of the spent solutions.

A concentrate is prepared from the following ingredients in the amounts specified:

Material	Grams per Liter
CrO_3	40.0
ZnO	7.6
HNO_3, 38° Bé	68.0
H_2SiF_6 as a 23% solution	91.2
Molybdic acid as 84% MoO_3	9.5
Water, balance	

From this concentrate a bath is prepared by diluting the concentrate with water to make a 5% (by volume) solution. The final solution pH was about 1.5.

Example 1: In Example 1, 6 sets of aluminum panels made from 3003 alloy were subjected to the following treatment sequence.

Sequential Steps	Process Step
Stage 1	Alkaline cleaning step
Stage 2	Tap water rinse step
Stage 3	Treatment at about 100°F with bath composition
Stage 4	Tap water rinse step

Following Stage 4 the sets of panels treated in the abovedescribed manner were painted and subsequently tested for performance. The results are set forth in Table 1.

Table 1: Bath Composition

(1) Panel Designation	(2) Panel Composition (alloy)	(3) Zn^{++} (g/ℓ)	(4) Cr^{+6} (g/ℓ)	(5) Free Acid as Equivalent (10^{-4})	(6) Coating Weight (mg/ft²)	(7) Spray Time (sec)
A-1	3003	0.0	0.9	0.9	7.2	15
A-2	3003	0.1	0.9	0.8	24.3	15
A-3	3003	0.2	0.9	1.0	31.2	15
A-4	3003	0.3	0.9	0.9	39.6	15
A-5	3003	0.4	0.9	0.9	37.8	15
A-6	3003	0.5	0.9	0.9	27.3	15
A-7*	3003	**	0.6	–	14.4	5
A-8	3003	used as blank (cleaned only)			0.3	not sprayed
A-9*	3003	**	0.6	–	40.5	15

*Standard.
**Ferricyanide treated.

A-7 and A-9, used as comparison reference panels, designate panels treated according to the method of U.S. Patent 2,988,465. These panels are standard ferricyanide treated panels. Panel A-8, also used as a reference panel, was cleaned in a similar manner as the other test panels but received no conversion coating of any type and thus performs as a blank.

The six sets of panels (A-1 through A-6) were prepared with varying amounts of zinc to demonstrate the accelerating effect of various levels of zinc in producing coatings on the panels treated in accordance with the process.

Table 2 contains data collected from the testing of the panels described in connection with Table 1.

Table 2

(1)	(2)	(3)	(4)	(5)	(6)	(7)	(8)	(9)	(10)
Identification and Type Metal	Zinc Content of Bath (g/ℓ)	Acid Salt Spray 5% at 95°F							
		500 Hours				1,000 Hours			
		Untaped Area		Taped Area		Untaped Area		Taped Area	
		Scribed	Unscribed	Scribed	Unscribed	Scribed	Unscribed	Scribed	Unscribed
A-1	0	slight	7.0	slight	*	moderate	3.0	moderate	moderate
A-2	0.10	*	10	*	*	slight	3.0	slight	moderate
A-3	0.20	*	10	*	*	moderate	3.0	moderate	moderate
A-4	0.30	*	10	*	10	moderate	3.0	moderate	slight
A-5	0.40	*	10	*	10	moderate	3.0	moderate	moderate

(continued)

(1)	(2)	(3)	(4)	(5)	(6)	(7)	(8)	(9)	(10)
Identification and Type Metal	Zinc Content of Bath (g/ℓ)	Acid Salt Spray 5% at 95°F							
		500 Hours				1,000 Hours			
		Untaped Area		Taped Area		Untaped Area		Taped Area	
		Scribed	Unscribed	Scribed	Unscribed	Scribed	Unscribed	Scribed	Unscribed
A-6	0.50	*	10	*	10	moderate	3.0	moderate	moderate
A-7	–	slight	10	slight	10	moderate	3.0	moderate	moderate
A-8	–	slight	7.0	slight	slight	heavy	2.0	moderate	moderate
A-9	–	*	10	*	10	moderate	3.0	moderate	moderate

*Very slight.

Columns 3 through 10 indicate results from a standard acid/salt spray of the particular panel in taped and untaped areas for both a scribed and unscribed portion of the panel. This particular test refers to a method as described in detail in the *American Society of Testing Materials Bulletin,* ASTM B-287.

In performing the aforementioned standard test the panels to be tested, after being treated by the composition and method of this process together with standards and blank, are then coated with paint. The particular paint system utilized in this test is known as PPG Duracron 630 (Pittsburgh Plate Glass). This paint is an acrylic paint applied and baked and is used as a standard for this purpose in this test.

Panels were rated both numerically and descriptively. Numerical rating of the panels was done visually on a scale ranging from 1 to 10 with 10 representing the best result. Descriptions such as slight, moderate, etc., indicate that the loss of paint was observed to be slight, moderate etc.

As can readily be seen from the results in Table 2, panels coated in accordance with the composition and method of this process give results overall which are as good as ferricyanide-activated solutions.

For Aluminum and Its Alloys: Phosphate and Fluoride Ions

N.J. Newhard, Jr., U.S. Patent 4,131,489; December 26, 1978; assigned to Amchem Products, Inc. provides a composition for coating the surface of a metal selected from the group consisting of aluminum and alloys thereof in which aluminum is the principal ingredient, comprising an aqueous acidic solution which consists essentially of from about 0.005 to 0.2 g/ℓ of CrO_3, from about 0.02 to 0.4 g/ℓ of phosphate ion, and from 0.005 to about 0.04 g/ℓ of fluoride ion, the pH of the solution being less than about 3.5.

It has been found that by using the compositions, aluminum surfaces having excellent properties in terms of corrosion resistance and paint adhesion may be obtained, notwithstanding the fact that the coating-forming ingredients in the composition are present in amounts considerably below the concentrations used in the prior compositions. By operating with such low concentrations the problem of waste removal from the coating process is largely overcome, since the concentration of hexavalent chromium in any wastewater will be very low and will only require a minimum of treatment to reduce the chromate concentration to within acceptable limits.

Example: *Coating composition* – A coating composition was prepared having the following composition: 0.1 g/ℓ of chromic acid (calculated as CrO_3), 0.098

g/ℓ of phosphoric acid and 0.02 g/ℓ of hydrofluoric acid. Deionized (DI) water is added to make 1 ℓ of composition. The pH of the composition was 2.59.

Formation of coating on aluminum cans using DI water baths – Using coatings as given in the table, alloy 3004 (beer) aluminum cans were cleaned and coated according to the following procedure:

(a) The cans were prewashed with tap water at 160°F by spraying at 15 psi for 60 sec.

(b) The cans were cleaned using a sulfuric acid/fluoride cleaner (≈ 1% by weight in tap water) at 125°F. The amount of fluoride in the cleaner solution was about 20 ppm.

(c) Some of the cans were then rinsed with tap water.

(d) The cans were spray-treated with the coating composition for 20 sec at 100°F and 5 to 6 psig.

(e) The cans were rinsed using DI water copiously applied from a plastic squirt bottle.

(f) The cans were oven-baked at 205°C for 2 min.

Sections from the resulting coated cans were then tested as follows:

Muffle Test – Sections taken from the sidewalls of the cans were heated in a muffle furnace at 900°F for 5 min. The color of the resulting metal was observed. The presence of a light gold to brown color on the surface of the metal after treatment in the muffle furnace evidences coating formation, whereas the absence of such color (no color) indicates that a coating was not formed.

Blackening Resistance Test – The exterior domes of the cans, (i.e., the exterior of the bottom portion of the cans) were boiled in tap water for 15 min and then examined for any discoloration of the metal. No darkening should be observed on a can that has been effectively coated.

Adhesion Test – Sections of the coated cans were painted with either a white ink or a base coat as follows:

(1) A white ink was applied to the coated metal surface. The inked surface was thereafter varnished and then heat cured for 6 min at 350°F.

(2) A white base coat of paint was applied to the coated metal surface using a roller-coat process. The resulting paint layer was cured at 400°C for 2½ min.

The painted can sections were immersed in a boiling 1% (by volume) solution of Joy dishwashing detergent in deionized water. The sections were then dried, and portions of the sections were scribed through the coating layers to bare metal.

The scribing and taping were performed as follows: Scribing was done very precisely. A cutting tool was employed to cut parallel lines through the painted surface which were approximately 1½ inches long and 1/16 inch apart. A second set of parallel lines was cut over the first set at a 90° angle to produce 100 squares of painted surface, separated by bare metal.

Run No.	Rinse*	CrO_3 as g/ℓ Cr	H_3PO_4 (g/ℓ)	HF (g/ℓ)	pH	Blackening Resistance**	Muffle Test	Test White Ink Adhesion**	Test Base Coat Adhesion***
Control	R	–	–	–	–	poor	no color	no pickoff	fail, moderate pickoff
Comparison	R	0.0	0.098	0.020	2.95	poor	light gold	slight pickoff	fail, heavy peeloff
Comparison	NR	0.0	0.098	0.020	2.95	poor	light gold	slight to moderate pickoff	fail, heavy peeloff
1	R	0.005	0.098	0.020	2.93	good, ND	light gold	no pickoff	no pickoff
2	NR	0.005	0.098	0.020	2.93	good, ND	light gold	no pickoff	no pickoff
3	R	0.010	0.098	0.020	2.90	good, ND	light gold	no pickoff	no pickoff
4	NR	0.010	0.098	0.020	2.90	good, ND	light gold	no pickoff	no pickoff
5	R	0.100	0.098	0.020	2.59	good, ND	light gold	no pickoff	no pickoff
6	NR	0.100	0.098	0.020	2.59	good, ND	light gold	no pickoff	no pickoff
Comparison	R	0.001	0.098	0.020	–	brownish, not acceptable	light gold	no pickoff	slight pickoff
Comparison	NR	0.001	0.098	0.020	–	brownish, not acceptable	light gold	no pickoff	slight pickoff
7	R	0.010	0.020	0.020	3.09	no darkening	light gold brown	pass, no pickoff	pass, no pickoff
8	NR	0.010	0.020	0.020	3.09	no darkening	light gold brown	pass, no pickoff	pass, no pickoff
9	R	0.010	0.039	0.020	2.99	no darkening	light gold brown	pass, no pickoff	pass, no pickoff
10	NR	0.010	0.039	0.020	2.99	no darkening	light gold brown	pass, no pickoff	pass, no pickoff
Comparison	R	0.020	0.039	0.004	3.04	fail, severe	gold-brown	pass, no pickoff	pass, no pickoff
Comparison	NR	0.020	0.039	0.004	3.04	fail, moderate to severe	gold-brown	pass, no pickoff	pass, no pickoff
11	R	0.020	0.039	0.008	3.02	mediocre, mild blackening	gold-brown	pass, no pickoff	pass, no pickoff
12	NR	0.020	0.039	0.008	3.02	pass, slight darkening	gold-brown	pass, no pickoff	pass, no pickoff
13	R	0.020	0.039	0.016	2.94	pass, no darkening	gold-brown	pass, no pickoff	pass, no pickoff
14	NR	0.020	0.039	0.016	2.94	pass, no darkening	gold-brown	pass, no pickoff	pass, no pickoff
15	R	0.020	0.039	0.020	2.98	pass, no darkening	gold-brown	pass, no pickoff	pass, no pickoff
16	NR	0.020	0.039	0.020	2.98	pass, no darkening	gold-brown	pass, no pickoff	pass, no pickoff
Comparison	R	0.050	0.098	0.0	2.77	fail, severe	light gold brown	fail, moderate pickoff	fail, massive pickoff
Comparison	NR	0.050	0.098	0.0	2.77	fail, severe	gold-brown	pass, no pickoff	pass, no pickoff

Note: ND means no darkening.

*R denotes rinse, NR indicates that no rinse was used between coating formation and baking. **Average of three runs. ***Average of two runs.

Scotch tape was applied over the scribed area and pressed firmly onto the metal. The tape was then rapidly peeled off in order to determine the extent of removal of paint or ink coating from the metal surface. The sections were then assessed for pickoff which is removal of discrete pieces and peeloff which is removal of the paint or ink as a continuous film.

The results are presented in the table. Comparison was made against an untreated can (see control) and against cans treated with compositions outside of the scope of the process. Note that effective coating was obtained using low CrO_3 concentrations of 0.005 g/ℓ (see runs 2 and 3), low concentrations of phosphate ion of 0.020 g/ℓ (see runs 7 and 8), and low concentrations of fluoride ion of 0.008 g/ℓ (runs 11 and 12) with effective coating. The pH range in these runs using deionized water was from 2.59 (see runs 5 and 6) to 3.09 (see runs 7 and 8). The results are presented in the above table.

For Aluminum and Its Alloys: Sodium Nitroferricyanide and Inorganic Fluoride

Compositions containing hexavalent chromium, sodium or potassium ferricyanide, and fluoride ion in acidified aqueous solutions have been used for producing corrosion-resistant conversion coatings on aluminum and its alloys.

E. Simon; U.S. Patent 4,036,667; July 19, 1977 discloses that substitution of sodium nitroferricyanide for sodium or potassium ferricyanide in a conversion solution composition of a nitric-acid-acidified solution of dichromate and a soluble fluoride, reduces the deposition of loosely adherent reaction products and improves the appearance of the conversion coatings without adversely affecting the corrosion resistance characteristics. This enhances its efficacy for commercial applications and as bonding substrates for finishes such as coatings, adhesives, and sealants.

The compositions are applied at ambient temperatures, within the range of 65° to 85°F, for immersion times of 5 to 10 min depending on the requirement of color, coating thickness, and corrosion resistance.

Representative compositions are comprised of from 0.75 to 1.25 g of the dihydrate of sodium dichromate, from 0.075 to 0.125 g of the dihydrate of potassium fluoride, from 0.075 to 0.125 g of the dihydrate of sodium nitroferricyanide [$Na_2Fe(CN)_5NO \cdot 2H_2O$], water to make 100 ml solution, and nitric acid (70%), if required, to provide a pH of 1.4±0.3. The addition of from 0.075 to 0.125 g of the tetrahydrate of sodium vanadate ($NaVO_3 \cdot 4H_2O$) appears to enhance the efficacy of the process, particularly as it relates to improved corrosion resistance in saltwater media of the treated aluminum and aluminum alloy substrates.

Example: *Characteristics of conversion coatings from sodium nitroferricyanide and potassium ferricyanide solutions* – Aluminum and aluminum alloy substrates were immersion-treated for 10 min at approximately 70°F in the solutions shown in the following table:

	A	B
$Na_2Cr_2O_7 \cdot 2H_2O$, g	1.00	1.00
$KF \cdot 2H_2O$, g	0.10	0.10
$Na_2Fe(CN)_5NO \cdot 2H_2O$, g	0.10	–
$K_3Fe(CN)_6$, g	–	0.10
H_2O (dist), to make 100 ml		
HNO_3 (70%) to about pH 1.5, ml	0.3	0.3

A-coated substrates were uniformly coated with reflective, adherent, gold-colored reaction products, whereas solution B coatings were matte, rust-colored with a nonadherent surface layer. Prior to the corrosion-resistance testing, both the A and B coatings were wet-rubbed with water; the A-coated specimens showed very little surface removal as compared to the B-coated specimens in which the excessive, and apparently more porous, build-up was easily removed. Corrosion protection to the base metals after total and partial immersion in seawater and 3% sodium chloride in tap water for greater than 3 months was very satisfactory, showing essentially no pitting; in contrast, controls of the same metal substrates that were not conversion-coating treated, were severely attacked, some within a couple of days.

For Zinc-Iron Alloy: Perchlorate and Fluoride Ions

W.C. Glassman, S.H. Melbourne, M.S. Morson and U. Soomet; U.S. Patent 4,141,758; February 27, 1979; assigned to Dominion Foundries and Steel, Limited Canada provide a composition for use in production of a protective coating on zinc/iron alloy surfaces, the chromium ion being the only metallic ion present in the solution.

The pH of the solution may be in the range of 1.20 to 0.80 (preferably 1.0 to 0.9). The chromium ion may be present in the amount of not less than 20 and not more than 36.4 g/ℓ, corresponding to the addition of 38.5 to 70.0 g/ℓ anhydrous chromium trioxide to the solution, (preferably 20 to 24 g of ion per liter). The perchlorate ion may be present in the amount of 0.4 to 1.2 g/ℓ, corresponding to the addition of 0.41 to 1.22 ml of 60% perchloric acid to each liter of solution, (preferably 0.5 to 1.0 g of ion per liter). The fluoride ion may be present in the amount of 0.1 to 0.35 g/ℓ, corresponding to the addition of about 0.19 to 0.65 ml/ℓ of 48% hydrofluoric acid to the solution, (preferably 0.12 to 0.25 g/ℓ).

The process includes applying a composition as specified above to a zinc/iron alloy surface for a period of from 1 to 10 sec at a temperature of from 150° to 195°F (65.5° to 90.5°C).

For a description of apparatus suitable for carrying out the process, reference may be made to the U.S. Patent 3,857,739, the disclosure of which is incorporated herein by reference.

The films produced by the application of the process are found to be unexpectedly much more protective than those of known prior methods. One test known as the humidity cabinet test involves the use of a Cleveland Condensing Test Cabinet having its interior maintained at 100°F (37.7°C) wherein strips of the iron-zinc-alloy-coated steel are suspended above a water bath, the side exposed to the bath being at the temperature of 100°F (37.7°C), while the other side is at the ambient temperature, usually about 70°F (21°C). The humidity of the cabinet interior is 100% and the water that condenses on the inner surface trickles back into the bath, giving the effect of a constant exposure to a rainy atmosphere.

The strips are maintained under this condition until visual inspection shows that storage staining has begun with the formation of black corrosion spots on the surface of the iron-zinc alloy. The use of conventional dilute chromic acid solutions typically provides a conversion coating which protects the alloy for approximately 200 hr before such black spots form in this test. The solution described in the abovementioned U.S. Patent 3,857,739, as applied to iron-zinc alloy coatings, will on average provide about 370 hr of protection before black storage stain spots form. Conversion coatings produced with the solution of the process provide an average of about 475 hr protection in this same test. It is believed that such a high degree of protection against storage stain formation is not provided by any other economical solution or treatment.

In another more severe test known as the Salt Fog Test (ASTM B117-62) the test panels are placed in a fine mist of 5% salt solution at 100°F. After 64 hr of exposure to this salt fog, iron-zinc alloy panels with no conversion coating show advanced corrosion, (i.e., red rust) over their entire surface areas. Conversion coatings produced by the use of conventional dilute chromic acid solutions offer protection such that, on average, only 60% of the surface area of a test panel shows corrosion after 64 hr. The solution described in U.S. Patent 3,857,739 will produce an improved conversion coating such that, on average 50% of the surface area of a test panel is corroded after 64 hr. After the same 64 hr test, test panels treated with the solution of the process show on average corrosion stains over only 6% of their surface area.

Black Chromate Coating on Zinc or Cadmium

There are commercially available single-dip immersion processes for producing chromate conversion coatings including jet black chromate conversion coatings on metal substrates such as zinc or cadmium, zinc or cadmium die-castings, zinc or cadmium plate, aluminum or copper. Such coatings can afford high corrosion and abrasion resistance, resist stains and fingerprints, and serve as an excellent base for paint and other organic finishes. Black chromate coatings are expected to have utility on a large scale as radiation adsorbers in solar energy devices. Zinc plate properly coated with a black chromate coating will withstand in excess of 100 hr of standard salt spray solution without white corrosion.

However, a disadvantage of such processes and solutions is that the coating and other conditions need to be rather carefully controlled in order to obtain optimum results.

J.R. House; U.S. Patent 4,065,327; December 27, 1977; assigned to Imasa Limited, England provides a method for treating a black chromate-coated zinc or cadmium metal surface, which comprises contacting the coated surface with an aqueous solution or dispersion of thiosulfate, thioglycolate or thiourea. Mixtures of such sulfur-containing compounds may be used with advantage.

A preferred sequence of operations is as follows:

(1) Provide an aqueous solution of hexavalent chromium at a concentration of from 0.1 M to 3.0 M and silver at a concentration of from 0.0003 M to 0.1 M and at a pH of from 0.5 to 4.

(2) Contact the zinc or cadmium surface to be treated with the aqueous solution, e.g., by dipping, until a black chromate coating has been formed. This generally requires from 1 to 3 min, 2 min being optimum.

(3) Rinse the coated product with cold water.

(4) Contact the black chromate-coated zinc or cadmium surface with an aqueous solution of a sulfur-containing compound as described above.

(5) Rinse.

(6) Dry at not more than 60°C.

The method provides the following advantages. It will be understood that not all these advantages will necessarily be observed at any one time.

(a) The corrosion resistance of black chromate-coated metal surfaces can be increased many times. Coatings having excellent corrosion resistance can be obtained more reliably than was hitherto possible.

(b) The coated surface can be dried at elevated temperatures, where it was previously necessary to dry at ambient temperature with a consequent delay and expenditure in energy for air circulation.

(c) The light fastness of the treated products can be better than that of prior chromate-coated products.

Example 1: A good quality black chromate finish will withstand 100 hr salt spray testing (BSS 1224/1970, ASTM B-36868). However, to illustrate the effectiveness of the postblackening dip process, articles were processed such that without the postblackening treatment the zinc coating would only withstand 48 hr salt spray.

Similar samples were processed in the same black chromating solution for 2 min and washed thoroughly. The control samples were then dried at room temperature.

The remaining samples were dipped in sodium thiosulfate (200 g/ℓ) for 2 min and then thoroughly washed. The samples were divided into two batches, and were dried at either room temperature (20°C) or at 50°C. The results are indicated below.

Test	Sample Treatment	Result
Salt spray BBS 1224/1970		
Control	No thiosulfate dip	48 hr maximum
Dried at 20°C	Postdipped in sodium thiosulfate	>120 hr
Dried at 50°C	Postdipped in sodium thiosulfate	>120 hr
Daylight		
Control	No thiosulfate dip	Fading after 4 wk exposure
Dried at 20°C	Postdipped in sodium thiosulfate	No fading after 3 mo
Dried at 50°C	Postdipped in sodium thiosulfate	No fading after 3 mo
UV Radiation		
Control	No thiosulfate dip	Fading after 1 wk exposure*
Dried at 20°C	Postdipped in sodium thiosulfate	No fading after 1 mo
Dried at 50°C	Postdipped in sodium thiosulfate	No fading after 1 mo

*24 hours per day.

Example 2: Black chromate-coated zinc panels were immersed for various times in thiourea solutions of varying concentrations with the following results.

Concentration (g/ℓ)	Immersion Time (sec)	Salt Spray Test (hr)
Saturated	30	200+
80	60	200+
40	120	200+
20	240	200+
10	480	200+
Control	–	50-70

USING TRIVALENT CHROMIUM

For Zinc or Zinc Alloy: H_2SO_4 and Peroxide Oxidizer

It has been recognized that surfaces of zinc and zinc-based alloys can be protected against corrosion by treatment with an acid solution containing hexavalent chromium.

Among the disadvantages of hexavalent chromium-type solutions is one in the area of waste disposal. Recent emphasis on water pollution problems has drawn attention to the fact that chromates are serious pollutants. In order to satisfy water quality standards, it frequently is necessary to subject the wastewater to a multistage purification sequence in order to remove chromates from the effluents.

C.V. Bishop, T.J. Foley and J.M. Frank; U.S. Patent 4,171,231; October 16, 1979; assigned to R.O. Hull & Company, Inc. found that a highly desirable clear to light blue chromate finish on all types of zinc plate which imparts superior corrosion resistance to the zinc surface can be obtained with an aqueous acidic coating solution comprising trivalent chromium as substantially the only chromium ion present, fluoride ion, an acid other than nitric acid and an oxidizing agent. Preferably, the acid is sulfuric acid, the oxidizing agent is a peroxide, and the solution also contains a small amount of a cationic wetting agent.

The trivalent chromium solution may be prepared by reducing an aqueous solution of hexavalent chromium with sufficient reducing agent to reduce all of the hexavalent chromium to trivalent chromium. The aqueous acidic coating solutions of the process have been found to achieve a satisfactory single dip chromate finish on all types of zinc plate over a wide operating range whether the zinc plate has been deposited by a cyanide or noncyanide-type zinc plating bath. Metal articles having zinc or zinc alloy surfaces which have been treated with the aqueous acidic coating solutions of the process exhibit the desired clear to light blue finish and are characterized by superior corrosion resistance.

The aqueous acidic coating solutions generally will contain from about 0.1 to 1 g/ℓ and preferably from about 0.3 to 0.7 g/ℓ of trivalent chromium ion, sufficient acid, preferably a mineral acid other than nitric acid, to lower the solution pH to between about 1 to 4 and preferably between about 1 to 3. The amount of oxidizing agent included in the coating solutions is an amount which is sufficient to oxidize the trivalent chromium to hexavalent chromium at the interface of the zinc surface and the coating solution where the pH is greater than the pH of the bulk of the solution. The concentration of the oxidizer is determined by the appearance of the treated zinc plate which preferably is a blue-white finish.

Example 1: A solution is prepared by mixing the following in the amounts and order indicated: 1.1% v/v of a Cr(III) compound formed by reacting 94 g/ℓ chromic acid with 86.5 g/ℓ potassium metabisulfite and 64 g/ℓ sodium metabisulfite in water; 3 cc/ℓ 96% sulfuric acid; 3.6 g/ℓ ammonium bifluoride; 0.25 ml/ℓ of an organic addition agent which is a solution of 32 cc/ℓ Armohib 25, (amine wetting agent, Akzona Chemicals) in water; and 2% v/v 35% hydrogen peroxide.

Example 2: Freshly plated zinc panels are immersed in the solution of Example 1 for about 15 to 30 sec whereupon a blue color appears on the surface. The panels are removed from the solution, rinsed with water and allowed to dry over a period of 48 hr at room temperature. The dried panels are subjected to a 5% neutral salt spray environment. At the end of 24 hr, the panels showed only 0 to 10% of white corrosion product, and at the end of 50 hr of salt spray environment, some panels still showed no white corrosion.

When the same procedure is carried out on the same type of freshly plated zinc panels except that the coating solution does not contain any hydrogen peroxide (oxidizing agent), and the treated panels are subjected to the same neutral salt spray environments, the panels showed 80 to 100% white corrosion at the end of 24 hr.

For Zinc or Zinc Alloy: Alum and Vanadate

The process of *K. Aoki; U.S. Patent 4,126,490; November 21, 1978; assigned to Caterpillar Mitsubishi Ltd., Japan* relates to a chromate coating for the reduction of corrosion of zinc metal or zinc alloys.

The composition does not contain any hexavalent chromium compounds and does not form any. Thus, the zinc metal can be coated without subjecting people or animals to environmentally undesirable hexavalent chromium.

The composition for forming a chromate coating on the surface of zinc metal or zinc alloy comprises water as a solvent and contains dissolved therein 1 to 50 g/ℓ of a trivalent chromium compound, 3 to 130 g/ℓ of an alum, 0.2 to 10 g/ℓ of a vanadate, 0.5 to 25 g/ℓ of a mineral acid and optionally an amount of a surfactant.

Example: In the test samples each specimen is prepared, treated, tested and estimated in a manner as noted below.

(1) A rectangular (1.0 x 50 x 75 mm) JIS G3141 cold-drawn steel specimen (bright-finished) having a 2 mm hole at each corner was plated with zinc to a thickness of 4 to 4.5 μ. The plated specimen was immersed in a chromate treatment liquor. After the completion of the treatment the specimen was removed from the liquor, immediately and thoroughly washed with tap water, and dried by an air stream.

(2) The specimen so treated was subjected to a continuous saltwater spraying in accordance with ASTM B117-54T for a period of 24 hr.

(3) Based on the total surface area of the tested specimen the percent of the sum of those areas where white rust had developed was determined, and depending upon the determined

percent the corrosion resistance of the specimen was rated as follows.

Percent of White Rust Areas (~%)	Rating
0	1
1	2
3	3
8	4
14	5
20	6
30	7
50	8
100	9

Tests were carried out with the typical composition and varying treatment conditions as indicated in the table, wherein A is chromium nitrate, B is aluminum potassium sulfate, C is ammonium metavanadate, and D is hydrochloric acid.

Test No.	Composition of Treatment Liquid in Water (g/ℓ) A	B	C	D	Treatment Conditions Time (sec)	Temperature (°C)	Corrosion Resistance Rating
1	10	30	2.25	5.1	5	ambient	2
2	10	30	2.25	5.1	10	ambient	1
3	10	30	2.25	5.1	20	ambient	2
4	10	30	2.25	5.1	30	ambient	2
5	10	30	2.25	5.1	40	ambient	3
6	10	30	2.25	5.1	60	ambient	3
7	10	30	2.25	5.1	10	42	1
8	10	30	2.25	5.1	10	50	1
9	10	30	2.25	5.1	10	60	2
10	10	30	2.25	5.1	10	68	2
11	10	30	2.25	5.1	10	81	2

The chromate treated test specimen from Test No. 3 and a similar test specimen which had been obtained using a commercially available chromate treatment liquid containing a hexavalent chromium compound were tested for paint adhesion. Each specimen was sprayed with a primer to a thickness of about 25 μ and allowed to stand for 48 hr. A cross incisure about 12.7 by 25.4 mm (0.5 x 1.0 inch) was scarred on the primer coating with a knife, and an adhesive tape (Scotch tape No. 610', 3 M Corp.) was adhered so as to completely cover the incisure. The tape was then suddenly peeled off, and the state of peeling of the primer in the vicinity of the incisure was examined with the naked eye. If any peeling of the primer was noted the paint adhesion of the test specimen was rated as being poor whereas if no peeling of the primer was observed the paint adhesion was rated as being good. The results indicated no peeling in the process sample or the conventionally treated one.

For Zinc, Zinc Alloy and Cadmium: Green Cr(III), Blue Cr(III), Fluoride Ions

The aqueous acidic coating solutions of *D.J. Guhde and D.M. Burdt; U.S. Patent 4,263,059; April 21, 1981; assigned to Rohco, Inc.* which are useful for treating a zinc or zinc alloy surface comprise a mixture of trivalent chromium as substantially the only chromium ion present, fluoride ion and an acid. As a source of trivalent chromium solutions, solutions of chromium(III) sulfate or chromium(III) nitrate, for example, may be used, but the preferred trivalent chromium solutions are those prepared by reduction of an aqueous hexavalent chromium-containing solution. The solutions are made by mixing a green and a blue trivalent chromium solution. They also can contain peroxide compounds.

Example 1: *Blue Solution* – The solution of this example is prepared by mixing the following ingredients in the amounts and order indicated: 1.1% v/v of a Cr(III) compound formed by reacting 94 g/ℓ of chromic acid with 86.5 g/ℓ of potassium metabisulfite and 64 g/ℓ of sodium metabisulfite in water; 3 cc/ℓ of 96% sulfuric acid; 3.6 g/ℓ of ammonium bifluoride; and 0.25 ml/ℓ of an organic addition agent which is a solution of 32 cc/ℓ of Armohib 25 (amine wetting agent, Akzona Chemicals) in water.

Example 2: *Blue Solution* – The procedure of Example 1 is repeated except that the sulfuric acid is replaced by an equivalent amount of nitric acid (67%).

Example 3: *Blue Solution* – To 28.4 parts of water is added 4.2 parts of chromium trioxide and 24.4 parts of a 25% solution of sodium metabisulfite in water while maintaining the temperature at about 125°F (52°C). To this solution is added 40 parts of nitric acid (67%) and 3 parts of ammonium bifluoride. A blue trivalent chromium solution results having a pH <1.

Example 4: *Green Solution* – A mixture of 71.4 parts of water, 4.2 parts of chromium trioxide and 24.4 parts of a 25% solution of sodium metabisulfite in water is stirred with cooling to maintain the temperature at about 125°F. The resulting solution is a green trivalent chromium solution with a pH of between 3 and 4.

The solutions of the process are prepared by mixing a blue trivalent chromium solution with a green trivalent chromium solution. The amount of the blue and green chromium included in the solution may be varied over a wide range although the weight ratio of blue to green chromium generally will be between about 1:10 to 10:1. Although the precise chemical nature of the two chromium forms is not understood it has been found that the properties of coatings deposited from the solutions such as corrosion resistance, are improved when compared to the properties of coatings deposited from either the green or blue chromate solutions used alone.

Example 5: The following illustrate the solutions of the process, and all parts and percentages are by volume.

	Parts by Volume
Example A	
Solution of example 3	1.5
Solution of example 4	1.5
Water	97
Example B	
Solution of example 3	10
Solution of example 4	1.5
Water	88.5
Example C	
Solution of example 3	1.5

(continued)

	Parts by Volume
Solution of example 4	10
Water	88.5

In the coating operation in which the method is used, the zinc or cadmium surface usually is first cleaned by chemical and/or physical means to remove any grease, dirt or oxides, although such treatments are not always required. After rinsing the surface with water, the surface then is treated with the aqueous acidic coating solutions. Treatment may be by any of the commonly used techniques such as spraying, brushing, dipping, roller-coating, reverse roller-coating and flow coating. The coating compositions are particularly useful in a dipping system.

The pH of the coating solutions during application generally is from about 1 to to 4 and preferably between about 1 to 2.

The temperature of application of the coating solutions to the metal surface usually is between about 10° to 50°C and is preferably between about 20° to 35°C. When the method of application is by dipping or immersion, a dipping or immersion time of about 10 to 30 sec and preferably of about 10 sec is utilized.

Example 6: Freshly plated zinc panels are immersed in the solution of Example A for about 15 to 30 sec whereupon a blue color appears on the surface. The panels are removed from the solution, rinsed with water and allowed to dry over a period of 48 hr at room temperature. The dried panels are subjected to a 5% neutral salt spray environment, and are inspected for corrosion.

For comparison purposes the same procedure is carried out on the same type of freshly plated zinc panels using the following coating solutions: the solution of Example 3, and the solution of Example 4. The treated panels are subjected to the same neutral salt spray environment, and after 24 hr the panels are inspected for evidence of corrosion.

The results of the 24 hr salt spray test are summarized in the following table. These results demonstrate the improved corrosion resistance obtained with the solutions of the process.

Bath of Ex. No.	Heavy Corrosion (%)	Total Corrosion (%)
3	15	70
4	15	60
A	10	25

Similarly good corrosion resistance is obtained when freshly plated zinc panels are coated with the solutions of Examples B or C.

For Iron, Zinc, Aluminum: Phosphate and Silica

K. Lampatzer and W. Rausch; U.S. Patent 4,169,741; October 2, 1979; assigned to Oxy Metal Industries Corporation describe a hexavalent chromium-free process for treating a metal surface to prepare the surface for the application of lacquers, adhesives and other synthetic or resinous materials. It comprises contacting the surface with an aqueous acidic composition consisting essentially of trivalent chromium, phosphate and dispersed silicic acid wherein the molar ratio of

$Cr(III):PO_4:SiO_2$ equivalent is 1:0.3-30:0.5-10 in an amount sufficient to yield a dry film weight of up to 0.6 g/m^2 of metal surface and thereafter drying the film without rinsing the surface.

The composition additionally comprises at least one component selected from the group consisting of acetate ions, and maleate ions wherein the molar ratio of Cr(III):acetate and/or maleate ion is 1:0-5.

The composition additionally contains at least one component selected from the group consisting of zinc ions and manganese ions wherein the molar ratio of Cr(III):zinc and/or mangenese ion is 1:0-3.

The solutions used contain the components in amounts such as to produce a residue from evaporation of between 5 and 150 g/ℓ. The film of liquid used for wetting is preferably between 2.5 and 25 ml/m^2 of workpiece surface. Satisfactory technical results may be obtained, for example, with a dried layer weighing between 0.03 and 0.6 g/m^2 of workpiece surface. The film of solution is then dried on the surface of the metal. Although this may be done at room temperature, better results are obtained with a higher temperature, preferably with the specimens at a temperature of between 70° and 300°C.

Example 1: The solution contained 2.1 g/ℓ Cr(III) (trivalent chromium); 48 g/ℓ PO_4 (phosphate); 2.1 g/ℓ CH_3CO_2 (acetate); and 10 g/ℓ SiO_2. The molar ratios were: (a) $Cr(III):PO_4:CH_3CO_2:SiO_2$ = 1:12.6:0.9:4.3; and (b) the ratio $Cr(III):(PO_4 + CH_3CO_2)$ = 1:13.5. Residue from evaporation: ~65 g/ℓ.

Example 2: This solution contained 3.6 g/ℓ Cr(III); 29 g/ℓ PO_4; 3.5 g/ℓ CH_3CO_2; 10 g/ℓ SiO_2. The molar ratios were: (a) $Cr(III):PO_4:CH_3CO_2:SiO_2$ = 1:4.4:0.9:2.5; and (b) $Cr(III):(PO_4 + CH_3CO_2)$ = 1:5.3. Residue from evaporation: ~50 g/ℓ.

Example 3: This solution contains 5 g/ℓ Cr(III); 9.7 g/ℓ PO_4; 4.9 g/ℓ CH_3CO_2; 5.5 g/ℓ $(:CHCO_2)_2$ (maleate); 10 g/ℓ SiO_2. The molar ratios were as follows: (a) $Cr(III):PO_4:[CH_3CO_2 + (:CHCO_2)_2]:SiO_2$ = 1:1.04:1.4:1.7; and (b) the ratio $Cr(III):[PO_4 + CH_3CO_2 + (:CHCO_2)_2]$ = 1:2.44. Residue from evaporation: ~35 g/ℓ.

Example 4: This solution contains 3.6 g/ℓ Cr(III); 29 g/ℓ PO_4; and 10 g/ℓ SiO_2. The molar ratios were (a) $Cr(III):PO_4:SiO_2$ = 1:4.4:2.5; and (b) $Cr(III):PO_4$ = 1:4.4. Residue from evaporation: ~32 g/ℓ.

Example 5: This solution contains 5 g/ℓ Cr(III); 9.7 g/ℓ PO_4; 4.9 g/ℓ CH_3CO_2; 10 g/ℓ SiO_2 and 1.7 g/ℓ Zn. Molar ratios were (a) $Cr(III):PO_4:CH_3CO_2:SiO_2$ = 1:1.1:0.9:1.7; (b) $Cr(III):PO_4$ = 1:1.1; and (c) Cr(III):Zn = 1:0.27. Residue from evaporation: ~32 g/ℓ.

Example 6: This solution contains 5 g/ℓ Cr(III); 9.7 g/ℓ PO_4; 4.9 g/ℓ CH_3CO_2; 10 g/ℓ SiO_2; and 1.4 g/ℓ Mn. Molar ratios were (a) $Cr(III):PO_4:CH_3CO_2:SiO_2$ = 1:1.1:0.9:1.7; (b) $Cr(III):PO_4$ = 1:1.1; and (c) Cr(III):Mn = 1:0.26. Residue from evaporation: ~32 g/ℓ.

With the exception of Example 4, the Cr(III) was introduced into the solutions in the form of basic chromium acetate, the PO_4 in the form of thermal phosphoric acid, the SiO_2 in the form of pyrogenic finely divided silicic acid, the Mn in the form of MnO and the Zn in the form of ZnO. The maleic acid was introduced as such.

In each example, the solutions were applied, by means of a roll-frame with counterrotating rolls, to sheets of metal previously subjected to alkaline spray degreasing, rinsing in water and squeezing between rubber rolls. The sheets were raised to a temperature of 80°C by placing them for 17 sec in a 220°C furnace; they were then raised to a temperature of 200°C by placing them for 90 sec in a 240°C furnace. The coatings produced weighed between 0.1 and 0.2 g/m^2. The color of the coatings on steel was blue-gray and, on aluminum and galvanized steel, gray.

The test pieces thus pretreated were coated with an acrylate lacquer and a polyester coil-coating lacquer and were tested for adhesion by means of a bend test and for resistance to corrosion by means of the ASTM B 117 salt spray test. These tests produced technological values showing results with this method at least equivalent to, and in some even rather better, than those obtained with solutions based upon the known Cr(VI)/CR(III)/SiO_2.

OTHER METHODS

Chromium Chromate Corrosion-Inhibiting Pigment

In accordance with the process of *L. Schiffman; U.S. Patent 4,161,409; July 17, 1979* corrosion-inhibiting pigments are provided in the form of finely divided solid particles of a chromium chromate, essentially free of other extraneous metal cation and free of acidic anions which might otherwise tend to promote metal corrosion.

The pigments are produced by partial reduction of a hexavalent chromium compound such as chromic acid in aqueous solution, with an oxidizable organic water-soluble compound containing an active hydroxyl, aldehyde or carboxyl group, evaporating the resulting aqueous reaction product to dryness, and grinding the residual solid mass to particle size suitable for use as pigment.

The pigments thus obtained when applied in conventional vehicles for coating metal surfaces have a passivating effect on such surfaces and impart pronounced capacity for resisting corrosion.

A wide selection of oxidizable organic compounds is applicable for partial reduction of the chromic acid; e.g., formaldehyde, methanol, ethanol, citric acid, sucrose, furfural, etc. From the standpoint of convenience, economy and performance, the preferred reducing agents are methanol, formaldehyde and furfural.

Example 1: 150 lb (= 68.04 kg) of chromium trioxide (CrO_3) are dissolved in 42 gal (= 159 ℓ) of water. To this solution is added, slowly and with vigorous stirring, a solution prepared by diluting 16 lb (= 7.26 kg) methanol with 15 gal (= 57 ℓ) of water. The rate of addition is controlled so that excessive reaction is avoided. This can be monitored by keeping reaction temperature preferably below 85°C. After the reaction has largely subsided, the obtained liquid is transferred to evaporating trays for drying, which is preferably carried out below

150°C. The dried product is then ground to pigment particle size in a pebble mill.

Example 2: A chromium chromate liquid concentrate is produced by dissolving 150 lb (68.04 kg) of chromium trioxide in 40 gal (151.4 ℓ) of water. To this is added, slowly with stirring, 42 lb (19.05 kg) of commercial formalin (36.6%) in 18 gal (68.1 ℓ) water. The addition rate is controlled so that the reaction temperature does not exceed 85°C. When the reaction subsides, water is added to make a total of 100 gal (378.5 ℓ) of concentrate. This concentrate contains 1.5 lb of total chromium oxides per gallon (180 g/ℓ). The chromium is present in both hexavalent and reduced state. For convenience, this liquid concentrate is designated Solution A.

To 50 gal (189.25 ℓ) of Solution A there is added 110 lb (49.9 kg) of Emtal 41 talc, which obtains a ratio of 0.35/1 chrome pigment (calculated as Cr) to talc extender in the final pigment product. The slurry of talc and Solution A is stirred to assure complete and uniform admixture. Excess water is removed by heating under continued agitation.

The obtained moist cake is dried in a kiln at 175° to 180°C and then ground in a pebble mill to pigment size. For convenience, this is hereinafter referred to as Pigment B.

Example 3: *Preparation of a baking primer* – Solution A concentrate as prepared in Example 2 above, is evaporated and dried, then ground to pigment size. The product is designated for convenience Pigment C.

Pigment C is formulated in a baking primer (Formula 2) and tested against a reference control (Formula 1) as tabulated below.

Formula 1

Component	Pounds per 100 Gallons	Grams per Liter
Beckosol 13-412*	100	119.8
Red oxide	92	110.2
Barytes	275	329.5
Water (demineralized)	193.5	231.8

*A water-soluble alkyd.

The above components are ground in a pebble mill and there is admixed therewith, the following:

Beckosol 13-412	162.5	194.7
Water (demineralized)	313.0	375.0
6% manganese naphthenate	1.8	2.2
4% calcium naphthenate	4.6	5.5

Formula 2

Component	Pounds per 100 Gallons	Grams per Liter
Beckosol 13-412	100	119.8
Red oxide	68	81.5
Pigment C	24	28.8
Barytes	275	329.5
Water (demineralized)	193.5	231.8

The foregoing components are ground in a pebble mill and there is admixed therewith, the following:

Beckosol 13-412	162.5	194.6
Water (demineralized)	313	375.0
6% manganese naphthenate	1.8	2.2
4% calcium naphthenate	4.5	5.5

Steel panels were cleaned and coated with the respective compositions of Formulas 1 and 2, baked for 30 min at 149°C and tested in accelerated laboratory corrosion test. The test procedure involved longitudinal scratching of the surface of each panel followed by partial immersion of panel in an aqueous solution containing 3% sodium chloride and 1% (by volume) of 30% hydrogen peroxide. The panel treated with Formula 1 showed evidence of corrosion within 10 min exposure. The panel treated with the Formula 2 composition showed no evidence of corrosion until after 45 min exposure to test conditions.

Electrodeposition of Chromite Coating

A method of producing chromium conversion coatings is described by *J.J.B. Ward and C. Barnes; U.S. Patent 4,137,132; January 30, 1979; assigned to BNF Metals Technology Centre, England.* The coatings are chromite, i.e., Cr_2O_3 coatings rather than the conventional chromate, i.e, CrO_3 coatings. The process includes an electrolyte for depositing chromite layers. The electrolyte is an aqueous solution containing Cr^{3+} ions, a weak complexing agent for Cr^{3+} ions and a poison for the electrodeposition of chromium metal. The electrolyte preferably also contains conductivity salts and may include other additives such as fluoride ion and boric acid. Examples of poisons are Cr^{6+} ions, peroxide, nitrate, polyamines, phosphates and formaldehyde. The chromite conversion coatings can be improved by aging and can be subsequently painted or lacquered. The electrolytes of the process are much less corrosive than Cr^{6+} electrolytes and thus the substrates which can be coated include materials which cannot readily be chromate coated because they are reactive towards Cr^{6+} electrolytes.

Example 1: Using 0.7 M Cr (as sulfate), 4 M NH_4^+, 0.4 M boric acid, and 0.7 M sodium hypophosphite, at pH 3.0 and temperature of 30°C, constituted as above, and operated as an electrolyte, at 1,000 A/m^2 cathode current density, decorative chromium plate was deposited. 5 g/ℓ sodium dihydrogen phosphate was added to the electrolyte. No chromium metal was deposited at any current density and careful examination revealed the presence of a transparent film. A variety of substrate metals were cathodically treated at 200 A/m^2 for 1 min in this solution. The results were as follows:

Substrate	Test	Results
Copper		
Untreated	Immersion in polysulfide	Immediate blackening
Cathodically filmed	solution for 5 min	Retained original appearance
Nickel		
Untreated	Exposed to humid corrosive	Became dull and tarnished
Cathodically filmed	environment for 3 mo	Retained original appearance
Silver		
Untreated	Immersed in polysulfide	Yellowed
Cathodically filmed	solution for 15 min	Retained original appearance

Example 2: The procedure was repeated as in Example 1 but with 10 g/ℓ hexamine in place of sodium dihydrogen phosphate. Identical results to Example 1 were obtained.

Example 3: The procedure of Example 1 was repeated but with 20 ml of 40% formaldehyde solution in place of sodium dihydrogen phosphate. Identical results were obtained.

Example 4: With an electrolyte of Example 1, copper panels were cathodically treated at 200 A/m^2 for 30 sec. Immersion in polysulfide solutions caused the copper to slowly blacken. Other copper panels, cathodically treated in the same way were oven-dried at 50°C for 16 hr. No blackening occurred when immersed in a polysulfide solution.

Example 5: Copper panels were cathodically treated in an electrolyte of Example 1 at a current density of 200 A/m^2 for a time of 1 min. After drying, the panels were sprayed with a clear lacquer. When the lacquer was dry one panel was cut in half. Examination showed that there was no flaking of the lacquer along the edges of the cut. For comparison, a copper panel was sprayed directly with lacquer. After cutting in half, some microflaking of the lacquer was detected.

Other copper panels, prepared as described above, were scribed to give a single long scratch penetrating to the copper. The panels were exposed to a humid, corrosive environment. After one month panels with the cathode film plus lacquer only showed corrosion along the length of the scratch. Lacquered panels without the cathode film showed corrosion spreading from the scratch underneath the lacquer.

Resinous Coating Containing Mixed Chromium Compounds

The process of *J.W. Davis; U.S. Patent 4,183,772; January 15, 1980; assigned to Amchem Products* represents a modification and extension of the teaching of U.S. Patent 3,185,596 in that it has been found that coating formulations can be made that impart outstanding corrosion resistance to metal surfaces such as iron, aluminum and zinc and at the same time are generally universally useful under almost all types of paints and paint systems. This improvement is achieved by blending hexavalent chromium, trivalent chromium, phosphoric acid, soluble polyacrylic acid and water-dispersible acrylic emulsion polymer in specific ratios.

In order to achieve optimum performance and widest universality under paint, the partial reduction of the hexavalent chromium is preferably controlled so that from about 46 to 50% of the hexavalent chromium is reduced to the trivalent state.

The preferred formulation involves a concentrate containing about 10 pbw of the mixed chromium compounds (calculated as CrO_3) in about 200 pbw of solution, which in use can be further diluted.

The phosphoric acid desirably should be present at about 3 to 4 pbw of the phosphoric acid (100% H_3PO_4) to above 10 parts of the mixed chromium compounds.

The treating solution should contain about 4 to 5 pbw polyacrylic acid solids to

10 parts of mixed chromium compounds in order to obtain optimum performance and universality of use under paint. For this purpose it is preferred to use Acrysol A-1 (an aqueous solution containing 25% by weight polyacrylic acid solids).

The formulation also includes a rather large concentration of acrylic emulsion polymer solids, namely about 17 to 20 pbw of such emulsion polymer solids to 10 pbw of the mixed chromium compounds.

A useful acrylic emulsion polymer is Rhoplex AC-73, an emulsion with a nonionic surfactant containing about 46.0 to 47.0% polymer solids of a pH of 9.0 to 10.0 and that is intended for use in neutral to alkaline applications for forming hard films.

Example 1: This example will describe the preferred method of making a concentrate which can be used as such or further diluted before being used to treat metal.

41.5 g of chromium trioxide were dissolved in about 250 to 300 ml deionized water together with 14.5 g phosphoric acid (100% H_3PO_4). The solution was agitated and heated to about 130°F and 4.4 g formaldehyde (100% HCHO) was slowly added. While stirring about 1 hr, the solution was maintained at a temperature between about 185° to 195°F. Upon completion of the formaldehyde addition, heating was continued for 2 hr and the solution temperature maintained at or near boiling. About 300 to 350 ml of deionized water was added to the solution and then 76 g Acrysol A-1 solution (25% solids polyacrylic acid, weight per volume) were introduced with stirring.

The resulting solution was diluted with deionized water to a volume of about 800 ml, thoroughly mixed and allowed to set for 72 hr. In a separate container 160 g Rhoplex AC-73 emulsion (46 to 47% solids weight per volume) was prepared as a 75% volume per volume solution in deionized water by adding sufficient water to make 200 ml total. The diluted acrylic emulsion was slowly added to the aged reduced chrome-acrylic acid solution with stirring. This gave a concentrate which can be used in its concentrated form to form corrosion-resistant coatings where a paint is not going to be applied thereto or it can be further diluted to as much as 20 ℓ volume with deionized water for use in underpaint application.

Example 2: This example illustrates the improved adhesion and corrosion resistance provided by the coating compositions of this process under various paint systems and also shows the universality of the coatings, i.e., their usefulness under different paint systems.

Test panels are prepared by using treating solutions of the process prepared in accordance with Example 1 at 20% dilution and for comparison panels were also treated with a formula in accordance with U.S. Patent 3,185,596 but not containing acrylic emulsion polymer (hereinafter designated as Formulation A).

Using a laboratory spin technique, coatings were applied to clean aluminum panels (0.024 inch thickness) at room temperature to a coating weight of about 20 mg/ft^2 in each case. After application of the coating, each panel was dried with hot air using a heat gun. The metal temperature during the drying reaches about 110° to 120°F. After cooling, the panels were painted with standard coil coating paints in accordance with the manufacturer's specification using three different paint systems: (a) standard single-coat polyester system, (b) vinyl paint system, and (c) a two-coat plastisol paint system using two different primer cure temperatures.

The cured painted panels were then subjected to standard tests to determine paint adhesion and corrosion resistance as follows:

(1) ambient taped reverse impact adhesion tested using ⅝" ball at 48 inch-pounds;

(2) cold taped reverse impact adhesion tested as above with the panel at -10°F at the time of impact;

(3) Cleveland condensing humidity test (ASTM D 714 rating);

(4) MEK solvent resistance test in which the number of double wipes with a cheesecloth soaked with methyl ethyl ketone that is required to remove paint from the metal surface is determined; and

(5) Pencil hardness test

The results are shown in the table below.

	...............Results...............	
Paint System and Tests	**Formulation A**	**Formulation of Ex. 1**
Polyester		
Ambient impact	No paint loss	No paint loss
Cold impact	30% paint loss	No paint loss
Cleveland humidity	No. 8 blisters at 72 hr	No blisters at 240 hr
MEK	57 double rubs	100+ double rubs
Pencil hardness	HB	HB
Vinyl		
Ambient impact	No paint loss	No paint loss
Cold impact	5% paint loss	No paint loss
Cleveland humidity	Few No. 8 blisters at 240 hr	No blisters at 240 hr
Pencil hardness	HB	H to 2H
Wet impact*	5% paint loss	No paint loss
Primer cured at 380° to 390°F**		
Ambient impact	30% paint loss	No paint loss
435°F**		
Ambient impact	No paint loss	No paint loss
Two-coat water-based paint		
Wet impact*	—	No paint loss

*Same as ambient impact test with the panel soaked in deionized water for 2 hr at room temperature and impacted immediately upon removal.
**Plus plastisol topcoat.

Emulsion plus Water-Soluble Chromium

The work of *Y. Hirasawa and H. Yamamoto; U.S. Patent 4,170,671; October 9, 1979; assigned to Nippon Paint Co., Ltd., Japan* relates to a method for the treatment of a metal surface to prepare it to receive a coating finish. The treatment of a metal surface comprises applying a treating liquid comprising (a) an emulsion prepared by emulsion-polymerizing an α,β-monoethylenically unsaturated monomer in the presence of a specific water-soluble polymer as an emulsifier and (b) a water-soluble chromium compound containing trivalent and 30 to 90% hexavalent chromium and optionally (c) a water-insoluble white carbon, the treating liquid containing substantially no alkali metal ion.

The emulsion is prepared by emulsion-polymerizing an α,β-monoethylenically unsaturated monomer in the presence of an emulsifier selected from a polyacrylic acid and a copolymer of acrylic acid and at least one monomer selected from the group consisting of methacrylic acid, acrylamide, methacrylamide and a hydrophilic monomer of the formula:

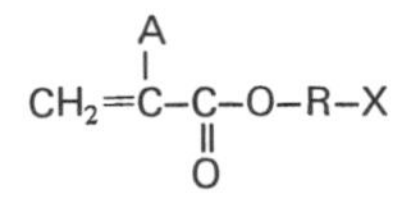

wherein A is hydrogen atom or methyl, R is a substituted or unsubstituted alkylene group having 2 to 4 carbon atoms and X is a functional group having at least one of oxygen atom, phosphorus atom and sulfur atom, the emulsifier being used in an amount of 20 pbw or more (in the solid content) to 100 pbw of the α,β-monoethylenically unsaturated monomer.

The emulsifier, (i.e., the water-soluble polymer) is polyacrylic acid and a copolymer of acrylic acid and at least one compound selected from methacrylic acid, acrylamides, (e.g., acrylamide and N-methylolacrylamide), methacrylamides, (e.g., methacrylamide and N-methylolmethacrylamide) and a hydrophilic monomer of the above formula.

The proportion of the acrylic acid to the other hydrophilic monomer in the above copolymer may be appropriately selected so as to make the content of the acrylic acid in the whole monomer 60% by weight or more.

The α,β-monoethylenically unsaturated monomers used include acrylic esters, methacrylic esters, and acrylonitrile; methacrylonitrile; vinyl acetate; vinyl chloride; vinyl ketone; vinyltoluene, and styrene, which may be used alone or in a mixture of two or more thereof.

The emulsifier is used in an amount of 20 pbw or more, preferably 20 to 50 pbw, (in the solid content) to 100 pbw of the α,β-monoethylenically unsaturated monomer.

The most suitable example of the chromium compound is chromic anhydride (CrO_3). It is important that the chromium compound contain 30 to 90% by weight, preferably 40 to 60% by weight, of hexavalent chromium based on the total chromium content.

The water-insoluble white carbon includes the following fine grain compounds:

(1) fine grain silicic acid anhydride, which has 98% by weight or more of SiO_2 content and contains little adhesive moisture and bound water, for example, a fumed silica prepared by a gaseous phase method, (e.g., Aerosil, Degussa Co.; or Carb-O-Sil, Cabot Co.); a silica prepared by an arc method, (e.g., Fransil, Fransol

Co.; or ArcSilica, PPG Industries Inc);

(2) fine grain silicic acid hydrate, which has 80 to 98% by weight of SiO_2 and contains a large amount of adhesive moisture and bound water, for example, a silica prepared by a wet process, i.e., by hydrolyzing a silicate with an acid and purifying the resulting silicic acid hydrate, (e.g., Hi-Sil, PPG Industries Inc.; Ultrasil, Degussa Co.; Tokusil, Tokuyama Soda Co. Ltd.; or Carplex, Shionogi Pharmaceutical Co.); and

(3) fine grain silicate, such as calcium silicate or aluminum silicate.

Among these, preferred white carbons are (a) a water-insoluble white carbon having the primary particles of 0.1 to 3 μ in particle size which is mostly present in the form of primary particles without aggregation in the treating liquid, and (b) a water-insoluble white carbon wherein the primary particles have large aggregation properties and the particles are present in the form of aggregated particles of 0.1 to 3 μ in particle size in the treating liquid.

Example 1: Into a flask provided with a stirrer, a reflux condenser, a thermometer and two dropping funnels, there are charged 150 parts of deionized water and 120 parts of water-soluble copolymer obtained by copolymerizing acrylic acid and 2-hydroxyethyl methacrylate in the ratio of 8:2 by weight (25% aqueous solution, $\overline{M}w$ = 66,000); and the mixture is heated to 60° to 65°C with stirring.

Then, a monomer mixture consisting of 35 parts of methyl methacrylate, 15 parts of styrene, 10 parts of 2-hydroxyethyl methacrylate and 40 parts of n-butyl acrylate and a catalyst solution consisting of 2 parts of ammonium persulfate and 50 parts of deionized water are separately and simultaneously added dropwise from the separate dropping funnels over a period of 8 hr. After completion of the dropwise addition, the resultant mixture is kept at 60° to 65°C for about 2 hr to complete polymerization reaction to give an emulsion having 30.1% solid content.

Preparation of treating liquid – 8.1 parts of the above emulsion, 7.4 parts of the aqueous solution of chromium compound (solid content 16.5%) which is obtained by reducing about 50% of the amount of the hexavalent chromium to trivalent chromium by adding 5 parts of formalin (37% aqueous solution) to 95 parts of 17% aqueous solution of chromic anhydride and 20 parts of a dispersion (solid content 10%) of fine grain silicic acid anhydride (Aerosil TT 600, Degussa Co.) in deionized water are mixed at room temperature, to which deionized water is added to prepare a treating liquid having 2.3% solid content.

Metal surface treatment and coating – The above treating liquid is applied with a roll coater to the surface of a galvanized steel (a plate having a thickness of 0.35 mm) degreased with an alkali degreasing agent (Ridoline No. 72, Nippon Paint Co., Ltd.) and the resultant is immediately dried at 100°C for 40 sec to give an even coating film having a film weight of 293 mg/m^2. Thereafter, the thus-surface-treated galvanized steel is coated with coating formulation for galvanized steel, i.e., with a primer formulation (Superlac DIF-TX-1, Nippon Paint Co., Ltd.) (baking at a furnace temperature of 220° to 240°C for 45 sec, film thickness 3 μ) and then with a topcoating formulation (Superlac DIF-A-55, Nippon Paint Co., Ltd.) (baking at a furnace temperature of 210° to 230°C for 60 sec, film thickness 11 μ) to give a coated plate.

Example 2: *Preparation of emulsion* – In the same manner as described in Example 1 except that 200 parts of polyacrylic acid (25% aqueous solution, $\overline{M}w$ being 59,000) is used instead of 120 parts of acrylic acid/2-hydroxyethyl methacrylate copolymer and the monomer mixture consists of 35 parts of methyl methacrylate, 15 parts of styrene, 10 parts of ethyl methacrylate and 40 parts of n-butyl acrylate, the emulsion-polymerization is carried out to give an emulsion having 30.8% solid content.

Preparation of treating liquid – 7.9 parts of the above emulsion are admixed with 7.4 parts of the same aqueous solution of chromium compound (solid content: 16.5%) as used in Example 1 and 20 parts of the same dispersion of white carbon (solid content: 10%) as used in Example 1 and thereto is added deionized water to give a treating liquid having 2.3% solid content.

Metal surface treatment and coating – In the same manner as described in Example 1 except that the above treating liquid is used, there is obtained a coated plate (the film weight of the coating film 278 mg/m^2).

Reference Example: A commercially available zinc phosphate treating liquid for chemical conversion treatment is applied to a galvanized steel previously degreased as in Example 1 by a spray coating for 2 min. The resultant is immediately washed with a large amount of water, dipped in an aftertreating liquid, and then dried at 100°C for 1 min to form an even phosphate coating film (film weight 1,500 mg/m^2). The treated galvanized steel is subjected to the coating finish as in Example 1 to give a coated plate.

The coated plates prepared in the above examples and reference examples are subjected to the following tests. The results are shown in the table.

Test	Example 1	Example 2	Reference Example
Scratch resistance	A	A	B
Bending resistance	10	10	8
Resistance to saline solution	5	5	4
Moisture resistance	5	5	4

Scratch resistance – The surface of the coated plate is scratched with a coin and the degree of the injury to the surface is observed. The results are evaluated as follows: A, superior (not injured); B, good (little injured).

Bending resistance – The coated plate (width: 5 cm) is bent at an angle of 180° and vised, and a pressure-sensitive adhesive tape is adhered onto the bent surface and then is peeled off. The results are evaluated by a 10-point method ranging from 10 (no trouble) to 1 (fully peeled off).

Resistance to saline solution – A 5% saline solution is sprayed onto the coated plate which is scribed with a knife at 35±1°C for 1,000 hr, and the width of blistered paint film from the scribe line is measured. The results are evaluated as follows: 5, no trouble; 4, <0.5; 3, 0.5 to 1.5 mm; 2, 1.5 to 2.5 mm; and 1, >2.5 mm.

Moisture resistance – The coated plate is kept under the atmosphere of a humidity of 98±2% and a temperature of 50±1°C for 1,000 hr; and then the blister density is measured (cf ASTM D714-56). The results are evaluated as follows: 5, no trouble; 4, few; 3, medium; 2, medium dense; and 1, dense.

SURFACE TREATMENT OF ALUMINUM

ANODIC OXIDATION

Use of Halogenated Amine-Metal Halide Complex as Bath Additive

Of the numerous finishes for metals, and particularly aluminum, none are as versatile as the electrochemical oxidation and anodizing process. There are numerous types of anodizing electrolytes that have been employed to produce an oxide coating with useful properties. However, sulfuric acid anodizing is the most common in this country. Many millions of pounds of aluminum products for applications requiring attractive appearance, good corrosion resistance and superior wearing quality are finished by this method.

Since the anodizing process is a balance between the competitive dissolution and oxide formation processes, an improvement in the efficiency of coating formation would result in a saving of time, material and energy as well as decreasing the volume of waste bath to be discarded or treated to make it environmentally acceptable.

A bath composition for surface finishing on metal surfaces is provided by *S. Kessler; U.S. Patent 4,023,986; May 17, 1977*. The coating bath provides a chemically converted surface which is more dense and organized and provides significant increase in efficiency of coating deposit rate. The anodizing baths of the process may be subjected to much higher current density without causing objectional burning of the film. Efficiency and uniformity of dissolution are also provided in etching baths containing the additive. Colored films are found to be lustrous, bright, dense, and uniform, to have good abrasion resistance and to be very smooth.

The chemical surface finishing bath composition comprises an aqueous vehicle containing an inorganic oxidant-etchant and an effective amount of the reaction product of a metal halide and a polyhalo-substituted alkarylamine. The metal surfaces are processed in a conventional manner, suitably after preliminary cleaning treatment and surface brightening or roughening, if desired for special effect. The part is immersed in the bath and is maintained in the bath until the desired thickness and quality of coating or etching have been effected. The article is then removed and subjected to conventional aftertreatment such as sealing, waxing or dyeing and is then ready for service.

The additive is generally present in the bath and in an amount from 0.1 to 50, preferably 1 to 20 g/ℓ and is formed from a combination of ingredients which react to form a fluoro-, chloro-, bromo- or iodo-substituted hydrocarbon amine-metal halide complex capable of improving deposition rate and coating characteristics. It is believed that the additive causes an organization of the layer that forms which permits the metal oxide or salt molecules to organize in a faster manner and to form a more organized, denser deposit providing a harder, denser, smoother, more abrasion and corrosion resistant deposit having more even color.

The first ingredient utilized in forming the additive material is an at least trihalogenated compound of fluorine, bromine, iodine or chlorine, and a metal, particularly Groups I-B, II, III-A, IV-B, V-B, VI-B and VIII metals such as copper, magnesium, boron, aluminum, titanium, vanadium, niobium, chromium and tungsten. A preferred material is boron trifluoride and especially in a stabilized form as a complex with a lower alkyl ether such as diethyl ether.

The other necessary ingredient is an alkarylamine, particularly a fluorinated alkarylamine having a relatively high content of available and active fluorine atoms which are reactive with the metal halide. A suitable material is α,α,α-trifluoro-m-toluidine. The presence of an amino group is believed to relieve stress in the deposited film in a manner analogous to the action exhibited by sulfonamides in electrodeposition or anodizing of aluminum.

Example 1: An additive was prepared from the following ingredients:

Component	Amount, ml
Tetrachloroethylene	900–960
Boron trifluoride etherate	50–20
α,α,α-Trifluoro-m-toluidine	50–20

The toluidine and tetrachloroethylene were combined and a cloudy suspension was formed. When the metal halide etherate was added, globules of a fluffy, waxlike, white precipitate was observed in copious volume after storage at room temperature. A maximum volume of waxlike solid of over one-half of the initial volume of the mixture was obtained after several days. The waxlike solid was separated by filtration and washed with methanol and water.

The reaction could be accelerated by heating the mixture to a higher temperature. The waxlike material was heated to 575°F and no decomposition or melting of the material was observed. Since the formation of a waxy solid is observed, a chloro-fluoro-boro-substituted hydrocarbon polymer is believed to be formed.

In the known processes of anodizing metals such as aluminum, the metal body is placed in a bath of suitable electrolyte and connected as an anode in a direct current electrical circuit which includes the electrolyte bath. When current is passed through the bath, an oxide layer is formed on the surface of the aluminum body that is characterized by being thicker than an oxide that would form in air.

The following table provides typical conditions for practicing anodizing aluminum in accordance with the process.

Ingredient	Range
H_2SO_4 (93%), g/ℓ	5-400
Boro-fluoroamine additive, g/ℓ	0.5-20
Current density, A/dm^2	5-200
Temperature, °C	-20-100
Time, minutes	2-120

Example 2: *(A)* – A 1 liter bath containing 185 g/ℓ of 93% H_2SO_4 was formed containing 1.2 g/ℓ of the additive of Example 1. The bath was contained in a stainless steel tank which was connected as cathode and a flat 1" square specimen of aluminum 3003-H14 alloy was connected as anode and inserted into the bath. The bath temperature was adjusted to 0°C and after 15 minutes at 100 A/dm^2, a thick, uniform, dense, hard coating of anodic aluminum oxide was formed on the specimen. The additive of the process causes at least a 40% increase in deposition rate as well as permitting much higher current densities without deterioration of the film.

(B) – The procedure of (A) was repeated on the same alloy specimen under the same conditions except that the additive was not present in the bath. The deposition thickness for equivalent times was only 60% of that achieved for the bath composition of (A). Furthermore, the coating was not as organized nor as dense. The color on the specimens was less uniform than that achieved on the specimen treated according to (A).

The hardness of the anodic deposits of Example 2(A) and (B) was compared by the conventional commercial scratch test which indicated that the anodic aluminum oxide deposit on the specimen of Example 2(A) was significantly harder than the deposit on the specimen of Example 2(B).

Addition of Lignin to Electrolyte Bath

J.L. Woods; U.S. Patent 4,270,991; June 2, 1981 found that if lignin and its acids and salts of acids are added to an acidic anodizing electrolytic bath the deposition of metal oxide on the aluminum surface will be thick and hard. It has also been found that when using a lignin or lignin derivative, the anodizing process may be carried out by a constant current.

The hard anodizing may be carried out without the use of surfactants or wetting agents in the anodizing electrolyte solution.

During laboratory testing it was shown that concentrations of sulfuric acid varying from 125 to 350 g/ℓ performed satisfactorily with a lignin derivative concentration varying from 0.5 to 5.0 g/ℓ. The 5 g figure was an arbitrary limit as there does not seem to be any maximum concentration inside the bounds of solubility at which the lignin derivative will function. It has been noted, however, that concentrations below about 0.25 g/ℓ do not function in a satisfactory manner.

All of the tests were conducted in a 12 gallon stainless steel tank using the tanks as cathodes. Aluminum samples to be tested were 4" x 4" square and ⅛" thick.

They were connected to the anode of a direct current power supply by attaching them to 8" long pieces of 1100 alloy aluminum wire.

Example 1: *Conventional technique* – An electrolyte bath containing 175 g of sulfuric acid was employed without any lignin additive. The anodizing was started using a current of 2.0 A which was gradually increased over a 3-minute time period to 8.5 A which equates to 36 A/ft^2. The electrolyte was maintained at 2°C. Almost immediately, with the voltage constant, the current drastically increased. The power supply was immediately shut off and the aluminum plate was taken from the anodizing cell and examined. The plate had started burning on three of the four corners. Metal was eroded away with no appreciable anodic coating being formed.

Example 2: The conditions noted in Example 1 were repeated with the exception that 0.25 g/ℓ of sodium lignosulfonate were added to the electrolyte. The anodizing operation progressed satisfactorily for about 19 minutes at which time the current started to increase at a rapid rate, which is an indication of burning. The power supply was shut off and the aluminum sample was examined. The anodic coating was satisfactory with a minor indication of erosion beginning on one corner. A hard coating thickness of 0.8 mil had been formed.

Example 3: The concentration of sodium lignosulfonate was raised to 1.0 g/ℓ and the sulfuric acid in the electrolyte was raised to 350 g/ℓ. The anodizing process proceeded as described. The experiment was terminated when the power supply ran out of voltage at 98 volts. The anodizing time was 78 minutes and the anodic coating had a thickness of 3.8 mils. The coating was very hard and there was no indication of burning.

Example 4: The procedure of Example 3 was repeated using 5 g/ℓ of sodium lignosulfonate and 250 g/ℓ of sulfuric acid. The only variation noted was that the anodizing time was 79 minutes and the coating had a thickness of 3.9 mils.

Example 5: Anodizing was carried out as in Example 1. The electrolyte used contained 250 g/ℓ of sulfuric acid and 1.0 g/ℓ of sodium lignosulfonate. The experiment utilized aluminum alloy 7075 which consists of 0.5% Si, 0.7% Fe, 1.1 to 2.0% Cu, 0.3% Mn, 2.1 to 2.9% Mg, and 5.1 to 6.1% Zn, with the remainder being aluminum as the test plate. The power supply reached a terminal voltage of 98 volts after 89 minutes. The coating formed was hard and had a thickness of 4.4 mils.

In all experiments except Example 5, aluminum alloy 2024 was used since the 2024 alloy is extremely difficult to hand anodize with conventional anodizing processes. Aluminum alloy 2024 consists of 0.5% Si, 0.5% Fe, 3.8 to 4.9% Cu, 0.6% Mn and 1.2 to 1.8% Mg with the remainder being aluminum.

Double Anodizing Treatment

According to *G.K. Creffield, A.J. Wickens, A.C.H. Dowd and V.F.F. Henley; U.S. Patent 4,278,737; July 14, 1981; assigned to United States Borax & Chemical Corporation* there is provided a method of anodizing aluminum which comprises subjecting it to electrolysis first in a bath of electrolyte containing an aqueous soluble borate and then in a bath of a conventional electrolyte containing, for example, sulfuric acid.

The electrolyte in the first bath is preferably an alkali metal borate, such as sodium borate or borax, at a concentration of about 3 to 5% by weight of the electrolytic solution. It is necessary to adjust the pH of the first bath within the range of 9 to 11, preferably 9.2 to 10.5, and this can be achieved by addition of a suitable alkali, for example, sodium hydroxide solution, to the first bath until the pH attains the desired alkalinity. Electrolysis in the first, borate, bath is generally carried out at an elevated temperature such as in the range of about 50° to 80°C, preferably about 60° to 70°C, and is continued until the depth of the oxide coating formed on the aluminum is at least about 2 μ and preferably between 5 and 10 μ.

The first anodizing process can be carried out over a wide range of operating conditions. Both alternating and direct current may be used, or ac superimposed on dc may be employed. The voltage can range between about 20 and 75 volts. Particularly suitable conditions are application of a dc voltage of about 25 to 40 volts, especially about 30 to 40 volts, with a 5% by weight borax electrolyte at a pH of 9.5 to 10 with a temperature of 60° to 70°C.

The electrolyte used for the second step in the process can be, for example, a conventional electrolyte such as sulfuric acid in a concentration of about 16% by weight.

After anodizing in a borate electrolyte, the aluminum article may be directly placed in the bath containing the sulfuric acid as the presence of small amounts of borate do not appear to be detrimental to the second anodizing process. However, on a continuous basis, it may be desirable to rinse the aluminum article after the first anodizing stage. This may be effected using water, dilute aqueous sulfuric acid solution (5 or 6% by volume H_2SO_4) or a solution based on the second electrolyte.

In the following examples, the aluminum samples were cleaned, etched and desmutted using conventional methods before anodizing.

Example 1: A sheet of H9 aluminum alloy was immersed in an electrolytic cell wherein an aqueous solution containing 5% borax adjusted to pH 9.5 with sodium hydroxide was the electrolyte, and as anode, was subjected to an electric current density of 20 mA/cm^2 for 30 minutes at 28 volts at 70°C. This formed an oxide coating about 6 μ thick on the surface of the aluminum. The partly anodized article was then rinsed with dilute aqueous sulfuric acid. The rinsed aluminum article was then placed in an electrolytic cell containing 16% by weight aqueous sulfuric acid as electrolyte and anodized at 18°C and 18 volts for 30 minutes. It was necessary to initially raise the voltage to 36 volts to recommence anodizing. The anodized product was sealed by immersing in boiling distilled water for 1 hour. The alloy sheet had an oxide coating about 25 μ thick. The coating had a similar appearance to anodized aluminum produced by the conventional sulfuric acid process.

Example 2: A section of aluminum alloy (H9) was anodized first in a solution containing 5% borax plus sodium hydroxide to adjust the pH to 9.5 at 70°C. Anodizing was carried out for 40 minutes at 30 volts with an initial current density of 13 mA/cm^2. After 40 minutes, the voltage was reduced to 15 volts over a period of 30 seconds. It was then held at 15 volts for a further 30 seconds. The current was switched off and the sample removed from the borax anodizing bath. It was rinsed in water, then transferred immediately to the second anodizing bath containing 10% (v/v) H_2SO_4 at 18°C. Anodizing was then recommenced by raising the voltage to 18 volts, giving a current density of 16.5 mA/cm^2. This second anodizing stage was continued for 30 minutes, after which time the current was switched off and the sample removed. The sample was rinsed in cold water and then the anodic layer was sealed in the usual way by immersing it in boiling water for 60 minutes. The combined borax plus H_2SO_4 produced anodic layer was 25 μ thick.

SEALING OF ANODIZED ALUMINUM

Solutions Containing Agar-Agar and Gelatine

Anodizing is a surface treatment which is widely used with aluminum and its alloys. The oxide layer produced in this treatment considerably improves the resistance of the surface towards corrosive media. The natural or subsequent purposeful coloring of the oxide can also be used to provide a decorative finish to the part in question. One particularly important improvement from the technical point of view is that the oxide layer produced by anodizing increases the wear resistance of the surface of the aluminum.

Oxide layers produced by anodizing do not, however, satisfy all requirements without some further treatment. The layers are porous, do not offer sufficient corrosion protection, and if colored, may have the colorant washed out of them again. The oxide, therefore, has to be sealed. This so-called sealing process is usually carried out in hot or boiling water, if necessary, with certain additions made to the water. As a result of this process, the pores are closed off, providing improved corrosion resistance and entrapping the colorant securely in the oxide.

The hydration of the aluminum oxide during the sealing process not only causes the pores to be closed off, but also results in the formation of a velvet so-called sealing deposit on the surface. This deposit from the hydrated oxide impairs in particular the decorative appearance of dark colored oxide layers. Because the specific surface area is increased, this deposit also impairs the corrosion resistance and leads to discoloring of the surface. This sealing deposit is usually removed by mechanical polishing which, of course, entails extra labor and, therefore, considerable costs.

F. Schneeberger, W. Zweifel and H. Weber; U.S. Patent 4,235,682; November 25, 1980; assigned to Swiss Aluminium Ltd., Switzerland provide a sealing solution for anodically oxidized aluminum or aluminum alloys comprising basically an aqueous solution and a process which prevents the formation of deposits during the sealing of anodized aluminum.

The sealing solution contains an addition of 0.1 to 2.5 g/ℓ comprising: (a) at least 10% agar-agar; and (b) 0 to at most 90% gelatine. The pH of the bath solution lies between 5 and 7. The addition can be solely agar-agar, the optimum concentration range being between 0.5 and 1.5 g/ℓ, preferably between 0.7 and 1.1 g/ℓ.

In accordance with a particularly advantageous version of the sealing bath the addition comprises both agar-agar and gelatine with the agar-agar constituting 30 to 90%. The presence of agar-agar and gelatine simultaneously in the sealing bath has surprisingly, compared with the individual implementation of these substances, a considerable improvement with respect to the prevention of sealing deposits, the sealing quality, residues on drying and the long term corrosion properties, apparently due to a synergistic effect of these substances.

Extensive trials in production have shown that it is advantageous to use an addition containing 40 to 70% agar-agar in a concentration range of 0.2 to 0.9 g/ℓ, preferably 0.3 to 0.6 g/ℓ.

Independent of the concentration of the addition made, the pH of the sealing bath is to be adjusted to 5.5 to 6.0. The process is carried out at temperatures between 90°C and boiling point, preferably at least at 95°C.

Example: The material used for investigation purposes was sheet material of the aluminum base alloy AlMg 1.5 in the half-hard condition and measuring 100 x 50 x 2 mm. The sheets were given a 1 minute caustic etching treatment in an aqueous solution of 150 g/ℓ NaOH at 50°C. They were then anodized in the classical direct current-sulfuric acid process at a constant current density of 1.5 A/dm^2. The electrolyte, an aqueous solution of 176 g/ℓ H_2SO_4 with an aluminum content of 7 g/ℓ, was kept constant at a temperature of 20°±1°C. The duration of the anodizing treatment was 40 minutes, which corresponds to an oxide layer thickness of about 20 μm. A part of this colorless anodized material was then colored black in a conventional manner using an organic coloring agent (Sanodal MLW, black), and another part colored dark brown, electrolytically in a metal salt solution (Colinal 3100) using alternating current.

The anodized and colored samples were then placed in deionized water, the pH adjusted to a value between 5.5 and 6.0 using acetic acid or ammonia, kept there for 40 minutes–corresponding to 2 min/1 μm of layer thickness–at a temperature of 98°±2°C to seal the oxide layer. The composition of the additions in wt % were increased in steps of 20% from 100% agar-agar + 0% gelatine to 0% agar-agar + 100% gelatine. The concentrations of the additive in the solution were 0.1, 0.2, 0.4, 0.6, 0.8, 1.0, 1.2, 1.5, 2.0 and 2.5 g/ℓ.

To judge the quality of the sealed sheet, the criteria listed in Table 1 were employed, along with the testing methods also listed there. Some of these results are shown in Tables 2, 3 and 4 which follow.

Table 2: Deposit on the black colored samples. The sheets were rubbed on one place with a rough black cloth. The assessment of the coating was based on the visual appearance of the sheet and on the residues on the cloth. The values used for the rating range from 0 (no deposit) to 5 (very pronounced deposit). In order to assess the sheets with respect to deposits after drying, the sheets were dipped briefly in deionized water after sealing.

Table 3: Evaluation of the color drip test as in ISO-R 2143. The rating scale ranges from 0 (very good) to 5 (very bad).

Table 4: Evaluation of an accelerated corrosion test as in DIN SO 947. The rating ranges from 0 (no visible corrosive attack) to 5 (pronounced pitting).

Table 1

Criterion	Test
Sealing deposit	Visual inspection, rubbing with a black cloth (residue)
Drying residues	Visual inspection
	Apparent conductivity as in ISO-DIS 2931
	Weight loss as in dissolution test (ISO-DIS 2932)
Sealing quality	Color drip test as in ISO-R 2143
	Accelerated corrosion test as in DIN 50 947
	Accelerated corrosion test as in LN 29 596 (Kersternich)
Resistance to light	Ultraviolet radiation
Long term corrosion behavior	Outdoor exposure to a mild industrial atmosphere

Table 2: Deposit on Black Samples

Addition Concentration (g/ℓ)	Composition of the Addition in Weight Percent ....Agar-Agar (A)/Gelatine (G)................					
	100A/0G	80A/20G	60A/40G	40A/60G	20A/80G	0A/100G
0.1	3	1	1	1	1	4
0.2	2	0	0	0	0	4
0.4	1	0	0	0	0	2
0.6	0	0	0	0	0	1
0.8	0	0	0	0	0	0
1.0	0	0	0	0	0	0
1.2	0	0	0	0	0	0
1.5	0	0	0	0	0	0
2.0	0	0	0	0	0	0
2.5	0	0	0	0	0	0

Table 3: Color Drip Test

Addition Concentration (g/ℓ)	Composition of the Addition in Weight Percent ....Agar-Agar (A)/Gelatine (G)................					
	100A/0G	80A/20G	60A/40G	40A/60G	20A/80G	0A/100G
0.1	0	0	0	0	0	0
0.2	0	0	0	0	0	0
0.4	0	0	0	0	0	0
0.6	0	0	0	1	0	0
0.8	0	0	0	0	0	0
1.0	0	1	0	0	0	2
1.2	0	0	0	1	1	3
1.5	2	0	0	0	0	5
2.0	2	1	0	0	1	5
2.5	3	0	0	0	2	5

Table 4: Accelerated Corrosion Test

Addition Concentration (g/ℓ)	Composition of the Addition in Weight Percent ...Agar-Agar (A)/Gelatine (G)...					
	100A/0G	80A/20G	60A/40G	40A/60G	20A/80G	0A/100G
0.1	0	0	0	1	0	0
0.2	0	0	0	0	1	1
0.4	0	0	0	0	0	0
0.6	1	0	0	0	0	0
0.8	0	1	1	0	1	1
1.0	0	1	0	1	0	2
1.2	1	0	0	0	0	3
1.5	0	0	0	0	1	3
2.0	2	0	0	1	2	5
2.5	2	1	1	0	2	4

Use of Ammonia Vapor in Sealing

Anodized aluminum is usually sealed to render it impervious to elements which could adversely affect the aluminum substrate, because, in many instances, especially in architectural applications, it will be exposed to the atmosphere for many years. It is therefore imperative that the seal be of very high quality to ensure satisfactory service.

High quality seal, as used herein, is defined as a sealed, smudge-free anodized aluminum, which, after being immersed for a period of 15 minutes in a solution containing 2.0 wt % chromic acid and 3.5 wt % phosphoric acid at a temperature of 100°F, has a weight loss of not more than 2 mg/in^2. This seal quality evaluation is known as the "acid dissolution test".

Sealed anodized aluminum is resistant to staining. Thus, sealing quality can be determined qualitatively by a dye stain test known as "Standard Method for Measurement of Stain Resistance of Anodic Coatings on Aluminum" (ASTM B136-72). In this method, after conditioning the sealed, anodized surface with a nitric acid treatment, a dye test solution is placed thereon for a period of about 5 minutes, then this test area is washed with water and rubbed with pumice powder. Staining of the anodized finish after this treatment indicates a poorly sealed anodic coating. Conversely, absence of stain indicates a satisfactory seal.

A. Alexander; U.S. Patents 4,103,048; July 25, 1978; and 4,031,275; June 21, 1977; both assigned to Aluminum Company of America has found that a high quality, smudge-free seal can be formed at a low temperature on anodized aluminum by exposing it to a source of ammonia vapors.

Anodized aluminum which can be sealed in accordance with the process may be anodized by conventional methods.

The temperature of the sealing media containing ammonia vapors can be in the range of 55° to 155°F, with a preferred range being 60° to 135°F, and most preferred, 80° to 125°F.

A preferred time period for sealing according to this process is 10 to 50 minutes, with a most preferred time being 20 to 40 minutes at slightly above ambient temperature, i.e., about 105°F.

A suitable amount of ammonia gas is that which develops about 2.0 to 250 lb/in^2 of pressure in the vaporous sealing media with a preferred pressure range being 5.0 to 75.0 lb/in^2 in the temperature ranges indicated above.

A convenient source of ammonia gas is an ammonium hydroxide solution which has a concentration of at least 10 wt % and more preferably 20 to 60 wt %.

Example: Specimens of a conventional Anoclad 11 sheet alloy (an Aluminum Association Alloy No. 110 clad with the same alloy) were anodized using conventional practices to produce a bronze colored oxide coating in a sulfophthalic acid/sulfuric acid electrolyte. The specimens were sealed by suspending them in ammonia vapors above a solution of 27 to 29 wt % ammonium hydroxide in sealed containers. Four specimens were exposed to the vapors at 82°F for 30, 60, 90 and 120 minutes and four were exposed to the vapors at 105°F for 30, 45, 60 and 75 minutes. Five specimens were treated in the vapors at 125°F for 15, 30, 45, 60 and 90 minutes. The results, as determined by the Dye Stain Test (ASTM B136-72) and acid dissolution test, of the sealing procedures are as shown below.

Temperature of Sealing	Seal Time	Dye Stain (ASTM B136-72)	Acid Dissolution (weight loss, mg/in^2)
82°F	30	Failed	18.5
	60	Failed	17.5
	90	Failed	2.4
	120	Passed	0.7
105°F	30	Failed	8.1
	45	Passed	0.5
	60	Passed	0.6
	75	Passed	0.4
125°F	15	Failed	9.2
	30	Passed	0.85
	45	Passed	0.30
	60	Passed	0.30
	90	Passed	0.30

Thus, it can be seen that the process provides a high quality seal on anodized aluminum. Also, as the temperature and the concentration of ammonia gas is increased, the time required to provide a high quality seal is shortened considerably. That is, by extrapolation of this data, it can be seen that a high quality seal can be obtained in about 5 minutes by increasing the temperature and concentration of the vapors above the ammonium hydroxide solution.

Use of Nickel Acetate, Triethanolamine and Soluble Sulfate

E.G. Remaley, R.W. Baker and R.J. Meyer; U.S. Patent 4,045,599; August 30, 1977; assigned to Aluminum Company of America found that anodized aluminum can be sealed at a lower temperature and the smudge formed during sealing removed with mineral acid. The sealing comprises treating anodized aluminum with an aqueous sealing solution containing 2 to 6 g/ℓ of a hydrolyzable metallic salt, 5 to 20 ml/ℓ of an ethanolamine and 50 to 2,000 mg/ℓ soluble sulfate ($SO_4^=$), at a temperature of at least 140°F to provide the seal. A suitable temperature is in the range of 150° to 170°F. Nickel acetate and nickel sulfate are preferred hydrolyzable metallic salts. Preferably, the ethanolamine used is triethanolamine; however, mono- or diethanolamine or a combination thereof can be quite suitable.

The aqueous media and the above ingredients should be combined to form a sealing solution having a pH in a range of 6.5 to 7.5, and preferably within 6.8 to 7.2.

The nickel acetate to triethanolamine ratio by weight should be at least 0.18 and preferably at least 0.27.

Another important aspect of the process is the order in which the ingredients are added to the deionized or distilled water. It is preferred that the soluble sulfate ($SO_4^=$) be added before the other constituents and thereafter, either the hydrolyzable metallic salt or the triethanolamine added. If either or both of the latter ingredients are added before the soluble sulfate, the sealing solution has been found to be less effective in providing a high quality seal for reasons which are not completely understood.

Anodized aluminum can be sealed in the sealing solution in a time period preferably not greater than 45 minutes, with a suitable seal being effected in a time period of 25 to 35 minutes.

The sealing system is considered to be unique in that its minimum temperature for sealing can vary with the type of anodic coating. The types of anodic coatings referred to are those providing integral color on aluminum upon anodization. More particularly, it has been discovered that integrally colored anodic coatings on aluminum, characterized by having colors ranging from very light to dark bronze, depending on the anodization conditions, can be treated by the sealing solution of the process to produce a high quality seal at a temperature in the range of 140° to 155°F.

Example 1: A specimen of a conventional Anoclad 11 sheet alloy was anodized using conventional practices to produce a light bronze colored oxide coating, thickness 0.7 mil, in a sulfophthalic acid/sulfuric acid electrolyte. The specimen was sealed for a period of 30 minutes in a sealing solution, pH of 6.9, temperature of 150°F and containing 4 g/ℓ nickel acetate, 5 ml/ℓ triethanolamine and 50 mg/ℓ soluble sulfate. Smudge formed during sealing was removed by immersing the specimen in room temperature (75°F) sulfuric acid, 15 wt %, for 3 minutes. In accordance with the acid dissolution test, quality of seal was then determined by immersing the specimen in chromic/phosphoric acid at a temperature of 100°F for 15 minutes. Weight loss was 0.34 mg/in^2 indicating a high quality seal.

Example 2: A specimen of Anoclad 11 was anodized to a medium bronze color in a sulfophthalic acid/sulfuric acid electrolyte and sealed according to the conditions in Example 1, except the sealing solution contained 500 mg/ℓ sulfate ($SO_4^=$).

The oxide coating thickness was 1.0 mil. Sealing smudge was removed as in Example 1. A high quality seal was obtained. Weight loss in the acid dissolution test was 1.8 mg/in^2.

Primary Chromium Phosphate and Emulsified Organic Resin

Anodizing processes consist of the formation of a layer of alumina by anodic oxidation of the metal surface and followed by a sealing process of the porosities of the formed layer.

The porosity of the layer is useful for coloring the metal surface by means of certain dyestuff solutions. Final sealing of these porosities is absolutely necessary to improve corrosion resistance and metal appearance.

The composition of *J.B. Fabregas; U.S. Patent 4,084,014; April 11, 1978* includes a primary chromium phosphate $(PO_4H_2)_3Cr$, solution and an organic resin emulsion, the final product being water-soluble. The composition must be entirely solubilized in such a way that trivalent chromium salts are able to penetrate into porosities of the alumina layer, imparting to this its protective properties.

The method of sealing surface porosities of a surface layer of aluminum oxide formed by anodic oxidation on an article comprised of aluminum comprises contacting the surface layer with an aqueous solution which includes phosphate ions, trivalent chromium ions, and an emulsion or dispersion of organic resin selected from the group consisting of polymers or copolymers of acrylic, vinyl-acrylic and styrene-acrylic, and thereafter causing the surface layer to dry. The ingredients are present in substantially the following concentrations: phosphate ions (expressed as P_2O_5), 5 to 150 g/ℓ; trivalent chromium (expressed as Cr_2O_3), 0.01 to 20 g/ℓ; and resin (solids), 2 to 300 g/ℓ.

Example 1: 50 g of commercial phosphoric acid (85%) are mixed with 1 g of chromium carbonate. After the reaction is completed, which is accelerated by heat, 70 g of acrylic resin emulsion Vinacryl are added. This solution is diluted to 1 liter with demineralized water.

Example 2: 2 g of sodium dichromate dissolved in 20 g of water, are added to 70 g of commercial phosphoric acid (85%). Afterwards, 1.5 g of ethylene glycol should be added, a redox reaction takes place and hexavalent chromium is converted into trivalent chromium, the color of the solution becoming green. After reaction is accomplished, 50 g of acrylic resin Vinacryl are added. The solution is then diluted to 1 liter with demineralized water.

NONELECTROLYTIC OXIDES

Treatment with Difficultly Soluble Magnesium Compound at pH 7 to 9

The oxidation process of *G.A. Dorsey, Jr.; U.S. Patent 4,113,520; September 12, 1978; assigned to Kaiser Aluminum & Chemical Corporation* for aluminum, when controlled to form a thin oxide coating, provides an aluminum surface which is particularly suitable for resistance welding and when controlled to form thicker oxide coatings generate integrally colored oxide coatings.

An aluminum workpiece is treated with a hot aqueous solution at neutral or slightly alkaline condition and with difficultly soluble magnesium compounds. Generally suitable are those magnesium compounds which are soluble in distilled or deionized water at 20°C in amounts less than 30 g/ℓ. Suitable magnesium compounds include magnesium carbonate, magnesium hydroxide, magnesium sulfate, magnesium silicate and the like. Magnesium carbonate is preferred.

Magnesium concentrations can range generally from about 0.001 to point of saturation (expressed as $MgCO_3$). Preferably, a suspension of the magnesium compound is maintained to assure saturation. The temperature of the aqueous bath should range from about 65°C to the boiling point of the solution, preferably from about 75° to 95°C. The pH of the solution should be about 7 to 9, preferably about 7.5 to 8.5. A high pH allows for the use of low bath temperatures, whereas a low pH usually requires high bath temperatures.

Treatment times generally are less than about 1 minute for thin colorless oxide coatings but usually are in excess of 5 seconds for coatings of any significant thickness. For integrally colored coatings, treatment times generally range from about 1 to 60 minutes. With a particular set of bath conditions, treatment times depend to a certain extent upon the alloy composition and the oxide characteristics desired in the final product.

The process should be preceded with an etching step to remove the natural oxide surface and other surface contaminants so that a clean aluminum surface is presented to the treatment solution. To uniformly generate the oxide coating over the aluminum surface, a nonionic or cationic wetting agent in amounts up to 1% by weight can be incorporated into the oxidizing solution.

The process forms an aluminum oxide coating having a thickness ranging up to about 5,000 Å units depending upon treatment times and conditions. The oxide coating is essentially all aluminum oxide except for a trace quantity of magnesium hydroxide or other magnesium salt which lies on the surface of the aluminum oxide.

Example 1: A group of aluminum alloys indicated in the table below were treated in an aqueous solution containing about 1 g/ℓ $MgCO_3$ (saturated solution) and about 1 g/ℓ of a nonionic wetting agent (Tergitol NPX) with the pH controlled to about 8.0 and the temperature controlled to about 80°C. The treatment times and the resultant colors of the oxide coatings are also indicated below.

Alloy*	Treatment Time (min)	Color of Oxide Coating
1199**	20	None
1100	10	Medium tan
1100	20	Dark amber
5086	20	Dark amber
6061	20	Dark amber
5005	10	Medium tan
5005	20	Dark amber
5005	10	Medium tan
5005	20	Amber
5252	20	Dark tan
7075	20	Gold
2024	20	Dark tan

*Aluminum Association alloy designations.
**High purity (99.99%) aluminum.

Example 2: A group of aluminum panels was treated for about 30 seconds in a bath which was essentially the same as that described above in Example 1 to develop a relatively thin aluminum oxide coating of about 100 Å and an electrical resistance of about 75 to 100 μohm. The panels were removed from the treatment bath and water-rinsed. One portion of the panels was stabilized in a hot aqueous alkaline solution containing stearic acid and a second portion of the panels was stabilized in an identical aqueous alkaline solution except that it contained isostearic acid instead of stearic acid. The bath temperatures were about 85°C, the pH from 9.3 and the carboxylic acid content about 1 g/ℓ. The specimens having the coating formed in the alkaline solution of stearic acid allowed over 3,000 spot welds before the electrode needed replacement. The specimens treated in the aqueous alkaline solution of isostearic acid allowed approximately 2,000 spot welds before electrode replacement. This represents an increase of about 10 to 100 times in the number of spot welds on aluminum with the same electrode obtained in commercial practice.

The electrical resistance described herein was measured by means of a surface resistance analyzer Model No. VT-11A (C.B. Smith Company).

Pseudoboehmite-Type Aluminum Oxide

The work of *G.A. Dorsey, Jr.; U.S. Patent 4,028,205; June 7, 1977; assigned to Kaiser Aluminum & Chemical Corporation* is directed to a readily controllable process for forming a hydrophilic coating on an aluminum surface which develops a tenacious bond with curable organic matter, such as paints and adhesives. Moreover, the solution used in the process contains no chromates, fluorides, phosphates or other deleterious compounds which are difficult to dispose of without special treatment.

The aluminum workpiece is treated for a relatively short period of time with a hot caustic alkali solution containing small effective quantities of a nonionic or anionic surfactant to form an aluminum oxide coating which develops a tenacious bond with paints, adhesives and the like. The aluminum oxide coating formed is believed to contain pseudoboehmite-type aluminum oxide which provides the improved adhesive properties.

The concentration of the surfactant ranges from preferably about 0.01 to 0.50% by weight. A broad spectrum of surface-active wetting agents can be utilized in the process. Generally nonionic or anionic types of surfactants are preferred with the nonionic type being most desirable. Use of the terms "nonionic" or "anionic" refers to the fact that the surfactant is nonionic or anionic under conditions of treatment.

The pH of the aqueous solution should range from about 8.5 to 10.0, preferably 9.5±0.5, which is maintained by a caustic alkali.

The solution temperature is from about 50°C up to and including the boiling point of the solution, preferably from about 75° to 95°C.

Example 1: Cleaned 3105 aluminum alloy panels (Aluminum Association alloy designation) were treated for 30 seconds on an aqueous solution of sodium hydroxide containing about 0.2% Tergitol NPX, a nonionic surfactant. The pH of the solution was about 9.5 (NaOH concentration about 2 mg/ℓ) and the temperature was about 90°C. The treated panels were rinsed with deionized water, dipped in a 50% nitric acid solution to desmut, again water-rinsed and then air dried. After treatment, the sheets were tested with No. 600 Scotch brand cellophane tape and in each instance a heavy tape adhesive residual was left on the aluminum surfaces when the tape was rapidly removed, indicating that the desired coating had been formed. The treated sheets were painted with a single coat (0.85 mil thick) of low gloss white Duracron 100 acrylic paint and cured in a conventional manner.

Example 2: A second group of clean 3105 aluminum alloy panels were prepared and tested in the manner set forth above in Example 1. The treated panels were electrocoated with a paint primer RF-2874A (PPG, Incorporated), top coated with low gloss white Duracron 100 acrylic paint (0.99 mil total thickness) and then both coats were cured in a conventional manner simultaneously.

Example 3: A third group of clean 3105 aluminum alloy sheets were treated with an aqueous solution of Alodine 1200 so as to form a chemical conversion coating thereon of approximately 25 mg/ft^2. After rinsing and drying, the sheets were painted with a single coat (0.84 mil thick) of low gloss white Duracron 100 acrylic paint and cured in a conventional manner.

Example 4: A fourth group of clean 3105 aluminum alloy panels were treated in an aqueous solution of Alodine 401-45 so as to form a chemical conversion coating thereon of about 25 mg/ft^2. After rinsing and drying, the panels were electrocoated with RF-2874A paint primer (PPG, Incorporated), top coated with low gloss white Duracron 100 acrylic paint (0.96 mil total thickness) and then both coats were cured in a conventional manner simultaneously.

The panels described above in Examples 1 through 4 were subjected to Water Fog tests (Method 6201 of Federal Test Method 141), Salt Fog test (ASTM Test Method B117), CASS test (ASTM Method B-368-65), SWAACT (Navy Ship Specs. AA-A-00250/20), and ASFC test–Acidified Salt Fog Cyclic test (Military Specifications MIL-A-897A). After testing, the panels were rated with a rating of 10 (no effect) to 2 (large-scale corrosion). The results of these are set forth in the table below.

	.... Example Number.........			
	1	2	3	4
Water Fog, 1,000 hr	10	10	10	10
Salt Fog, 1,000 hr	10	10	10	10
ASFC, 150 hr	8	9	9	9
ASFC, 300 hr	7	8	8	8
SWAACT, 100 hr	4	8	6	8
SWAACT, 200 hr	2	7	4	7

Wedge bend and reverse impact tests were generally equivalent between the specimens of Examples 1 and 3 and specimens of Examples 2 and 4.

From the data above, it can be seen that the treatment of the process is equivalent in almost all respects to a two-coat (ED primer) paint system with an Alodine 401-45 pretreatment and almost as good as a single-coat paint system with an Alodine 1200 pretreatment. Treatment of spent solution is simple–merely neutralize.

HYDROPHOBIC AND OLEOPHILIC COATINGS FOR SHORT TERM PROTECTION

Long Chain Aliphatic Carboxylic Acids

Many processes are available for coating aluminum surfaces, such as anodizing, plating, chemical conversion coatings, painting and the like. The coatings, although designed for long life, require extensive surface pretreatments and are quite expensive. However, frequently, only short-term protection is needed or desired, for example, in shipping or storing semifabricated aluminum products, such as coiled sheet and the like, to prevent the formation of water stain or other oxidation products. On other occasions, it is desirable to prevent the gradual buildup of natural oxide on the aluminum surface, for example, in welding applications and adhesive bonding applications because the buildup of natural oxide can interfere with these types of operations. However, no simple and inexpensive process is known which will give short-term protection without interfering with subsequent fabrication or surface treatments, particularly when lubricants must be applied to the surface.

In accordance with the process of *G.A. Dorsey, Jr.; U.S. Patent 4,004,951; January 25, 1977; assigned to Kaiser Aluminum & Chemical Corporation* an aluminum surface is treated with an aqueous alkaline solution containing a long chain aliphatic carboxylic acid, an equivalent alkali metal salt thereof or a compound which generates a long chain aliphatic carboxylate anion in an alkaline solution at elevated temperatures greater than 60°C. Treatment times usually will be about one second for a clean surface, but extended treatment times do not seem to detrimentally affect the coating.

The surface coating is hydrophobic and usually highly oleophilic. Moreover, the coating is not usually affected by mineral acids, such as nitric acid, hydrochloric acid or sulfuric acid or by common polar solvents, such as acetone or ethyl alcohol. The coating formed is very difficult to analyze because under most circumstances, it appears to be a monomolecular layer about 100 Å thick. The carboxylate anion generating compound in the alkaline solution is apparently either reacting with the aluminum surface to form a type of aluminum soap or at least strongly associating with the aluminum surface.

The pH of the alkaline treating solution is preferably about 9 to 10. For optimum results, the temperature is maintained at about 85°±5°C.

The long chain carboxylate anion should have from 10 to 20 carbon atoms, preferably 12 to 18 carbon atoms.

Example 1: A clean 3004-H19 aluminum alloy sheet was treated. The treating solutions, which were maintained at 85°±5°C, were prepared by adding 1.0 g/ℓ of the noted acid to deionized or distilled water and then adjusting the pH to 9.0±0.1 with NaOH if needed. Treatment time in each case was 30 seconds. Each treated specimen was checked for water wettability after treatment in the alkaline solution, after a 30-second dip in a 35% (by weight) nitric acid solution and then after an acetone-ethyl alcohol rinse.

(A) – Myristic acid (C_{14}) formed a hydrophobic, oleophilic surface which remained hydrophobic after a 30-second dip in the nitric acid. However, after the

nitric acid treatment, the acetone-ethyl alcohol solution apparently removed the hydrophobic coating because the treated surface could then be wet with water.

(B) – Palmitic acid (C_{16}) formed a hydrophobic, oleophilic surface which remained so after both the nitric acid dip and the acetone-ethyl alcohol rinse.

(C) – Stearic acid (C_{18}) formed a hydrophobic, oleophilic surface which remained so after both the nitric acid dip and the acetone-ethyl alcohol rinse.

Example 2: *(A)* – A clean 3004-H32 aluminum alloy sheet was treated for 5 seconds in an aqueous alkaline solution maintained at 80°C which contained 1.0 g/ℓ sodium stearate. The pH of the solution was 9.3. Initially, a burst of effervescence occurred but the effervescence quickly subsided and the desired hydrophobic, oleophilic coating formed. The treated sheet withstood 20 hours of continuous water-fog exposure with no evidence of water stain or other surface defects. The coating was fully compatible with various metal-working lubricants, such as are used in rolling, forging, drawing and ironing, shaping, stamping and the like. Initially, the treated surface had an electrical resistance of 16 μohm/cm^2 and after 6 weeks of laboratory exposure (23°C and 70% humidity) had an electrical resistance of only 30 μohm/cm^2.

(B) – A plurality of clean, closely packed 3004-H32 aluminum sheets were treated in the manner set forth in (A) above except that 2.0 g/ℓ of a polyoxyethylene sorbitan trioleate (Tween 85) was added to the solution as a wetting agent. The coating formed was fully equivalent to the coating formed in (A). The wetting agent allowed the solution to penetrate in between closely packed aluminum sheets and react with the surfaces thereof.

Example 3: Clean 3004-H32 aluminum alloy sheets were treated for 30 seconds in a hot aqueous alkaline solution containing 1 g/ℓ of sodium stearate. The pH of the solution was 9.5 and the temperature was 80°C. After treatment, the sheets were rinsed and then separate sheets were treated for 5, 15, 25 and 35 seconds in a second hot, aqueous alkaline solution containing 2 g/ℓ of a polyoxyethylene sorbitan trioleate (Tween 85). The pH and temperature of the second solution were also 9.5 and 80°C, respectively. The coatings formed were hydrophobic and highly oleophilic.

Treated sheets were then evaluated for compatibility with mineral oil. The evaluation was conducted by placing a drop of mineral oil on a treated surface inclined about 70° from the horizontal and then determining the time required for the drop of oil to travel 3" on the inclined surface. Longer times indicate greater wettability and thus greater compatibility with the lubricant. The results are as follows:

Duration of Second Treatment, sec	Compatibility Test Time, sec
0	8-10
5	25-35
15	60
25	90
35	90

Results with a drop of 30 vol % o/w emulsion of Texaca 591 are similar.

Treatment with Isostearic Acid

The process of *G.A. Dorsey, Jr.; U.S. Patent 4,101,346; July 18, 1978; assigned to Kaiser Aluminum & Chemical Corporation* relates to the treatment of an aluminum surface to form a tenacious, hydrophobic and oleophilic coating which not only protects the underlying aluminum surface from significant oxidation but also facilitates the application of lubricant during subsequent fabrication of the aluminum workpiece. As used herein, aluminum refers to pure aluminum, commercially pure aluminum and aluminum alloys. Numbered aluminum alloy identifications herein refer to Aluminum Association alloy designations.

The method of preparing a hydrophobic and oleophilic protective coating on an aluminum workpiece consists essentially of developing a fresh aluminum surface on the workpiece and then treating the fresh aluminum surface under alkaline conditions with a liquid selected from the group consisting of nonaqueous liquids and aqueous liquids with a pH less than 10, the liquids containing an effective amount of a branched chain aliphatic carboxylic acid or an equivalent carboxylate compound having from 12 to 22 carbon atoms.

The tenacious surface coating formed is both hydrophobic and highly oleophilic. Moreover, the coating is not significantly affected by short-term exposure to mineral acids, such as nitric acid, hydrochloric acid or sulfuric acid at room temperatures and is not removed by common polar solvents, such as acetone or ethyl alcohol. The coating formed is very difficult to analyze because under most circumstances, it appears to be a monomolecular layer of about 100 Å thick. Coating formation is attributed to the carboxylic acid anion under alkaline conditions either reacting with a fresh aluminum surface to form a type of aluminum soap or at least strongly associating with the fresh aluminum surface.

Large amounts of the branched chain carboxylic acid are not necessary, and amounts down to 1 ppm of the solution have been found functional. However, usually an amount far in excess of that soluble is added to the alkaline solution to avoid depletion of the carboxylate component over extended periods of operation. Operational levels normally can range from about 0.01 to 10 grams carboxylate compound per liter. Commercially available acids include Emersol 871 and 875 which are isostearic acids.

Several modes of coating formation have been developed, all of which require alkaline conditions. For example, the aluminum surface can be treated with an aqueous alkaline solution containing the branched chain fatty acid. Alternatively, the aluminum surface can be deformed or abraded in such a manner to disrupt the natural oxide coating and thereby expose the underlying aluminum surface to a branched chain carboxylic acid under alkaline conditions. In both cases, the natural oxide coating is apparently removed or disrupted and the underlying nascent aluminum is immediately oxidized. The newly formed oxide is more reactive than aged natural oxide and an aluminum surface having a newly formed oxide coating is that which has been referred to as a "fresh" aluminum surface.

Treatment times are usually longer than 0.1 second, preferably longer than 1 second. Excessively long treatment periods, e.g., 60 minutes, do not detrimentally affect the coating formed. The treating solution preferably has a pH of 9 to 10. For optimum results, the temperature is maintained at about 85°±5°C.

If desired, wetting agents, such as Emsorb 6903, Tween 85 and Ultrawet can be added to the aqueous alkaline solution in an amount up to 3% by weight to facilitate the wetting of the surface by the alkaline solution during treatment.

Example 1: An aqueous emulsion of 1% by weight isostearic acid (Emersol 871) was prepared at a temperature of 80°C and the pH of the aqueous phase was adjusted to 9.5 by additions of NaOH. Panels of 5657 aluminum alloy were submerged in the aqueous emulsion for 30 seconds, rinsed with distilled water, then air dried. The panels had a tenacious, hydrophobic and oleophilic coating which was highly resistant to polar organic solvents.

Example 2: An aqueous bath containing 0.1% by weight of isostearic acid (Emersol 871) was prepared at a temperature of 85°C and the pH of the aqueous phase was adjusted to 9.5 by additions of NaOH. Panels of 3004 aluminum alloys were submerged in the bath for 30 seconds, rinsed with distilled water and then air dried. The panels had a tenacious, hydrophobic and oleophilic coating which was highly resistant to polar organic solvents.

Example 3: A 3004 aluminum alloy sheet was cold rolled on a small laboratory rolling mill with an alkaline oil-based lubricant consisting of 4% by weight isostearic acid (Emersol 871), 3% by weight of a mixture of C_{14}, C_{16} and C_{18} alcohols (Alfol 1418) and the remainder a base hydrocarbon oil (Somentor 43). About 0.2 g/ℓ of KOH was included with the alcohol mixture so that the lubricant exhibited a pH of 9.3 when mixed 1:1 with distilled water. The coating formed during rolling was hydrophobic, oleophilic and highly resistant to polar organic solvents.

The treated panels of the above examples exhibited a more highly oleophilic surface and the surfaces thereof were more readily wet with commercial lubricants than the panels which had been treated in the same manner but with straight chain carboxylic acids.

Coating Formed Simultaneously with Fabrication

In accordance with the process of *G.A. Dorsey, Jr.; U.S. Patent 4,099,989; July 11, 1978; assigned to Kaiser Aluminum & Chemical Corporation* an aluminum surface is mechanically deformed or upset under aqueous alkaline conditions in the presence of a long chain aliphatic carboxylic acid, an equivalent alkali metal salt thereof or a compound which generates a reactive long chain aliphatic carboxylate anion during the deformation of the aluminum surface. The surface coating which forms under these conditions is hydrophobic and usually highly oleophilic. Moreover, the coating is not significantly affected by short-term exposure to mineral acids, such as nitric acid, hydrochloric acid or sulfuric acid at room temperatures or by common polar solvents, such as acetone or ethyl alcohol. The coating formed is very difficult to analyze since under most circumstances, it appears to be a monomolecular layer about 100 Å thick.

The carboxylic acid can be most readily applied to the exposed fresh surface by incorporating the acid or its equivalent into the metal-working lubricant. This allows the coating to be formed while the aluminum workpiece is being shaped or reduced in cross section, thereby eliminating an additional processing step. Oil-based lubricants or oil-in-water emulsified lubricants can be used.

Particularly effective carboxylic acids are those described in U.S. Patent 4,101,346 which have at least one short secondary alkyl group on or near the nonpolar end of the carboxylic acid, such as isostearic acid.

The hydrophobic coating has an electrical resistance initially of about 10 μohm which remains relatively stable for at least 7 weeks. This indicates that essentially no oxidation of the underlying aluminum surface is occurring. These resistance levels are to be compared with a natural oxide coating which has an initial resistance of about 10 μohm and which can gradually increase to well over 1,000 μohm in a matter of days.

Example 1: A 3004 aluminum alloy sheet was cold rolled on a small laboratory rolling mill with an alkaline, oil-based lubricant consisting of 4% by weight isostearic acid (Emersol 817), 3% by weight of a mixture of C_{14}, C_{16} and C_{18} alcohols (Alfol 1418) and the remainder a base hydrocarbon oil (Somentor 43). About 0.02 g/ℓ of KOH was included with the alcohol mixture so that the lubricant exhibited a pH of 9.3 when mixed 1:10 with distilled water. The coating formed during rolling was hydrophobic, oleophilic and highly resistant to polar organic solvents.

Example 2: A 3004 aluminum alloy sheet was cold rolled on a small laboratory rolling mill with an alkaline oil-based lubricant consisting of 4% by weight stearic acid, 3% by weight of a mixture of C_{14}, C_{16} and C_{18} alcohols (Alfol 1418) and the remainder a base hydrocarbon oil (Somentor 43). About 0.2 g/ℓ of KOH was included with the alcohol mixture so that the lubricant exhibited a pH of 9.3 when mixed 1:10 with distilled water. The coating formed during rolling was hydrophobic and oleophilic, although less oleophilic than the coating formed in Example 1.

Example 3: A solution of stearic acid and a small quantity of ethyl alcohol saturated with KOH was prepared with the pH of the solution, when mixed 1:10 with distilled water, at 9.0. This solution was placed at the interface between two 3004 aluminum alloy sheet specimens and then the specimens were briskly rubbed together by hand for a few moments to mechanically abrade the oxide surfaces at the interface. After removing the residual solution, the abraded surfaces of the specimens exhibited the hydrophobic and oleophilic coating of the process, although the surface was less oleophilic than those of Examples 1 and 2.

Although cold rolling is described as the primary mode of surface deformation, forging, extruding and the like can be used. Moreover, surface abrading in any suitable manner has been found adequate to remove the oxide coating and expose a fresh aluminum surface.

USE OF TANNINS IN COATINGS

Ternary Composition of Zirconium, Tannin and Fluoride of pH <3

In the past, satisfactory corrosion resistance and concomitant good adhesion of the organic finish were provided by typical metal treatment solutions composed of about 1% solution of a mixture of hexavalent chromium, phosphoric acid and fluoride. Greater sensitivity to preserving a clean, safe environment has spurred the development of acceptable coatings which also provide adequate protection.

G.L. Tupper; U.S. Patent 4,277,292; July 7, 1981; assigned to Coral Chemical Company found that a satisfactory corrosion resistant and adhesion enhancing coating for aluminum surfaces can be obtained at pH values of 2.3 to 2.95. It is totally unexpected that satisfactory coatings can be produced at this range of pH. The composition contains zirconium and fluoride ions and tannin in an effective amount. The zirconium ion concentration varies between 1.4 and 4.8 g/ℓ in the concentrate solution and 0.014 to 0.096 g/ℓ in the treatment bath. This may also be expressed as 14 to 96 ppm in the treating bath. The fluoride, both free and complex, varies between 0.020 and 0.140 g/ℓ or 20 to 140 ppm in the treating bath, and the tannin between 0.060 and 0.240 g/ℓ, or 60 to 240 ppm in the treating bath.

The coating compositions can be applied to aluminum substrates, particularly aluminum cans which have been pretreated with an acid cleaner (environmentally acceptable) and water rinses.

The coating solution can be applied to the aluminum in a variety of methods employed by the industry. For example, the solution can be applied by spraying, immersion, or flow-coating techniques. Spraying is effective and economical.

The temperature of the coating solution is between 100° and 130°F. Generally, it is 110°F. The aluminum is in the bath for about 30 seconds.

Before immersing the aluminum, it is cleaned. Available acid or alkaline cleaners, well-known in the art, are used. After cleaning, the surface is water rinsed and finally deionized water rinsed. No chromium or phosphate treatment is used.

After the coating has been applied, it is dried, for example, in forced hot air circulating ovens at 350°F for 5 minutes.

Subsequently, the decorative organic coating of lacquer, ink or resin is applied and, in the case of cans used for foods or beverages, a sanitary interior coating is applied.

The coatings have a general thickness of 200 to 750 Å and a weight within the range of 10 to 35 mg/ft^2.

Preferred is a coating composition of a ternary variety of zirconium ions of 30 to 45 ppm, fluoride ions of 30 to 70 ppm, and vegetable tannin of about 100 ppm. The pH is preferably 2.4 to 2.8.

Example 1: *Comparative example* – This example is not illustrative of the process. A treating solution is prepared as follows:

Treating Bath	Grams	Percent by Weight
Water	955.0	95.5
Nitric acid, 42° Bé	28.0	2.8
Hydrofluosilicic acid (30%)	3.0	0.3
Ammonium zirconyl carbonate	14.0	1.4

The drawn and ironed aluminum cans are cleaned in a 1% solution of a hot aqueous sulfuric acid cleaner and rinsed with cold water to a no-water-break surface. The cans are immersed in a 1½% v/v solution (40 to 60 ml in 4,000 ml) of the above for 30 seconds at 120°F. The pH is 2.6. A cold tap water rinse follows. The cans are dried in the oven for 5 minutes at 350°F.

The following tests are done on the bottom of a can, cut from side wall.

A. Reynold Metals TR3 Test: The bottom is heated at 150°F for 30 minutes, immersed in a solution of 0.5 g sodium chloride, 1.0 g sodium bicarbonate and water to make 1 liter. The surface is then observed for discoloration.

   Result: The can bottom is brown.

B. The can bottom is heated in a muffle furnace at 1000±50°F for 5 minutes. A yellow to brown color indicates the presence of a coating.

   Result: No color.

Example 2: *Process examples* – These examples are illustrative of the process. (A) The following solution is prepared:

	Grams
Water	941.5
Nitric acid, 42° Bé	28.0
Hydrofluosilicic acid (30%)	3.0
Ammonium zirconyl carbonate (20%)	14.0
Mallinckrodt tannic acid	10.0
Hydrofluoric acid (70%)	3.5

The treating bath is made at 1% v/v (40 ml treating solution diluted to 4,000 ml). The pH is 2.95.

TR3 result: No discoloration after 30 minutes.

Furnace: Develops bright golden color.

(B) The following solution is prepared:

	Grams
Water	913.0
Nitric acid, 42° Bé	46.0
Hydrofluosilicic acid (30%)	5.0
Ammonium zirconyl carbonate (20%)	21.0
Tannic acid	10.0
Hydrofluoric acid (70%)	5.0

A 2% v/v solution is used as the treating bath. The pH is 2.3.

TR3 result: No discoloration.

Furnace: Light yellow.

Solution of Phosphate, Tannin, Titanium and Fluoride at Acidic pH

Y. Matsushima, H. Oka, K. Noji and H. Nakagawa; U.S. Patent 4,017,334; April 12, 1977; assigned to Oxy Metal Industries Corporation found that aluminum cans may be successfully treated to provide a paint receptive noncorrosive, colorless coating with a contact time of 30 seconds or less. The clean aluminum surface is contacted with an aqueous solution containing phosphate, a tannin, titanium and fluoride at an acidic pH value prior to application of the organic finish.

The term "aluminum" as used herein is meant to include alloys of at least 90% aluminum which are commonly employed in can manufacture. Such alloys may contain elements such as magnesium, manganese and zinc, for example, 3000, 5000 and 6000-type aluminums are suitable examples.

Experimental work has shown that hydrolyzable, condensed, and mixed varieties of vegetable tannins may all be used. Quebracho extract and tannic acid in accordance with Japanese Industrial Standard K8629 have been found very effective.

The following examples demonstrate the process. In all cases, the aluminum cans were pretreated as follows: (1) 15-second hot water rinse; (2) 30-second spray cleaning using a sulfuric acid-based cleaner; (3) 15-second hot water rinse; (4) spray application of treating solution; (5) 15-second cold water rinse; and (6) 3-minute oven dry at 350°F.

Coke red ink (Acme Ink Co. alkyd-based) was then applied using rubber rolls. Next, clear overvarnish (Clement Coverall Co., Code No. P-550-G alkyd polyester) was applied over the wet ink using a No. 5 drawdown bar. The cans were then baked 5 minutes at 350°F followed by 3 minutes at 410°F to cure.

Comparative Example A: A solution was prepared in accordance with Example 1 of U.S. Patent 2,502,441 to contain 13.5 g/ℓ NaH_2PO_4, 0.15 g/ℓ MoO_3, 0.08 g/ℓ quebracho, and pH adjusted to about 5.0.

The solution was used as the treating solution in the above procedure at a temperature of about 55°C for either 20 seconds or 1 minute. The coating obtained appeared dull and nonadherent. After the organic finish was applied as above, the can was subjected to the pasteurization and adhesion tests. In the pasteurization test, the surface was grossly discolored and in the adhesion test, almost complete paint removal was observed indicating unacceptable adhesion.

Comparative Example B: To the above solution was added 0.4 g/ℓ of fluoride as HF. When used to treat aluminum cans for either 20 seconds or 1 minute, a coating was obtained which was dusty-brownish and nonadherent. Adhesion and pasteurization test results were unacceptable. Again, almost complete paint removal occurred in the adhesion test.

Example 1: To the above fluoride containing solution was added 0.2 g/ℓ of Ti as $Ti(SO_4)_2$ and cans treated for either 20 seconds or 1 minute. The resulting coating was colorless, nondusty and acceptable pasteurization and adhesion results (essentially no paint removal) were obtained.

Example 2: A solution was prepared to contain 0.4 g/ℓ phosphate as H_3PO_4, 0.7 g/ℓ Ti as H_2TiF_6, 2 g/ℓ quebracho, 1.7 g/ℓ F as H_2TiF_6, and a pH of 1.2 to 5.5.

Aluminum cans were treated for 20 seconds as above. The coating was shiny and colorless. Adhesion and pasteurization results were excellent.

Two-Stage Vegetable Tannin Treatment

P.F. King and G.A. Reghi; U.S. Patent 4,054,466; October 18, 1977; assigned to Oxy Metal Industries Corporation describe a composition and process useful for the treatment of an aluminum surface comprising contacting the surface with an aqueous solution containing at least 0.000025% by weight of a vegetable tannin, which solution exhibits a pH of at least 3. Best results can be obtained via a two-stage tannin treatment in which the treating solution of the first stage also contains titanium and fluoride in dissolved form.

The following tests have been employed in the examples to evaluate the quality of the treated surface.

Bend Adhesion: This test is a measure of the ability of a finish to withstand cold deformation after painting. A standard test panel is bent 180° about a mandrel. The radius of curvature at the bend is a function of the mandrel thickness which thickness is expressed in terms of multiples of the test panel thickness. The most severe condition is encountered when no mandrel at all is employed and the panel is simply bent back upon itself (O-T Bend). Paint adhesion is then determined by application and removal of Scotch brand transparent tape (No. 610) from the bend and the proportion of paint remaining is rated from 10 (100% adhesion) to 0 (0% adhesion).

Impact: This test is designed to show the effect upon paint adhesion of an impact deformation. A ⅝" diameter tool is impacted on the unpainted side of a panel. The force of the impact is approximately 2,000 times the panel thickness (e.g., 50 in-lb for a panel 0.025" thick). The standard impact test is performed shortly after the paint is cured and at ambient temperatures. A "Cold Impact" is performed on a painted panel which has been refrigerated to a temperature of 15°F or less. A "Delayed Cold Impact" is performed on a panel at least 5 days after painting. In any impact test, adhesion is measured by the application and removal of Scotch brand transparent tape to the deformed surface and the proportion of paint remaining on the surface is rated from 10 (100% adhesion) to 0 (0% adhesion).

MEK Resistance: This test is employed by paint manufacturers as a measure of the degree of cure of a paint. A cloth soaked with methyl ethyl ketone is rubbed briskly back and forth over the painted surface. The number of back and forth rubs required to completely remove the paint over a 10 mm length is recorded. 100 or more rubs are normally required for acceptability.

Example: A basic processing cycle for aluminum panels was established as follows: (1) 10-second alkaline cleaner, 120°F; (2) 10-second hot water rinse; (3) 5-second spray application of treating solution, 120°F, pH 5; (4) 10-second cold water rinse; and (5) 3-second post-treatment (0.025% quebracho tannin, pH 4.5).

Identical aluminum panels were processed through four variations of the above process cycle as follows:

(A) Clean only – Panels are cleaned, water rinsed and painted, omitting steps 3, 4 and 5.

(B) Single-stage tannin treatment – Panels are treated omitting steps 3 and 4.

(C) Two-stage tannin treatment – Panels are treated according to the basic process employing an aqueous solution of 0.015% by weight chestnut tannin extract in Step 3.

(D) Modified two-stage tannin treatment – Panels are treated according to the basic process employing in Step 3 an aqueous solution of 0.015% by weight chestnut tannin, 0.014% Ti (added as $TiOSO_4$) and 0.1% F (added as HF).

Separate sets of the thus-treated panels were then painted with the following paints: Polyester (#71308 Poly-Lure 2000, Glidden); acrylic enamel (Duracron Super 630, PPG); modified epoxy (8-C-2002, Technical Coatings Co.); and vinyl (1401-3706-11, Bradley-Vrooman).

The thus-painted panels were then subjected to physical testing in accordance with the Bend, Cold and Delayed Cold Impact and MEK tests. The panels which had been cleaned only (Variation A) failed in almost all physical testing. Where a tannin solution was employed (Variations B, C and D), results improved markedly. Two-stage tannin treatment (Variations C and D) exhibited further improvement with best results obtained in the modified two-stage treatment (Variation D). The results of these tests are shown in the following table.

	Paint Type							
							Vinyl	
Process Variation	Polyester O-T Bend	Polyester MEK	Acrylic O-T Bend	Acrylic MEK	Epoxy O-T Bend	Epoxy MEK	O-T Bend	Delayed Cold Impact
A	9.5	26	0	65	8	100+	1	1
B	10	100+	10	100+	10	100+	9	4
C	10	100+	9.5	100+	10	100+	9.8	5
D	10	100+	9.5	100+	10	100+	10	10

Soluble Lithium Compound

J.K. Howell, Jr.; U.S. Patent 4,063,969; December 20, 1977; assigned to Oxy Metal Industries Corporation found that the corrosion resistance imparted to an aluminum surface by an aqueous tannin containing composition can be improved by including a soluble lithium compound in the treatment composition. The presence of the lithium compound improves corrosion resistance without detrimentally affecting the adhesion of a subsequently-applied organic finish. The lithium compound should be present in an amount, at least 0.001 g/ℓ, sufficient to improve corrosion resistance imparted by the treating solution. The improved results obtained from the addition of lithium are not evident when other alkali metals or ammonium are employed in equal amounts.

Example: A treating solution of pH 5.0 was prepared to contain the following:

Component	Grams per Liter
Ti (in $TiOSO_4$)	0.13
F (in 70% HF)	0.95
PO_4 (in 75% H_3PO_4)	0.094
Chestnut extract	0.155
NaOH	0.0075
Li (in $LiOH \cdot H_2O$)	0.013
NH_3	0.796

Aluminum (5050 alloy) panels were then processed as follows: (1) alkaline cleaner, 160°F, 10 seconds; (2) hot water rinse, 10 seconds; (3) treating solution, 120°F, 5-second spray; (4) cold water rinse, 5 seconds; (5) aqueous tannin post-treatment, 3 seconds (0.25 g/ℓ quebracho extract, pH 5); and (6) squeegee and air dry.

Panels were also run using steps 1, 2 and 6 only, giving cleaned only control panels. Sets of both treated and cleaned only panels were painted immediately and another set of treated panels was aged three months prior to painting. The panels were painted with Mobil's S-9009-105 vinyl-based paint and subjected to testing.

Panels were immersed in boiling deionized water for 10 minutes. After crosshatching of the surface and drying, Scotch brand transparent tape (No. 610) was applied and removed from the crosshatched surface. The results were then rated from 10 (no paint removal) to 0 (complete paint removal).

Similarly immersed panels were bent 180° without a mandrel and tape pulled along the flat panel surface adjacent the bend. The results were as follows:

Test	. . .Treated Panels. . .		Cleaned Only
	Fresh	Aged	
Adhesion	10	10	5
Bend adhesion	10	10	5

Prior Treatment with Iron Ions and Complexing Agent

Y. Nagae and T. Utsumi; U.S. Patent 4,163,679; August 7, 1979; assigned to Oxy Metal Industries Corporation found that the surface of aluminum and its alloys can be provided with corrosion resistant and paint receptive properties without the use of chromium compounds by treating the surface with an aqueous alkaline solution containing iron ion and a complexing agent and having a pH of higher than 10, rinsing the treated surface with water, and then treating the surface with an aqueous acidic solution containing an organic tannin.

Example 1: Alloyed aluminum panels (Material No. 5052) having a size of 50 x 100 x 0.3 mm were treated at 65°C for 6 seconds by spraying with an aqueous alkaline solution prepared by dissolving 70 g of sodium hydroxide, 2 g of ferric ion in the form of ferric sulfate and 18 g of sodium gluconate in 10 liters of water. After rinsing with water, the panels were treated at 55°C for 6 seconds by spraying with an aqueous solution prepared by dissolving 50 g of tannic acid (Chinese gallotannin) in 10 liters of water and a pH value of 3.5 followed by rinsing with water, rinsing with demineralized water and drying.

The thus treated panels were then subjected to the salt spray test according to JIS-Z-2371 and the humidity test according to JIS-Z-0228. In addition, another identically prepared set of panels was painted with an epoxy paint (Kancoat XJL165L Clear, Kansai Paint Co.) to a thickness of from 5 to 6 μ and baked at 205°C for 10 minutes. The painted panels were then subjected to the salt spray test and paint adhesion test. Excellent paint adhesion results were obtained for the treated panels of the process. Tables 1 and 2 show the results obtained in the salt spray and humidity testing. The comparison examples were also subjected to testing.

Comparative Example 1A: Panels identical to those of Example 1 were treated at 65°C for 6 seconds by spraying with a strong alkaline cleaning solution prepared by dissolving 70 g of sodium hydroxide and 18 g of sodium gluconate in 10 liters of water, rinsed with water and then demineralized water and dried.

Comparative Example 1B: Panels identical to those of Example 1 were treated at 65°C for 6 seconds by spraying with an aqueous alkaline solution prepared by dissolving 70 g of sodium hydroxide, 18 g of sodium gluconate and 2 g of ferric ion in the form of ferric sulfate in 10 liters of water, rinsed with water and then with demineralized water and dried.

Comparative Example 1C: Panels identical to those of Example 1 were treated with the same strong alkaline cleaning solution as in Comparative Example 1A at 65°C for 6 seconds by spraying and then rinsed with water. The treated sheets were then treated at 55°C for 6 seconds by spraying with an aqueous solution prepared by dissolving 50 g of a tannic acid (Chinese gallotannin) in 10 liters of water and adjusting the pH to a value of 3.5, rinsed with water and then with demineralized water and dried.

Comparative Example 1D: Panels identical to those of Example 1 were chromated with a conventional bath containing chromium phosphate so that a chromate coating of 20 mg/m^2 was provided.

Table 1: Unpainted Panels

Ex. No.		48 Hr Salt Spray Test	48 Hr Humidity Test
		(% white rust)	
1	Process	5	0
1A	Clean only	90	90
1B	No tannin treatment	70	90
1C	No iron treatment	70	70
1D	Conventional chromate	0	5

Table 2: Painted Panels

Ex. No.		500 Hr Salt Spray Corrosion Width (mm)
1	Process	0
1A	Clean only	0-1
1B	No tannin treatment	0-1
1C	No iron treatment	0-1
1D	Conventional chromate	0

Aqueous Acidic Fluoride-Containing Cleaner

G.A. Reghi and S.T. Farina; U.S. Patent 4,111,722; September 5, 1978; assigned to Oxy Metal Industries Corporation found that when a vegetable tannin-containing composition is employed as the primary treatment in place of the conventional chromate-phosphate treatment for an aluminum surface, the cleaner employed is very critical to obtaining acceptable improvements in corrosion resistance and paint receptivity. A marked improvement in corrosion resistance is obtained by including fluoride ion in an aqueous acidic cleaning composition prior to treatment with a vegetable tannin-containing composition.

The precise minimum and maximum effective fluoride concentrations suitable for use in the cleaner cannot be stated without reference to parameters such as the particular cleaner and treating formulations employed; processing conditions such as contact time, method and temperatures of treatment; and the quality desired of the final product. In general, however, effective fluoride concentrations of from about 0.01 to 0.5 g/ℓ have been found effective with concentrations of from 0.01 to 0.2 g/ℓ being preferred.

Tannins are generally characterized as polyphenolic substances having molecular weight of from about 400 to 3,000. They may be classified as "hydrolyzable" or "condensed" depending upon whether the product of hydrolysis in boiling mineral acid is soluble or insoluble, respectively. Often extracts are mixed and contain both hydrolyzable and condensed forms. No two tannin extracts are exactly alike. Principal sources of tannin extracts include bark such as wattle, mangrove, oak, etc; woods such as quebracho, chestnut, oak, etc.; fruits such as myrobalans, valonia, divi-divi, etc.; leaves such as sumac and gambier; and roots such as canaigre and palmetto.

The term "vegetable tannins" is employed to distinguish organic tannins such as those listed above from the mineral tanning materials such as those containing chromium, zirconium and the like. Experimental work has shown that hydrolyzable, condensed, and mixed varieties of vegetable tannins may all be suitably used in the process. Quebracho and chestnut have been found to be very effective condensed tannins and myrobalan an effective hydrolyzable tannin.

The following tests were employed to evaluate the corrosion resistance and organic finish receptivity of the treated aluminum surface:

Pasteurization: This test is a measure of the resistance to discoloration of a substrate which has been treated but to which no organic finish has been applied. The treated surface is immersed in tap water at 140° to 160°F (60° to 70°C) for 45 minutes. The surface is then observed for discoloration and rated "Acceptable" (colorless), "Marginal" (slight brown color) or "Unacceptable" (brown colored).

Tape Adhesion: This test is a measure of the adhesion between an organic finish and a treated substrate. The painted surface is subjected to a standard 1% detergent solution (Joy) at boiling for 30 minutes, rinsed in tap water, crosshatched (approximately 64 squares/in^2), and dried. Scotch brand transparent tape (No. 610) is then applied to the crosshatched area and the amount of paint removed by the tape is observed. Results are rated "Excellent" (100% adhesion), "Good" (95+% adhesion) or "Poor" (less than 95% adhesion).

Example: An aqueous tannin treatment bath was prepared to contain the following: Chestnut tannin extract, 0.15 g/ℓ; Titanyl sulfate, 0.14 g/ℓ as Ti; HF (70%), 1.0 g/ℓ as F; H_3PO_4, 0.1 g/ℓ as PO_4; NH_4OH to pH 5.1; and water, the balance.

Cleaner A was prepared to contain: H_2SO_4, 6.3 g/ℓ; $(NH_4)_2SO_4$, 2.1 g/ℓ; Triton CF-10 (alkylaryl polyether surfactant, Rohm & Haas Co.), 1.9 g/ℓ; Surfactant AR-150 (polyethylene glycol ester of rosin, Hercules, Inc.), 1.9 g/ℓ; and fluoride (as HF), 0 to 0.1 g/ℓ.

Cleaner B was prepared to contain: H_2SO_4, 6.2 g/ℓ; Antarox LF 330 (aliphatic polyether surfactant, GAF Corp.), 1.3 g/ℓ; Surfactant AR-150, 1.3 g/ℓ; and Fluoride (as HF), 0 to 0.5 g/ℓ.

The following process sequence was employed to spray-treat aluminum cans: (1) clean for 30 seconds; (2) water rinse for 5 seconds; (3) tannin treatment, 105° to 120°F, 20 seconds; (4) cold water rinse, 5 seconds; (5) deionized water rinse, 5 seconds; and (6) oven dry at 350°F, 3 minutes.

Transparent ink (Acme Ink Co.) was then applied to the can exterior using rubber rolls. Next, clear overvarnish (Clement Coverall Co., Code No. P-550-G, alkyd polyester) was applied over the wet ink using a No. 5 drawdown bar. The cans were then baked 5 minutes at 350°F. A sanitary interior lacquer (Mobil S-6839-009, vinyl-based) was then applied to the interior followed by 3 minutes at 410°F to cure.

Both the interior and exterior surfaces were then tested for Tape Adhesion and the exterior can bottom was subjected to the Pasteurization test for discoloration of the unpainted surface. Both Cleaner A and Cleaner B were employed at temperatures of 120°F and 180°F with either no fluoride or with an effective fluoride concentration of 0.1 g/ℓ. In every instance, the presence of fluoride improved the Pasteurization test results from "Unacceptable" to either "Marginal" or "Acceptable".

Twenty-one tests were run varying the effective fluoride concentration of Cleaner B at 120°F from about 0.01 to 0.5 g/ℓ. One test rated "Marginal" on Pasteurization while the other twenty were "Acceptable". When the fluoride-free cleaner was employed, "Unacceptable" Pasteurization was observed. As the fluoride concentration approached 0.4 to 0.5 g/ℓ etching of the cans began to occur which is normally undesirable. Further, with the particular cleaner and tannin treatment employed, paint adhesions appeared consistently "Poor" for fluoride concentrations above about 0.2 g/ℓ.

It should be understood that the maximum desirable fluoride concentration will be a function of parameters such as the particular cleaner and treating formulations employed, processing conditions such as contact time, method and temperature of treatment, and the quality desired of the final product. Suitable fluoride levels may be selected by simple experimentation once these parameters have been determined.

Melamine-Formaldehyde Resin and Vegetable Tannin

J.K. Howell, Jr. and L. Kulick; U.S. Patent 4,174,980; November 20, 1979; assigned to Oxy Metal Industries Corporation disclose a single-application nonelectrolytic method for the treatment of bare metal surfaces to form a corrosion-resistant coating. The clean metal surfaces are treated with an aqueous composition containing a melamine-formaldehyde resin and a vegetable tannin. The resulting dried coating exhibits superior qualities for a single-application coating process when used as a paint base and the process does not require the use of environmentally objectionable chromium compounds. Application to ferrous, zinc or aluminum surfaces may be by any known technique designed to provide a deposited layer of desired uniformity.

Very small quantities of the tannin material, when included in combination with a melamine-formaldehyde resin, have been found very effective in increasing the anticorrosion properties imparted by the treating solution. It is desirable to include at least 0.01 g/ℓ of the vegetable tannin in the solution. Most preferably, the weight ratio of the resin to the tannin is in the range of 1:30 to 30:1 with a resin concentration of at least 0.01 g/ℓ. For many paint systems a ratio of resin:tannin of at least 1:1, preferably at least 3.75:1, and most preferably at least 7.5:1 is suitable.

Humidity corrosion resistance was measured in accordance with the procedure of ASTM 2247-64T. The panels were rated in terms of the number and size of the blisters; F for few, M for medium and D for dense, and from 9 for very small size to 1 for very large. 10 represents no blisters.

Salt Spray corrosion resistance was measured in accordance with the procedure of ASTM B117-61. The panels were rated in terms of the amount of paint loss from a scribe in 1/16" increments (N for no loss of paint at any point). The principal numbers represent the general range of the creepage from the scribe along its length. Thus, 2 to 7 means representative creepage varied from 2/16" to 7/16".

Example 1: An aqueous concentrate solution was prepared to contain the following:

Component	Percent by Weight
Melamine-formaldehyde resin*	24.5
Tannin (nonbisulfited)**	1.5
NaOH	0.25
Water	Balance

*Resimene X714, Monsanto Company
**Quebracho extract, Arthur C. Trask Corp.

The quebracho was added as an aqueous solution containing a small amount of NaOH for solubilizing.

A treating solution was prepared of the above in Detroit tap water at a concentration of 16.0 g/ℓ corresponding to approximately 4 g/ℓ resin and 0.25 g/ℓ tannin. The pH was adjusted to 3.0 with phosphoric acid (25%). SAE 1010 cold rolled steel panels were then processed according to the following sequence: (1) alkaline cleaner, 1 oz/gal, 150°F, 1 minute spray; (2) warm water rinse, ½ minute spray; (3) treating solution, ambient temperature, ½ minute spray; and (4) dry-off oven, 5 minutes at 350°F.

The panels were painted with Dulux 704-6731 white alkyd-based paint and subjected to the Salt Spray and Humidity tests for 336 hours.

As controls, identical panels were treated with a conventional iron phosphating bath containing approximately 1% PO_4 and 0.5% chlorate, water rinsed, and post-treated with a conventional dilute (0.1% CrO_3) hexavalent chromium rinse. The results were as follows:

Test	Resin-Tannin Treatment	Conventional Treatment
Humidity	10*	10*
Salt Spray	2-3**	3-5**

*No blisters
**Creepage in 1/16" increments

The results show that the corrosion resistance for resin-tannin treated surfaces as measured by the Salt Spray and Humidity tests is as good as or better than that of conventionally treated surfaces.

Example 2: An aqueous treating solution was prepared to contain the following:

Component	Grams per Liter
Melamine-formaldehyde resin*	12.6
Quebracho extract (nonbisulfited)	1.7
NaOH	0.16
Triton CF 54	0.4

*Cymel 7273-7

The above room temperature solution was roll coated onto aluminum (3003 alloy) panels after the panels had been cleaned with the alkaline cleaner of Example 1. The panels were then oven dried at 400°F for 20 seconds (metal temperature 150° to 180°F). The coating weight was about 10 mg/ft^2. Groups of the panels were then separately painted with polyester, acrylic and vinyl-based paints. Thereafter, the panels were subjected to the Salt Spray, Acetic Acid Salt Spray, MEK Resistance and Bend Adhesion tests.

The resin-tannin treatment provided a paint base of acceptable quality.

AQUEOUS ACIDIC COATING SOLUTIONS

Zirconium and/or Titanium, Fluoride and Phosphate Compounds

It is known to coat aluminum surfaces with aqueous coating solutions that are effective in forming thereon coatings which are corrosion resistant and thereby protect the surface from degradation due to attack by corrosive materials. In general, the coatings formed from such coating solutions should also have properties such that overlying coatings which are applied thereto adhere tightly and strongly. Such overlying coatings are decorative or functional in nature and are formed from materials such as paints, lacquers, inks, etc.

An acidic aqueous coating solution for forming on an aluminum surface a coating which is corrosion resistant and to which overlying coatings adhere excellently is disclosed by *T.L. Kelly; U.S. Patent 4,148,670; April 10, 1979; assigned to Amchem Products, Inc.* The coating solution contains compounds of zirconium and/or titanium, fluoride and phosphate in dissolved form, and optionally, polyhydroxy compounds having 6 or fewer carbon atoms. The coating solution is capable of forming on an aluminum surface a uniformly colorless and clear coating so that the coated surface has the appearance of the underlying metal surface, that is, the coating can be formed without changing the appearance of the metal surface.

The pH of the coating solution should be within the range of about 1.5 to 4.0.

Example 1: Unless stated otherwise, the aluminum surfaces treated with the solutions identified in the tables below were drawn and ironed aluminum cans which were first degreased, as necessary, in an acidic aqueous cleaner containing sulfuric acid and detergents. Unless stated otherwise, the coating solutions were applied by spraying for about 25 seconds at a temperature of about 110°F. After treatment with the solutions, the aluminum surfaces were rinsed in deionized water and dried in an oven for 2 minutes at about 400°F.

Thereafter, the aluminum surfaces were tested for corrosion resistance by subjecting them to a pasteurization test. This test consists of immersing the aluminum surface in water having a temperature, as indicated, and for a period of time, as indicated. A cleaned-only aluminum surface, when subjected to the pasteurization test, turns black after a few minutes. The results of the tests were rated as follows: 5, perfect, no blackening; 3+, acceptable; and 0, total failure, severe blackening.

Aluminum surfaces treated with the solutions described were tested also for paint adhesion. After the treated surface was dried, as described above, a portion of the surface was painted with a white base coat (No. 12W100A white polyester, H.C.I.) and the other portion of the surface was painted with an interior vinyl-lacquer (Modified Vinyl Epoxy Lacquer C-5054, Mobil).

After the paint was cured, the painted surface was immersed either in boiling water-detergent or water-NaCl solution. After removing the painted surface from the solution, it was rinsed in water, and the excess water was removed from the surface by wiping. The painted surface was then crosshatched, using a sharp metal object to expose lines of aluminum which showed through the paint or lacquer, and tested for paint adhesion. This test included applying cellophane tape firmly over the crosshatched area and then drawing the tape back against itself with a rapid pulling motion. The results were rated as follows: 10, perfect, when the tape did not peel any paint; 8, acceptable; and 0, total failure.

In Table 1, the solutions were adjusted to pH 2.7 with HNO_3 or NH_4OH.

Table 1: Composition

Solution Number	$(NH_4)_2ZrF_6$	HBF_4	H_2SiF_6	H_2TiF_6	HF	H_3PO_4
	....	....	........(g/ℓ)	....	....	....
1	0.240	–	–	0.164	0.050	0.294
2	–	–	–	0.164	0.050	0.294
3	–	0.264	–	0.164	0.050	0.294
4	0.240	0.264	–	0.164	0.050	0.294
5	0.240	0.264	–	–	–	0.294
6	0.240	0.264	–	0.164	–	0.294
C-1*	0	0	0	0	0	0
C-2	0.240	–	–	–	0.050	–
C-3	0.240	0.264	–	–	0.050	–
C-4	–	0.264	–	–	0.050	0.294

(continued)

Table 1: (continued)

Solution Number	$(NH_4)_2ZrF_6$	HBF_4	H_2SiF_6	H_2TiF_6	HF	H_3PO_4
			(g/ℓ)			
C-5	0.240	–	0.442	–	0.050	–
C-6	–	–	0.442	–	0.050	0.294
C-7	0.240	–	–	0.164	0.050	–

*Cleaned only, no treatment.

Table 2: Test Results

Solution Number	Pasteurization Test 212°F 15 min	Pasteurization Test 160°F 45 min	Paint Adhesion Test* White Base Coat	Paint Adhesion Test* Interior Vinyl Lacquer
1	3+	4–	10, 9+, 9+, 9	10, 10, 10, 10
2	3	3	10, 9+, 9+, 9	10, 10, 10, 10
3	2	3–	9, 9, 9, 8	10, 10, 10, 10
4	3	3+	9+, 8+, 8, 8	10, 10, 10, 10
5	4+	4+	10, 10, 9+, 9	10, 10, 10, 10
6	4+	4+	10, 10, 10, 10	10, 10, 10, 10
C-1**	0	0	0, 0, 0, 0	10, 10, 9+, 8
C-2	0	0	9, 8+, 8, 6	10, 10, 10, 10
C-3	0	0	6, <5, <5, 0	10, 10, 10, 10
C-4	0	0	0, 0, 0, 0	10, 10, 10, 9+
C-5	0	0	0, 0, 0, 0	10, 9+, 9+, 9+
C-6	0	0	0, 0, 0, 0	10, 10, 10
C-7	0	0	7, 6, <5, <5	10, 10, 10, 9+

*0.7% Orvus-K detergent, 212°F, 15 min.
**Cleaned only, no treatment.

Example 2: This example illustrates the improved ink adhesion obtained by the use of polyhydroxy compounds in the coating solution, specifically, the use of various concentrations of sodium gluconate in a coating solution containing ammonium fluozirconate and/or fluotitanic acid, phosphoric acid, and hydrofluoric acid.

The coating solutions were applied by spraying for 15 seconds at a temperature of 90°F and the thus coated aluminum cans were then coated with a white base polyester ink [M61513 (Schlitz white), Acme Printing Ink Co.]. Prior to curing the ink, an alkyd-amine overvarnish was applied to the wet ink coating. Curing was effected for 6 minutes at 375°F.

In the table below all solutions were adjusted to a pH of 2.7 with HNO_3.

Solution Number	$(NH_4)_2ZrF_6$	H_2TiF_6	H_3PO_4	HF	Sodium Gluconate	Pasteurization Test 212°F, 15 min	Adhesion 1% Joy, 212°F, 15 min
			(g/ℓ)				
7	0.120	0	0.098	0.010	0	5, 5	0, 0
8	0.120	0	0.098	0.010	0.04	4+	8, 8
9	0.120	0	0.098	0.010	0.08	4+	8, 9+
10	0.120	0	0.098	0.010	0.195	4+, 4+	9, 9+

(continued)

Solution Number	$(NH_4)_2ZrF_6$	H_2TiF_6	H_3PO_4 (g/ℓ)	HF	Sodium Gluconate	Pasteurization Test 212°F, 15 min	Adhesion 1% Joy, 212°F, 15 min
11	0.120	0	0.098	0.010	0.40	4	9+, 9+
12	0.120	0.041	0.098	0.010	0	4, 4	5, 9
13	0.120	0.041	0.098	0.010	0.195	4, 4	9, 9+
14	0.120	0.082	0.098	0.010	0	4–, 3+	7, 9
15	0.120	0.082	0.098	0.010	0.195	3+, 3+	9+, 9+

Ammonium Fluozirconate and Gluconic Acid

T.L. Kelly; U.S. Patent 4,273,592; June 16, 1981; assigned to Amchem Products, Inc. provides an aqueous treatment or coating solution which contains as essential ingredients a zirconium and/or hafnium compound, a fluoride compound, and a polyhydroxy compound having no more than 7 carbon atoms. Such solution can be used to treat a bright shiny aluminum surface in a manner such that the bright shiny appearance of the surface is not changed, while forming on the surface a uniformly colorless and clear coating which is corrosion resistant and to which overlying coatings adhere excellently.

A particularly preferred coating solution has a pH within the range of about 3.4 to 4.0 and contains the following:

Ingredient	Approximate Concentration in Mols per Liter
Zr	0.50×10^{-3} to 1.75×10^{-3}
Polyhydroxy compound	0.30×10^{-3} to 1.75×10^{-3}
Available fluoride	0.50×10^{-3} to 2.50×10^{-3}

The preferred source of Zr in the above composition is ammonium fluozirconate, and the preferred polyhydroxy compound is gluconic acid. Preferably hydrofluoric acid is used as the source of available fluoride, and nitric acid is used to adjust the pH.

When utilizing hafnium, it is preferably used in an amount of from 0.5×10^{-3} to 1.75×10^{-3} mols/liter. The preferred source of hafnium is HfF_4. Other of the preferred ingredients and amounts thereof are described immediately above for the preferred Zr-containing solution.

Unless stated otherwise, the aluminum surfaces treated with the solutions identified in the examples were drawn and ironed aluminum cans which were first degreased, as necessary, in an acidic aqueous cleaner containing sulfuric acid, fluoride and detergents. Also, unless otherwise stated, the coating solutions were applied by spraying for about 20 seconds at the temperatures set forth below. After treatment with the solutions identified in the examples, the aluminum surfaces were rinsed in deionized water and dried in an oven for 3.5 minutes at about 400°F.

Thereafter, the aluminum cans were tested for corrosion resistance by subjecting them to a water stain resistance test simulating can exposure during commercial pasteurization processes. The test consisted of immersing the cans for a period of 30 minutes in a hot solution of distilled or deionized water containing 0.220 g/ℓ of sodium bicarbonate, 0.082 g/ℓ of sodium chloride, and 2.180 g/ℓ of a water conditioner (Dubois 915, a proprietary product, which exhibits a total alkalinity of 5.8% Na_2O and on analysis contains $NaNO_3$, carbonate, triethanolamine and dodecylphenyl polyethylene glycol). The solution was maintained at 150°±5°F during the test. After immersion, the cans were rinsed with tap water, dried with a paper towel and then examined for staining. A cleaned-only aluminum surface, when subjected to this test, turns black or brown after a few minutes.

The results of the tests were rated as follows: 5, perfect, no blackening; 3+, acceptable; and 0, total failure, severe blackening.

Aluminum cans treated with the solutions described in the examples were tested also for paint adhesion. After the treated surface was dried, as described above, a portion of the surface was painted with a waterborne white base coat (No. CE3179-2 white polyester, PPG Industries Inc.) and the other portion of the surface was painted with a waterborne overvarnish (Purair S145-121).

After the paint was cured, the painted surface was immersed in boiling water for 15 minutes. After removing the painted surface from the solution, it was crosshatched, using a sharp metal object to expose lines of aluminum which showed through the paint or lacquer, and tested for paint adhesion. This test included applying Scotch transparent tape No. 610 firmly over the crosshatched area and then drawing the tape back against itself with a rapid pulling motion such that the tape was pulled away from the crosshatched area. The results of the test were rated as follows: 10, perfect, when the tape did not peel any paint from the surface; 8, acceptable; and 0, total failure.

Example 1: This example shows the effect of gluconic acid concentration on water stain resistance, as well as on the adhesion of waterborne siccative coatings, at two different pH and temperature levels. Zirconium was present in each solution in the form of ammonium fluozirconate [$(NH_4)_2ZrF_6$] at a concentration of 1.25×10^{-3} mols/liter, and the pH of each solution was adjusted by the addition of concentrated nitric acid. Two cans were employed in determining the paint adhesion rating while the water stain resistance rating represents the average rating of six cans.

Sample No.	Gluconic Acid ($M \times 10^{-3}$)	pH	Temperature (°F)	Water Stain Resistance	. . . . . Adhesion.	
					CE3179-2	S145-121
1	0	3.5	125	1	10, 10	10, 7
2	0	3.5	135	$2^2/_3$	10, 10	10, 8
3	0	4.25	125	4	10, 10	10, 10
4	0	4.25	135	$2^2/_3$	10, 8	10, 8
5	0.5	3.5	125	5	10, 10	10, 10
6	0.5	3.5	135	5	10, 10	10, 10
7	0.5	4.25	125	$4^2/_3$	10, 10	10, 10
8	0.5	4.25	135	$4^2/_3$	10, 9	10, 10
9	1.25	3.5	125	$4^1/_6$	10, 10	10, 10
10	1.25	3.5	135	5	10, 10	10, 10
11	1.25	4.25	125	5	10, 10	10, 10
12	1.25	4.25	135	5	10, 10	10, 10
13	2.5	3.5	125	$1^1/_6$	10, 10	10, 10
14	2.5	3.5	135	$2^2/_3$	10, 8	10, 10

Example 2: In order to demonstrate that aluminum surfaces coated with a coating solution containing gluconic acid, zirconium and fluoride undergo the so-called "muffle test", while aluminum surfaces coated with a like coating solution free of gluconic acid do not, a number of aluminum cans were coated with solutions having the compositions shown below. The coated cans were then heated at a temperature of 900°F for 5 minutes and the color of the cans was observed. The solutions employed all had a pH of 4.25, obtained by the addition of concentrated nitric acid, and were applied at the temperatures shown.

Muffle Test Results of Coated Aluminum Surfaces

Sample No.	$(NH_4)_2ZrF_6$ ($M \times 10^{-3}$)	Gluconic Acid ($M \times 10^{-3}$)	Temperature (°F)	Surface Color After Heating at 900°F for 5 min
1	1.25	0	125	silver
2	1.25	0	135	silver
3	1.25	0.5	125	light golden brown
4	1.25	0.5	135	light golden brown
5	cleaned only	–	–	silver

Polyacrylic Acid, H_2ZrF_6, H_2TiF_6 or H_2SiF_6

D.Y. Dollman and T.J. O'Grady; U.S. Patent 4,191,596; March 4, 1980; assigned to Amchem Products, Inc. provide a composition for coating the surface of a metal selected from the group consisting of aluminum and alloys thereof in which aluminum is the principal ingredient. This comprises an aqueous acidic solution which consists essentially of:

(1) from about 0.5 to 10 g/ℓ of at least one polymer selected from the group consisting of polyacrylic acid and esters thereof; and

(2) from about 0.2 to 8 g/ℓ of at least one acid selected from the group consisting of H_2ZrF_6, H_2TiF_6 and H_2SiF_6.

The pH of the solution is less than about 3.5 (preferably less than 2.5).

The preferred composition is one that consists essentially of from about 1.5 to 6 g/ℓ of component (1) and from about 0.75 to 4 g/ℓ of component (2).

After application of the coating, the aluminum surface can be dried at a temperature of from about 60° to 500°F, thereby avoiding the production of process waste effluent.

Examples 1 through 4: Several coating solutions were prepared using polyacrylic acid (Acrysol A-1) and H_2ZrF_6 (Example 1), H_2TiF_6 (Examples 2 and 3), or H_2SiF_6 (Example 4), in the amounts specified in Table 2 below. The formulations of Comparisons B and C were prepared using only polyacrylic acid. In addition, a comparative formulation was prepared using polyacrylic acid and ammonium fluozirconate (Comparison D). All formulations were made up using tap water.

Panels of aluminum extrusion measuring 0.1" x 3" x 12" were cleaned with a mild alkaline phosphate cleaner. They were then spray-coated on both sides with the abovedescribed coating solutions. The panels were dried without rinsing and then heated in a warm air oven at 190°F for 5 minutes. After cooling to room temperature, the panels were spray-painted with metallic bronze paint. (Note that Comparison A is a control which was not spray-coated prior to painting). The painted panels were then baked in a hot air oven for 20 minutes at 300°F.

The painted panels were tested using AAMA (Architectural Aluminum Manufacturers Association) method and ASTM method given in Table 1 below.

In addition, the following adhesion tests were used.

Bend, 180° – This is a test for paint loss when a painted panel is bent back on itself to form a sharp crease. Visual observation is made of the extent of paint loss that results on the surface of the crease.

Pucker – A few drops of a paint stripper (Ensign Epoxy Stripper 803) are placed on a painted test panel. Visual observation is made of the time it takes for the paint to pucker.

Table 1: Adhesion Test Methods

Test	AAMA Method	ASTM Method
Detergent, 72 hr	6.6.3.1	
Salt spray, 1,000 hr	6.7.2.1	B-117-73
Cross hatch (film adhesion)		
Dry	6.4.1.1	
Wet	6.4.1.2	

The test results are given in the following Table 2.

Table 2

	Comparison				Example			
	A	B	C	D	1	2	3	4
Formulation, g/ℓ								
Polyacrylic acid*	–	4.11	2.05	2.67	4.11	4.11	2.05	4.11
$(NH_4)_2ZrF_6$	–	–	–	1.66	–	–	–	–
H_2ZrF_6**	–	–	–	–	1.23	–	–	–
H_2TiF_6	–	–	–	–	–	1.942	0.971	–
H_2SiF_6	–	–	–	–	–	–	–	0.85
Metal ion	–	–	–	0.730	0.540	0.567	0.284	0.166
Solution pH	–	2.8	3.02	~3.9	~2.1	~2.1	~2.2	~2.1
Adhesion Tests***								
Bend, 180°	100% loss	10% loss	20% loss	no loss	trace loss	no loss	no loss	trace loss
Detergent, 72 hr	failed slight loss	failed excessive loss	failed moderate loss	failed excessive loss	no loss	no loss	no loss	no loss
Salt spray, 1,000 hr	NR	pass–field blisters	pass–field blisters	NR	pass	pass	pass	pass–medium blisters

(continued)

Table 2: (continued)

	Comparison				Example			
	A	B	C	D	1	2	3	4
Pucker	paint loss after 21 sec	paint loss after 35 sec	paint loss after 35 sec	paint loss after 90 sec	no loss after 10 min	no loss after 10 min	no loss after 10 min	no loss after 10 min
Cross hatch wet	failed excessive loss	failed slight loss	failed moderate loss	NR	no loss	no loss	no loss	no loss
Cross hatch dry	no loss	no loss	no loss	NR	no loss	no loss	no loss	no loss

Note: NR denotes that the test in question was not performed, i.e., not run.

*Added as Acrysol A-1.

**Prepared by mixing 1 mol of ZrF_4 with 2 mols of HF.

***Conducted using metallic bronze paint.

The results as presented in Table 2 clearly show the improved results that are obtainable using the coating solutions of the process as compared to conventional coating solutions.

Organic Film-Forming Polymer and Soluble Titanium Compound

K. Muro, K. Yashiro, H. Kaneko and K. Yamazaki; U.S. Patent 4,136,073; January 23, 1979; assigned to Oxy Metal Industries Corporation disclose a composition and process useful for improving the corrosion resistance and/or paint adhesion of an aluminum surface. The composition is an aqueous acidic one containing an organic film-forming polymer and a titanium compound in a weight ratio of polymer:Ti of 100:1 to 1:10, preferably 20:1 to 1:1. The aluminum surface is contacted with the composition at a temperature not in excess of 70°C.

For most polymers, the preferred pH ranges from 1.2 to 5.5, more preferably from 1.5 to 4.0.

The polymer concentration is 0.1 to 60% by weight and the Ti concentration is 0.01 to 5% by weight.

The film-forming polymer is selected from the group of polymers and copolymers consisting of vinyl, acrylic, aminoalkyd, epoxy, urethane polyester, styrene, and olefin polymers and copolymers and natural and synthetic rubbers.

In order to promote the film-forming reactions, there may be used an oxidizing agent such as $NaNO_3$, HNO_3, $NaClO_3$, HNO_2, H_2O_2, hydroxylamine and the like which have been conventionally used in the phosphating, chromate conversion coating and similar processes.

The presence of fluoride in the bath is advantageous because fluoride promotes the chemical reaction with the aluminum surface and also because it aids in solubilizing the titanium present.

Example: A treating bath was prepared as follows. 5 g of titanium ammonium fluoride $(NH_4)_2TiF_6$ was dissolved in 100 ml of pure water, to which was added 25 g of an aqueous 40% solution (by weight) of commercially available acrylic ester emulsion (Toacryl N-142) and made up to 1 liter by diluting with pure water. The resulting solution was adjusted to a pH of about 2.8 by addition of 1.2 g of phosphoric acid.

The treating bath thus prepared was maintained at 30°C and a clean aluminum panel (2S) having a size of 0.6 x 70 x 150 mm and which had been cleaned by rinsing with a conventional alkaline degreasing agent was immersed therein for 3 minutes, followed by rinsing with water and drying.

When the surface-treated aluminum panel was subjected to the salt spray test according to JIS-Z-2371, no development of white rust was observed even after the test for 120 hours. The development of white rust after the test for 240 hours was less than 5% as shown in the table below. The aluminum panel was surface-treated in the same manner as mentioned above, followed by rinsing with water and dried. The surface-treated plate was then coated with an acrylic paint (KP 2406 Enamel Light Blue, Kansai Paints Co.) in a thickness from 18 to 20 μ and then baked at 290°C for 45 seconds to provide a pencil hardness from H to 2H. The table below shows the results of paint adhesion testing.

As controls, aluminum panels of the same grade were prepared and treated with one bath free from the acrylic emulsion (Control 1) and another bath free from titanium ammonium fluoride (Control 2) under the same conditions, followed by coating with the same paint and the same adhesion tests. The table also shows results of test specimens cleaned only (free from any treatments) (Control 3).

The test method used for adhesiveness of coating is the bend test which is performed as follows. The painted test panel was bent 180° around one (1T) or three (3T) panels the same panel thickness as the specimen. The painted surface was then subjected to the friction pull test at the bend with Scotch tape to measure the adhesion of paint thereto. Because of the smaller radius of curvature, the 1T bend is a much more severe test than the 3T bend.

Specimen	Corrosion Resistance 240 hr Salt Spray Percent White Rust	Paint Adhesion: As Painted 1T	3T	Immersion in Boiling Water 1T	3T	Moisture Proof Test for 1,000 hr 1T	3T
Example	<5	9.5	10	9.5	10	7.5	9.5
Control 1	100	0	7.5	0	7.5	0	0
Control 2	100	0	7.5	0	0	0	0
Control 3	100	0	0	0	0	0	0

The results were evaluated as follows: 10 = no change; 9.5 = peeling of less than 5%; 7.5 = peeling of less than 25%; and 0 = peeling of more than 25%.

Inositol Phosphate plus Titanium Fluoride

K. Yashiro, Y. Ayano and Y. Miyazaki; U.S. Patent 4,187,127; February 5, 1980; assigned to Nihon Parkerizing Co., Ltd., Japan provide a surface treatment of an aluminum or aluminum alloy surface to impart excellent surface characteristics which are similar to those of the chromate treatment without disadvantages of the chromate treatment. This is attained by treating a surface of an aluminum or aluminum alloy surface with a surface processing solution comprising 0.1 to 10 g/ℓ of one or more compounds selected from the group consisting of inositol di- to hexaphosphates and water-soluble salts thereof, such as alkali metal, alkaline earth metal and ammonium salts thereof and 0.1 to 10 g/ℓ as Ti of a titanium fluoride.

Suitable inositol di-~hexaphosphates include myoinositol phosphates such as myoinositol diphosphate, myoinositol triphosphate, myoinositol tetraphosphate, myoinositol pentaphosphate, and myoinositol hexaphosphate and other inositol phosphates.

Preferred salts include Na, K, Li, Mg, Ca, Sr or Ba salts of myoinositol phosphates.

Myoinositol hexaphosphate means phytic acid. Myoinositol di-~pentaphosphates are mainly prepared by a hydrolysis of phytic acid and accordingly, phytic acid is especially important.

Phytic acid occurs widely in grains (cereals) and is nontoxic because of natural substance.

Suitable titanium fluorides include K_2TiF_6, Na_2TiF_6, $(NH_4)_2TiF_6$ and TiF_4.

A concentration of the inositol phosphate is usually in a range of 0.1 to 10 g/ℓ and preferably 0.3 to 3 g/ℓ. A concentration of the titanium fluoride is usually 0.1 to 10 g/ℓ and preferably 0.3 to 3 g/ℓ as Ti.

A ratio of the inositol di-~hexaphosphate or a water-soluble alkali metal or alkaline earth metal salt thereof to the titanium fluoride as Ti is usually in a range of 1:10 to 10:1 by weight, preferably 1:2 to 4:1 by weight.

The surface processing solution is contacted with the surface of the aluminum or aluminum alloy substrate at 20° to 80°C for 10 seconds to 5 minutes to form a layer by a conventional method such as a spraying method and an immersing method, and then the surface is rinsed with a city water and with a deionized water and dried at 80° to 150°C for 1 to 10 minutes in an oven such as a hot air oven.

An amount of Ti component adhered on the surface of the aluminum or aluminum alloy substrate after drying is in a range of 1 to 85 mg/m^2.

Example: In a 15 liter stainless steel tank, 30 g of 50% aqueous solution of phytic acid was diluted with 5 liters of water and then 30 g of ammonium titanium fluoride was dissolved in the solution with stirring. After dissolving it, 28% ammonia water was added to the solution to prepare a surface processing solution having pH of 3.8.

A sample panel made of AC4C aluminum alloy having a size of 75 x 150 x 0.5 mm was treated by spraying a mild alkaline degreasing solution (10 g/ℓ) (Fine Cleaner No. 359, Nihon Parkerizing Co., Ltd.) at 60°C for 2 minutes and the degreasing solution remaining on the sample panel was removed by spraying with city water for 1 minute. The surface processing solution was heated to 40°C and sprayed under a pressure of 0.5 kg/cm^2 (gauge) for 1 minute to form a coating. The surface processing solution remaining on the surface was removed by spraying city water for 1 minute then rinsed by mist-spraying deionized water having an electroconductivity of 15 μV-cm^{-1} for 15 seconds. The surface was then dried at 120°C for 5 minutes in a hot air oven.

The amount of Ti component adhered on the treated surface of the sample was determined by a fluorescent x-ray analytical method to be 5.5 mg/m^2. An acryl type powder paint (Powdux A40 Clear) was coated to a thickness of 80±10 μ on the treated surface of the sample panel by electrostatic powder coating equipment (Gema 720) applying -70 kV, and it was baked at 180°C for 30 minutes in a hot air oven.

After the baking, a paint adhesion test, a hot water immersion test and a salt spray test were carried out on the coated sample panels. As a control, another sample panel was treated in the same manner as described in the example, but omitting the treatment with the surface processing solution and the following water rinse, and it was also tested. Results are shown in the table below.

In the table the following test methods were used:

(1) Paint adhesion test: The paint film of the sample panel was crosshatched with a knife edge with each gap of 2 mm and the number of nonpeeled crosses/100 crosses was counted.

(2) Hot water immersion test: The coated sample panel was immersed in hot water at 40°C for 240 hours. The paint adhesion test (1) was then carried out.

(3) Salt spray test: In accordance with Japanese Industrial Standard Z-2371, the salt spray test, sample panels with crosscuts in diagonal were exposed to salt spray for 240 hours, and a degree of blister development measured (both sides, mm).

	Example	Control
Paint adhesion test, crosses/100 crosses	100	88
Hot water immersion test	100	60
Salt spray test, mm	<0.5	5

ENHANCING PAINT ADHESION

Use of Epoxy Functional Silanol Adhesion Promoter in Paint

A method is disclosed by *A.F. Hofstatter; U.S. Patent 4,208,223; June 17, 1980; assigned to Superior Industries* for the painting of aluminum surfaces that provides a highly superior coating than prior methods. The method comprises a controlled oxidation of the aluminum surface, the application to the oxidized surface of an epoxy functional silanol, and coating of the aluminum surface with paint by conventional methods. These silanols can be applied to the oxidized aluminum surface by dissolving them, or their corresponding trialkoxysilanes, in an aqueous solution which is applied to the surfaces and permitted to dry or can be incorporated in the coating composition.

Referring to Figure 4.1, the method of treating an aluminum surface, such as an aluminum wheel, is described. The wheels, after casting, are conditioned for the application of a paint which contains a preselected color pigment. Prior to treatment, the wheels have received the cleaning, heat treatment and other customary treatments in preparation of a finished product. The wheels are then cleaned by treatment with an alkaline etching cleaner in a first treatment step **30**. This treatment is performed by immersing or spraying the wheels with an aqueous solution of from 2 to about 10, preferably about 5 volume percent of an alkali metal phosphate, typically trisodium orthophosphate having a pH of about 13.5. Suitable commercial products are available for this purpose such as Oakite STC. Aluminum is etched by this material at a rate of about 0.5 in/yr. The wheels are maintained in the treatment bath for a limited period of time, typically from 2 to about 5, preferably about 3.5 minutes, and the treatment cleans the wheels and etches the surfaces.

Figure 4.1: Block Diagram of Treatment Steps

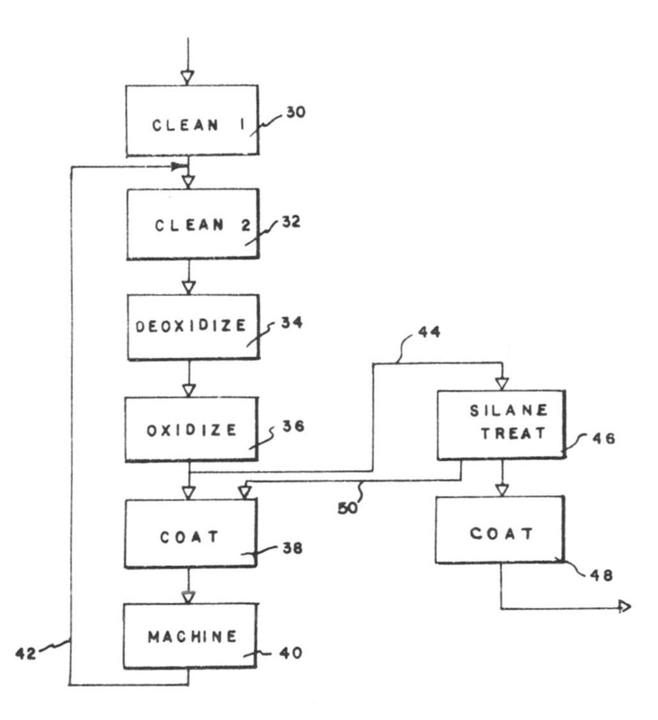

Source: U.S. Patent 4,208,223

The wheels are rinsed by immersing or spraying in a rinse tank and thereafter sprayed with rinse water. The wheels are passed to treatment step **32** where they are cleaned in an aqueous solution of a nonetching, nonsilicate cleaner. A suitable cleaner for this purpose is Oakite aluminum cleaner 166, which is a nonetching alkaline cleaner comprising a mixture of alkali metal phosphates and borates with a surface active agent and a corrosion inhibitor to prevent etching. Typically, the cleaner has a concentration of from 2 to about 10% by weight. The wheels are cleaned by immersing or spraying with the aqueous solution of this cleaner for a short period of time, from 1 to about 5, preferably for about 2.5 minutes. Following this treatment, the wheels are rinsed by immersing or spraying in a tank of rinse water and thereafter spray rinsed.

The cleaned wheels are then passed to treatment step **34** for a controlled deoxidation and removal of aluminum oxide and any discoloration. This treatment comprises immersing or spraying the wheels for a period of time, from 1 to about 5, preferably about 2.5 minutes in an aqueous solution of an inorganic reducing agent such as an aqueous acidic solution of ferrous sulfate. A material which has been found suitable for this deoxidation treatment is Oakite deoxidizer LNC which is used at a concentration from 10 to 20 volume percent and has a pH value from about 0.8 to 1.0%. The wheels removed from the deoxidation treatment tank are then rinsed with a rinse water spray and are passed to the controlled oxidation treatment in step **36**.

A suitable oxidation step comprises the contacting of the cleaned and freshly deoxidized aluminum surfaces with an aqueous solution of chromic acid having a pH no greater than about 4.5, typically from 2 to about 4.0. A commercially available material useful for this treatment is Alodine 1000. The solution can be buffered by the presence of an alkali metal, e.g., sodium or potassium. The chromic acid is used in an aqueous solution having a concentration from 0.2 to about 2% by weight, preferably about 0.6% by weight.

The treatment is conducted at ambient temperatures, e.g., from about 60° to 100°F, preferably at about 80°F, and is performed by immersing or spraying the aluminum object such as the wheel with a bath of the abovedescribed aqueous chromate solution. The aluminum object is maintained in the bath for a period of from 1 to about 5, preferably 1 to 3 minutes and is thereafter removed, and rinsed by immersion or spray with deionized water. The object then proceeds through one of two alternate courses. If it goes directly to step **38**, coat, it is permitted to drain, and dried in an oven maintained in a flowing air stream at a temperature of from 200° to 250°F, preferably 200°F. This treatment provides a controlled and uniform oxide surface on the aluminum object that insures maximum adhesion of the coating composition.

The aluminum wheels which have been oxidized and provided with a controlled oxide coating in accordance with the process are thereafter subjected to treatment with the silanol adhesion enhancer in step **38**. The silanol adhesion improver used has the following formula:

$$\underbrace{CH_2CH}_{O}(CH_2)_n{-}O{-}(CH_2)_mSi(OH)_3$$

where n is a whole number integer of 1 through 3 and m is a whole number integer of 1 through 5.

The silane can be added directly to the paint composition. A very suitable silane for this purpose is A-187 (Union Carbide Corporation) which is γ-glycidoxypropyltrimethoxysilane. The material can be used at a concentration of from 1 to about 10, preferably from about 3 to 7% by weight in a coating composition.

The silane can be incorporated in a variety of coating compositions. The coating composition can be a solvent base composition, typically a polyurethane base comprising a mixture of color base, polyurethane resin and solvent or thinner. A catalyst is added immediately prior to use of the composition. Suitable paint compositions for this purpose are available as Durathane DMU. Another suitable coating composition which can be employed in combination with the silane for the combined surface treatment and coating step can be a water base composition such as a water-soluble emulsion of a melamine-acrylic-styrene polymer. Typically, the solvent employed for this composition is approximately 20 volume percent organic solvent and 80 volume percent water. The coating composition comprises a mixture of the aforementioned solvent, color base and resin base. A suitable water base coating composition is WPB-8805-3 (Spraylat).

The other alternate course is shown as path **44** to step **46**. After the oxidize, step **36**, the object is rinsed in deionized water then treated with the silane, step **46**, then on to path **50** to coat step **38**. In this alternate course, no silane is added to the coating, but the silane is introduced directly to the surface in an aqueous solution.

The silanol can be obtained and used directly as an aqueous solution or, alternatively, the silanol can be obtained by dissolving a trialkoxysilane having alkoxy groups of from 1 to about 4 carbons in water since these silanes readily hydrolyze in aqueous solution to form the silanol active adhesion promoter.

The silanol obtained from the silane is contacted with the oxidized aluminum surfaces separately or jointly with the contacting of the surfaces by the coating composition. When the silanol is separately contacted with the surfaces, it is preferably employed as a spray of an aqueous solution containing from 0.05 to about 3% by weight, preferably from 0.1 to about 0.5% by weight of the aforementioned silanol.

Priming Layer of Aluminum Hydroxyoxide

It is known that metals such as aluminum and alloys therewith contain an oxide surface thereon, resulting from the metal oxidizing in normal atmospheric environment. This oxide has a very low adhesion itself for organic coatings, and also effectively prevents the buildup of a boehmite or hydroxyoxide coating on the metal surface. For this reason, such barrier oxides must be removed from the metallic surface so as to allow the proper boehmite formation.

K.L. Craighead and V. Mikelsons; U.S. Patent 4,149,912; April 17, 1979; assigned to Minnesota Mining and Manufacturing Company describe a process for providing a priming layer of aluminum hydroxyoxide on aluminum and aluminum alloy surfaces. This comprises treating the surface with a chemical reagent having a solvent action for the naturally-occurring oxides which form on such surfaces until the oxides are dissolved, and contacting the surfaces with steam while a film of the reagent is still present on the metal surface. The priming layer provides for excellent adhesion of subsequent organic coatings to the metal surface.

Chemical reagents which are suitable for removal of the oxide coating include those compounds which in essence are capable of dissolving aluminum oxides. Solutions of strong bases such as sodium hydroxide or potassium hydroxide are particularly effective because of their capability to rapidly dissolve the naturally-occurring oxide film on the metal surface.

With the barrier oxide layer removed, the bare aluminum metal is now susceptible to hydrolysis by the action of steam, thereby forming a hydroxyoxide film on the aluminum surface.

One critical and unique feature of the process is that the aluminum surface must remain covered with a thin layer of the chemical reagent when exposure to the boehmite-forming steam begins. If the chemical reagent utilized to dissolve the barrier oxide layer is removed from the metal surface, such as by rinsing, prior to exposure of the surface to steam, it has been ascertained that an insufficient film of the hydroxyoxide will be formed in an allotted reaction time, i.e., about 10 seconds of exposure to steam, which will therefore result in inadequate adhesion of the organic coating thereto, and high speed processing is therefore not applicable.

One of the distinct advantages of the process, in addition to the reduced time consumption, is that the chemical reagents utilized and reaction products formed are essentially nonpolluting.

To minimize the hydroxyoxide-forming treatment time, the concentration level of the reagent should be optimized. If the reagent concentration is too low, the barrier oxide layer may not be completely removed, and rapid formation of the boehmite layer will be prevented, although elevated temperatures, i.e., above about 50°C, will assist the reagent action. If the reagent concentration is too high, etching of the metal surface may continue during exposure thereof to steam, again preventing effective hydrolysis. Reagent concentrations of from about 0.3 to 5.0% by weight have been found acceptable as a general range, depending on the particular reagent chosen and the treatment temperature utilized.

In practice, because of the diversity of aluminum oxide solvents available, the surface is exposed to an aluminum oxide solvent-containing solution for a time, at a concentration, and at a temperature sufficient to allow for substantially complete removal of the barrier oxide layer.

Example: A 14 x 15 cm panel of 0.024 cm thick Alloy 5352 (containing approximately 2.5% by weight magnesium as the major alloying additive) was cleaned by dipping for 1 minute in a solution consisting of 95 g sodium borate, 95 g sodium pyrophosphate, 6.0 g Maprofix NEU, a sodium lauryl sulfate surfactant, 2.0 g Ultrawet DS, a sodium linear alkylate sulfonate surfactant, 2.0 g Igepal CO 730, a nonylphenoxypoly(ethyleneoxy)ethanol surfactant, and 4,000 ml deionized water at 70°C.

After cleaning, the panel was rinsed by spraying with deionized water at room temperature for 10 seconds and was then dipped for 5 seconds in a 2.5% solution of NaOH in water at 50°C. The excess liquid was drained from the aluminum surface for 5 seconds and the panel immersed in a chamber containing steam at 92°C. After 10 seconds of exposure to steam the panel was withdrawn, rinsed with deionized water and dried with a jet of warm air.

The effectiveness of this treatment for improving the adhesion of organic coatings was tested by the following procedure, which is useful for nonbrittle coatings. The primed surface was coated with a vinyl-epoxy lacquer, DAV 210 (Mobil Chemical Co.), using a No. 30 Meyer bar. After drying in air at room temperature, the coated panel was cured for 105 seconds at 210°C, cooled and cut into 5 x 15 cm strips. The thickness of the cured lacquer layer was about 1.8 x 10^{-3} cm. To test the adhesion of the coating, each strip was bent to a sharp crease, and 3M No. 610 high tack tape was applied over the bend area and peeled back at a 180° angle. The width of the zone from which the lacquer was removed by the tape was then measured starting from the bend area. The adhesion of the lacquer has an acceptable level if the failure zone is less than 0.08 cm wide. In this case, the zone was 0.04 cm wide, i.e., the lacquer adhesion was good.

MISCELLANEOUS TREATMENTS

Protection Against Water Staining

R.J. Sturwold, W.E. Utz, N. Christ, W.R. Ford, Jr. and G.P. Koch; U.S. Patent 4,153,464; May 8, 1979; assigned to Emery Industries, Inc. and Reynolds Metals Company found that a blend or reaction product of a polybasic acid and polyol partial ester can be applied to aluminum and aluminum alloys to protect against water staining. The protective compositions can be used in neat form or combined with an inert diluent and applied as a protective oil.

The protective compositions are a blend or the complete or partial reaction product of (1) aliphatic or cycloaliphatic polybasic acids containing 18 or more carbon atoms and 2 to 4 carboxyl groups per molecule, and (2) polyol partial esters derived from aliphatic or ether polyols containing from 3 to 30 carbon atoms and 3 to 22 hydroxyl groups and a C_{12-22} aliphatic monocarboxylic acid. Application of a continuous film of the compositions to the surface of aluminum products provides effective protection against water staining even under severe conditions.

Useful polybasic acids for preparation of the protective compositions contain usually, 21 to 54 carbon atoms. Especially useful polybasic acids are the commonly called dimer acids obtained from the polymerization of unsaturated C_{16-26} monocarboxylic acids. Superior results are obtained when C_{36} dimer acids containing less than 25% by weight trimer and higher polymer acids are employed. Polyol partial esters of aliphatic polyols and ether polyols containing 3 to 12 carbon atoms and having 3 to 10 hydroxyl groups esterified with a C_{14-20} fatty acid or mixture are also preferably used.

The weight ratio of polybasic acid to polyol partial ester will range between about 20:1 and 1:4 and the protective compositions will have hydroxyl values greater than about 15, and more preferably from about 25 to 200. If the components are blended the weight ratio preferably ranges from 5:1 to 1:2 whereas if the components are reacted the preferred range is from 3:1 to 1:3. At least about 0.5 mg of the protective composition is required per square foot of metal surface to obtain effective water stain protection. The protective compositions exhibit excellent solubility in aliphatic and aromatic hydrocarbons and protective oils containing as little as 0.1% by weight of the protective composition can be advantageously utilized. Protective oils prepared from aliphatic and aromatic hydrocarbons having 100°F viscosities of 20 to 100 SUS and which contain 0.5 to 10% by weight of the protective composition are particularly useful for application to the metal surface by spraying.

Blends or partial reaction products of C_{36} dimer acid containing less than 25% trimer and higher polymer acids and polyol partial esters derived from glycerol, trimethylolpropane, pentaerythritol, di-, tri- and tetraglycerol and di-, tri- and tetrapentaerythritol and C_{18} fatty acids and fatty acid mixtures form an especially preferred variation of the process.

Example: A protective composition was prepared by blending 75 parts C_{36} dibasic acid (Empol 1018 dimer acid containing about 15% trimer acid) and 25 parts glycerol monooleate. The resulting blend had an acid value of 136, hydroxyl value of 49.5 and 100°F and 210°F viscosities of 791 and 43.5 cs, respectively. 1%, 2%, 4% and 6% solutions (identified as A, B, C and D, respectively) of the protective composition were prepared by diluting with mineral seal oil (100°F viscosity 38.6 SUS; boiling range 520° to 610°F).

6" x 3" aluminum panels were coated with each of the protective oils. Prior to application of the oil the panels were cleaned with trichlorotrifluoroethane to remove any residual rolling oil. Two methods of application were used to coat the panels. In the first method, hereinafter called the spray method, panels were sprayed with the solution and allowed to stand in a vertical position overnight. This method applies a relatively heavy film of the protective oil to the test panels.

In the second method, hereinafter identified as the weighed film method, several drops of the oil were placed on one side of each previously weighed panel and spread uniformly over the surface with a lint-free tissue. Each panel was reweighed and the film weight adjusted, if necessary, until 5 to 10 mg of the oil was present on the sheet. The panels were then allowed to stand at ambient temperature and humidity for at least 24 hours to allow the mineral seal oil to evaporate, thus leaving only a thin film of the protective composition on the surface of the aluminum, and reweighed.

To demonstrate the ability of the protective composition to prevent water staining the treated aluminum sheets were subjected to a steam test. For this test 3" x 1.5" coupons were cut from the panels and suspended at a distance from 0.5" to 0.75" in front of the sidearm of a 500 ml filtration flask containing vigorously boiling water. After 5 minutes exposure the coupon was removed and visually examined. Water stained coupons were then rated from 1 to 5 depending on the diameter of the stain as follows: 1 = <⅛"; 2 = ⅛" to ¼"; 3 = >¼" to ½"; 4 = >½" to ¾"; and 5 = >¾".

Results obtained with duplicate samples tested using the abovedescribed procedure are listed below. NS indicates no visible stain.

	Water Stain Rating	
Protective Oil	Spray Method	Weighed Film Method
A	NS	NS
	NS	1
B	NS	NS
	NS	NS
C	NS	NS
	NS	NS
D	NS	NS
	NS	NS

The above results demonstrate that even with very low concentrations of the protective composition it is possible to prevent water staining on the surface of the aluminum even under severe conditions. An untreated control panel as well as a panel treated with the straight mineral seal oil had a water stain rating of 5.

Improved Weathering

The method of *M. Hasegawa, S. Segawa, Y. Koise, H. Okada, S. Gomyo and Y. Kudo; U.S. Patent 4,115,607; September 19, 1978; assigned to Yoshida Kogyo KK, Toa Paint Co., Ltd. and Shin-Etsu Chemical Co., Ltd., Japan* comprises coating the newly extruded aluminum material while its surface is maintained at temperatures between about 550° and 250°C following extrusion with an organopolysiloxane-based material having a limited composition as set forth hereinafter. The thus coated aluminum material may be subjected to annealing at a temperature between about 170° and 200°C for about 2 to 6 hours in order to further improve the mechanical strengths of the coatings.

The mechanism by which the effects of the method are obtained may be as follows. The surface of a newly extruded aluminum material remaining at a temperature not lower than 250°C is so reactive that a strong bond can be formed between the surface and the coating composition. In this case, the coating composition must be an organopolysiloxane composition, and any substitutes which are made from organic resins are not suitable for the purpose. This is because those organic resins have a relatively poor thermal stability and are not capable of forming a strong bond with the surface of the newly extruded aluminum material kept at 250°C or above due to thermal decomposition and denaturation with coloration.

The organopolysiloxane coating composition once applied on the surfaces of aluminum material rapidly cures and hardens, so that the subsequent handling of the coated aluminum material is quite easy.

Examples 1 through 5: The organopolysiloxane (A) and the diorganopolysiloxane (B) used in the examples were prepared by the following procedures.

Organopolysiloxane (A): A mixture of 5.5 mols of phenyltrichlorosilane, 2.0 mols of methylvinyldichlorosilane and 2.5 mols of dimethyldichlorosilane was subjected to cohydrolysis. The resultant cohydrolyzate in a 40% by weight solution in toluene was polymerized by condensation in the presence of a small amount of potassium hydroxide as the alkali catalyst, followed by removal of the toluene and other volatile matter by distillation, to finally produce the organopolysiloxane having the desired resin structure.

Diorganopolysiloxane (B): A mixture of 3.8 mols of dimethyldichlorosilane, 3.2 mols of diphenyldichlorosilane, 2 mols of methylvinyldichlorosilane and 1 mol of trimethylchlorosilane was subjected to cohydrolysis and subsequently to condensation in the presence of an alkali catalyst, followed by removal of volatile matter, to finally produce the desired diorganopolysiloxane, having both chain ends terminated with trimethylsilyl groups and a viscosity of 480 cs at 25°C.

The various test values or results were determined in accordance with the following. Thickness, adhesion and pencil hardness of coating film are each determined in accordance with Japanese Industrial Standard (JIS) A 4706. Impact strength was determined by the Du Pont impact tester to be expressed in cm, the falling distance, with 12.7 mm x 500 g. Contact test with acid solution or with alkali solution was conducted by keeping the coated surface in contact with a 5% sulfuric acid solution or with a 1% aqueous solution of sodium hydroxide, respectively, for 48 hours and examined in accordance with JIS A 4706, to determine whether each test piece should pass the test.

Spraying with saline solution was evaluated by the manner such that an aqueous salt solution containing 5% sodium chloride and 0.026% copper(II) chloride was sprayed on the coated surface for 48 hours, and the state of the resulting surface was observed and recorded by ratings in cardinal in accordance with JIS H 8601.

Accelerated weathering was determined by subjecting the coated surface to weathering in accordance with JIS A 4706 for 250 hours to be recorded with respect of discoloration, adhesion and gloss retention of the coatings.

Appearance of the coated surface was visually examined immediately following the annealing step.

A bullet of aluminum-based alloy 6063 (AA specification) was continuously extruded through a die attached to an extruder machine at a velocity of 30 m/min to form a continuous bar having the same cross-sectional shape as the die opening. The tip of the extruding bar was held and pulled by a puller so that it advanced afloat in the air through a coating machine. In the coating machine the extruding bar was spray-coated while its surface temperature remained at about 420°C, using an airless sprayer or hot-melt applicator Model VII (Nordson Co.). The bar having been coated and held by the puller was put to compulsory cooling by a cooler to a temperature below 100°C and then it was cut by a cutter to form a piece having a length approximate to what was desired exactly.

The thus obtained piece of the coated bar was then taken off the coating and cooling line and then subjected to stretching for stress compensation by means of a pair of stretchers. The resulting piece was cut by another cutter in the exact desired length, and then kept in an annealing oven at 180°C for 5 hours to effect the annealing of the metal component and the curing of the coating composition to produce the finished product which was useful as a window sash.

In the above spray-coating procedure, the coating composition was heated at 180° to 200°C and sprayed all over the surfaces of the bar using four spraying rate-adjustable nozzles positioned around at a distance of about 25 cm from the bar.

The coating composition used in each of Examples 1 through 5 was a mixture of organopolysiloxane (A) and diorganopolysiloxane (B) in a ratio indicated in the table below.

Tests were conducted to determine the various properties of the coatings formed on the finished product and coated surfaces, and the results are also shown in the table below.

	Example Number				
	1*	2	3	4	5
Organopolysiloxane, % by wt	10	20	30	40	50
Diorganopolysiloxane, % by wt	90	80	70	60	50
Thickness of coating film, μm	15	16	15	15	16
Adhesion of coating film	100/100	100/100	100/100	100/100	91/100
Pencil hardness	H	3H	3H	3H	4H
Impact strength, minimum, cm	40	50	50	50	40
Contact test with acid solution	Passing	Passing	Passing	Passing	Passing
Contact test with alkali solution	Passing	Passing	Passing	Passing	Passing
Spraying with saline solution	10	10	10	10	10
Accelerated weathering					
Discoloration	None	None	None	None	None
Adhesion	100/100	100/100	100/100	100/100	100/100
Gloss retention, %	>95	>95	>95	>95	>95
Appearance of coated surface	Smooth	Smooth	Smooth	Smooth	Smooth

*Control.

Providing Corrosion Resistance

According to *R.G. Newell and D.C. Perry; U.S. Patent 4,202,706; May 13, 1980; assigned to Minnesota Mining and Manufacturing Company* the surfaces of articles made of aluminum, such as anodized aluminum and bare aluminum, are treated with N-alkylfluoroaliphaticsulfonamidophosphonic acids or salts thereof to impart corrosion resistance to their surfaces. Preferably the compound is selected from the group consisting of the following:

diethyl 3-(N-butyltrifluoromethanesulfonamido)-propanephosphonate;

3-(N-butyltrifluoromethylsulfonamido)-propanephosphonic acid;

diethyl 3-(N-ethylperfluorobutanesulfonamido)-propanephosphonate;

3-(N-ethylperfluorobutanesulfonamido)-propanephosphonic acid;

4-(N-ethylperfluorooctanesulfonamido)-propanephosphonic acid;

4-(N-methylperfluorooctanesulfonamido)-butanephosphonic acid;

3-(N-ethyltrifluoromethanesulfonamido)-propanephosphonic acid; and

3-(N-butylperfluorooctanesulfonamido)-propanephosphonic acid.

Example 1: A mixture of N-butyltrifluoromethanesulfonamide (62.0 g, 0.302 mol), potassium carbonate (45.54 g, 0.330 mol) and methanol (250 ml) was refluxed with stirring for 2 hours. Allyl bromide (39.93 g, 0.330 mol) was then added and this mixture stirred under reflux for 24 hours, cooled, filtered, and the solvent evaporated in vacuo. The residue was fractionally distilled yielding 37.0 g (boiling point of 53° to 58°C at 0.07 torr) of the desired compound, N-allyl-N-butyltrifluoromethylsulfonamide.

Over a 60-minute period, a solution of the above prepared N-allylsulfonamide (24.60 g, 0.10 mol) and diethylphosphite (15 g) was added dropwise concurrently with a solution of di-tert-butyl peroxide (0.90 g) in diethylphosphite (5 g) to diethylphosphite (80 g) being stirred at 150°C with an argon purge. The resulting mixture was stirred for an additional 1 hour under these conditions. The excess diethylphosphite was distilled at 10 to 20 torrs and the residue fractionally distilled to give 19.64 g (boiling point of 150° to 153°C at 0.20 torr) of the desired compound, diethyl 3-(N-butyltrifluoromethanesulfonamido)-

propanephosphonate. A mixture of the phosphonate prepared above (15.36 g, 0.040 mol) and bromotrimethylsilane (13.46 g, 0.088 mol) was stirred under a reflux condenser and calcium chloride drying tube for 96 hours. Then water (75 ml) was added and the resulting mixture stirred for 15 minutes followed by extraction with three 50 ml portions of diethyl ether. The ether portions were combined, dried with $MgSO_4$ and evaporated in vacuo to give 12.25 g of the desired compound, 3-(N-butyltrifluoromethanesulfonamido)-propanephosphonic acid.

Example 2: The procedure of Example 1 was repeated with the exception that N-ethylperfluorobutanesulfonamide was used in place of N-butyltrifluoromethanesulfonamide to make N-allyl-N-ethylperfluorobutanesulfonamide (boiling point of 50° to 55°C at 0.12 to 0.10 torr). The latter was reacted with diethylphosphite to produce the desired compound, diethyl 3-(N-ethylperfluorobutanesulfonamido)-propanephosphonate (boiling point of 148° to 152°C at 0.4 to 0.5 torr). A solution of the phosphonate prepared (30.0 g, 0.062 mol) and bromotrimethylsilane (20.0 g, 0.130 mol) was stirred under a reflux condenser and calcium chloride drying tube for 24 hours, water (75 ml) was added, and then the mixture stirred for 15 minutes. The resulting white precipitate was filtered and washed with water (50 ml) and diethyl ether (100 ml), then dried in an oven at 100°C for 24 hours at 10 to 20 torrs, to yield the desired compound, 3-(N-ethylperfluorobutanesulfonamido)-propanephosphonic acid (melting point, 138.5° to 140°C).

Example 3: Following the procedure of Example 1, N-allyl-N-ethyltrifluoromethanesulfonamide (boiling point of 395°C at 1.50 to 0.35 torrs) was reacted with diethylphosphite to produce diethyl 3-(N-ethyltrifluoromethanesulfonamido)-propanephosphonate (boiling point of 147° to 155°C at 0.25 torr).

Following the procedure of Example 2, this phosphonate was hydrolyzed to produce the corresponding phosphonic acid (melting point 137.5° to 139.5°C).

Example 4: A freshly cleaned bare aluminum panel (Al 7072, typical of that used in the fabrication of aluminum radiators for automobiles) was immersed for about 1 hour at 25°C in a bath of 50 vol % aqueous ethylene glycol solution containing 0.1 wt % $C_8F_{17}SO_2N(C_2H_5)(CH_2)_3P(O)(OH)_2$ dissolved therein. The bath contained "corrosive" water, viz., one hundred times the chloride, sulfate, and bicarbonate concentration specified in ASTM D 1384.

During the immersion, the corrosion rate was measured by the linear polarization method, an electrochemical technique well-known in the art of metal protection. As explained in detail in *Adv. Corr. Sci. Technol.* 6, 163 (1976), one can use this method to measure the current generated during a corrosion process. The magnitude of that corrosion current, denoted by i_{corr} and expressed as a current density, in amperes per unit area, is directly proportional to the corrosion rate. Thus the action of an effective corrosion inhibitor is reflected by a reduced value for i_{corr} in its presence. The polarization was run under nitrogen at 25°C.

For purposes of comparison, another bare aluminum panel (Al 7072) was similarly immersed in an aqueous ethylene glycol bath which did not contain any corrosion inhibitor, and the corrosion rate was likewise determined by polarization. Results are set forth in the table below.

Run	Corrosion Inhibitor	i_{corr}, $\mu A/cm^2$
1	None	3.6
2	$C_8F_{17}SO_2N(C_2H_5)(CH_2)_3P(O)(OH)_2$	0.3

Water-Resistant Bond Between Aluminum and Polysulfide

There has been a need in the industry for improving the adherence of polysulfide-based elastic sealing materials to aluminum. This adherence usually tends to weaken upon exposure to water and heat and, therefore, there has been a particular need to improve the adhesiveness of such materials in a moist and warm environment.

T. Borresen and N.U. Harder; U.S. Patent 4,138,526; February 6, 1979 found that a strong and water-resistant bond between aluminum and a polysulfide material can be achieved by applying to the aluminum metal a solution of a strongly basic-reacting inorganic alkali metal compound.

As the inorganic alkali metal compound or base of the process, there is preferably used one or more of the strongly basic reacting hydroxides of lithium, sodium or potassium. Also, sodium phosphate, potassium phosphate, sodium silicate and potassium silicate may be employed. Preferred inorganic bases also are carbonates of alkali metals, more particularly, lithium, sodium and potassium.

As a solvent for the inorganic base, there can be used, in principle, any liquid which will dissolve to a sufficient degree the inorganic compound to be included in the primer. It is, however, an advantage of the process that water can be used as a solvent. Also, lower alcohols can be used as a solvent for the inorganic base or, if desired, mixtures of such alcohols with water.

The concentration of base contained in the primer can be quite low and is desirably less than 5% by weight, preferably about 1% by weight.

The primer has proved to be particularly effective in connection with a polysulfide jointing compound based on a polymercaptan polymer.

Example: Aluminum extrusions which had been electrolytically oxidized were primed at room temperature by immersion in a solution as indicated below (% by weight). The last three tests were included for comparison with Tests 1 through 3 which were carried out according to the process.

	Test Number					
	1	2	3	4	5	6
Water	98	99	–	98	–	–
Ethylene glycol	–	–	98	–	98	–
Na_2CO_3	2	–	–	–	–	–
NaOH	–	1	2	–	–	–
$Ca(OH)_2$	–	–	–	–	2	–
H_2SO_4	–	–	–	2	–	–
No priming	–	–	–	–	–	x

After drying for one-half hour (in air at 28°C), the aluminum extrusions were joined to glass by means of a polysulfide jointing compound, PRC 408 P, intended for the production of insulating windows. The test samples were then left to cure for one week at room temperature, after which they were stored in water at 70°C for 4 weeks. After this treatment the adherence between the jointing compound and the metal was tested, the results of which follow.

In Water at 70°C, wk	Test Number 1	2	3	4	5	6
1	K	K	K	A	K	K
2	K	K	K	A	5A	K
3	K	K	K	A	50A	10A
4	10A	K	K	A	A	A

Note: A is rupture by adhesion failure to aluminum; K is rupture by cohesion failure of the jointing compound; and 50A is 50% rupture by adhesion failure to aluminum.

Reducing Transfer Etch

Aluminum polishing compositions of the type which comprise a mixture of phosphoric and nitric acids and which additionally contain sulfuric acid are known.

Typically the essential ingredients are phosphoric and nitric acids, but because of the high cost of phosphoric acid it has often been found commercially advantageous to substitute cheaper sulfuric acid for a part of the phosphoric acid.

It would be economically advantageous to increase the proportion of sulfuric acid, but a particular problem, common to phosphoric/sulfuric/nitric acid polishing compositions has prevented the commercial introduction of any composition containing more than about 24% by weight of sulfuric acid, i.e., one part by weight of commercial concentrated (SG 1.84) sulfuric acid to three parts concentrated (SG 1.75) phosphoric acid. This problem is called "transfer etch".

Transfer etch occurs when the polished work is removed from the polishing bath and drained preparatory to being transferred to the next treatment stage (usually a rinsing stage). If the work is allowed to drain for too long, an unsightly, white, etched effect mars the surface of the work. In baths containing a high proportion of phosphoric acid the onset of transfer etch is generally sufficiently slow for it to be practical to transfer the work before significant etching can occur. However, if the proportion of sulfuric acid is increased, the onset of transfer etch becomes more rapid, shortening the permissible time available for transferring the work until eventually it is impossible in practice to polish the work without a quite unacceptable degree of etching.

T.R. Rooney; U.S. Patents 4,251,384; February 17, 1981; and 4,116,699; September 26, 1978; both assigned to Albright & Wilson Ltd., England found that certain aromatic organic compounds have a beneficial effect in reducing the occurrence of transfer etch in aluminum polishing solutions. The presence of such etch inhibitors therefore permits the proportion of sulfuric acid in an aluminum polishing solution to be substantially increased.

The process provides an aluminum polishing solution comprising phosphoric acid, nitric acid, sulfuric acid and dissolved copper, which additionally comprises as an etch inhibitor, an organic compound comprising an aromatic ring having at least two heteroatoms conjugated therewith. The aromatic ring is preferably a benzene ring but may alternatively be a naphthalene ring or a pyridine, pyrazine or other heteroaromatic ring. The heteroatoms are preferably nitrogen, oxygen or sulfur atoms having electron pairs conjugated with the aromatic ring.

Effective etch inhibitors are readily identified by the presence of an aromatic ring system (usually, but not essentially, a six carbon ring) which is stable in the highly acidic medium, and at least two heteroatoms conjugated or conjugable with the ring. An aromatic system is essential for stability in the aggressive polishing solution. Compounds lacking an aromatic ring system, such as thiazole, thiadiazole, dimercaptothiadiazole or triazole are ineffective, probably due to instability in the medium. At least two heteroatoms, preferably nitrogen, oxygen or sulfur, especially nitrogen, stabilized by conjugation with the ring, are necessary, probably to provide chelating power.

Provided that the essential aromatic nucleus is present, together with the conjugated heteroatoms, the only other necessary limitation is that the compound should be soluble in the bath. This generally implies some limitation of the size of the molecule. The etch inhibitors usually contain a total of from 3 to 25 carbon atoms, preferably 4 to 20.

The etch inhibitor is preferably present in a proportion of from 0.05% by weight up to 0.7% or higher. Preferably the proportion of 100% nitric acid is 1.6 to 3.5% by weight and most preferably between 2.4 and 3.1%.

It is preferred to employ proportions of phosphoric to sulfuric acid of less than 3:1, e.g., 1.5:1 to 1:1.5, preferably 1.2:1 to 1:1.2, typically 1:1. Sulfuric acid and phosphoric acid together usually constitute at least 90%, preferably at least 93%, e.g., at least 95% of the weight of the composition.

The baths of the process contain copper as an essential ingredient, e.g., in a proportion of up to 0.2% by weight, most preferably 0.1 to 0.15%. Polishing baths may optionally contain ammonium or substituted ammonium ions, in order to reduce fuming.

Example: The composition of the polishing solution employed is as follows.

Constituent	Percent by Weight
H_3PO_4 (SG 1.75)	56.0
H_2SO_4 (SG 1.84)	38.5
HNO_3 (SG 1.50)	3.4
$CuSO_4 \cdot 5H_2O$	0.25
H_2O	1.85

The SG after aging was 1.80.

Samples of this composition were aged, i.e., their aluminum contents were raised to 30 g/ℓ Al, a typical concentration found in working aluminum chemical polishing solutions.

A sample of the aged polishing solution was heated to 105°C and adjusted to the optimum nitric acid content of 3% w/w SG 1.50 acid. Test pieces of an aluminum alloy suitable for chemical polishing (BA 211) were treated for 2 minutes by immersion in the solution while gently agitated. These test pieces were drained in air for (a) 1 second and (b) 30 seconds before rinsing. The short draining time was too short for the transfer etch to manifest itself and was taken as a standard that the particular solution sample was performing satisfactorily. A transfer time of 30 seconds is the longest used in commercial practice and in solutions of the above composition produced a complete coating of light grey transfer etch over the whole surface of the test piece.

The compound to be tested was added to the sample in increments of 1 g/ℓ and between each addition, after complete dissolution, test pieces were treated as above and drained in air for 30 seconds before rinsing in water. The efficiency of the compound at each concentration was estimated by visual estimation of the proportion of the area of the test piece covered with transfer etch to the nearest 100%.

Additions were carried out until (1) 100% removal of transfer etch was obtained; (2) the transfer etch reached a minimum which was not reduced by subsequent additions; and (3) no effect was observed in reducing transfer etch and additions totalled 10 g/ℓ.

1,2,3-benzotriazole itself has been tested up to 50 g/ℓ without any further effect upon performance being observed after complete suppression of transfer etch at 5 g/ℓ.

The compounds are listed below in decreasing order of image clarity (specular brightness) of the finish and increasing order of transfer etch.

Compound	(g/ℓ)	Reduction in Transfer Etch (%)
1,2,3-Benzotriazole	5	100
Benzofuroxan	2	100
2,1,3-Benzothiadiazole	2	100
2-Mercaptobenzothiazole	1	90
2-Mercaptobenzothiazole	1	80
1,2,4-Triazole	10	0
2,5-Dimercapto-1,3,4-thiadiazole	10	0

Inhibiting Fatigue Corrosion

Aluminum corrosion inhibitors are often addressed in terms of reducing the uniform surface corrosion rate of aluminum. However, in addition to surface corrosion, fatigue corrosion is an important contributing factor toward shortening the life of aluminum structural members, especially in cyclical high stress conditions which are encountered in aircraft.

Aluminum, aluminum alloys and other metals are elastic and will, although to an extent much less than encountered in a highly elastic material such as a rubber band, "stretch" and "compress" in reaction to external tensile or compressive forces.

The term "fatigue" embraces the general sequence of events which occur in reaction to external stress being applied to a metal. In other words, a metal fatigues when it, in reaction to external mechanical forces, develops slip planes and cracks. Fracture of the metal part due to external mechanical forces forming slip planes and cracks in the metal, in absence of chemical changes in the composition of the metal, is termed fatigue failure.

R.L. Crouch; U.S. Patent 4,261,766; April 14, 1981; assigned to Early California Industries, Inc. describes a method of inhibiting the fatigue of aluminum comprising the steps of immersing aluminum in an aqueous solution of a water-soluble cyanide compound at room temperature and continuously maintaining the aluminum in contact with the aqueous solution. The aqueous solution is substantially free of chromium.

It is generally found that a concentration of about 0.25% by weight of the cyanide compound in water, ethyl alcohol or acetone will provide the desired degree of corrosion inhibition of aluminum or aluminum alloys.

Example: This example illustrates the improvement in fatigue and fatigue corrosion characteristics of aluminum which results from contacting the metal with a composition of water and a cyanide component.

A test specimen of aluminum alloy (2024-T3) measuring 0.25" x 0.50" x 14" is oriented in the long transverse direction, notched at the center, degreased and inserted through slits cut in the side wall of a polyethylene bottle. The slits are sealed around the test beam with silicone caulking and the bottle is filled with the corrosion inhibiting composition of deionized water containing 0.25% by weight sodium ferrocyanide. The ends of the specimen are then attached to the vice and the crank of a Fatigue Dynamics VSP-150 plate bending machine and the loading is adjusted to 6,800 psi.

The test beam is then stressed at 100 cycles/min at 70°F until the specimen breaks.

With only deionized water in the polyethylene bottle, the test specimen breaks at 720,000 cycles. With a solution of deionized water containing 0.25% by weight sodium chromate, the test specimen breaks at 854,000 cycles. Duplicate tests with the bottle filled with deionized water containing 0.25% by weight sodium ferrocyanide were conducted and the following data obtained:

Test Number	Cycles to Failure
1	1,010,100
2	1,404,000
3	1,203,100

In all cases the aluminum bar did not fail.

As is known, the oxide layer which normally forms on the surface of aluminum is very resistant to ordinary water. Similarly, it is known that cyanide is an equally effective corrosion inhibitor for aluminum. Thus, when the aluminum bar was immersed in the aqueous solution of deionized water containing a sodium ferrocyanide, the aluminum bar was placed in a solution which would cause minimal corrosion. The failure of the aluminum bar was therefore predominantly due to the effects of fatigue.

Pretreatment for Resistance Welding of Aluminum

One of the major problems in the resistance welding of aluminum products is the large and frequently erratic variation in electrical resistance of the oxidized aluminum surface. A freshly cleaned aluminum surface has an electrical resistance of about 10 μohm. This resistance is usually too low for effective resistance welding due to the rapid deterioration of the electrode which it frequently causes. Only about 20 to 400 spot welds can usually be made before the electrode needs replacement.

The work of *G.A. Dorsey, Jr.; U.S. Patent 4,097,312; June 27, 1978; assigned to Kaiser Aluminum & Chemical Corporation* relates to a method of pretreating an aluminum surface for the resistance welding thereof wherein an oxide coating is formed on the aluminum surface having a controlled electrical resistance and then the oxidized surface is stabilized by treatment with a hot aqueous alkaline solution containing long chain carboxylic acids.

In accordance with the process, the aluminum surface is oxidized to a controlled level of electrical resistance and then the freshly oxidized aluminum surface is treated with an aqueous alkaline solution containing long chain aliphatic carboxylic acids as described in U.S. Patents 4,004,951 and 4,101,346. The hydrophobic coatings formed by this latter treatment prevent further oxidation of the aluminum substrate and the aging of the controlled oxide layer formed so that the electrical resistance of the surface remains relatively constant over long periods of time. The hydrophobic coating formed by the latter processing step does not significantly increase the electrical resistance of the coating nor does it detrimentally affect the resistance welding of the aluminum substrate.

Generally, the oxide layer having a controlled level of electrical resistance can be developed by any convenient manner.

The oxidation of the aluminum surface should be controlled so that the initial electrical resistance of the fresh oxide is less than 500 μohm, preferably less than 300 μohm but greater than 25 μohm, preferably greater than 50 μohm.

One attractive method for developing an oxide coating described above is set forth in U.S. Patent 4,113,520. This process generally comprises treating the aluminum surface with a hot aqueous solution at neutral or slightly basic conditions which is essentially saturated with a difficultly soluble magnesium compound.

Example: A group of 2036-T4 aluminum alloy panels were first etched in a concentrated aqueous solution of sodium hydroxide, water rinsed and then treated for about 30 seconds in an aqueous solution containing about 1 g/ℓ $MgCO_3$ (a saturated solution) with pH controlled to about 8.0 and the bath temperature controlled to about 80°C. The solution also contained a nonionic wetting agent (Tergitol NPX) at a level of about 0.1 ml/ℓ of solution. This treatment formed a relatively thin aluminum oxide coating of about 100 Å with an electrical resistance of about 75 μohm.

The panels were removed from the treatment bath and rinsed with water. One portion of the panels was treated in a hot aqueous alkaline solution containing stearic acid in accordance with U.S. Patent 4,004,951 and a second portion of the panels was treated in a hot aqueous alkaline solution containing isostearic acid (the methyl branched chain isomer of stearic acid) under the exact same conditions in accordance with U.S. Patent 4,101,346.

Each group of specimens was then subjected to numerous spot resistance welding with essentially the same electrode pressure and electrical current until the electrode tip of the welding machine was no longer suitable for producing acceptable welds. The specimens having the coating formed in the aqueous alkaline solution of stearic acid allowed over 3,000 spot welds before the electrode tip needed replacement. The specimens having the coating which had been formed in the aqueous alkaline solution of isostearic acid allowed approximately 2,000 spot welds before the tips needed replacement. This represents an increase of 10 to 100 times the number of spot welds on aluminum heretofore obtained in commercial practice with the same electrode.

The electrical resistance measurements of the aluminum surfaces described herein were measured by means of a surface resistance analyzer Model No. VT-11A (C.B. Smith Company). All discussions herein of surface electrical resistance refer to surface electrical resistance as measured by the aforementioned surface resistance analyzer or its equivalent.

Chromate conversion coatings for aluminum substrates are discussed in Chapter 3. The coloring of aluminum can be found in Chapter 8.

SURFACE TREATMENT OF FERROUS METALS

ANTICORROSIVE OXIDE LAYER ON STEEL

Heating in Nitrogen or Air Followed by Superheated Steam

A process for forming an anticorrosive, oxide layer on steel by subjecting the steel surface to superheated steam is known from the German Patent 1,621,509. According to this process, a corrosion-preventing, protecting layer of Fe_3O_4 is formed by conducting superheated steam, of at least 250°C, through pipelines, apparatus and vessels of steam power plants. The protective layer formed in this manner, however, does not withstand all chemical influences. Furthermore, undesirable hydrogen embrittlement can occur.

M. Pfistermeister, H. Krapf and E. Coester; U.S. Patent 4,141,759; February 27, 1979; assigned to URANIT (Uran-Isotopentrennungs-GmbH), Germany provide a method for forming an anticorrosive, oxide layer on steel, in which the steel surface is subjected to superheated steam for a period of at least 1 hour, and which comprises heating a cleaned steel surface in a nitrogen, air or oxygen atmosphere to at least 200°C, and then subjecting the heated steel to a further heating period in which a temperature between 450° and 520°C is reached and maintained, with superheated steam under steam flow-through conditions in which the steam flow is turbulent or should have a Reynolds number greater than 900.

Normal cleaning of the steels is only done if the surfaces are contaminated with oils, greases or other substances arising from the manufacturing process. On the other hand, if the steel is initially contaminated with easily oxidizable products, the initial heating process is performed with air or oxygen only. Should the steel be contaminated with a thin oxide layer in the range 500 to 1000 Å, then the steel is heated to over 400°C in an atmosphere of N_2 and H_2 (4-5:1) or in an atmosphere of N_2 and NH_3 (4-5:1). Normally oxide layers up to 500 Å need not be removed and so far initial oxide layers thicker than 1000 Å have not been encountered.

The clean steel surface is heated in a nitrogen, air or oxygen atmosphere to a temperature in the range of 200° to 250°C. The heating in the nitrogen, air or oxygen atmosphere can bring the steel to a temperature from about 25° to 250°C. Generally, the steel is subjected to the nitrogen, air or oxygen treatment for a period of time from 20 to 60 min. During this initial nitrogen, air or oxygen treatment, the steel is not subjected to superheated steam.

Upon reaching the desired temperature of over 200°C, the steel is then subjected to treatment with superheated steam. The superheated steam raises the temperature of the steel to the range of 450° to 520°C, generally in about 1 to 3 hours, and the steel is maintained by the superheated steam at this temperature. Preferably, the steel is maintained at a temperature of about 450° to 520°C by the superheated steam for a period of from about 1 to 5 hours.

During the treatment with the superheated steam, the flow conditions should be as turbulent as possible. Generally, the Reynolds number of the flow must be at least 900 but the optimum range is from about 2,100 to 2,500.

After the steam treatment, the steel is cooled to about 100°C by subjecting the steel, for several hours, to a stream of air having a temperature which can be adjusted in the range of 10° to 30°C. In the case where hydrogen or ammonia plus nitrogen were used for cleaning purposes, it is desirable to use nitrogen instead of air for the cooling down process. This prevents further oxidation of the oxide layer.

The entire process, including the initial heating in nitrogen, air or oxygen, the superheated steam treatment, and the cooling can take place in an accurately regulatable fluidized bed furnace. The steam used during the superheated steam treatment may have added to it nitrogen, air or oxygen in the ratios $H_2O:N_2$ or air or O_2 of no more than 4:1 or 5:1. The best results are obtained, however, with steam which is completely free of nitrogen, air or oxygen, respectively.

Example: A clean maraging steel sample (Ni-Co-Mo) was heated to 480°C in an accurately regulatable fluidized bed furnace during a period of 60 minutes. In this 60-minute period, the heating initially took place in air to bring the steel to a temperature of 200°C (20 min) and upon reaching 200°C, heating then took place with 21 Nm^3/hr superheated steam (Reynolds number is 2,100) to bring the heated steel to a temperature of 480°C. This steam treatment was continued for 3 hours at 480°C. Thereafter, the steel was cooled to 100°C in a stream of air of greater than 20 Nm^3/hr during a period of time of about 3 hours. A microcrystalline firmly adhering mixed oxide layer was formed which had a thickness of about 1 μ, and which consisted of mixed and pure spinels of the type $Fe(Fe_2O_4)$, $Ni(Fe_2O_4)$, $Ni(Co_2O_4)$, $Co(Co_2O_4)$, and $FeMoO_4$, with an average lattice constant of 8.4 Å.

The mechanical properties of the heated sample, such as tensile strength, modulus of elasticity and coefficient of expansion, remained fully unchanged. The H_2 content of the treated sample was less than 1 ppm.

Heating in Gaseous Formic Acid Followed by Superheated Steam

In spite of careful cleaning of the steel surfaces, the formation of undesirable thin oxide layers before the steam treatment in the method described in U.S. Patent 4,141,759 above could not be prevented in all cases. These thin oxide layers interfere with the actual protective layer formation that occurs during the treatment with the superheated steam.

E. Coester, H. Krapf, M. Pfistermeister, B. Sartor and H. Mohrhauer; U.S. Patent 4,153,480; May 8, 1979; assigned to URANIT (Uran-Isotopentrennungs-GmbH), Germany have developed a method for forming an anticorrosive oxide layer on steel in which the steel surface is subjected to superheated steam for a period of at least 1 hour. This comprises heating a steel surface in a gaseous formic acid atmosphere to a raised temperature up to 480°C, and then subjecting the heated steel to a further heating period in which a temperature between 450° and 520°C is maintained, with superheated steam under steam flow-through conditions in which the steam flow has a Reynolds number greater than 900.

The steel surface is heated in a formic acid atmosphere to a raised temperature up to 480°C. The raised temperature achieved during the formic acid treatment preferably is in the range of 400° to 480°C, especially when the steel surfaces initially contain a thin oxide layer of 500 to 1000 Å.

Generally, the steel is heated to a temperature of at least 450°C during the formic acid treatment. The heating in the formic acid atmosphere can raise the steel from an ambient temperature of, for example, about 25°C to the desired temperature range of, for example, 400° to 480°C. Generally, the steel is subjected to the formic acid treatment for a period of time from 20 to 60 minutes. During this initial formic acid treatment, the steel is not subjected to steam treatment. The formic acid preferably is mixed with an inert gas, such as nitrogen or argon, which serves as a carrier gas. The formic acid atmosphere can contain 4 to 5 volume parts of the carrier to 1 volume part of formic acid.

Upon reaching the desired raised temperature of up to 480°C, the heated steel is then subjected to treatment with superheated steam. If the initial treatment with formic acid has raised the temperature of the steel to 450°C or higher, the steam treatment is preferably selected to maintain the steel temperature achieved during the formic acid treatment or to raise the steel to a still higher temperature within the range up to 520°C. If the initial treatment with formic acid has not raised the temperature of the steel to 450°C, the treatment with superheated steam raises the temperature of the steel to the range of 450° to 520°C.

Once the steel has reached the desired temperature for the steam treatment within the range of 450° to 520°C, the steel is maintained by the superheated steam at this temperature. Preferably, the steel is maintained at a temperature of about 450° to 520°C by the superheated steam for a period of from about 1 to 5 hours.

During the treatment with the superheated steam, the flow conditions should be as turbulent as possible. Generally, the Reynolds number of the flow must be at least 900, but the optimum range is from about 2,100 to 2,500. The treatment with superheated steam under flow conditions where the Reynolds number is greater than 900 prevents hydrogen embrittlement since no equilibrium state then can form which would permit successive penetration of the hydrogen into the steels. During the steam treatment, an oxide layer is formed which can consist of the mixed and pure spinels of the type $Fe(Fe_2O_4)$, $Ni(Fe_2O_4)$, $Ni(Co_2O_4)$, $Co(Co_2O_4)$, and $Fe_4Mo_6O_{16}$.

After the steam treatment, the steel is cooled to about 100°C by subjecting the steel, for several hours, to a gas stream having a temperature which can be adjusted in the range of 10° to 30°C. It is desirable to use nitrogen instead of air for the cooling down process to prevent further oxidation of the oxide layer.

Example 1: Normal maraging steel samples which may have blue and/or yellow annealing colors as well as rust pits, were heated from 20° to 480°C in a regulatable fluidized-bed furnace within 40 to 60 minutes, by means of a mixture of nitrogen/formic acid in a volume ratio of 5:1 and with a flow speed of 5 ℓ/min. Upon reaching 480°C, the treatment was continued for 30 minutes with a gas mixture of nitrogen/formic acid in a volume ratio of 4:1 and a flow speed of 7 ℓ/min. Thereafter, the formic acid content was adjusted to 0 with a gradual switchover to superheated steam (10 to 15 min). The steam treatment took place under turbulent conditions at 480°C, during 2.5 hours. Cooling to 100°C took place in the nitrogen stream. Compared with untreated samples, the thus-treated samples distinguish themselves in the corrosion test by a very high protection factor of about 90 to 110. The layer thickness was about 0.8 to 0.9 μ.

Example 2: A clean maraging steel sample (Ni-Co-Mo) was heated to 480°C in an accurately regulatable fluidized-bed furnace during a period of 60 minutes. In this 60-minute period, the heating initially took place in a formic acid/nitrogen atmosphere with 3 vol % formic acid to bring the steel to a temperature of 200°C within 20 minutes. Afterwards the formic acid concentration in nitrogen was raised to 20 vol % and heating was continued to reach 460°C in another 40 minutes. After having reached this level heating then took place with 21 Nm^3/hr superheated steam (Reynolds number = 2,100) to bring the heated steel to a temperature of 480°C.

The changeover from the nitrogen/formic acid mixture to superheated steam was carried out firstly by the exclusion of formic acid gas and secondly by the gradual displacement of nitrogen by superheated steam. This steam treatment was continued for 2 hours at 480°C. Thereafter, the steel was cooled to 100°C in a stream of nitrogen greater than 20 Nm^3/hr during a period of time of about 3 hours. A microcrystalline firmly adhering mixed oxide layer was formed which had a thickness of about 0.8 μm, and which consisted of mixed and pure spinels of the type $Fe(Fe_2O_4)$, $Ni(Fe_2O_4)$, $Ni(Co_2O_4)$, $Co(Co_2O_4)$, and $Fe_4Mo_6O_{16}$, with an average lattice constant of 8.4 Å.

The mechanical properties of the heated sample, such as tensile strength, modulus of elasticity and coefficient of expansion, remained fully unchanged. The H_2 content of the treated sample was less than 1 ppm. A number of steel samples were treated in a similar manner, and the H_2 content of the treated samples was always less than 1 ppm. The samples also distinguished themselves by a high protection factor of 120 to 130. The average layer thickness was determined by weighing the samples and gave a value of 0.7 to 0.8 μm.

METALLIC CARBIDE LAYER ON FERROUS ALLOY

Carbide Layer of V-B Group Element or Chromium

There are several methods for coating metallic articles or for forming a metallic carbide layer thereon. Prior coating methods have a drawback. They use a molten treating bath containing metal particles. The metal particles need a relatively long time to dissolve in the bath, and undissolved metal particles deposit in the formed carbide layer, making a rough surface on the treated articles.

The work of *N. Komatsu and T. Arai; U.S. Patent 4,158,578; June 19, 1979;*

assigned to KK Toyota Chuo Kenkyusho, Japan relates to a method of forming a carbide layer of a V-B Group element of the Periodic Table or of chromium on the surface of the ferrous alloy article.

A fine and uniform carbide layer of a V-B Group element or of chromium is formed on the surface of a ferrous alloy article in a molten treating bath consisting essentially of boric acid or a borate in addition to an oxide of a carbide-forming element (CFE), such as a V-B Group element of the Periodic Table and chromium (U.S. Patent 3,719,518 and U.S. Patent 3,671,297), and a boron-supplying material. The method is highly productive.

The carbide of a V-B Group element, such as vanadium carbide (VC), niobium carbide (NbC) and tantalum carbide (TaC), and chromium carbide (CrC) is very hard. Therefore, the formed carbide layer is extremely hard and has superior wear resistance. It is highly suitable for surfaces of molds, such as dies and punches; of tools, such as pincers and screw-drivers; of parts for many kinds of tooling machines; and of automobile parts subjected to wear.

Upon intensive investigation of the mechanism for forming a carbide layer on the surface of an article by diffusing a V-B Group element or chromium (all references in this text to CFE are to these elements) from the molten treating bath composed mainly of (a) boric acid or borate, (b) a CFE oxide and (c) boron-supplying material, it was found that the main source of CFE came from the dissolved element in the molten treating bath rather than directly from an undissolved solid metal particle.

To dissolve CFE in the molten treating bath (to form a carbide layer having a smooth surface), CFE oxide was introduced into the molten bath rather than metal powder. The CFE oxide easily and quickly dissolves in a molten bath of boric acid or of borate. However, the resulting molten bath fails to form a carbide layer on an article (composed of carbon-containing ferrous alloy) immersed in the bath containing dissolved CFE oxide. A boron-supplying material (wherein the boron is not bonded to oxygen) must be incorporated in the molten bath together with the CFE oxide. The boron-supplying material reduces the CFE oxide, facilitates dissolving the CFE in the bath and enables the bath to form a carbide layer on the surface of the article immersed therein.

With regard to the ratio of the boron-supplying material to the CFE oxide, too much boron-supplying material results in forming (on the article) a boride layer composed of FeB or Fe_2B due to the excess boron dissolved in the bath. On the other hand, too little boron-supplying material results in the formation of no layer. Generally speaking, the weight of boron in the boron-supplying material is between 7 and 40% of the weight of the CFE oxide.

Example: Dehydrated borax ($Na_2B_4O_7$) was introduced into a crucible made of heat-resistant steel and heated in an electric furnace to melt the borax. A bath of 950°C was prepared. A treating bath was made by adding granular Nb_2O_5 and then B_4C powder (–325 mesh) little by little while stirring the prepared bath. Many kinds of baths were prepared in the same manner by changing the composition ratio of Nb_2O_5 and B_4C. Test pieces made of JIS SK 4 (carbon tool steel) with a diameter of 7 mm were immersed into each of the treating baths and kept therein for 2 hours, taken out therefrom and cooled in oil baths. Any treating material adhering to the surface of the test pieces was removed by washing with hot water. After cutting the test pieces, cross sections of each were observed micrographically.

A bath containing 10% of Nb_2O_5 and 3% of B_4C (2.4% of B) formed a layer of NbC with a 0.7 μ thickness, and a bath containing 10% of Nb_2O_5 and 5% of B_4C (3.9% of B) formed a 0.6 μ thick layer consisting of two layers, the uppermost one being a NbC layer and the other one (formed beneath the NbC layer) being a Fe_2B layer. However, baths containing 10% of B_4C (7.8% B) or 20% of B_4C (15.6% B) did not attain the object of the process because the baths formed exclusively a Fe_2B layer having a thickness of 40 μ and 63 μ, respectively. Strictly speaking, the NbC layers used in the specification are layers of Nb(C,B) in which a part of C is replaced with B. The content of B increases with an increase in the content of B_4C. The surfaces of all treated test pieces were smooth, and no powder adhesions were observed.

The test pieces with a NbC layer which were obtained by the method described above were placed in an air atmosphere in an electric furnace of 550°C, kept there for 10 minutes, and then cooled in the air while observing the surface of each piece. These steps were repeated. Also, for comparative data, a test piece was made of JIS SK 4 to form a NbC layer of 9 μ thickness by dipping it in a treating bath composed of molten borate and (as additive) 20% by weight of ferroniobium (Fe-Nb) powder. It was tested by the same previously noted steps. The results are shown below.

Comparative Test Data

No. of Cycles	Bath Composition		
		10% Nb_2O_5 Powder plus...	
	20% Fe-Nb Powder	3% B_4C	5% B_4C
40	no peeling	no peeling	no peeling
60	partial peeling	partial peeling	no peeling
100	complete peeling	partial peeling	no peeling

The comparative piece was gradually discolored by oxidation of NbC as the cycles increased, and peeling occurred before the 60th cycle; the layer was completely peeled off by the 100th cycle.

Compared with this result, although peeling occurred, with test pieces treated in the bath containing 10% by weight of Nb_2O_5 and 3% by weight of B_4C (according to the process) complete peeling did not occur even after the 100th cycle. The test pieces reflected greater oxidation resistance. Especially, in a test piece of the process treated in a bath containing 10% of Nb_2O_5 and 5% of B_4C, no peeling was found even after the 100th cycle.

Other treating baths were prepared by using Ta_2O_5 as CFE oxide, and by using B_4C as the boron-supplying material, or by using Fe-B. Suitable treating-bath compositions have a ratio of B (from the boron-supplying material) to Ta_2O_5 of from about 7 to 24% in the former case and of from about 7 to 35% in the latter case.

Additional data was obtained by using Cr_2O_3 as CFE oxide, and using B_4C as the boron-supplying material or by using Fe-B. Suitable treating-bath compositions have a ratio of B (from the boron-supplying material) to Cr_2O_3 of from about 7 to 26% in the former case and of from about 7 to 32% in the latter case.

Mixed Carbide Layer of V-B Group Elements

N. Komatsu, T. Arai and H. Fujita; U.S. Patent 4,202,705; May 13, 1980; assigned to KK Toyota Chuo Kenkyusho, Japan provide a method for forming a mixed carbide layer of V-B Group elements on the surface of a ferrous alloy article in a molten treating bath. The mixed carbide layer of V-B Group elements has the combined merits of a plurality of single-carbide layers of each of the V-B Group elements.

The carbide layer is conveniently formed on such a ferrous alloy article by immersing the article in a molten borate or boric acid bath containing a V-B Group element in a suitable form. Although oxides of V-B Group elements are soluble in these molten baths, it appears that the free-metal form is needed. However, V-B Group elements in free-metal or alloy form lack the requisite solubility and tend to impair the surface of articles immersed in molten baths containing them.

Dissolving a V-B Group element (in metal or alloy form) in the molten bath is somewhat facilitated by concurrently incorporating in the same bath one or more oxides of the V-B Group element. By having the V-B Group element (in metal or alloy form) as a finely divided powder (particle size, e.g., from −100 to −325 mesh), solution is further assisted. Mesh is based on Tyler Standard.

Carbide layers of different V-B Group elements impart different advantageous properties, e.g., hardness, smoothness and resistance to corrosion, wear, peeling, cracking and oxidation, to surfaces of ferrous alloys. To obtain a combination of advantageous properties, two or three different V-B Group elements are incorporated in a molten bath to form a suitable treating bath. The V-B Group elements so incorporated are not all initially in the same form. At least one is in metal or alloy form and at least one different V-B Group element is in the form of an oxide. For all combinations of V-B Group elements, any one or two are optionally in metal or alloy form and a different one or two are optionally in the form of one or more oxides.

The weight percent of oxide of V-B Group element(s) in the molten treating bath ranges from about 1 to about 35 to 40%. The weight percent of V-B Group metal or alloy similarly ranges from about 1 to about 35 to 40%. The total of such metal or alloy and such oxide is preferably not more than about 80% by weight. Illustrative bath compositions are those in which each of the elements vanadium (V), tantalum (Ta) and niobium (Nb) is in powdered (particle size no larger than 100 mesh) metal form. Rather than powdered metal form, each of V, Ta and Nb is optionally in the form of a powdered alloy (having a corresponding particle size) of such element.

Example 1: 5 kg of borax were introduced into a crucible made of heat-resistant alloy and heated up to 950°C in an electric furnace. Then, 0.33 kg (5% of the composition) of V_2O_5 particles having a diameter of several millimeters and 1.4 kg (20% of the compositions) of ferroniobium (Fe-Nb) powder (including about 50% of Nb), which was not more than 100 mesh, were added to the borax to prepare a treating bath.

Next, test pieces made of carbon tool steel (JIS SK 4) with a diameter of 8 mm and a length of 50 mm were dipped in the bath, kept there for 4 hours at 950°C and then taken out of the bath. The test pieces were washed with water after being cooled in air. The test pieces were cut to observe the cross section of the treated surface layer. The thickness of the layer was measured, and the structure of the layer was analyzed under a microscope, an x-ray microanalyzer and an x-ray diffraction meter.

The cross-sectional structure was observed to have one uniform surface layer with a thickness of 9 μ on each surface of test pieces. This surface layer was identified to be (Nb, V)C, i.e., a mixture of niobium and vanadium carbide or, more precisely, a structure of niobium carbide in which some niobium atoms are replaced by vanadium atoms, by x-ray diffraction and was measured by an x-ray microanalyzer to determine that the ratio of Nb to V was about 4 to 1 by weight and that the V was mainly distributed near the mother material in the layer.

Example 2: 5 kg of borax were introduced into a crucible made of heat-resistant alloy and heated up to 950°C in an electric furnace. Then 0.17 kg (3% of the composition) of Nb_2O_5 particles and 0.57 kg (10% of the composition) of ferrovanadium (Fe-V) powder (including about 40% of V) which was not more than 100 mesh were added to the borax to prepare a treating bath.

Next, test pieces which had the same form, composition and quality as those used in Example 1 were dipped in the noted bath, kept therein for 4 hours at 950°C and then taken out of the bath. The test pieces were washed with water after being cooled in air. In the same manner as in Example 1, the structure of the treated test pieces was analyzed by a microscope, an x-ray microanalyzer and an x-ray diffraction meter. The surface layer produced was observed to have one uniform surface layer which was identified to be (Nb, V)C with a thickness of 8 μ on each test piece surface. The ratio of Nb to V in this carbide layer was about 1 to 1 by weight.

Mixed Carbide Layer of Chromium and V-B Group Elements

Here, too, *N. Komatsu, T. Arai and H. Fujita; U.S. Patent 4,230,751; Oct. 28, 1980; assigned to KK Toyota and Chuo Kenkyusho, both of Japan* provide a method for forming a mixed carbide layer of chromium and of V-B Group element on the surface of a ferrous alloy article. The method is characterized by the ferrous alloy article, which comprises at least 0.1% by weight of carbon and by immersing and keeping such article in a molten borate and/or boric acid bath which contains (a) chromium oxide and one or more V-B Group elements in powdered metal or powdered alloy form or (b) chromium powder in metal or alloy form and of one or more V-B Group elements in powdered oxide form.

It is convenient to use chromium in oxide form when at least one V-B Group element is in metal or alloy form and to use chromium in metal or alloy form when at least one V-B Group metal is in oxide form. When two different V-B Group elements (one in oxide form, e.g., Ta_2O_5, and one in metal or alloy, e.g., ferro-vanadium having about 50 wt % of vanadium, form) are incorporated in the molten boric acid and/or borate bath, added chromium is optionally in oxide form or in metal or alloy form.

The oxides, metals and alloys are solids. Since the metals and alloys are not readily soluble in molten boric and and/or molten borate, they should be incorporated therein in powder form or, possibly, in flake form. As the oxides are more soluble in the molten bath, flake form is more suitable for them, but powder form is still preferred.

The minimum amount of compounded metal or alloy additive is 1% by weight and that of oxide is also 1% by weight. The maximum amount of compounded metal or alloy additive is from 35 to 40% by weight and that of oxide is also from 35 to 40% by weight (leaving only from 20 to 30% by weight of boric acid and/or of borate). For the purpose of these percentages the entire amount of the metal, the alloy or the oxide is counted.

Exemplary makeup components for 5-kg molten treating baths are presented in the table. Each of the 11 treatment-bath formulations is compounded by melting the boric acid and/or borate components in a suitable vessel and incorporating the oxide and the element components in the resulting melt. The element and oxide components are preferably in powder form when added to the melt. When an element is incorporated in the melt in alloy form, the actual amount of alloy used must be increased sufficiently to provide the resulting melt with the indicated quantity of the element.

The molten medium for the treating bath is provided by boric acid or a borate or any suitable mixture. Alkali metal borates are conveniently used for this purpose, particularly borax.

When chromium or one or more V-B Group elements are introduced into the molten bath, they are generally in particulate, e.g., powder or flake form. A V-B Group element or chromium in metal or alloy form is preferably introduced into a molten bath as a fine powder, i.e., having a particle size finer than 100 mesh.

Bath Compositions

	1	2	3	4	5	6	7	8	9	10	11
Ingredients						(kilograms)					
V	0.25	–	–	–	–	–	–	–	–	–	–
Ta	–	0.50	–	–	–	–	–	–	–	0.25	–
Nb	–	–	0.50	–	–	–	–	–	–	–	–
Cr	–	–	–	0.5	1.0	1.5	0.75	1.5	0.75	0.50	0.5
Cr_2O_3	0.13	0.25	1.0	–	–	–	–	–	–	–	–
V_2O_5	–	–	–	0.5	–	–	–	–	–	–	–
VO_2	–	–	–	–	–	–	0.25	–	–	–	–
Ta_2O_5	–	–	–	–	0.5	–	–	0.5	–	0.50	–
Nb_2O_5	–	–	–	–	–	0.25	–	–	–	–	–
NbO_2	–	–	–	–	–	–	–	–	0.25	–	–
NbO	–	–	–	–	–	–	–	–	–	–	0.25
$Na_2B_8O_{13}$	2.00	–	–	–	–	–	–	–	–	–	–
$Na_2B_6O_{10}$	–	–	–	–	–	1.25	–	–	–	–	–
$Na_2B_4O_7$	2.62	4.25	–	–	2.0	–	–	1.5	–	3.75	2.0
$Na_2B_2O_4$	–	–	3.50	–	–	2.0	–	–	–	–	–
$K_2B_4O_7$	–	–	–	–	1.5	–	–	2.5	–	–	2.25
$Li_2B_4O_7$	–	–	–	4.0	–	–	4.0	–	–	–	–
CaB_4O_7	–	–	–	–	–	–	–	–	4.0	–	1.2

RUST

Rust Treatment Preparation

B.E. Svenson; U.S. Patent 4,289,638; September 15, 1981 provides a rust treatment preparation comprising a strong acid, in an emulsion comprised of wool fat, wool grease and/or derivatives thereof, optionally together with a solvent therefor, and/or a stabilizing and/or absorbing agent.

The preparation may contain the following ingredients in preferred concentrations: concentrated phosphoric acid (95%)–5 to 35% v/v (e.g., 25%); wool fat or wool grease and/or derivatives thereof–25 to 35% w/v (e.g., 25%); kaolin–10 to 15% w/v (e.g., 10%); and methylated or industrial spirits to 100% volume.

Once the preparation has dried (e.g., overnight), the wool fat, wool grease and/or derivatives thereof, contained therein, form a protective barrier against moisture penetration, thus minimizing further corrosion at the applied location. Additionally, the preparation allows the phosphoric acid to remain in close contact with the rust until all the reaction has ceased. The water-repellant nature of the wool fat, wool grease and/or derivatives thereof also prevents the preparation being washed away, thus lengthening the time between applications.

Example: To produce 1 ℓ of the preparation, 250 g of commercial anhydrous wool grease was placed in a container. A mixture of 330 ml of 95% concentrated phosphoric acid and 500 ml of methylated spirits was prepared and added to the wool grease. The heat of reaction of the acid and methylated spirits was sufficient to dissolve all the wool fat. Once the wool fat had dissolved, 100 g of light or colloidal kaolin was added to the mixture. If necessary, methylated spirits may be added to the liter volume mark. The mixture was placed in a homogenizer so as to further stabilize the emulsion and finally allowed to cool.

The emulsion is suitable for application by hand. This is a convenient technique in instances where rust is not visible but can be felt by the hand and hence the composition can be spread over the affected surface. There are numerous situations where application by the hand is more accurate than other techniques (e.g., spraying). Because of the acidic nature of the composition prolonged contact with the skin is not advocated but no adverse effects, e.g., burning of the skin, are encountered provided that this preparation is washed off the hand once application is completed.

Rust Removal Composition

Removal of corrosion from metal surfaces has been a problem since there has been metal, particularly, the removal of rust from iron-containing metals. *G.B. Anderson; U.S. Patent 4,014,804; March 29, 1977; assigned to Gultex, Inc.* provides a composition for the effective removal of corrosion which does not damage the metal surfaces containing such corrosion.

The composition comprises 2 to 10% by weight of a sulfate of copper, 0.05 to 1.0% by weight of an alkali metal bichromate, 0.05 to 1.0% by weight of a surfactant, 15 to 35% by weight of phosphoric acid (based on contained phosphoric acid) and 50 to 85% by weight of water, the total of all components equaling 100%.

The method for preparing a corrosion-removing composition comprises heating water to a temperature of 140° to 160°F, dissolving therein while agitating, a sulfate of copper and an alkali metal bichromate, adding a surfactant, allowing to cool and dispersing therein phosphoric acid, the respective amounts of each of the ingredients added being sufficient to produce a composition containing 2 to 10% by weight of a sulfate of copper, 0.05 to 1.0% by weight of an alkali metal bichromate, 0.05 to 1.0% by weight of a surfactant, 15 to 35% by weight of phosphoric acid and 50 to 85% by weight of water, the total of all components being 100%.

Preferably the sulfate of copper is cupric sulfate, the alkali metal bichromate is sodium bichromate or the equally effective potassium dichromate, and the surfactant is one selected from the group consisting of the alkyl phenyl ethers of polyethylene glycol, dodecylphenol adducts with ethylene oxide, trimethylalkyl ethers of polyethylene glycol, polyalkylene glycol ethers, alkyl mercaptan adducts with ethylene glycol, secondary alcohols and their ammonium sulfate salts with ethylene oxide, alkylaryl polyether alcohols and mixtures thereof.

For optimum results, the rust-coated surfaces are first freed of dirt, grease or oil. After application of the corrosion-removing composition, the composition is allowed to remain in contact with the metal surface for a period of 5 to 20 minutes, preferably 10 to 15 minutes or until dryness occurs, whichever first occurs. The treated area of the metal surface is then wiped clean. For severe rusting, two or more applications of the corrosion-removing composition may be required.

Rust-Preventive Compositions

Phosphate esters are finding increasing use as fire-resistant lubricating and hydraulic fluids. These ester lubricants have the desirable properties of low flammability, high lubricity and long service life.

New machinery or machinery in storage or transport which is designed for use with phosphate esters is frequently rust-proofed with conventional petroleum-based compositions or other formulations which are not compatible with phosphate esters. This situation often necessitates an extensive cleaning of rust-preventive-treated machinery before operation with phosphate ester based fluids may be begun.

J.F. Anzenberger; U.S. Patent 4,263,062; April 21, 1981; assigned to Stauffer Chemical Co. provide a method of rust-proofing machinery by the application of phosphate ester containing rust-preventive compositions.

The first essential component of the rust-preventive composition is a liquid aryl phosphate ester. Examples of suitable triaryl phosphates are triphenyl phosphate, tricresyl phosphate, trixylyl phosphate, cresyldiphenyl phosphate, isopropylphenyldiphenyl phosphate, diisopropylphenyl/diphenyl phosphate, tert-butylphenyl/diphenyl phosphate or di-tert-butylphenyl/diphenyl phosphate. A mixture of phosphate esters may be used if desired. The aryl phosphate ester component constitutes about 50 to about 95 wt % of the rust-inhibiting composition. Preferably, the aryl phosphate ester constitutes from 65 to 90 wt % of the rust-preventive composition.

The second essential ingredient in the rust-preventive composition is an oil-soluble calcium sulfonate. Oil-soluble calcium sulfonates are detergent additives having a molecular weight of from about 350 to about 550. These sulfonates are formed by reacting petroleum sulfonic acid with a 10 to 100% excess of calcium carbonate or calcium hydroxide neutralizing agent. The oil-soluble calcium sulfonate ingredient should constitute from about 0.1 to about 5.0 wt % of the rust-preventive composition.

The third essential ingredient of the rust-preventive composition is a liquid polyolester having a viscosity of 4.30 to 4.70 cs at 100°C. The polyolester is the reaction product of a polyhydric alcohol and a monocarboxylic acid. A preferred polyolester is the reaction product of pentaerythritol with butyric and heptanoic acids.

The liquid polyolester ingredient constitutes from about 5 to about 50 wt % of the rust-preventive composition. Preferably, the proportion of liquid polyolester is from about 20 to about 30 wt % of the composition.

The three essential ingredients previously described are mixed to form a single phase. The essential ingredients should comprise in combination at least 80 wt % of the rust-preventive composition. The balance of the composition may, if desired, include minor amounts of optional ingredients such as dyes, perfumes, or diluents. Other corrosion-inhibiting agents may be mixed with the rust-preventive compositions if desired.

Metal surfaces (especially, ferrous metal surfaces) are protected from corrosion by applying a coating of the rust-preventive composition. The composition may be applied by any conventional means such as spraying, dipping, brushing, flushing, etc. Since the rust-preventive composition has the ability to adhere to metal surfaces, it is only necessary to contact the metal with the composition to deposit a rust-preventive effective coating.

Example: This example illustrates the preparation and use of the rust-preventive composition. In addition, this example compares the composition of this process to control compositions and compositions containing known rust-preventive additives.

Test procedure – A steel paint panel (approximately 7.62 x 12.7 cm) was completely immersed in a rust-preventive formulation. The panel was allowed to drip dry for 30 minutes. The panel was then laid flat in a container slightly larger than the panel and 80 ml of distilled water was poured into the container to completely cover the panel. The container was allowed to stand at room temperature until all of the water evaporated. The test panel was then examined and evaluated for rust formation.

Rust-preventive compositions (all percentages by weight) –

Sample A: Trixylyl phosphate (Fyrquel 220) 79 wt %; oil-soluble calcium sulfonate detergent additive (Tergol 8BH) 1%; liquid polyolester, a reaction product of pentaerythritol with C_7 and C_4 alkanoic acids (Base Stock 874) 20%.

Sample B: Trixylyl phosphate (ingredient used in Sample A) 78%; oil-soluble calcium sulfonate detergent additive (ingredient used in Sample A) 2%; liquid polyolester (ingredient used in Sample A) 20%.

Sample C: Trixylyl phosphate (ingredient used in Sample A) 100%.

Sample D: Trixylyl phosphate (ingredient used in Sample A) 99%; calcium sulfonate detergent additive (Tergol 180H) 1%.

Sample E: Trixylyl phosphate (ingredient used in Sample A) 99%; calcium sulfonate detergent additive (ingredient used in Sample A) 1%.

Sample F: Trixylyl phosphate (ingredient used in Sample A) 99%; dilauryl acid phosphate rust inhibitor (Ortholeum 162) 1%.

Sample G: Trixylyl phosphate (ingredient used in Sample A) 99%; fatty imidazoline tertiary amine inhibitor (Unamine C) 1%.

Experimental results are displayed below:

Sample	Test Results
A*	No rust
B*	No rust
C	Heavy rust
D	Medium rust
E	Light rust
F	Medium rust
G	Heavy rust

*These samples conform to the rust-preventive compositions of this process.

The combination of the three essential ingredients of the rust-preventive composition of this process show superior corrosion-resisting properties in comparison to selected individual ingredients or other known anticorrosion agents.

Treatment of Oxidized Surfaces Without Removal of Rust

The physical structure of an oxidized surface, particularly a metal surface such as a ferrous surface, normally involves selective pitting of the surface. It seems that in the presence of moisture and oxygen, the metal surface is progressively oxidized. In the case of a ferrous material, there are various iron oxides formed in a generally porous structure; sometimes, however, following a first pitting, the active center of the rust formation is at the deepest portion of the formed pitting.

I.B. Schafer; U.S. Patents 4,264,377; April 28, 1981; and 4,170,493; October 9, 1979 has found that if an oxide-reducing substance is applied to the surface of rust without any removal of this rust, and the material is left long enough, it can eventually reach the deepest portion of the pits of rust. The surfaces appear to be very porous but if any sealing effect that is inhibiting is used, the material will block its own path.

However, if a substance, while being eminently effective for reducing the oxide, has nonetheless included therewith a second substance which will retard the mobility of the molecules or in some other way simply reduce significantly the rate of the action, then applying such a substance to the surface of even untreated rust, will allow the substance to penetrate deeply before closing its own passage and inhibiting its own access to the deepest portions of the rust.

Some acids are particularly effective for the first substance. Other substances can effectively limit the mobility of the action of the molecules and especially in one case, by reason of an action that appears to be a catalytic action, the reaction products are formed into an insoluble complex which grows directly from and about the site of the reduction action thus, in effect, providing a coating that grows at the apparent susceptible area of the corroded surface. This has been found to be the case particularly where a reasonably strong solution of phosphoric acid, i.e., orthophosphoric acid, has added thereto a small proportion of urea. In this case, the urea is found to slow down the mobility of the phosphoric acid which, however, by being in strong concentration, has ample capacity to substantially reduce significant quantities of oxide.

Adding small quantities of metals as soluble salts where the metals are one or more of transitional metals helps to create a somewhat stronger and more dense insoluble deposit deep within the porous rust. If required, such deposit will envelope a substantial portion of the surface treated where there is an adequate proportion of oxidized material. Cobalt has been found to be a useful addition to strengthen this layer. It has also been found to be useful to add nickel and this has been added as a nickel sulfate.

Concentrations of the reduction agent, such as phosphoric acid, should be up in the concentration percentages of 55% and in practical terms probably not less than 40% should be used. It is generally difficult to maintain concentrations greater than 75 to 80% considering that additives in solution must also be added.

A range of between 5 to 15% of urea might be considered a reasonable range although ideally experiments should be conducted in relation to the specific concentration of reduction agent used and the materials should be selected specifically for the particular product to be treated.

Example: 400 g of dry urea are dissolved in 1,600 ml of water at ambient temperature and to this is added 200 ml of cobalt sulfate solution (200 g cobalt $CoSO_4 \cdot 7H_2O$ dissolved in 1,100 ml of water) and then to this mixture is added 3,200 ml of 82% phosphoric acid (technical grade). This provides approximately 5 ℓ of product that is now suitable for either brushing on or otherwise applying to the oxidized surfaces. The above solution contains by weight: urea, 5.3%; cobalt sulfate, 0.5%; phosphoric acid (H_3PO_4), 58.0%; and water, 36.0%.

The mixture as prepared in one experiment was applied to a sheet of mild steel. The mild steel surface exhibited moderate rusting with a loose surface deposit. As a comparison, a commercially available product was applied to separate identical mild steel sheets. The product was Rust Dissolver (Selleys, Australia).

In the control cases, it was an instruction to first wire brush the surface of the mild steel, and this was done; but it was not done for one of the portions of the surface to which the process solution was applied.

The sheets were in each case covered according to the appropriate instructions with a film of the substance and in each case were left for 24 hours and each of the strips was then placed outdoors in an exposed position and from that time until 14 days later were wet thoroughly twice daily to induce rusting. Photomicrographs of the treated surface were taken after the 24-hour indoor and the 14-day outdoor exposure.

Test results (ferrous metal) – After 24 hours the mild steel surface treated with the process solution had a hard, glossy crystalline deposit of black/blue color formed over the entire surface of application. The deposit appeared identical on both the wire brush cleaned portion and the uncleaned portion of the plate and there was no sign of rust in any of the treated areas.

Selleys rust dissolver: A hard grey layer had formed on the surface of the steel and the rust had appeared visually to be removed in the treated areas.

The following effects were noted after the 14 days outdoor exposure: The crystalline deposits had flaked and there was evidence of the formation of a white powder on most flakes. In explanation of this, it appears that some excess of the solution not used in the basic process can be subsequently removed by dissolving. The revealed substrate was grey in color, and in a few places very light rusting could be seen. The results were identical in both cases of cleaned and uncleaned mild steel surface.

Selleys rust dissolver: The grey surface film formed initially was still substantially retained over most of the treated area, with some flaking evident. The surface revealed beneath the flaked area was also of grey color but there were generally signs of rusting evident in a few areas.

Test results (nonferrous metals) – Observation of the brass, copper and aluminum surfaces after application of the process solution under the conditions outlined earlier revealed that the tarnish was removed and a shiny surface was presented. After 14 days indoors, all 3 surfaces remained untarnished and no other surface effect could be detected.

Rust Transforming Composition with Synthetic Binder

E. Hengelhaupt and L. Peier; U.S. Patent 4,086,182; April 25, 1978; assigned to Noverox AG, Switzerland provide a method of converting an oxide film on a body composed of iron or steel into a corrosion-resistant protective coating. The composition for converting a surface film of iron oxide on an iron body into a corrosion-resistant film containing complexed compounds of iron comprises an aqueous dispersion or emulsion of a synthetic binding agent together with at least one polymeric esterification product of an aromatic oxycarboxylic acid having phenolic hydroxyl groups with a substance selected from the group consisting of acid anhydrides and substituted acid anhydrides.

Preferably, the dispersion or emulsion of the synthetic binding agent is mixed with a synthetic plastics material which is in solution. The quantity of the solution of the synthetic plastics material present may be up to 50 wt % of the total content of plastics material. The presence of the synthetic plastics material produces a significant increase of the corrosion stability of the resulting protective film.

The ratio of the quantities of the dispersed or emulsified synthetic binding agent to the synthetic plastics material is most preferably between 1.5:1 and 1:1.5. The dispersed or emulsified synthetic binding agent may, for example, be a styrene-butadiene copolymer, a styrene-butadiene-acrylonitrile rubber, or a vinyltoluene-butadiene resin. The synthetic plastics material which is employed as a solution may, for example, be a long or medium oil chain alkyd resin, polyvinyl chloride, a chlorinated rubber, an acrylate, a cyclic rubber, a styrene-butadiene rubber or any other suitable synthetic rubber, or a styrene-acrylate copolymer. The synthetic plastics material can be used in the form of a solution in an organic solvent, particularly an aromatic hydrocarbon such as benzene or xylol. It is also possible to use the same synthetic compound, for example, a styrene-butadiene copolymer, both as the dispersed or emulsified synthetic binding agent and as the synthetic plastics material. Furthermore, the solution of the synthetic plastics material may additionally contain corrosion-protective oils such as fish oil, safflower oil or wood oil.

The oxycarboxylic acid with phenol characteristics from which the esterification product is formed may, for example, be natural gallic acid, tannin or synthetic organic oxycarboxylic acids with phenol characteristics, particularly a dioxy- or trioxycarboxylic acid.

Example 1: A polyester of a styrene-maleic acid copolymer containing gallic acid units was prepared by esterifying gallic acid with a styrene-maleic anhydride copolymer. The esterification was carried out by refluxing the styrene-maleic anhydride copolymer and the gallic acid in a molar ratio of about 10:1 in acetone at 60°C for a period of about 1 hour using p-toluenesulfonic acid as catalyst. The product had a ratio of gallic acid units to maleic acid units of about 1.25:1 and was isolated as a powder after removal of the acetone. A rust-transforming composition was formed by mixing together the following ingredients:

	Weight Percent
Styrene-maleic acid-gallic acid copolymer*	5
Styrene-butadiene rubber dispersion**	30
Copolymer of styrene and an acrylate in solution in 50% white spirit***	25
Water	34.5
Isopropanol	5
Formic acid	0.5

*Powder prepared as described above.
**Solids content 50% (containing 47% polymer and 3% monomer).
***43.2% polymer; 1.8% monomer.

The styrene-acrylate copolymer solution is emulsified into the styrene-butadiene solution. By means of formic acid (which acts as a catalyst in the subsequent rust transformation), the product is adjusted to pH 3.5. Iron parts treated with the rust-transforming composition prepared as described have the yellowish film of surface rust immediately transformed into a black corrosion-resistant coating containing complexed iron compounds in the synthetic binding agent.

Example 2: A polyester is prepared by esterifying phthalic anhydride with gallic acid in ethylene glycol at about 100°C for a period of 1 hour using p-toluenesulfonic acid as catalyst to give a polymeric product containing phthalic acid anhydride units to gallic acid units in the ratio of about 1:1. Removal of the ethylene glycol gave the required polyester in the form of a powder. The following ingredients are mixed together to give a rust-transforming composition:

	Weight Percent
Polyester*	5
Vinyltoluene-butadiene resin dispersion**	30

(continued)

	Weight Percent
Styrene-acrylate copolymer***	25
Isopropanol	5
Water	34.5
Phosphoric acid	0.5

*Powder prepared as described above.
**Containing 48 wt % polymer and 2 wt % monomer.
***In solution in heavy benzene 50%.

The styrene-acrylate copolymer solution is emulsified into the vinyltoluene-butadiene resin dispersion. By means of the phosphoric acid (which acts as a catalyst for transforming the rust), the product is adjusted to a pH of 3. The resulting composition when applied to rusty iron parts converts the film of surface oxide into a black rubberlike corrosion-protective coating.

PRETREATMENT TO PASSIVATE SURFACE

Pretreatment with Ascorbic Acid-Containing Composition and Molybdenum

It is known that chromic acid pretreatments passivate and improve the corrosion resistance and coating properties of metals, particularly ferrous metal surfaces. However, chromic acid pretreatments are undesirable because chromic acid is toxic and its effluent creates serious pollution problems.

A method of providing corrosion resistance to a metal surface is provided by *N.T. Castellucci; U.S. Patent 4,120,996; October 17, 1978; assigned to PPG Industries, Inc.* The method comprises pretreating the metal surface with an ascorbic acid-containing composition free of β-diketones, and preferably in combination with elemental molybdenum to passivate the surface, followed by directly coating the passivated metal surface.

The pretreatments render the metal surfaces resistant to corrosion without the application of a coating detectable by eyesight or weight change of the pretreated metal. Thus, passivating pretreatments of the process are distinguished from coating metal surfaces with corrosion-inhibiting primer which are visually detectable and do result in a weight gain for the coated metal. Also, the treated metal surfaces obtained in accordance with the process are electrically conductive such that they can be subsequently electrocoated.

The weight ratio of ascorbic acid to molybdenum should be at least 5:1 and usually within the range of from about 5 to 50:1. For application to the metal surface, the molybdenum usually in powder form and ascorbic acid are usually mixed with a compatible liquid. Preferably, water constitutes about 90 to 100% by weight of the liquid. In general, the treating composition will contain preferably 0.05 to 5% by weight of molybdenum, preferably 5 to 24% by weight of ascorbic acid, and at least 75 and preferably 75 to 94% by weight of liquid, the percentages by weight being based on total weight of molybdenum, ascorbic acid and liquid.

Where ascorbic acid is used without molybdenum, it is used in a concentration of at least 5 and preferably from about 5 to 24% by weight of the treating composition, the percentage by weight being based on total weight of liquid and ascorbic acid.

The preferred way of treating the metal substrate is to form a liquid solution or dispersion of the treating ingredients and then immerse the metal to be treated. Previously treated iron phosphated steel only requires a 6 to 12 second immersion, whereas cold-rolled steel requires about 2.5 to 5 minute immersion for acceptable corrosion pretreatment. In general, the metal article should be immersed for at least 6 seconds, usually from about 8 seconds to 12 minutes, followed by removal of the article from the bath and rinsing. The temperature of the immersion bath is also important and usually is at least room temperature, preferably between 35° to 50°C.

After drying, the metal surface is then directly coated with a film-forming material. The coating can be an adhesive coating or a protective coating such as a layer of paint.

Example: A series of untreated and previously treated iron phosphated steel panels were dipped in a pretreatment bath containing 10% by weight ascorbic acid and 1% by weight of molybdenum metal powder.

Alkaline-rinsed, cold-rolled (untreated) and previously treated iron phosphated steel panels were dipped in the acid bath at 40°C for various times as is reported in the table below. The steel panels were then removed from the bath and further processed as described in the table below.

After drying, the pretreated panels were topcoated with a thermosetting acrylic coating composition, Duracron 200 (PPG Industries, Inc.). The topcoating was accomplished by drawing down to approximately 1 mil thickness with a drawbar. The topcoated sample was then baked for 8 minutes at 400°F (204°C), scribed with an "X" and placed in a salt spray chamber at 100°F (38°C) and 100% relative humidity atmosphere of a 5% by weight aqueous sodium chloride solution for 1 week. The panels were then removed from the chamber, dried, the scribe mark taped with masking tape, the tape pulled off at a 45° angle and the creepage from the scribe mark measured. The results are reported in the table below. Creepage is a rusted darkened area of the panel where the coating has lifted from the panel surface.

No.	Previous Steel Panel Treatment	Ascorbic Acid, Elemental Molybdenum Pretreatment Conditions	Scribe Creepage (mm)
*	Untreated	None	**
*	Iron phosphated	None	2.5
1	Untreated	Dip 5 min, rinse with deionized water, blow dry	4
2	Untreated	(Same as 1), panel baked 5 min at 400°F (204°C) after blowing dry	5
3	Untreated	Dip 5 min, blow dry, bake 5 min at 400°F (204°C)	31
4	Iron phosphated	Dip 6 sec, rinse with deionized water, blow dry	5
5	Iron phosphated	(Same as 4), panel baked 5 min at 300°F (149°C)	4
6	Iron phosphated	(Same as 4), panel baked 5 min at 400°F (204°C)	7

(continued)

No.	Previous Steel Panel Treatment	Ascorbic Acid, Elemental Molybdenum Pretreatment Conditions	Scribe Creepage (mm)
7***	Iron phosphated	(Same as 4), panel drip dried at room temperature rather than blowing dry	1.2
8	Iron phosphated	(Same as 7), panel baked 5 min at 300°F (149°C) after dripping dry	1.2
9	Iron phosphated	(Same as 7), panel baked 5 min at 400°F (204°C) after dripping dry	2.8
10	Iron phosphated	Dip 6 sec, blow dry, bake 5 min at 400°F (204°C)	**

*Control.
**Complete delamination.
***Preferred embodiment of the process.

The results indicate that rinsing the pretreated panels is necessary to obtain good results. Also, baking the pretreated panels can be detrimental. Further, it was noted that without a rinse, the molybdenum powder mixed with the Duracron coating during drawdown to give a smeary gray-white surface topcoat.

Pretreatment with Onium Salt

N.T. Castellucci and J.F. Bosso; U.S. Patent 4,053,329; October 11, 1977; assigned to PPG Industries, Inc. provide a method of improving corrosion resistance of metal substrates, particularly ferrous metal substrates. The method comprises first passivating the surface of the substrate by pretreatment with an onium salt followed by direct coating of the pretreated metal surface. Examples of onium salts are those selected from the class consisting of: $(R)_4N^+A^-$, $(R)_4P^+A^-$ and $(R)_3S^+A^-$ where R are organic radicals and A is an anion of an acid which will not detrimentally attack the surface of the substrate.

Preferably, the onium salt is a monomeric phosphonium salt such as ethyl triphenyl phosphonium acetate and tetrabutyl phosphonium acetate. In a second preferred variation, the onium salt is a quaternary ammonium salt derived from a polyepoxide.

The term "passivating" means rendering the surface of the substrate resistant to corrosion without applying a visually detectable coating. The amount of pretreating material applied to the surface of the substrate is less than about 100 mg/ft^2 (328 mg/m^2).

The pretreating composition is applied to the surface of the metal substrate in any convenient manner such as by immersion, spraying or wiping the surface either at room temperature or at elevated temperature. The preferred way of pretreating the metal substrate is to form an aqueous dispersion or solution of the onium salt and then immerse the metal to be treated.

The times and temperatures of immersion can be critical, depending on the onium salt and its concentration, and on the identity of the metal substrate treated. In general, the metal article should be immersed at a bath temperature of about 25° to 50°C, preferably 40° to 45°C, for at least 5 seconds, usually for about 5 seconds to 5 minutes, followed by removal of the article from the bath, and optionally rinsing with deionized water. The article is then dried. Preferably, the article is dried with forced air, and then baked at elevated temperature.

After drying, the metal has sufficient corrosion protection so that it can be exposed to the atmosphere without danger of atmospheric oxidation on the surface. The metal substrate is then directly coated. The coating can be adhesive coating or a protective coating such as a layer of paint.

The process is particularly useful for treating ferrous metal substrates. The substrates can be untreated steel or steel which has been previously pretreated such as iron phosphated or zinc phosphated steel substrates.

Example: A series of steel panels, both untreated and iron phosphated, were dipped in a 5% solids pretreatment bath of ethyl triphenyl phosphonium acetate. Pretreatment conditions were varied as reported below. The pretreated panels were coated and the coating baked, scribed and exposed to a salt spray fog as described in U.S. Patent 4,120,996 above. After one week exposure, the scribe creepage was measured and the results are reported below.

No.	Steel Panel	Pretreatment Conditions	Scribe Creepage (mm)
1	Untreated	Dipped 5 min at bath temperature of 60°C, drip dried at room temperature	34
2	Untreated	Dipped 5 min at 60°C, rinsed with deionized water and blown dry	9
3	Untreated	Dipped 5 min at 60°C, rinsed with deionized water and blown dry, then baked at 400°F (204°C) for 5 min after blowing dry	2
4	Iron phosphated	Dipped 10 sec at 60°C, drip dried at room temperature	25
5	Iron phosphated	Dipped 10 sec at 60°C, drip dried at room temperature, then baked 5 min at 400°F (204°C) after drip drying	23
6	Iron phosphated	Dipped 10 sec at 60°C, rinsed with deionized water and blown dry	3
7	Iron phosphated	Dipped 10 sec at 60°C, rinsed with deionized water and blown dry, then baked 5 min at 400°F (204°C) after blowing dry	0.7
8	Untreated	Dipped 5 min at 25°C, rinsed with deionized water and blown dry	16
9	Untreated	Dipped 5 min at 60°C, rinsed with deionized water and blown dry, then baked at 400°F (204°C) for 5 min after blowing dry	20
10	Iron phosphated	Dipped 6 sec at 25°C, rinsed with deionized water and blown dry	3.8
11	Iron phosphated	Dipped 6 sec at 25°C, rinsed with deionized water and blown dry, then baked 5 min at 400°F (204°C) after blowing dry	2.3
12	Untreated	Dipped 1 min at 40°C, blown dry, rinsed with deionized water and blown dry	8.5

(continued)

No.	Steel Panel	Pretreatment Conditions	Scribe Creepage (mm)
13	Untreated	Dipped 1 min at 40°C, blown dry, rinsed with deionized water and blown dry, then baked 5 min at 400°F (204°C) after last blow dry	8.5
14	Iron phosphated	Dipped 6 sec at 40°C, blown dry, rinsed with deionized water and blown dry	2
15	Iron phosphated	Dipped 6 sec at 40°C, blown dry, rinsed with deionized water and blown dry, then baked 5 min at 400°F (204°C) after last blow dry	0.6

Embedding Ferrous Abrasive Particles in Surface Followed by Phosphating

Stainless steel is steel alloyed with chromium, nickel and other elements. Materials referred to as stainless steel generally contain a minimum of 12% chromium. Its outer surface exposed to air forms a protective oxide coating. However, contrary to popular belief, stainless steel will corrode. It needs oxygen to maintain its protective capability and if there is no oxygen, stainless steel will corrode.

If stainless steel is to be coated with a black coating for nonreflective purposes, the stainless steel must be activated before the black coating will form on it. This requires removal of the protective oxide coating. This may be done by dipping it into an acid bath, usually sulfuric or hydrochloric acid. The activated stainless steel is then coated by immersion in a blackening bath.

Cast stainless steel also requires corrosion-proofing. This steel is heat treated to get a desired high strength material. It corrodes readily due to the formation of an iron-rich as-cast surface that readily corrodes when exposed to air.

In accordance with the process of *J.T. Menke; U.S. Patent 4,194,929; March 25, 1980; assigned to U.S. Secretary of the Army*, stainless steel is activated by abrasive blasting with steel grit to remove the protective oxide coating. This grit is embedded into the surface, making the surface suitable for phosphating by spray or dip techniques. This treatment forms a protective passive coating over which supplementary coatings of paint, lubricating oils, or preservative compounds may be added, if desired. Cast stainless steel may be phosphated without steel grit blasting, since its surface is already iron-rich.

The embedded particles of ferrous metal abrasive generates a galvanic potential with the base metal and constitute local anodes when the treated steel is immersed in an electrolyte. A large cathode, small anode area relationship is established between the embedded metal anodes and the large corrosion-resistant steel cathode. When further treated in an acid phosphate solution or spray of heavy metal ions such as zinc, manganese or iron, a progressive depletion of the embedded ferrous metal anodes occurs, and a simultaneous electrolytically equivalent cathode reaction occurs, resulting in an extremely corrosion-resistant surface.

Example: To prepare the stainless steel panel with steel grit embedded in its surface, grit of 80 mesh size was applied with a force of 80 to 120 pounds per square inch (psi) pressure. These embedded particles generate a galvanic potential with the stainless steel and constitute local anodes when the treated steel is immersed in an electrolyte. When treated in an acid phosphate solution a progressive depletion of the embedded ferrous metal anodes occurs, and a simultaneous electrolytically equivalent cathode reaction occurs, resulting in an extremely corrosion-resistant surface. This reaction is self-terminating, ending when equilibrium is reached, i.e., when all anode and cathode reactions have been completed. A completely passive surface remains.

Phosphatizing processes for ferrous and other metals are described in Chapter 1.

LUBRICANT COATED STEEL SUITABLE FOR IRONING

Complex Coating

"Ironing" is a method which, after squeezing a sheet by the use of a suitable punch and die to form a cup, elongates the side wall of the cup by the use of a punch and die having a clearance between the die and the punch smaller than the thickness of the side wall of the squeezed cup, thus decreasing the thickness of the side wall to obtain a cuplike container. After an end plate is fixed thereto, the so-called two-piece can is obtained.

The ironing is a severe processing to a material. Accordingly, when a steel sheet is subjected to the processing by the use of a normal cutting or machine oil, a phenomenon of scorch occurs between the die and the material, which gives rise to deep linear scars on the surface of the body of the can.

According to *H. Asano, S. Maeda and Y. Oyagi; U.S. Patent 4,235,947; Nov. 25, 1980; assigned to Nippon Steel Corp., Japan,* there is provided a method for the manufacture of a steel sheet adapted for use in ironing having good lubrication property which is treated by applying to the surface of the steel sheet one aqueous solution selected from the group consisting of (a) an aqueous solution of ammonium phosphate, (b) an aqueous solution of ammonium molybdate, (c) a mixed aqueous solution of ammonium phosphate and ammonium molybdate, and (d) an aqueous solution of nickel salt; heating the steel sheet in an atmosphere of an inert or reducing gas to form a surface film caused by the thermal decomposition thereof; and thereafter applying thereto a lubrication oil consisting essentially of an animal or vegetable oil or fat or a mineral oil, as a basic oil, a high molecular compound, and a higher fatty acid.

A steel sheet which has preliminarily been degreased and washed is dipped in an aqueous solution of ammonium phosphate, ammonium molybdate or nickel acetate whereby the salt is coated on the steel sheet in a suitable amount by means of roll squeezing technique. Thereafter the sheet is dried by hot blast, and subjected to a heat treatment. The heat treatment is effected in the presence of a nonoxidizing gas such as N_2, H_2, mixture of N_2 and H_2, or Ar and the like. The heating temperature should preferably be between 200° and 750°C. As a result of the heat treatment, the ammonium phosphate, the ammonium molybdate or the nickel acetate is decomposed to form a film on the surface which is chiefly composed of phosphorus, molybdenum or nickel. The film thus formed will act to enhance the preservation of a lubrication oil having the abovementioned fundamental composition which is to be subsequently applied, and shows excellent resistance to the scar with an aid of a lubrication effect of the film itself.

The film amount of the aqueous solution should preferably be 2 to 200 mg/m^2 in the case of the aqueous solution of ammonium phosphate, 5 to 300 mg/m^2 in the case of the aqueous solution of ammonium molybdate and 5 to 300 mg/m^2 in the case of the aqueous solution of nickel. The optimum lubrication effect can be obtained in the range of the above amount.

Example: A cold-rolled steel sheet of 0.35 mm thickness before annealing was subjected to degreasing and washing. It was dipped into various treating solutions as shown in the table which consisted mainly of ammonium phosphate or ammonium molybdate, and then dried by hot blast after or without roll squeezing. It was subsequently heated at a temperature of 600°C or so in an atmosphere of N_2 plus H_2 mixed gas, so that the thermal decomposition of the coated agent and the annealing for removal of strain was concurrently effected. A film was thus formed. A skin-pass rolling was then carried out under a reduction rate of 1%. The surface-coated steel sheet thus obtained was further coated with high-molecule-containing lubrication oils shown in the table and thereafter subjected to ironing processing.

The ironing process was conducted by continuously making a lot of cans, using an Erichsen testing machine and the lubrication property was evaluated by the number of cans processed before which the scars appeared.

The conditions for ironing processing were as follows: cupping, one step; ironing, two steps; diameter of can, 50 mm; and processing rate for ironing (rate of decrease of thickness), 70%.

Treating Bath Composition (g/ℓ)		Treating Method*	Lubrication Oil Composition	Amount Applied (g/m^2)	No. of Cans Before Scars
Diammonium hydrogen phosphate	15	(A)	Tallow 70% + polypropylene (MW 10,000, atactic) 20% + lauric acid 10%	1	>200
Ammonium dodecamolybdate	20				
Diammonium hydrogen phosphate	10				
Ammonium dodecamolybdate	10				
Diammonium hydrogen phosphate	10	(B)	Spindle oil 70% + polypropylene (MW 12,000, atactic) 20% + oleic acid 10%	1	>200
Nickel acetate	10				
Ammonium heptamolybdate	10				
Chromium acetate	10				
Ammonium dihydrogen orthophosphate	10				
Nickel nitrate	10				
No treatment		–	Palm oil 70% polypropylene 20% lauric acid 10%	0.5	160
No treatment		–	Palm oil 70% polypropylene 20% lauric acid 10%	10	>200
No treatment		–	Cutting oil #620	10	<5

*(A) was dip coating, hot blast drying, heat-treatment, H_2 (10%) + N_2 (90%) gas, 600°C heating; (B) was dip coating, roll squeezing, heat treatment, H_2 (10%) + N_2 (90%) gas, 630°C heating.

In the above table, the "atactic" polypropylene means one of the three isomers of the polypropylene.

Composite Oil Coating

The work of *H. Asano and S. Maeda; U.S. Patent 4,027,070; May 31, 1977; assigned to Nippon Steel Corp., Japan* relates to a lubricating agent that provides luster to a material to be treated and also prevents scratching, for a special working treatment, such as ironing.

This is achieved with use of those lubricating oils which are produced from organic high molecular weight compounds (MW 500 to 30,000), e.g., polybutene, polypropylene, polyisobutylene and polyacrylic ester by diluting them with a mineral oil, such as, machine oil or an animal or vegetable oil, such as, tallow oil and palm oil. These oils are mixed, if necessary, with saturated or unsaturated monocarboxylic acids, such as, lauric, stearic, palmitic and oleic acids. These lubricating oils are liquid or semisolid at room temperature, depending on the amount and the average molecular weight of the high molecular substances added, and they are readily degreased by spraying a suitable commercially available degreasing agent. To increase the degreasing efficiency, if necessary, mono- or diglyceride of an aliphatic carboxylic acid or a commercial ionic or nonionic surfactant is added beforehand.

Further, the critical ironing rate can be made larger without sacrificing the surface appearance, if the steel plate to be treated is coated beforehand with phosphate of zinc, manganese or iron in a thickness of 0.2 to 10 mg/dm^2 before the abovementioned lubricating oils containing diluted high molecular substances is applied.

The optimum concentration of the high molecular substances to be added depends on type of material and the average molecular weight. Substances of higher molecular weight need a smaller amount to exhibit the effect. For instance, polypropylene of an average molecular weight 10,000 to 15,000 shows the highest effect when added in an amount of 20 to 30% (against tallow or spindle oil).

Example 1: A lubricating oil consisting of 70 parts of tallow (extra fancy class), 20 parts of polypropylene (atactic, MW 12,000) and 10 parts of lauric acid was applied to the thickness of 5 mg/dm^2. A 0.35 mm thick steel plate (tin-plated, 1-grade tempered) was subjected to ironing with an ironing rate of 50%. More than 100 cans were produced successively. Every can had a beautiful lustrous appearance without forming a scratch. For comparison, machine oil was applied to the same kind of material, when the appearance was beautiful and lustrous, but a scratch was formed from the third can and after, and many gashes were observed from the fifth or eighth can onward.

Example 2: A lubricating oil consisting of 80 parts of tallow, 10 parts of methyl polymethacrylate with the average molecular weight 28,000, 7 parts of stearic acid and 3 parts of monoglyceride of stearic acid was applied to the surface of a steel plate to the amount of 5 mg/dm^2. The steel plate thus treated was processed by ironing with an ironing rate of 50%. No scratching nor gashing occurred when more than 100 cans were produced successively. Appearance was good and beautiful.

Example 3: A lubricating oil consisting of 70 parts of tallow, 20 parts of polypropylene and 10 parts of lauric acid was applied to a steel plate which had been subjected to a zinc phosphate treatment so as to be covered with a coating of 1 mg/dm^2 and then processed by ironing.

No crack was observed even at an ironing rate of 70%, while a steel plate which was not treated with zinc phosphate made a crack at an ironing rate of 60%. However, both kinds of products showed no scratch nor gash and had good lustrous appearance.

PREPARING STEEL FOR FLUXLESS HOT DIP COATING

Forming Visible Iron Oxide Layer on Stock Surface

In the fluxless hot dip metallic coating of steel strip and sheet stock it is necessary to subject the surfaces to a preliminary treatment which provides a clean surface free of iron oxide scale and other surface contaminants, and which is readily wettable by the molten coating metal in order to obtain good adherence. Two types of in-line anneal preliminary treatments are in common use in this country, one being the so-called Sendzimir process or oxidation-reduction practice (disclosed in U.S. Patents 2,110,893 and 2,197,622), and the other being the so-called Selas process or high intensity direct-fired furnace line (disclosed in U.S. Patent 3,320,085).

In the Sendzimir process steel strip or sheet stock is heated in an oxidizing furnace to a temperature of about 370° to 485°C without atmosphere control, withdrawn into air to form a controlled surface oxide layer varying in appearance from light yellow to purple or blue grey, introduced into a reduction furnace containing a hydrogen and nitrogen atmosphere wherein the stock is heated to about 735° to 925°C and the controlled oxide layer is completely reduced. The stock is then passed into a cooling section containing a hydrogen and nitrogen atmosphere, brought approximately to the temperature of the molten coating metal bath, and then led beneath the bath surface while still surrounded by the protective atmosphere.

In the Selas process steel strip or sheet stock is passed through a direct fired preheat furnace section, heated to a temperature above 1315°C by direct combustion of fuel and air therein to produce gaseous products of combustion containing at least about 3% combustibles in the form of carbon monoxide and hydrogen, the stock reaching a temperature of about 425° to 705°C while maintaining bright steel surfaces completely free from oxidation. The stock is then passed into a reducing section which is in sealed relation to the preheat section and which contains a hydrogen and nitrogen atmosphere, wherein it may be further heated by radiant tubes to about 425° to 925°C and/or cooled approximately to the molten coating metal bath temperature. The stock is then led beneath the bath surface while surrounded by the protective atmosphere.

According to *A.F. Gibson and M.B. Pierson; U.S. Patent 4,123,292; October 31, 1978; assigned to Armco Steel Corp.*, there is provided a method of preparing the surfaces of steel strip and sheet stock for fluxless hot dip coating with molten metal. The method comprises the steps of passing the stock through a first heating section under conditions which form a visible iron oxide layer on the stock surfaces within the color range of dark straw through blue, continuing the heating of the stock in a second heating section isolated from the first heating zone in an atmosphere containing less than 5% hydrogen by volume, thereby preserving the oxide layer, and cooling the stock approximately to the temperature of the molten coating metal in a cooling zone isolated from the preceding heating zones, the cooling zone containing a reducing atmosphere comprising at least 10% hydrogen by volume, whereby to reduce the oxide layer completely to a metallic iron surface wettable by the coating metal. The temperature to which the stock is heated in the successive heating sections is not critical as long as the formation of a thick oxide layer is avoided.

Referring to Figure 5.1, strip **10** is to be treated. A direct fired preheat furnace is shown at **32**, which may be heated to a temperature of at least about 1205°C by direct combustion of fuel and air. A further heating section, which is preferably a radiant tube furnace is shown at **34**, and baffle means **36**, **36a** are provided between the preheat furnace **32** and radiant tube furnace **34**, thus isolating one from the other. An inlet for nitrogen into furnace **34** is shown at **38**, and baffle means **42** is provided isolating the cooling section from the radiant tube furnace. An inlet for hydrogen or a hydrogen-nitrogen mixture into section **40** is shown at **44**. The use of baffles **36**, **36a**, and **42** is optional with the configuration shown, as an atmosphere containing less than 5% hydrogen by volume in furnace **34** may be alternatively insured by providing a sufficiently large flow of nitrogen through inlet **38**.

Figure 5.1: Modified Selas Line for Use in Process

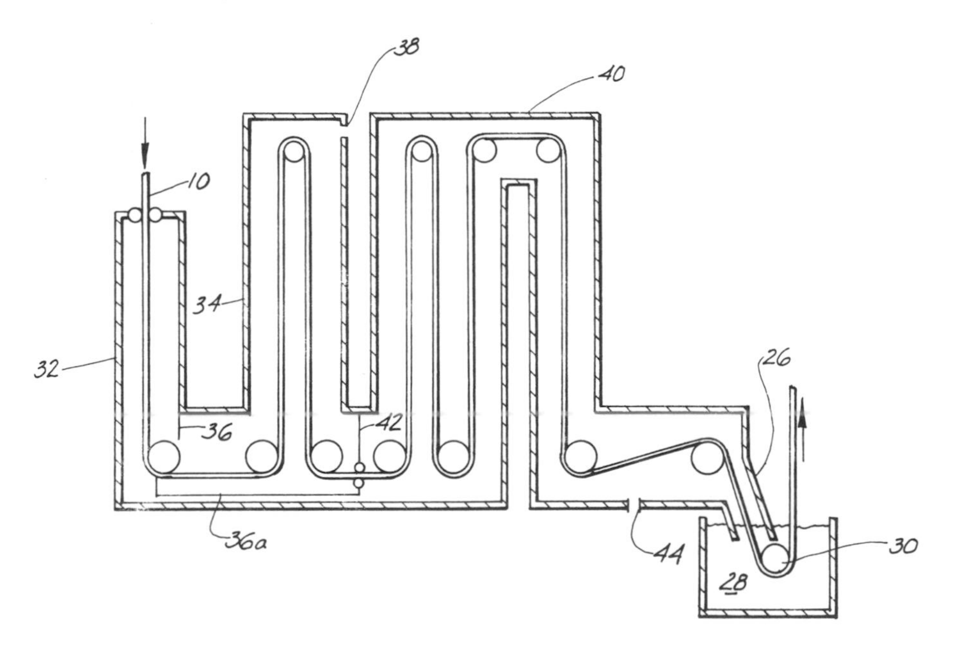

Source: U.S. Patent 4,123,292

Tests have been conducted on a production Selas-type line to compare conventional practice with that of the process. Conventional practice involved maintaining 3% by volume excess combustibles in the preheat furnace atmosphere (having five heating zones) and a 5% hydrogen-95% nitrogen atmosphere in the radiant tube heating section and cooling section. When conditions were altered to the practice of the process, i.e., approximately perfect combustion in the first four zones of the preheat furnace (0% excess combustibles and 0% oxygen), 1.5% oxygen in the fifth and final zone of the preheat furnace, less than 5% hydrogen in the radiant heating section, and 25% hydrogen-75% nitrogen by volume in the cooling section, it was found that the production rate was increased 10 to 20% in comparison to conventional practice. Complete removal of the oxide layer was effected in the cooling section in this test.

Further mill trials were conducted on low-carbon rimmed steel and aluminum-killed steel strip ranging in thickness from 0.054" to 0.099" (1.4 to 2.5 mm). In the preheat furnace conventional operating conditions were initially established as follows: zone 1 had 0.7% (by volume) excess combustibles; zone 2 had 0.4% excess combustibles; zone 3 had 0.6% excess combustibles; zone 4 had 0.2% free oxygen; zone 5 had 0.3% combustibles. The strip had a light straw color exiting zone 5, indicating an extremely thin oxide film, i.e., less than 10^{-5}" thickness.

Gas flow to the radiant tube furnace was then changed to pure nitrogen rather than the conventional nitrogen-hydrogen mixture, and zone 5 of the preheat furnace was adjusted to 1.3% excess oxygen in accordance with the method in this process. The strip exiting the preheat furnace then exhibited a reddish purple to blue iridescent color, indicating an oxide layer on the order of 10^{-5}" thickness. With this change in conditions the radiant tube furnace temperature began to drop (even though the firing rate was maintained at 100%), thus indicating greater heat transfer to the strip due to greater radiant energy absorptivity.

The strip exiting the radiant tube furnace was oxidized and ranged in temperature from 1350° to 1420°F (732° to 770°C), depending upon strip thickness and speed.

The cooling section was supplied with a 30% hydrogen-70% nitrogen mixture, and the oxide layer was reduced in the cooling section. The line speed was increased to 110% of scheduled speed (as a conservative measure) because of the higher strip temperature in the radiant tube section.

The trials were concluded by returning to conventional practice, and it was observed that the radiant tube section temperature increased and the strip temperature decreased therein, with the firing rate maintained constant at 100%.

The above trials show that close control of the initial heating zones in the preheat furnace is not essential so long as an oxidizing atmosphere (greater than 0% and up to 2% free oxygen) is maintained in the final zone or final two zones. The initial zones may thus be operated at perfect combustion or with up to about 1% by volume excess combustibles.

Forming Visible Sulfur and Oxygen Rich Layers in First Heating Zone

In all prior processes for preliminary treatment of steel strip and sheet surfaces which are exposed to atmospheres of direct fired furnaces, it has been considered that the presence of even small amounts of sulfur in the atmosphere would be highly deleterious. Accordingly, substantially sulfur-free fuel such as natural gas has been prescribed for use in such furnaces. However, natural gas shortages have made it necessary to consider alternative sources of fuel. In a steel mill having coke ovens, the use of coke oven gas as a fuel source would be an obvious choice except for the fact that raw coke oven gas ordinarily contains about 300 to 500 grains of sulfur per 100 cubic feet of gas, the sulfur being present primarily as hydrogen sulfide with a small amount of organic sulfur compounds.

J.L. Arnold, F.C. Dunbar, A.F. Gibson and M.B. Pierson; U.S. Patent 4,123,291; October 31, 1978; assigned to Armco Steel Corp. found that sulfur-bearing coke oven gas can be used as a fuel in direct fired furnaces for preliminary treatment of the surfaces of steel strip and sheet stock, and that greater increases in energy efficiency and/or production rates can be achieved in both the Sendzimir and Selas processes by increasing the radiant energy absorptivity of the steel stock. This absorptivity is increased by forming a film or layer rich in sulfur and oxygen on the stock surfaces in the initial direct fired (or preheat) furnace section, and by preserving this film throughout the heating sections.

It has been found that a film rich in sulfur and oxygen, which is thin and uniform, can be readily formed on the stock surfaces, and that this film can be easily reduced in a subsequent cooling section to produce a fresh ferrous surface which is readily wetted by molten coating metal, with resultant excellent adherence after solidification of the coating.

The temperature to which the stock is heated in the successive heating zones is not critical so long as the formation of a thick sulfur and oxygen containing scale is avoided. In general, the temperatures may be the same as those described above for conventional practice, i.e., for the Sendzimir process a range of about 370° to 485°C in the oxidizing furnace and about 735° to 925°C in the further heating zone.

Referring to Figure 5.2, steel strip to be treated is indicated at **10**, the direction of travel being shown by arrows. An oxidizing furnace is shown at **12**, which is heated to a temperature of, e.g., about 870°C by combustion of scrubbed coke oven gas. A second heating section, which may be a radiant tube furnace, is shown at **14**. An inlet for nitrogen into the second heating section is provided as shown at **16**. Baffle means **18** are provided between heating section **14** and cooling section **20**, which isolate the atmospheres in each section from one another.

Figure 5.2: Modified Sendzimir Line

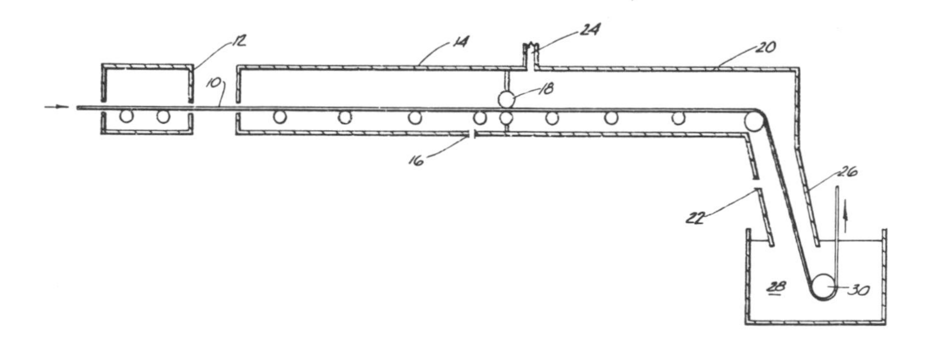

Source: U.S. Patent 4,123,291

A hydrogen inlet into cooling section **20** is shown at **22**, and a stack for flaring hydrogen is provided as shown at **24**. A protective snout **26** extends downwardly beneath the surface of a coating metal bath **28**, which surrounds strip **10** as it is conducted beneath the surface of the bath, around a reversing roll **30** and vertically upwardly out of the bath. Any conventional finishing means (not shown) may be used for metering and solidifying the metal coating.

Since the oxidizing furnace **12** and heating section **14** are separate, with heated strip **10** exposed to atmosphere therebetween, it is evident that the atmospheres of each are isolated from one another.

When practicing an anneal cycle the stock is brought to a temperature of about 427° to about 705°C in the preheat zone and to a maximum temperature of about 788°C in the radiant tube zone. When practicing a full hard cycle the stock is brought to a maximum temperature of about 565°C in the preheat zone and a maximum of about 538°C in the radiant tube zone. In the full hard cycle the hydrogen content in the cooling zone is preferably increased to about 40% by volume.

The amount of sulfur present in the coke oven fuel and in the atmosphere of the preheat furnace has been found to have little effect on the nature of the sulfur and oxygen rich film formed on the strip surfaces and may be varied from about 5 to about 1,600 grains per 100 cubic feet in the coke oven gas (about 0.007% to about 2.6% hydrogen sulfide by volume at standard temperature and pressure). Similarly, variations in sulfur content have little influence on coating metal adherence except in the practice of a full hard cycle wherein the maximum strip temperature is about 565°C. Under these conditions an increase in the hydrogen content in the cooling zone to about 40% by volume will result in improvement in coating adherence, as indicated above.

IRON-CHROMIUM ALLOYS

Surface Layer of Oxidized Chromium

The work of *G. Hultquist and C. Leygraf; U.S. Patent 4,168,184; September 18, 1979*, relates to a method of making a surface layer with improved corrosion properties on articles of iron-chromium alloys where the article is heated in an oxygen-containing atmosphere with vacuum.

The article is heated in an oxygen-containing gaseous atmosphere with a total vacuum of about 10^{-1} down to 10^{-8} mm Hg, preferably about 10^{-5} mm Hg, to a temperature at which diffusion of the most reactive alloying component or components, primarily chromium, with respect to its tendency of combining with oxygen, to the surface becomes perceptible, at which temperature the remaining properties of the alloy substantially are maintained. The abovedescribed condition for the article is maintained until a surface layer containing oxidized chromium has grown to a thickness of preferably 10^{-9} to 10^{-7} m, whereby the resulting surface layer with respect to structure and composition transforms by degrees from a condition on the surface to a condition in the matrix of the metal alloy.

This process thus, distinguished from conventional methods where in all cases a surface layer is applied externally to a substrate, relates to a method at which the surface layer is made by means of alloying components comprised in the substrate. The surface layer thereby is formed by heating the substrate, which is a chromium steel, in a gaseous atmosphere at vacuum containing a low and well-controlled oxygen content.

The temperature, to which the article is heated, is chosen so that diffusion of the most reactive alloying component or components to the surface becomes perceptible. A preferred temperature is about 500°C. Surface layers, thus, with improved corrosion properties have been observed after heating in gaseous atmosphere in the range 300° to 550°C, with optimum results about 475°C.

Example 1: (a) A plate of iron-chromium alloy containing 5% chromium was exposed on a surface of 20 mm^2 to tap water for 2 hours. After 2 hours a corrosion current of 10^{-5} A was obtained. (b) A plate similar to that in (a) was exposed according to the process to a treatment consisting of heating at 475°C for 6 hours at the total pressure 1.8 x 10^{-5} mm Hg. The resulting surface layer was characterized by a gradual transformation from a high chromium content farthest outwardly in the surface layer to the low chromium content in the matrix of the alloy. When 20 mm^2 of this treated surface were exposed to tap water, after 2 hours a corrosion current of 10^{-7} A was obtained. The one hundred times lower corrosion current was corresponding also to a hundred times smaller decrease in weight.

Example 2: Bright annealed plates of stainless steel SIS 2343 (as regards composition, see designation A in the table below) were exposed on surfaces of 20 mm^2 at room temperature to 3% neutral NaCl solution in the presence of a gap and during simultaneous potentiodynamic load, which in steps was increased by 50 mV/min. The potential, at which local attacks were initiated adjacent the gap, is hereinafter called crevice corrosion potential. The samples, which were exposed to a treatment according to the process comprising heating at 475°C for 2 hours at a total pressure of 5 x 10^{-5} mm Hg, showed a crevice corrosion potential, which on the average was 550 mV higher than for corresponding untreated bright annealed steel samples.

Example 3: Pickled plates of stainless steel SIS 2333 (as regards composition, see designation B in the table below) were subjected to a similar corrosion test as described in Example 2. The samples, which were subjected to a treatment according to the process comprising heating at 475°C for 2 hours at the total pressure of 7 x 10^{-6} mm Hg, showed a crevice corrosion potential, which on the average was 130 mm higher than for corresponding untreated pickled steel samples.

Chemical Analysis in Percent by Weight

Designation	Material	C	Si	Mn	Cr	Ni	Mo
A	SIS 2343	0.046	0.38	1.30	17.7	11.0	2.74
B	SIS 2333	0.040	0.47	1.67	18.8	9.1	0.53

Prevention of Sulfidation Attack by Oxidation Pretreatment

The attack of sulfur compounds on metals at elevated temperatures is an exceedingly important phenomenon in petroleum refining. Compounds naturally occurring in crude petroleum and other compounds formed during the processing of the oil may corrosively attack at various places within the processing equipment.

Z.A. Foroulis; U.S. Patent 4,017,336; April 12, 1977; assigned to Exxon Research and Engineering Co. discloses a method of producing a protective film on iron and iron alloys which may be used to significantly reduce the corrosion rate which would be otherwise experienced and thereby providing a longer useful service life for the equipment and reducing the cost of sulfidation attack.

It was found that resistance of carbon steel, iron-chromium alloys and iron-chromium-nickel alloys to sulfur compound attack is improved by pretreating the metal under controlled conditions to form an extremely thin oxide film which serves as a barrier to sulfur attack. Pretreatment is accomplished by heating at temperatures above that to be experienced in refining service and in the presence of various oxidizers, preferably air, with conditions being controlled so as to favor the formation of the barrier oxide film. The temperature of treatment will vary depending upon the nature of the alloy to be treated.

The effectiveness of the pretreatment is illustrated in the following table which shows corrosion data at 700°F for various steels which have been exposed to an Arabian light crude containing 1.58% total sulfur, with and without preoxidation treatment.

	Corrosion Rate, mils/yr				
		With Pretreatment, Temperature of Preoxidation in Air			
	Without Pretreatment	800°F	900°F	1000°F	1200°F
Carbon steel	101	93	100	17	231
2½ Cr steel	103	110	166	76	83
5 Cr steel	105	88	101	35	42
9 Cr steel	91	72	69	33	0
12 Cr steel	77	68	43	23	11.2
304 Cr-Ni steel	30	22	22	9.5	24
347 Cr-Ni-Cb-Ta steel	24	22	30.9	13.5	3.6

The above table illustrates that the corrosion rate is extremely high without the preoxidation treatment, illustrating also that adding chromium and nickel reduces the corrosive attack to acceptable but still high levels. Thus, chromium and chromium-nickel steels are ordinarily used in protecting refining equipment against sulfidation attack.

The table also illustrates that preoxidation treatment must be done at carefully preselected temperatures if the film formed is to be protective. It appears that carbon steel and low-chrome steels are best protected by preoxidation temperatures at about 1000°F, whereas somewhat higher temperatures, about 1200°F, are more effective for the higher chromium and chromium-nickel steels. The table illustrates the desirability of determining the precise conditions under which preoxidation may have its best effect. Precisely 1000°F for carbon steel and low chromium steels, or alternatively 1200°F for high chromium steels or stainless steels are not necessarily the optimum pretreatment conditions. It is believed that the nature of the oxide film may vary in structure depending upon the conditions which have been selected for pretreatment and that careful selection must be made in order to obtain the optimum performance.

With resulting oxide film being submicroscopic, it is difficult to analyze the effect of changes in the pretreatment conditions independently of their secondary effect in mitigating sulfidation attack. The significant reduction in corrosion rate, which is found with properly selected pretreatment conditions, may be very usefully applied in the specification of materials for use in refining equipment. The preoxidation treatment itself would ordinarily be done during the prestartup period prior to introducing crude oil into the processing units.

Treatment in Oxygen with Volatile Inorganic Boron Compound

The work of *G.O. Lloyd, J.E. Rhoades-Brown and S.R.J. Saunders; U.S. Patent 4,019,926; April 26, 1977; assigned to U.K. Secretary of State for Industry, England* relates to a process for the diminution of high-temperature oxidation of iron alloys containing chromium.

The expression "high-temperature oxidation" is defined as oxidation which occurs on surfaces of metal at temperatures in excess of about 400°C when the surfaces are in contact with air, carbon dioxide or other oxidizing gases, i.e., the gaseous products of combustion of coal or fuel oil.

This oxidation often forms protective layers by producing a barrier of oxide between the metal and the oxidizing environment. The effectiveness of this barrier in reducing oxidation varies widely with different metals. Thick layers of oxide result in changes in dimensions and loss of strength, because of the conversion of metal to much weaker oxide.

It was found that iron alloys containing chromium, particularly alloys containing from about 5 to about 20% chromium, may be given protection against high-temperature oxidation by enclosing or partially enclosing the alloy with a volatile inorganic boron compound, heating the boron compound and allowing the volatile constituents of the boron compound to contact at least one surface of the alloy.

The expression, "volatile boron compound" includes all compounds of boron which are known to produce a detectable deposit containing boron by transfer of material through the vapor phase in the conditions of the oxidation experiment. This expression includes borosilicate glasses, sodium metaborate, sodium tetraborate, boric acid, and boric oxide.

The processes confer excellent resistance to high-temperature oxidation, up to a temperature of at least 900°C. Various modifications of the process are known. In a first modification the alloy and the inorganic compound are heated together to a temperature in the range about 400° to about 900°C.

In another, the inorganic compound is heated by a supplementary heater to a higher temperature than the alloy. The alloy may initially be maintained at a lower temperature than the inorganic compound and be subsequently heated to its working temperature or may be heated to its working temperature from the start.

In either case, the enclosure may be the final assembly containing the parts to be protected in high-temperature surface or may be a temporary enclosure loaded with the parts for treatment.

The following examples give details of tests actually carried out. All the tests were carried out in air in laboratory furnaces, using an iron alloy containing 10% chromium. Thicknesses of scale formed were estimated by weighing the specimens to measure the weight of oxygen taken up. A weight gain of 100 $\mu g/cm^2$ corresponds to a thickness of oxide of approximately 0.6 μm.

Example 1: Strips of 10% chromium iron sheet were electropolished to a mirror finish and were exposed to flowing air at 600°C in a furnace tube of mullite in which a small quantity of Pyrex borosilicate glass (No. 774) had been melted at about 900°C to form a band about 1 cm wide around the inner surface in the central zone of the furnace.

The specimens were mounted in a silica frame, and were not allowed to touch the wall of the tube. Weight gains were on average about 10 $\mu g/cm^2$ in periods up to 1 hour, increasing to about 20 $\mu g/cm^2$ in 28 days; the largest gains recorded in the entire series of experiments were about 50 $\mu g/cm^2$. Oxide films of this thickness correspond merely to a dulling of the initial highly polished specimen.

Similar specimens oxidized in air in a furnace not treated with Pyrex glass gained about 4 mg/cm^2 in 6 days and 10 to 12 mg/cm^2 in 28 days. Complete destruction of the specimen corresponds to a gain of some 25 mg/cm^2.

Similar specimens oxidized in the protected furnace at 900°C gained about 3 mg/cm^2 in 24 hours. The oxide was adherent at temperature but spalled on cooling. A specimen in a nonprotected furnace was completely consumed within 24 hours.

Cold-rolled and abraded specimens of the same sheet material were exposed in flowing air in the protected furnace at 600°C. Weight gains in the first 24 hours averaged about 200 $\mu g/cm^2$. After 21 days, this increased to only 280 $\mu g/cm^2$.

Example 2: Strips of the same alloy were electropolished to a mirror finish and were oxidized in a silica frame surrounded by a cylinder of borosilicate glass (Pyrex glass No. 774) about 2 cm in diameter. The glass was flame-polished after cutting. It was secured by gold wire to the silica specimen holder, and was allowed to touch neither the specimens nor the mullite furnace tube.

Weight gains after 3 days in flowing air at 600°C averaged 60 $\mu g/cm^2$, the largest weight gain in a large batch of specimens being 230 $\mu g/cm^2$. Weight gains in the same furnace tube before the use of the Pyrex glass cylinder were about 480 $\mu g/cm^2$ in 3 days at 600°C.

After dismantling the specimen holder, and removing the glass cylinder, a further batch of specimens of the same alloy showed weight gains at least as low as those obtained in the presence of the glass. Baking out the furnace tube at 1300°C for 3 days restored the normal nonprotective oxidation behavior.

A particular advantage of this process is that it makes it possible to produce a protective oxide film on articles which have already begun to oxidize rapidly in a nonprotective atmosphere, without either removing the oxide already formed, or immersing in liquids. A further advantage is that it can be applied to treat large, complex or inaccessible equipment either after fabrication or after deterioration has occurred in service.

COATED ELECTRICAL STEEL SHEETS

Coating with Particle Separating Resin, Chromic Acid, and Compatible Resin

The work of *T. Irie, T. Tanda, T. Ichi and T. Sadayori; U.S. Patent 4,032,675; June 28, 1977; assigned to Kawasaki Steel Corporation, Japan* relates to a method for producing coated electrical steel sheets having excellent punchability, weldability, electrical insulation and heat resistance.

The process is characterized in that a treating dispersion is coated on an electrical steel sheet and then baked. The treating dispersion is obtained by compounding a resin emulsion which separates agglomerated particles having an average particle diameter of 3 to 40 μ when such a resin emulsion is compounded in an aqueous solution containing chromic acid and a resin having a compatibility to an aqueous solution containing chromic acid, to the aqueous solution containing chromic acid in such an amount that the total amount of the nonvolatile component of both the above resins is 5 to 150 pbw based on 100 pbw of chromic acid (CrO_3), and an amount of the nonvolatile component of the former resin emulsion to separate the agglomerated particles is at least 1 pbw among the abovedescribed total amount and is at least 5% by wt based on the total amount of the nonvolatile fraction of both the above resins.

Example 1: An acrylic resin having a monomer composition composed of 82 parts of methyl methacrylate, 10 parts of butyl acrylate and 8 parts of methacrylic acid was used as a particle separating resin. To 100 ml of an emulsion of this resin, whose concentration of nonvolatile matter was 42%, was added 100 ℓ of water to prepare a diluted resin emulsion containing 21% of nonvolatiles.

To 100 ℓ of a 32% aqueous solution of magnesium dichromate (concentration as CrO_3 is 26.6%) was added gradually 20 ℓ of the above diluted resin emulsion (the amount of nonvolatile matter of the particle separating resin emulsion is 12 parts based on 100 parts of CrO_3) under stirring to separate out particles, and further to the resulting mass were added 10 ℓ of a compatible acrylic resin emulsion having a concentration of nonvolatile matter of 50% (Voncoat 4001) (the amount of nonvolatile matter of the compatible resin emulsion is 14 parts based on 100 parts of CrO_3), 5 ℓ of ethylene glycol (16 parts based on 100 parts of CrO_3), 5 kg of boric acid (15 parts based on 100 parts of CrO_3) and 300 ℓ of water to prepare a treating dispersion.

An electrical steel strip of 0.5 mm thickness and 940 mm width containing 0.32% of Si and having a surface roughness of 0.9 μH_{max} was immersed in the treating dispersion at a rate of 60 m/min, squeezed by means of grooved rubber rolls and baked in a hot-air furnace kept at 400°C for 60 seconds to obtain a coating having no gloss and stripe pattern.

Properties of the resulting coating are shown in the following Table 1 together with properties of the coatings formed in the following example and comparative examples.

Example 2: A vinyl acetate-ethylene copolymer having a monomer composition of vinyl acetate:ethylene = 80:20 was used as a particle separating resin, in the form of an emulsion, whose concentration of a nonvolatile matter was 55%. To 100 ℓ of a 30% aqueous solution of calcium dichromate (concentration as CrO_3

is 23.4%) was added gradually 4 ℓ of the above resin emulsion (the amount of nonvolatile matter of the particle separating resin emulsion is 6 parts based on 100 parts of CrO_3) under stirring to separate out particles. Then, to the mass were added 40 ℓ of a compatible acrylic-styrene resin emulsion having a concentration of nonvolatile matter of 40% (Voncoat 4280) (the amount of nonvolatile matter of the compatible resin emulsion is 48 parts based on 100 parts of CrO_3) and 300 ℓ of water to prepare a treating dispersion.

An electrical steel strip of 0.5 mm thickness and 940 mm width containing 0.32% of Si and having a surface roughness of 1.4 μH_{max} was immersed in the treating dispersion at a rate of 80 m/min, squeezed by means of grooved rubber rolls and baked in a hot-air furnace kept at 500°C for 45 seconds to obtain a coating having no gloss and strip pattern.

Comparative Example 1: To 100 ℓ of a 30% aqueous solution of calcium dichromate were added, while stirring, 20 ℓ of the same compatible acrylic resin emulsion as used in Example 1 (the amount of the nonvolatile matter of the resin emulsion is 28 parts based on 100 parts of CrO_3), 7 ℓ of ethylene glycol (17 parts based on 100 parts of CrO_3) and 300 ℓ of water to prepare a treating liquid.

The resulting treating liquid was applied on an electrical steel strip (thickness 0.5 mm, width 940 mm, Si content 0.30%, surface roughness 1.4 μH_{max}) at a rate of 60 m/min by means of grooved rubber rolls, baked in a hot-air furnace kept at 400°C for 60 seconds to obtain a glossy coating having a uniform appearance.

Comparative Example 2: To 100 ℓ of a 18% aqueous solution of calcium dichromate were added 5 ℓ of ethylene glycol and a very small amount of a surfactant to prepare a treating liquid. The treating liquid was applied to the same electrical steel strip as used in Comparative Example 1 and treated in the same manner as described in Comparative Example 1 to obtain a uniform glossy coating.

Table 1

Properties of Coatings and Coated Steel Strips	Examples 1	Examples 2	Comparative Examples 1	Comparative Examples 2
Surface roughness, μH_{max}	4.3	3.4	1.3	1.5
Coating weight after baking, g/m^2 per side	2.6	2.7	2.6	2.5
Insulation resistance*, Ω-cm^2/layer				
Before annealing	53	32	9	7
After annealing at 750°C in N_2	23	18	1.8	0.7
Punchability**, no. of punching times	$>240 \times 10^4$	$>200 \times 10^4$	$>220 \times 10^4$	30×10^4
Weldability***, cm/min	130	100	20	50
Adhesion†, mm ϕ				
Before annealing	<10	<10	<10	<10
After annealing at 750°C in N_2	15	15	15	15
Corrosion resistance††, %	0	0	0	20
Space factor*, %	98.8	99.0	99.1	99.0

*According to ASTM A-344.
**The punchability is shown by the number of punching times until the burr height reaches 50 μ when a coated steel strip is punched by means of a tool steel die.
***Maximum welding velocity, at which no blowhole is formed.
†Maximum bending diameter, at which exfoliation of coating is observed.
††Rusted area after 10 hours in the salt spray test.

The conventional coating obtained by using an aqueous chromic acid solution compounded with only a compatible resin is very smooth, while the coating according to the process is rough. That is, the coating of Comparative Example 1 had a surface roughness of 1.3 μH_{max}, while the coatings of Examples 1 and 2 had surface roughnesses of 4.3 and 3.4 μH_{max}, respectively.

The coating according to the process contains a large number of deposited particles which are visible on the surface.

It can be seen from Table 1 that the coating according to the process has a very high insulation resistance, is excellent in the adhesion and in the heat resistance and does not exfoliate after stress relief annealing.

The coated steel strips of the process can be welded at a high speed and further are excellent in the punchability.

In the process, the weight of the coating should be 1.2 to 4.8 g/m^2 per one side. When the coating weight is less than 1.2 g/m^2, the punchability and insulation resistance of the coated steel strip are poor. While, when the coating weight exceeds 4.8 g/m^2, the space factor of the coated steel strip is deteriorated.

Further Protective Phosphate Coating for Silicon Steel Sheets

In the manufacture of silicon steel sheet material for use as so-called electrical sheets which have a grain orientation, after rolling and decarburization, the sheet material is conventionally heat treated at about 850° to 1350°C in order to achieve the necessary grain growth of the crystals, i.e., so that the sheet material will acquire the required magnetic properties. Before the heat treatment, however, the sheet material is usually coated with chemicals in order to produce, during the subsequent heat treatment, an electrically insulating protective coating.

According to the process developed by *C.A. Åkerblom; U.S. Patent 4,120,702; October 17, 1978; assigned to Asea AB, Sweden* silicon steel sheets which have a silicate protective coating are further protected by being first coated with an aqueous solution containing (1) phosphate ions, (2) silica grains, (3) iron and/or manganese ions, and (4) negative ions which convert to volatile products at temperatures below 400°C, and then heated to temperatures of between about 400° and 1100°C for periods of between about ½ and 10 minutes in order to form a further protective phosphate layer.

With respect to the amounts of the components useful in the phosphate solution, the following quantities per 100 pbw of silica are preferred (calculated as SiO_2 without water):

10 to 200 pbw, preferably 50 to 150 pbw of phosphate ions (calculated as PO_4^{3-});

1 to 30 pbw, preferably 2 to 20 pbw of iron ions or manganese ions, or both together;

0 to 25 pbw, preferably 2 to 20 pbw of aluminum ions or magnesium ions, or both together;

An amount of negative ions such that the solution contains a sufficient amount of metal ions to react with the phosphate ions of the solution, preferably an amount exactly equivalent to the amount of hydrogen ions in the solution or else which deviates therefrom by a maximum of 40% (preferably 25%); and

5 to 50 pbw, preferably 10 to 30 pbw of fillers if such are added.

The thickness of the applied layer of phosphate solution is 0.1 to 20 μ, preferably 0.5 to 5 μ, and most preferably 1 to 3 μ.

A sheet of silicon steel having a thickness of 0.3 mm is pretreated to have a grain orientation and has been decarburized at 720° to 900°C (preferably 820°C) in a wet hydrogen atmosphere. The sheet is drawn from a coil on a reel, and passes under a roll which rotates in a pan containing a suspension of the particulate material with which the sheet is to be coated. The suspension can, for example, be manufactured by suspending 90 pbw of magnesium oxide consisting of particles, 95% by wt of which have a grain size of less than 5 μ and the rest of which have a grain size of less than 25 μ, in 1,000 pbw water. After passing the pan, the sheet is passed between wiping rollers which are suitably rubber-clad, and into a furnace where it is dried at a temperature of about 100°C for about 30 seconds before it is coiled up on the reel after having passed transport rollers. Thereafter, the sheet is annealed (at high temperature) in a batch annealing furnace at around 1000° to 1350°C in a hydrogen atmosphere for several hours, during which time a protective coating of silicate with a thickness of 1 μ is formed on the sheet.

When the so-treated sheet has been liberated from excess coating by brushing, it is coated with phosphate in the apparatus according to Figure 5.3.

Figure 5.3: Apparatus for Additional Application of Protective Phosphate Coating

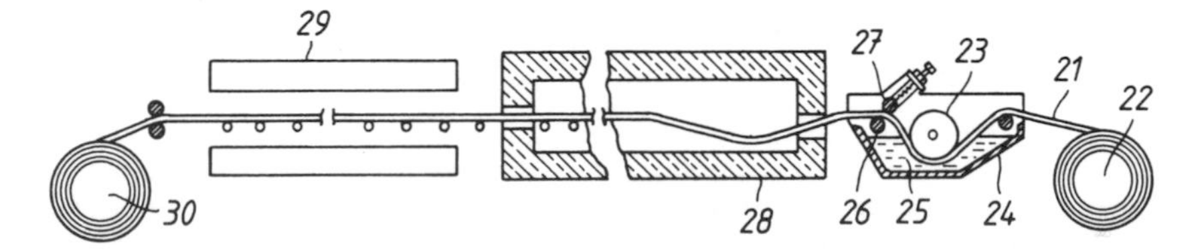

Source: U.S. Patent 4,120,702

More specifically, the sheet, designated **21**, is drawn from a reel **22** and passes under a roll **23** rotating in a pan **24** with a solution **25** of phosphate in water, possibly containing suspended filler. The sheet is then passed between the wiping rollers **26** and **27** which are suitably rubber-clad and into a furnace **28**, after which the sheet is cooled in a cooling device **29**, before it is coiled up on the reel **30**. The concentration of phosphate in the treatment liquid **25** is adjusted with regard to the profile of the rubber rollers **26** and **27** and to the roller pressure so that the desired thickness of the phosphate layer is obtained. For all the compositions of the solution **25** exemplified in the examples below, the furnace **28** had a temperature of 800°C and the time for the sheet to pass through the furnace was 2 minutes. The furnace atmosphere was air. The thickness of the phosphate layer in the exemplified cases was 2 μ. The preparations of the solution **25** are indicated by the following examples.

Example 1: A solution was prepared from 20 parts by weight of ferrous sulfate ($FeSO_4 \cdot 7H_2O$), 15 pbw of phosphoric acid (SG = 1.54), 100 pbw of water and 180 pbw of colloidal silica containing 300 g of SiO_2 per liter and with a particulate size of the silica of 100 to 200 Å and a specific surface of 250 m^2/g.

Example 2: A solution was prepared from 20 parts by weight of manganese sulfate ($MnSO_4 \cdot H_2O$), 15 pbw of phosphoric acid (SG = 1.54), 100 pbw of water and 180 pbw of colloidal silica of the kind described in Example 1.

Example 3: A solution was prepared from 6.7 parts by weight of ferrous sulfate ($FeSO_4 \cdot 7H_2O$), 13.3 pbw of manganese sulfate ($MnSO_4 \cdot H_2O$), 24 pbw of phosphoric acid (SG = 1.54), 40 pbw of aluminum phosphate solution (600 g $AlPO_4$ per liter, pH 2), 25 pbw of water and 180 pbw of colloidal silica of the kind described in Example 1.

Results: The electrically insulated silicon steel sheets produced as a result of the treatment sequences noted in conjunction with Figure 5.3 utilizing phosphate solutions according to Examples 1 through 3 were all insensitive to water and the insulation layers displayed excellent adhesion to the base silicon steel sheets. In addition, the magnetostrictive properties were all very good.

INSULATING COATINGS FOR SILICON STEEL SHEETS

Chromate-Organic Resin Insulating System

The work of *N. Morito, T. Sugiyama and Y. Obata; U.S. Patent 4,255,205; March 10, 1981; assigned to Kawasaki Steel Corporation, Japan* relates to a method of producing grain-oriented silicon steel sheets, and more particularly, to a method of producing grain-oriented silicon steel sheets, which can be easily separated from an annealing separator after final annealing and have substantially no glass film and have excellent punchability and magnetic properties.

In this method of producing grain-oriented silicon steel sheets having substantially no glass film, a cold rolled silicon steel sheet having a final gauge is subjected to a decarburization annealing to form subscale on the surface of the steel sheet, an annealing separator consisting mainly of Al_2O_3 is applied on the subscale and then the steel sheet is subjected to a final annealing. The process comprises using an annealing separator consisting of 5 to 30% of hydrated silicate mineral fine particles, 0.2 to 20%, calculated as strontium or barium, of a compound containing strontium or barium, 2 to 30% of CaO or $Ca(OH)_2$ and the remainder being Al_2O_3.

Since an annealing separator consisting only of Al_2O_3 does not shrink during the final annealing, gas can not flow smoothly through the space between adjacent coiled steel sheet layers, and it is difficult to remove S, Se and N during the final annealing, which S, Se and N would affect adversely the magnetic properties of the finally annealed steel sheet. When hydrated silicate mineral is contained in an Al_2O_3 series annealing separator, the annealing separator liberates water and shrinks during the final annealing, and gas flows smoothly in the space between adjacent coiled steel sheet layers, and as the result, the steel sheet can be easily purified and sticking of Al_2O_3 on the surface of steel does not occur.

The presence of a compound containing strontium or barium in the Al_2O_3 series annealing separator serves to float up oxide particles, which have been formed during the decarburization annealing, during the final annealing, and therefore the use of a compound containing strontium or barium is necessary in order not to form glass film.

When the amount of CaO or $Ca(OH)_2$ contained in the annealing separator of the process is less than 2%, sulfur or selenium can not be completely removed; while when the amount exceeds 30%, glass film is locally formed, and a strong pickling must be carried out. Therefore, the amount of CaO or $Ca(OH)_2$ contained in the annealing separator must be 2 to 30%.

Example: A silicon steel ingot containing 0.03% of C, 2.98% of Si, 0.055% of Mn, 0.018% of Sb and 0.020% of Se was hot rolled into a thickness of 3 mm, annealed at 970°C for 5 minutes, and subjected to two cold rollings with an intermediate annealing at 900°C therebetween to produce a cold rolled sheet having a final gauge of 0.30 mm. The cold rolled sheet was subjected to a decarburization annealing under a wet hydrogen atmosphere, and then subjected to a final annealing under hydrogen atmosphere. In the final annealing, the cold rolled sheet was firstly kept at 850°C for 50 hours and then kept at 1180°C for 10 hours. The magnetic properties and other properties of the finally annealed steel sheet are shown in the following table together with the oxygen content in the subscale formed during the decarburization annealing and the composition of the annealing separator.

Magnetic Properties and Other Properties of Finally Annealed Steel Sheet

Sample	Composition of Annealing Separator	Oxygen Content in Subscale** (g/m^2)	Magnetic Properties B_{10}(T)	$W_{17/50}$ (w/kg)	S*** (%)	Glass Film	Punchability* Insulating Coating System Phos-phate	Cr-Org. Resin
Comparative								
No. 1	MgO 100%	1.1	1.91	1.12	0.001	yes	5,000	–
No. 2	Al_2O_3 100%	0.6	1.92	1.11	0.010	yes	40,000	–
No. 3	Al_2O_3 + 10% serpentine	0.8	1.93	1.09	0.005	yes	40,000	70,000
No. 4	Al_2O_3 + 5% $Ba(OH)_2$	1.0	1.93	1.10	0.007	yes	30,000	–
Process								
No. 1	Al_2O_3 + 10% serpentine + 5% $Ba(OH)_2$ + 10% $Ca(OH)_2$	0.6	1.94	1.06	<0.001	none	70,000	300,000
No. 2	Al_2O_3 + 10% serpentine + 1% $Ba(OH)_2$ + 5% $Ca(OH)_2$	0.4	1.93	1.06	<0.001	none	–	–

*Punchability = The number of punching times until the height of burr reaches 50 μm when steel sheets are punched by means of a steel die having a diameter of 15 mm.
Cr-Org. = Chromate-organic resin mixture.
**On the surface of steel sheet after decarburization annealing; g/m^2 of one surface.
***Amount of S remaining in steel.

It can be seen that when a cold rolled silicon steel sheet is subjected to a decarburization annealing so as to form subscale containing 0.2 to 1.0 g of oxygen per square meter of one surface of the steel sheet, and the decarburized steel sheet is applied with an annealing separator consisting of 5 to 30% of hydrated silicate mineral fine particles, 0.2 to 20%, calculated as strontium or barium, of a compound containing strontium or barium, 2 to 30% of CaO or $Ca(OH)_2$ and the remainder being Al_2O_3, and then subjected to a final annealing, a grain-oriented silicon steel sheet having excellent magnetic properties and having substantially no glass film can be obtained. Moreover, when an organic resin system insulating coating is formed on the above obtained grain-oriented silicon steel sheet, the punchability of the coated steel sheet is remarkably superior to the punchability of a grain-oriented silicon steel sheet having a conventional phosphate system insulating coating.

Insulating Coating Dispersion for Oriented Silicon Steel Sheet

The process of *T. Ichida and T. Funahashi; U.S. Patent 4,238,534; December 9, 1980; assigned to Kawasaki Steel Corporation, Japan* relates to forming an insulating coating on a crystalline forsterite-ceramic film formed on an oriented silicon steel sheet. The coating is able to decrease the iron loss and suppress the magnetostriction of the steel sheet.

The heat-resistant insulating coating having a high adhesion can be formed on an oriented silicon steel sheet by the use of an aqueous coating dispersion containing colloidal silica dispersed therein, at least one of monobasic phosphates of Mg, Al, Sr, Ba and Fe, at least one compound selected from chromic acid anhydride, chromate and dichromate, and at least one fine particle oxide selected from SiO_2, Al_2O_3 and TiO_2 having a primary particle size of 70 to 500 Å and an apparent density of not higher than 100 g/ℓ.

Examples 1 through 3: An oriented silicon steel sheet rolled to a final gauge of 0.30 mm and containing 1 to 4% of Si was subjected to a decarburization annealing, and an oxide layer containing SiO_2 was simultaneously formed on the surface of the steel sheet. Then, a separator consisting mainly of an MgO-water slurry was applied to the sheet surface, and after the separator was dried, the steel sheet was wound into a coil shape and annealed at 1200°C for 20 hours under hydrogen atmosphere to form a forsterite-ceramic film on the surface of the oriented silicon steel sheet.

The oriented silicon steel sheet having the forsterite-ceramic film thereon was washed with water to remove unreacted separator, and the following aqueous coating dispersions, each having a specific gravity of 1.20 and the following composition, were applied to the steel sheet by means of a grooved roll, and the coated steel sheets were baked at 800°C.

The following table shows the characteristic properties of the resulting coated silicon steel sheets. The steel sheets are remarkably superior to conventional coated oriented silicon steel sheets in any of the magnetic properties, the magnetostriction under compression stress and the properties of coating. Particularly, the sheets do not at all stick to each other during the stress relief annealing, and have smooth surface and beautiful appearance after the stress relief annealing.

Example 1 (aqueous coating dispersion No. 1) – colloidal silica (20% aqueous dispersion), 100 ℓ; magnesium monobasic phosphate (35% aqueous solution), 50 ℓ; chromic acid anhydride, 3 kg; and fine particle Al_2O_3 (Aluminium Oxide C, Degussa Co.), 3 kg. The molar ratio of SiO_2 in colloidal silica/monobasic phosphate = 3.45.

Example 2 (aqueous coating dispersion No. 2) – colloidal silica (20% aqueous dispersion), 100 ℓ; magnesium monobasic phosphate (35% aqueous solution), 60 ℓ; chromic acid anhydride, 3 kg; and fine particle SiO_2 (Aerosil-100), 0.5 kg. The molar ratio of SiO_2 in colloidal silica/monobasic phosphate = 2.87.

Example 3 (aqueous coating dispersion No. 3) – colloidal silica (20% aqueous dispersion), 100 ℓ; magnesium monobasic phosphate (35% aqueous solution), 45 ℓ; chromic acid anhydride, 3 kg; and fine particle TiO_2 (Titanium Oxide P-25, Degussa Co.), 0.5 kg. The molar ratio of SiO_2 in colloidal silica/monobasic phosphate = 3.83.

Aqueous Coating Dispersion	Magnetic Properties B_{10} (T)	$W_{17/50}$ (w/kg)	Magnetostriction Under Compression Stress* 0	0.3	0.5	0.7 (kg/mm²)	.....Properties of Coating..... Insulation Resistance (Ω-cm²/sheet)	Space Factor (%)	Peeling Test by Bending (mm)
No. 1	1.93	1.11	0.8	0.7	1.0	5.3	200**	97.9	15
No. 2	1.92	1.12	0.9	0.7	1.5	6.0	200**	98.0	15
No. 3	1.93	1.09	1.0	0.7	1.5	6.2	200**	97.8	15

*x 10^{-6} **At least.

ORGANIC COATINGS

Polymeric Coating Hardened Under UV Light

H. Yamagishi, A. Murao and H. Tsutsumi; U.S. Patent 4,025,692; May 24, 1977; assigned to Nippon Kokan KK, Japan have developed a method for forming a corrosion-resistant coating on a steel sheet. This comprises applying a water solution on the surface of the steel sheet, followed by drying; and exposing the coated surface to ultraviolet rays having a main wavelength of 365 mμ in an air atmosphere to obtain a cured coating. The water solution contains a composition comprising (1) a prepolymer of an epoxy resin selected from the group consisting of bisphenol and novolak-phenol type reacted with an acid selected from the group consisting of acrylic acid and methacrylic acid, (2) a chelate-forming compound selected from the group consisting of a chromate and a dichromate, and (3) a water-soluble compound of molybdenum to stabilize the water solution.

Example 1: A prepolymer of bisphenol-type epoxy resin (Epikote 1001) was reacted with acrylic acid. Then the acrylate was further reacted with maleic anhydride, followed by neutralization with triethylamine. 50 pbw of ammonium dichromate were dissolved in pure water to provide 100 pbw of a treating solution.

After degreasing an electrolytically galvanized steel sheet using a common weak alkaline degreasing solution, followed by water-washing, the steel sheet was coated with the prepared treating solution by a grooved roller. The coated steel sheet was dried by hot air, and was irradiated by ultraviolet rays 5 seconds in an air atmosphere using an irradiating apparatus having two tubes each of 2 kW in order to polymerize and harden the coated layer.

The buildup of chromium coating was 50 mg/m². The coating became faintly yellow when not irradiated by ultraviolet rays, whereas colorless or faintly green when irradiated thereby. Furthermore, the sticky surface of the coated steel sheet appearing in the absence of the irradiation of ultraviolet rays vanished by irradiation. A test of exposing a coated steel sheet sample to salt spray showed that these coated steel sheet samples, indicated high enough corrosion resistivity to generate no white or red rust even 450 hours after the test.

A cylindrical deep-drawn article 50 mm in inner diameter prepared from the electrolytically galvanized steel sheet of this example 90 mm in blank diameter was tested for press-moldability and resistance to a degreasing agent, using a suitable kind of press oil. Then the deep-drawn article was cleaned for 3 minutes by a degreasing agent solution containing about 30 g/ℓ of a surfactant, the pH value and the temperature of the solution being 12 and 65°C, respectively. After being washed with water and dried, the article was tested for corrosion resistivity and paint adhesivity. The salt spray exposure test of 120 hours produced white and red rusts on the entire surface of the article which was not irradiated by ultraviolet rays, but did not give rise to any rust on the surface of the irradiated article. Further, the irradiated article using melamine resin was demonstrated by test to show high paint adhesivity.

Example 2: An electrolytically galvanized steel sheet previously treated with a common primary chromate solution was coated with a secondary treating solution of the same type as in Example 1. After being dried, the coated steel sheet was irradiated for 5 seconds by ultraviolet rays.

The secondary buildup of chromium coating (excepting the primary buildup thereof) was 30 mg/m^2. The coated steel sheet presented substantially an excellent moldability and paint adhesivity as in Example 1. The abovementioned two-step chromate treating was found to improve the corrosion resistivity of the coated steel sheet even if the buildup of the secondary chromium coating was somewhat small.

Example 3: A prepolymer of novolak-phenol type epoxy resin (Den, Dow Chemical Corporation) was made water-soluble in the same manner as in Example 1. 25 pbw of water-soluble prepolymer of novolak-phenol type epoxy resin acrylate, 25 pbw of the prepolymer further containing phosphoric acid groups, 16 pbw of ammonium bichromate and 3.2 pbw of ammonium molybdate were dissolved in pure water to provide 1,000 pbw of a treating solution.

The surface of an electrolytically galvanized steel sheet coated with this reacting solution was exposed to ultraviolet rays to have the coated layer hardened. The coating thus hardened indicated the same degree of resistivity to corrosion and a degreasing agent as in Example 1.

Example 4: A degreased cold-drawn steel sheet was coated with the same treating solution as in Example 3 using a grooved roller, followed by an irradiation of ultraviolet rays. The hardened coating indicated suitable physical properties including external appearance, corrosion resistivity and paint adhesivity for the subsequent application of paint on the surface of the coated steel sheet. In this case, the buildup of chromium coating was 60 mg/m^2.

Preparing Ferrous Metal for Coating with Aqueous Fluoropolymer

E. Vassiliou; U.S. Patent 4,226,646; October 7, 1980; assigned to E.I. Du Pont de Nemours and Company provides a process of preparing a ferrous metal substrate to be coated with an aqueous fluoropolymer coating wherein the substrate is first coated with triethanolamine in a volatile liquid carrier. Preferably, the triethanolamine is applied to the substrate as a 2 to 15%, preferably 4%, by weight solution in a volatile liquid carrier such as isopropanol. The carrier volatilizes quickly, leaving the triethanolamine.

The process can be used to prepare ferrous metal substrates for coating with aqueous coating compositions. It prevents the deleterious formation of iron oxide or flash rusting of the substrate that would occur without the triethanolamine treatment.

A carbon steel substrate can be coated with a solution of isopropanol containing 4% by wt of the total of triethanolamine. Enough coating is used to completely wet the surface, thereby preferably giving at least about a monomolecular layer of triethanolamine on the surface.

While other compounds such as ascorbic acid or vitamin C may give some useful effect in minimizing flash rusting, triethanolamine appears to be superior.

After the triethanolamine has been applied, the isopropanol quickly evaporates at room temperature, such as about 23°C. Then the thus treated substrate can be coated with, for instance, the aqueous fluorocarbon coatings, particularly those of the example of U.S. Patent 4,039,713.

Monobasic Organic Acid, Lubricant and Amine Mixture

The work of *A.J. Conner, Sr.; U.S. Patent 4,233,176; November 11, 1980* relates to a nonpetroleum based metal corrosion inhibiting composition consisting essentially of a solution of 1 pbw of an aqueous concentrate and up to 5 pbw of water. The aqueous concentrate comprises, per 100 pbw of the concentrate:

(a) 5 to 20 pbw of a monobasic organic acid having from 8 to 20 carbon atoms;

(b) 0.5 to 4 pbw of a lubricant;

(c) 0.5 to 4 pbw of an alkylaminoalkanolamine of the formula:

$$H_2NR^1-\underset{\underset{\displaystyle R^3}{|}}{N}-R^2-OH$$

where R^1 and R^2 are independently alkylidene of 1 to 4 carbon atoms, and R^3 is hydrogen or alkyl of 1 to 4 carbon atoms;

(d) 10 to 35 pbw of benzoic acid; and

(e) 5 to 20 pbw of an amine which forms a water-soluble salt with benzoic acid.

In a preferred modification, a mixture of tall oil fatty acids and rosin, because of its availability and cost and the properties of the resultant corrosion inhibiting composition, is used as the high molecular weight organic acid component of the composition.

The lubricant may be either a petroleum or a nonpetroleum product. Good results were achieved with 100 SUS viscosity petroleum oil. In lieu of a petroleum oil, esters such as butyl stearate, dioctyl sebacate, butyl benzoate, or any of the light alkyl esters with boiling ranges above 350°F may be used.

Aminoethylethanolamine is preferred because of its cost and the good results that it provides.

The metal corrosion inhibiting composition includes as an inhibitor a water-soluble salt of benzoic acid. Although it is believed that virtually any amine which forms a water-soluble salt with benzoic acid can be used, particularly good results have been obtained with the use of lower (C_{2-4}) alkanolamines and, particularly, monoethanolamine and diethanolamine.

The preparation of a typical 55 gallon batch of a concentrated solution of the nonpetroleum based corrosion inhibitor is as follows (approximate weights are in parenthesis).

Pump 30 gal of water (250 lb) at 120°F into tank and agitate. Add 10 gal of a tall oil fatty acid/rosin mixture (80 lb), Unitol-DT-40, and 1 or 2 gal of 100 SUS viscosity petroleum oil (7 to 14 lb). The oil will dissolve in the tall oil-rosin mixture, but neither the petroleum oil nor the tall oil fatty acid-rosin mixture will dissolve in the water. While agitating add 1 gal of aminoethylethanolamine (8 lb). An oil in water emulsion will form. This emulsion is milky and completely opaque. Add 8 gal of monoethanolamine (64 lb) and the mixture will become clear and stable. Add 100 lb of benzoic acid and the mixture will become hazy because of the portion of the benzoic acid which has not been neutralized to a soluble salt. To complete neutralization of the benzoic acid, add more monoethanolamine (or morpholine, cyclohexylamine, etc.) until the solution is completely clear and has a pH of 8.0 to 9.5. Continue mixing for 30 minutes and recheck pH. If pH drops below 8.0, add more monoethanolamine to bring pH to 9.0.

For use at the mills or manufacturing plants, one part of the above composition is diluted with up to five parts of water and applied as either a rust preventative or lubricant. The recommended dilution ratio is 1 part concentrate to about 4 parts water.

To illustrate the corrosion inhibiting properties of the nonpetroleum based corrosion inhibitor compositions when applied to steel, the following compositions were prepared according to the general procedures described above. In the compositions, percentages are by weight and the tall oil fatty acids-rosin mixtures employed are commercially available compositions in which the fatty acids are composed primarily of a mixture of oleic and linoleic acids.

Composition 1:

60% tall oil fatty acid (TOFA)-40% rosin–12 to 18%
100 SUS vis petroleum oil–2 to 4%
Amine mixture: 40% aminoethylethanolamine (AEE)-60% monoethanolamine (MEA)–5 to 10%
Benzoic acid–10 to 20%
Water–71 to 48%

Composition 2:

60% TOFA-40% rosin–12 to 18%
100 SUS vis petroleum oil–2 to 4%
Amine mixture: 40% AEE-50% MEA-10% morpholine–5 to 10%
Benzoic acid–10 to 20%
Water–71 to 48%

Composition 3:

70% caprylic acid-30% abietic acid—12 to 18%
100 SUS vis petroleum oil—2 to 4%
30% AEE-70% MEA—5 to 10%
Benzoic acid—10 to 20%
Water—71 to 48%

Composition 4:

80% TOFA-20% rosin—12 to 18%
Butyl stearate—2 to 4%
30% AEE-70% MEA—5 to 10%
Benzoic acid—10 to 20%
Water—71 to 48%

These compositions were evaluated for corrosion inhibiting properties according to the testing procedures described below.

Cold roll dry strips (1¼ inches wide by 4 inches long, dry, clean and rust-free) were used as test specimens. A 1/16 inch hole was punched ⅛ inch from the top and bottom, and ⅝ inch from one side. A hook, fabricated from galvanized wire, was used to hang the strips in a humidity cabinet. Each strip was marked for identification by embossing a number with a metal stamp about ¼ inch below the punched hole. To standardize the test, a strip as described above was dipped 2 inches in the solution to be tested and suspended by a metal hook with the dipped or coated portion of the strip at the bottom. The strip was allowed to dry or drain for one hour, and the hook transferred to the opposite end of the strip, which was then suspended on a rack in the humidity cabinet. The coated or dipped end was now on top, and the lower, uncoated end of the strip on the bottom. The conditions in the humidity cabinet were maintained at 100°F and 100% humidity.

Observations of the strip were made every 24 hours. The lower or dry parts of all strips were completely rusted after 24 hours. All tests were run for 120 hours. The coated parts were completely free of rust in all cases.

Inhibitor for Preventing Blistering in Autodepositing Compositions

A relatively recent development in the coating field is the provision of water-based coating compositions which are effective, without the aid of electricity, in forming on metallic surfaces immersed therein organic coatings that increase in thickness or weight during the time the surfaces are immersed in the compositions. (For convenience, a coating composition of this type is hereafter referred to as an autodepositing composition and a coating formed from such a composition is hereafter referred to as an autodeposited coating.)

Defects in autodeposited coatings formed on metal surfaces of the type described above can take various forms, depending on the particular application involved. For example, the defects may show up as pinholes, blisters and/or craters in the coating. (The term crater, as used herein, refers to a defect that looks like a collapsed blister.) In some applications, defects are not observable in the wet unfused coating, but they appear after the coating has been fused by subjecting it to an elevated temperature.

Bridging is another type of defect which can be encountered when coating certain types of metal articles. Bridging occurs in autodeposited coatings formed on articles having surfaces which do not lie in the same plane, but which intersect, and appears at the line of intersection. Bridging is evidenced by the pulling away of the coating from the underlying metallic surface at the aforementioned line of intersection.

In accordance with the process of *F.P. Lochel, Jr.; U.S. Patents 4,178,400; December 11, 1979; and 4,108,817; August 22, 1978; both assigned to Amchem Products, Inc.* an additive is included in an autodepositing composition for the purpose of reducing or preventing the tendency of the composition to form coatings which have defects such as pinholes, blisters, and/or craters. The process can be used also to prevent or deter a coating defect known as bridging.

The additive for use in the composition is referred to herein as a corrosion inhibitor or acid inhibitor or simply inhibitor and can be selected from a wide variety of compounds including, for example, aldehydes, ketones, amines, thiols, mercaptans, sulfides, thioureas, silicates, phosphates, carbonates, nitrites, oximes, alkanols, chromates and dichromates. Mixtures of inhibitors can be used also.

Among the corrosion inhibitors that can be used are amine compounds which are the product of a Mannich reaction which involves the reaction of a nitrogen compound having at least one active hydrogen attached to a nitrogen atom with an α-ketone and formaldehyde in the presence of an acid. The use of rosin amines in a Mannich reaction, as described in U.S. Patent 2,758,970, is exemplary, although other types of primary or secondary amines can be used.

A preferred inhibitor for use is propargyl alcohol. Particularly good results have been attained also by using an inhibitor composition prepared from thiourea and the product of a Mannich reaction.

In general, it is recommended that the amount of inhibitor included in the composition be in the range of about 100 to 5,000 ppm and adjustments be made as needed for the specific application.

When utilizing the preferred inhibitor composition prepared by the Mannich reaction, it is recommended that it be used in a concentration of about 100 to 3,000 ppm. When utilizing propargyl alcohol, it is recommended that it be used in a preferred concentration of about 100 to 500 ppm.

Example 1: *Basic coating composition –*

Ingredients	
Latex containing about 54% solids	190 g
Ferric fluoride	3 g
Hydrofluoric acid	2.3 g
Black pigment dispersion	5 g
Water	qs 1 ℓ

The resin of the latex used in the above composition comprised about 62% styrene, about 30% butadiene, about 5% vinylidene chloride and about 3% methacrylic acid. A film formed from the resin is soluble in refluxing chlorobenzene to the extent of about 13%. That the resin is crosslinked is indicated by its insolubility in Soxhlet extraction with chlorobenzene. The water-soluble content of the latex is about 2% based on the weight of dried resin, with the water-soluble content comprising about 10% sodium phosphate, about 13% sodium oleoyl isopropanolamide sulfosuccinate and about 75% sodium dodecylbenzene sulfonate, the first mentioned ingredient being a buffering agent used in preparing the latex, and the last 2 mentioned ingredients being emulsifiers. The pH of the latex was about 7.8 and the surface tension thereof about 45 to 50 dynes/cm. The average particle size of the resin was about 2000 Å.

The black pigment dispersion used in the above composition is an aqueous dispersion having a total solids content of about 36%. Carbon black comprises about 30% of the dispersion. It has a pH of about 10 to 11.5 and a specific gravity of about 1.17. The dispersion contains a nonionic dispersing agent for the solids (Aquablak 115). The inhibitors of the examples were added to the coating compositions, as abovedescribed.

Example 2: *Inhibitors prepared by a Mannich reaction* – Inhibitor A: This inhibitor is prepared by reacting, in the presence of HCl, acetophenone, acetone, formaldehyde, and a composition comprising stabilized abietylamines, predominately dehydroabietylamine of the formula:

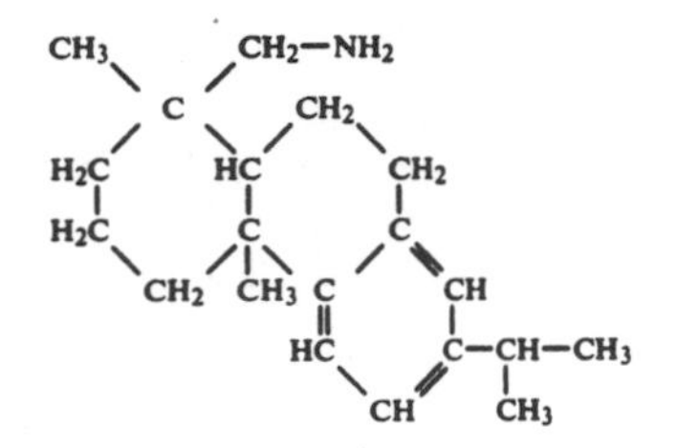

The aforementioned composition is a clear yellow viscous liquid sold as Amine 750 (Hercules Incorporated). The composition is derived from pine resin acids and is produced to contain a minimum amount of secondary amines.

A two-stage Mannich reaction is used in which, in the first stage, 2.3 parts of the amine composition are reacted with 0.8 parts of acetophenone and 1 part of 37 wt % aqueous solution of formaldehyde in the presence of HCl (20°Bé) at a temperature of about 80°C for about 24 hours. In the second stage of the reaction, 2.5 parts of acetone and 1 part of 37 wt % aqueous solution of formaldehyde are added to the product of the first reaction stage and the reactants are reacted for 24 hours at a temperature of about 60°C. After completion of the reaction, excess reactants (primarily acetone) are separated from the reaction mixture by distillation. To the reaction product (6.4 parts), there are added 1.4 parts of a surface active agent (nonylphenyl polyethyleneglycol ether sold as Tergitol NP 35), about 0.1 part of water and about 0.9 part of isopropanol. This product is Inhibitor A.

Inhibitor A': Inhibitor A' is prepared by admixing the following ingredients in the proportions stated.

Ingredients	Weight Percent
Inhibitor A	48.1
Thiourea	18.8

(continued)

Ingredients	Weight Percent
Isopropanol	5.0
Surface active agent, nonylphenoxy poly(ethyleneoxy)ethanol (Igepal CO-850)	24.2
HCl (20°Bé)	3.9

Inhibitor A'': This inhibitor is the same as Inhibitor A' except that the Inhibitor A portion is prepared in the presence of about 1 part of 70% hydroxyacetic acid instead of 20°Bé HCl and contains 3.9 wt % of 70% hydroxyacetic acid instead of 20°Bé HCl.

Unless stated otherwise, the metallic surfaces coated in the examples below were unpolished cold rolled steel panels (Q-panels) 3" x 4". All metallic surfaces were cleaned with a conventional alkali cleaner and rinsed with water prior to being coated.

In those examples in which the coated surfaces were rinsed with an aqueous chromium rinse solution, the Cr rinse solution comprised 3 wt % of an aqueous concentrate prepared from 150 g/ℓ of $Na_2Cr_2O_7 \cdot 2H_2O$ and an aqueous mixture of formaldehyde-reduced CrO_3 which contained 20 g/ℓ of reduced Cr, 30 g/ℓ of Cr^{6+}, and 25 g/ℓ of 75 wt % H_3PO_4.

Example 3: The applications discussed below show the formation of coatings with no pinholes on smooth panels and the effects that higher concentrations of inhibitors (Inhibitor A'') have on the thicknesses of the coatings formed. The panels were immersed in the composition for 60 seconds and the coated panels were then air dried for 60 seconds, water rinsed for 30 seconds, immersed in an aqueous chromium rinse solution for 30 seconds and baked at 170°C for 10 minutes. The baked coatings were then subjected to salt spray tests for 168 hours.

Inhibitor A''	. . . . Coating		Salt Spray Rating	
(ppm)	Thickness (mils)	Appearance	Scribe	Field
–	1.25	no pinholes	7	10
1,000	1.15	no pinholes	8	10
2,000	1.0	no pinholes	7+	10
3,000	0.75	no pinholes	8	9
4,000	0.3	no pinholes, transparent coating	0	0

Example 4: This group of samples shows the use of various inhibitors in coating compositions and their effect on reducing pinhole formation in coatings when used in the amounts indicated. Sanded panels were immersed in the coating compositions for 60 seconds and the coated panels were then air dried for 60 seconds, water rinsed for 30 seconds, and baked at 170°C for 10 minutes. The appearances of the baked coatings are set forth below.

. . . . .Inhibitor (ppm).		Coating Appearance
Thiourea	200	very few pinholes
Thiourea	500	no pinholes
Diethylthiourea	200	many pinholes
Diethylthiourea	500	very few pinholes
Sodium silicate	200	many pinholes
Sodium silicate	500	no pinholes

(continued)

........Inhibitor (ppm)........		Coating Appearance
Ammonium dichromate	200	many pinholes
Ammonium dichromate	500	many pinholes
Ammonium dichromate	800	very few pinholes
Ammonium dichromate	2,000	no pinholes
Glyoxime	200	many pinholes
Glyoxime	500	many pinholes
Glyoxime	800	few pinholes
Propargyl alcohol	200	no pinholes
A''	200	no pinholes
A'	200	no pinholes

Example 5: The next series of applications show the use of an inhibitor (Inhibitor A'') to reduce the tendency of coating compositions to form coatings which are bridged. The metallic articles coated were 2 inch pieces of ¼ inch diameter threaded steel rod. They were immersed in the coating composition for the periods of time noted below and further treated according to the process described in connection with the examples reported in Example 4. The appearances of the baked coatings are set forth below.

Inhibitor (ppm)	Coating Time (sec)	Coating Appearance
–	30	severely bridged
1,000	30	severely bridged
2,000	30	moderately bridged
3,000	30	no bridging
3,000	60	no bridging

OTHER METHODS

Facilitating Cold-Forming of Metals

The work of *K.-H. Nuss, K.-D. Nittel, H.-Y. Oei and G. Siemund; U.S. Patent 4,199,381; April 22, 1980; assigned to Oxy Metal Industries Corporation* relates to a process for facilitating the cold-forming of metals by phosphatizing and subsequent treatment with an aqueous lubricant having an alkali-soap base. The improvement comprises employing a lubricant consisting essentially of between 10 and 100 g/ℓ of alkali soap and between 0.1 and 10 g/ℓ of a complex-former of at least one compound selected from the group consisting of EDTA; NTA; trans-1,2-diaminocyclohexanetetraacetic acid; diethylenetriaminepentaacetic acid; N-hydroxyethylenediaminetriacetic acid; N,N-di(β-hydroxyethyl)glycine; sodium glycoheptonate and the alkali metal or ammonium salts of any of the foregoing. Preferably the aqueous, soap-containing lubricant comprises more than 40% of C_{18} alkali metal soaps. In the process, the soap deposit is preferably in excess of 5 g/m^2, and the soap coating is dried prior to cold-forming.

Examples 1 through 5: Steel stampings were treated as follows: pickling in 20% sulfuric acid at 65°C for 15 minutes, and rinsing with cold tap-water, 1 minute, and water at 80°C for another minute. The stampings were then phosphatized with a nitrate-accelerated zinc-phosphatizing system at 98°C for 10 minutes; the thickness of the phosphate coating was 15 μ. The process was continued by rinsing with cold water for 1 minute, followed by immersion in aqueous soap compositions as described in the table below at 73°C for 3 minutes, and then air drying.

Soap Compositions

Composition (g/ℓ)	1	2	3	4	5
Sodium soap with 50% C_{18} component	50	100	50	50	50
Ethylenediamine tetraacetic acid				2	
Nitrilotriacetic acid					2
pH value	10	10	11	10	10

After this treatment, the layer of soap was assessed visually and the deposit was determined by differential weighing after the layer of soap had been removed with boiling water and perchlorethylene vapor. The results are given below.

Deposited Coatings

Ex. No.	Soap Deposit (g/m²)	Visual Assessment
1	4.7	Grey, adheres well, uniform
2	17.2	White, does not adhere well, not uniform
3	10.5	White, does not adhere well, not uniform
4	17.4	White, adheres well, uniform
5	11.0	White, adheres well, uniform

The parts with their different layers of soap were then cold-extruded to form cups. The forming results are given below.

Cold Extrusion

Example	Forming Results
1	Poor (the formed parts had striations)
2	Good (but the press had to be cleaned after a few parts because of soap build-up)
3	Better than Example 1, but still unsatisfactory (additional lubrication in the press with MoS_2 was required in order to eliminate striations)
4	Good (no build-up on the press)
5	Good (no build-up on the press; no additional lubrication in the press with MoS_2 needed)

Phosphate coatings on iron and steel have previously been described in Chapter 1 of this book.

Pretreating Cold Rolled Steel Sheet for Annealing

The work of *N. Ohashi, M. Konishi, M. Nishida and Y. Inokuchi; U.S. Patent 4,042,425; August 16, 1977; assigned to Kawasaki Steel Corporation, Japan* relates to a process for pretreating cold rolled steel sheet for annealing so as to prevent graphite formation on steel sheet surfaces and nitrogenization in the course of the annealing.

The process is based on findings that both the graphite formation and the nitrogenization can be prevented by inactivating the steel sheet surfaces by causing an effective amount of inactivating element or elements, such as sulfur, aluminum, tin, arsenic, lead, antimony, bismuth, selenium, and/or tellurium, to exist on the steel sheet surfaces, because the graphite formation and the nitrogenization are a kind of surface reaction which takes place on the steel surfaces.

To effectively apply the element or elements on the steel sheet surfaces, an aqueous solution of a suspension containing at least one element selected from the group consisting of sulfur, aluminum, tin, arsenic, lead, antimony, bismuth, selenium, and tellurium is uniformly spread on the steel sheet surfaces, so as to achieve a uniform coverage at a density of not smaller than 2 g/m^2, and then the steel sheet is subjected to the recrystallization annealing.

Example: Commercial low-carbon steel was melted in a 200 ton converter for making ingots, and rolled to slabs; hot rolled to coils, pickled, and cold rolled (at a reduction ratio of 70%) in a conventional fashion.

The chemical composition of the low-carbon steel at ladle is: 0.04 to 0.05% carbon; 0.30 to 0.35% manganese; 0.007 to 0.01% phosphorus; 0.015 to 0.020% sulfur; and substantially the remainder being iron.

Ten coils, No. 1 to No. 10, were prepared by degreasing after the cold-rolling. Six coils, i.e., No. 1 to No. 6, were applied with 0.001 mol/ℓ aqueous solutions at $Na_2S_2O_3 \cdot 5H_2O$, $Al_2(SO_4)_3 \cdot 18H_2O$, $Na_2SO_4 \cdot 10H_2O$, $SbCl_3$, $SnCl_2$, and K_2TeO_3, respectively. These aqueous solutions were dripped onto the steel sheet surfaces after the degreasing but before the drying of rinsing water, and to achieve a uniform coverage at the amount of about 10 g/m^2 of aqueous solutions, the steel sheets were squeezed by rubber roll before drying the aqueous solutions with hot air. Four coils, i.e., No. 7 to No. 10, were similarly covered with the four mixtures by dripping and squeezing to achieve the amount of about 10 g/m^2 of aqueous solutions, as shown below. The four mixtures were those of $Al_2(SO_4)_3$ plus $Na_2S_2O_3$; $Na_2S_2O_3$ plus Na_2SO_4; $SbCl_3$ plus $Al_2(SO_4)_3$; and $SbCl_3$ plus $SnCl_2$. The concentrations of the two compounds in each of the four mixtures are shown in the table.

Suppression of Graphite Formation

Coil No.	Compound Applied to Steel Sheet Surface	Composition (mol/ℓ)	Rejection Rate (%)*
1**	$Na_2S_2O_3 \cdot 5H_2O$	0.001	0
2**	$Al_2(SO_4)_3 \cdot 18H_2O$	0.001	0
3**	$Na_2SO_4 \cdot 10H_2O$	0.001	0.6
4**	$SbCl_3$	0.001	0
5**	$SnCl_2$	0.001	0
6**	K_2TeO_3	0.001	0
7**	$Al_2(SO_4)_3 \cdot 18H_2O$ $Na_2S_2O_3 \cdot 5H_2O$	0.005 0.00005	0
8**	$Na_2S_2O_3 \cdot 5H_2O$ $Na_2SO_4 \cdot 10H_2O$	0.005 0.00005	0.4
9**	$SbCl_3$ $Al_2(SO_4)_3 \cdot 18H_2O$	0.00005 0.00005	0
10**	$SbCl_3$ $SnCl_2 \cdot 2H_2O$	0.00005 0.00005	0.5
11***	$Na_2S_2O_3 \cdot 5H_2O$	0.001	0
12***	$Al_2(SO_4)_3 \cdot 18H_2O$	0.001	0
13***	$Na_2SO_4 \cdot 10H_2O$	0.001	1.2
14***	$SbCl_3$	0.001	0
15***	K_2TeO_3	0.001	0
16†	Not applied	–	65.8
17†	Not applied	–	87.4
18†	Not applied	–	49.2

*Due to graphite formation.
**Applied to cold-rolled steel sheet after degreasing.
***Applied to cold-rolled steel sheet without degreasing.
†Without any treatment (only degreasing is applied).

Five coils, i.e., No. 11 to No. 15, were treated as cold rolled without degreasing; namely, 0.001 mol/ℓ aqueous solutions of $Na_2S_2O_3 \cdot 5H_2O$, $Al_2(SO_4)_3 \cdot 18H_2O$, $Na_2SO_4 \cdot 10H_2O$, $SbCl_3$, and K_2TeO_3 were dripped onto the surfaces of the steel sheets forming the five coils, so as to achieve a uniform coverage at the amount of about 10 g/m^2 of aqueous solutions.

All the coils thus treated were subjected to the recrystallization annealing in an HNX gas atmosphere in the tightly wound coil condition. The rejection rates (by weight) of the coils thus annealed, due to the graphite formation on the steel sheet surfaces, are shown. The preceding table also shows the corresponding rejection rates for reference coils, i.e., coils No. 16 to No. 18, which were identical with the coils No. 1 to No. 15 except that they were not treated by the process.

As shown, steel sheet surface defects due to the graphite formation thereon, as experienced in nontreated steel coils, can be almost completely eliminated by applying the process.

Chromizing of Steel by Gaseous Method

The process of *R. Leveque; U.S. Patent 4,242,151; December 30, 1980; assigned to Creusot-Loire, France* relates to an improvement in chromizing methods. The method of chromizing steels to a depth greater than 30 μ, usable for steels with a carbon content of at least 0.2%, especially for steels for construction work and steels for tools is characterized by the combination of three successive treatments. The first of these three treatments consists of an ionic nitriding of a surface layer between 100 and 350 μ thick, this ionic nitriding being realized in an atmosphere constituted by a mixture of nitrogen and hydrogen, at a temperature of between 450° and 650°C, for between 5 and 40 hours, so as to obtain between 1.5 and 2.5% nitrogen in the nitrided layer. The second of these treatments consists of a chromizing by gaseous method forming chromium carbides, lasting between 5 and 30 hours, and realized at temperatures of between 850° and 1100°C. The third of these three treatments is a thermal treatment comprising a quenching in oil of the chromized piece followed by a tempering at a temperature of between 600° and 650°C, lasting between 30 minutes and 10 hours, depending on the size of the piece treated.

Example: A chromium-molybdenum-vanadium steel, of the 35CDV12 type, and consequently with 0.35% carbon, is treated, in order to obtain a depth of chromizing of 50 μ.

The ionic nitriding which forms the first of the three successive treatments according to the process is here effected in a metal vessel provided with thermal shields and cooled by the circulation of water, which vessel constitutes the earthed anode. The electrical parameters are chosen in such a way that the current increases with the direct voltage produced by the generator and the sample to be nitrided which constitutes the cathode becomes covered by the glow discharge corresponding to the conditions of abnormal discharge. The gaseous ions are formed near the cathodic surface and accelerated towards the sample, causing it to heat up, this being continued until the temperature selected for realizing the thermochemical treatment is reached. Temperature regulation is obtained by means of a thermocouple protected by an aluminum casing and placed in the sample in conditions which allow the striking of arcs to be avoided.

The pressure at which the thermochemical treatment is carried out is generally between 2.5 and 8.0 millibars; a primary pump is sufficient to make the initial vacuum and then to allow the nitriding gas close to the sample to be renewed. The nitriding gaseous mixture is composed of nitrogen and hydrogen. The partial pressures P_N of nitrogen at which a solid solution of nitrogen is obtained in the lattice of the ferrite are between 0.1 and 0.6 of a millibar. The temperature is regulated to an average at 530°C and does not go beyond the range of between 510° and 530°C. Disregarding the rise in temperature and the lowering of the pressure of the atmosphere, the duration of the ionic nitriding at the correct pressure and temperature is 25 hours. Through this first treatment, the average nitrogen content of the steel at depths of between 50 and 200 μ reaches 2.1% and the nitride layer contains no iron nitrides or chromium nitrides.

The metal piece of 35CDV12 steel thus nitrided is then extracted from the ionic nitriding furnace and introduced into a cementation tank which is to effect the second treatment, which is chromizing by a gaseous method.

The cementation agent used is a powder constituted of 99.5% ferrochromium with 60 to 70% chromium and 0.5% ammonium chloride, with no aluminum or magnesium oxide. This powder presents a particulate size distribution comprised between 0.5 and 4 mm, with an average dimension of about 2.7 mm. This powder is disposed in the bottom of the cementation tank, which is shaped like a vertical cylinder, and is covered by a partition on which is placed the piece of steel to be chromized. On the upper part of the cementation tank, in a carrier, is a reserve of ferrochromium for direct regeneration of the active vapor of chromium chloride $CrCl_2$. Hydrogen introduced creates a reducing atmosphere.

The vessel is brought to an average temperature of 950°C, without leaving the range 920° to 980°C, for 20 hours.

The following phenomena occur in the cementation tank: On heating, the ammonium chloride dissociates. The chloride ion thus liberated acts on the chromium of the ferrochromium so as to form chromium chloride $CrCl_2$ in the vapor state, which produces the surface chromizing in accordance with the following reaction: $CrCl_2(g) + Fe(s) \rightarrow FeCl_2(g) + Cr(s)$.

The ferrous chloride vapors produced by this reaction react with the reserve of chromium placed at the upper part of the tank, which regenerates the gaseous chromium chloride $CrCl_2$ which plays a part in the chromizing according to the above reaction.

After 20 hours at 920° to 980°C, the chromized piece undergoes the third treatment, that is to say it is extracted from the cementation tank, immediately quenched in oil and then introduced into a tempering furnace kept at a temperature of the order of 625°C, for 2 hours.

After tempering, it is observed: that the surface layer containing the chromium carbonitrides is nearly 50 μ thick; that the chromium carbonitrides of this surface layer are almost exclusively of the $Cr_2(C,N)$ type; that the hardness of this layer is between 1,800 and 2,000 on the Vickers scale; and that it cracks under a load of 1 kg weight.

The chromized coating thus obtained in the process can be compared with that of a chromizing of known type, not preceded by an ionic nitriding. In this prior art process the surface layer containing the chromium carbides is nearly 15 μ thick with two phases of chromium carbide, $M_{23}C_6$ mainly at the surface and M_7C_3 towards the metal substrate (M is a metal such as Fe, Cr or Ni); the hardness of the surface layer is between 1,200 and 1,800 on the Vickers scale, with heterogeneities associated with surface porosities; and the load which causes cracks at the angles of Vickers indentations is 300 g.

Increasing Corrosion Resistance of Nitrided Structural Parts Made of Iron

H. Kunst and C. Scondo; U.S. Patent 4,292,094; September 29, 1981; assigned to Degussa AG, Germany describe a process for increasing the corrosion resistance of nitrided structural parts made of iron materials by an oxidation treatment in connection with the nitriding process which is usable for all iron material.

Satisfactory for the process are salt baths which consist of a mixture of alkali hydroxides, preferably with addition of 2 to 20% of alkali nitrate. Usually there is employed a mixture of sodium hydroxide and potassium hydroxide with potassium nitrate and/or sodium nitrate.

The oxidation treatment preferably takes place at a temperature of 250° to 450°C for 25 to 45 minutes.

Example 1: Bent sheet metal parts about 120 mm long and 60 mm wide made of steel type C15 were nitrided at 580°C for 45 minutes in a salt bath in which the bath composition was 37.6% cyanate and 1.8% cyanide, specifically the bath was composed of 37.6% CNO^-, 1.8% CN^-, 14.5% CO_3^{2-}, 37.2% K^+ and 8.9% Na^+. Subsequently the sheet metal parts were oxidizingly posttreated in a salt bath having the composition 37.4 wt % sodium hydroxide, 52.6 wt % potassium hydroxide and 10.0 wt % sodium nitrate.

Example 2: Rods 450 mm long and 18 mm in diameter made of steel type 42CrMo4 were nitrided for 120 minutes at 570°C in a gaseous mixture having the composition 50% ammonia and 50% endogas.

After withdrawal of the parts from the nitriding oven they were oxidizingly posttreated for 40 minutes at 400°C in a salt bath having the same composition as in Example 1. Subsequently they were cooled off in water to 30°C.

Example 3: Work pieces of cold worked steel were exposed to a glow discharge for 240 minutes under nitrogen at 530°C. Then they were oxidizingly posttreated in a salt bath of the composition set forth in Example 1.

In all the examples corrosion investigations were undertaken, which in all cases compared to the only nitrided composition, gave a substantially higher corrosion resistance.

Comparative Example: A salt spray test was carried out according to the customary process on samples of the steels C15 and 42CrMo4, the untreated salt bath nitrided and quenched according to the customary process and salt bath nitrided and posttreated according to this process.

The time in hours until the first occurrence of traces of rust was measured. The results are set forth in the table on the following page.

Type of Steel	C15	42CrMo4
Untreated	.... (hours)	....
Salt bath nitrided,	132	176
quenched in water or oil		
cooled in a salt bath*	132	176
Hard chromed	236	300
	190-220	190-220

*based on alkali hydroxide and nitrate, treated 35 minutes at 350°C.

Similar results were obtained in other corrosion testing processes (condensation water test, sea water test).

Steel Sheet for Forming Cans: Nickel Diffused Base plus Chromium Acid Treated Top Layer

The work of *H. Asano, Y. Oyagi and T. Egawa; U.S. Patent 4,035,248; July 12, 1977; assigned to Nippon Steel Corporation, Japan* relates to enhancing the anti-corrosion property of the steel sheet after can-forming work without injuring the good paint adhesion property which is the merit of the chromium-acid-treated steel sheet.

There is provided a method for the manufacture of a steel sheet material to be preferably used for a container or can which effectively combines the desirable formability of the Ni-diffused base layer with the desirable coatability or lacquerability of the Cr-acid-treated upper layer.

The process comprises (1) coating an aqueous solution containing Ni ion on a surface of steel sheet as cold rolled in the step of manufacturing a thin steel sheet, followed by drying, (2) thereafter subjecting the same to recrystallization annealing while simultaneously allowing the coated Ni compound to be reduced on the surface of the steel sheet, (3) and thereafter plating thereon a film of metallic chromium and/or hydrated chromium oxide by electrolysis using a bath consisting chiefly of a chromic acid.

The Ni compound may be nickel acetate, nickel nitrate, nickel formate, nickel oxalate, nickel carbonate, and the like. The range of concentration of the aqueous solution may be about 10 to 100 g/ℓ, depending upon the solubility of the particular Ni compound used. In this case it is preferable that the concentration be as large as possible in order to decrease the heat capacity for drying after coating with aqueous solution. The thickness of the diffusion of Ni may be about 1 μ or less, and the amount of Ni attached may be about 0.2 to 5.0 mg/dm^2. The cold rolling step including temper rolling to be held after the Ni treatment is intended to correct the strength or the shape of the material sheet. Thereafter, a treating bath containing a chromic acid as a main component is used to plate a film of metallic chromium and/or hydrated chromium oxide on the material according to an electrolytic process. The amount of metallic chromium attached may be about 0.3 to 1.0 mg/dm^2 while that of chromate may be about 0.2 mg/dm^2 or less.

The reason why the steel sheet treated with a bath consisting chiefly of chromic acid and having a Ni-diffused layer exert an excellent characteristic in this process is considered as depending upon the synergetic effect of the excellent lacquer adhesion property of the Cr-containing layer and the excellent formability of the Ni-diffused layer. The excellent formability of the Ni-diffused layer is caused by the fact that the layer does not produce any different phase in the surface of the steel sheet.

Example: In one test, a Ni plating is conducted on a steel sheet by means of an electric plating or by means of a conversion coating, followed by Ni diffusion by annealing. In another test, a Ni compound in aqueous solution is coated on a steel sheet, followed by drying, which is then subjected to annealing in an atmosphere of 6% H_2 and N_2 while the film is formed thereon. The results of these tests are shown in the table below, in which the samples are prepared as follows.

The steel sheet having a Ni-diffused layer formed in a manner as above is subjected to temper rolling under the reduction rate of 1.5%, which is then coated, in a plating bath consisting chiefly of a chromic acid, with 1.0 mg/dm^2 of metallic chromium and then with 0.1 mg/dm^2 (calculated as Cr) of hydrated chromium oxide thereon. An epoxy resin paint is coated thereon in an amount of 50 mg/dm^2, which is then subjected to a flanging work of 3 mm and 6 mm by the use of an Erichsen testing machine, followed by dipping in an aqueous solution of 1.5% citric acid and 1.5% salt. The resulting amount of dissolved iron is measured.

It is seen from the results shown in the table that the steel sheet obtained by the method of this process as represented by samples 6 to 8 shows much better properties than that of the other steel sheets.

Sample No.	Method of Ni-Plating	Ni Attached (mg/dm^2)	Condition of Annealing	Erichsen Flanging Test . . . Dissolved Iron . . . 3 mm	6 mm
1	None	0	Temperature kept at 680°C	15 ppm	80 ppm
2	Substitute-plating	1.0		3	15
3	Substitute-plating	2.0		2	7
4	Electric plating	1.0	Kept for 30 sec	3	10
5	Electric plating	2.0		2	6
6	Nickel acetate coating	1.0		1	3
7	Nickel nitrate coating	1.0	Annealing atmosphere*	1	4
8	Nickel formate coating	1.0	Annealing atmosphere*	1	3

*H_2, 6%; N_2 rest.

Treating Stainless Steel for Subsequent Enamelling

C.G.H. Brun and P.M.R. Tirmarche; U.S. Patent 4,012,239; March 15, 1977; assigned to Union Siderurgique du Nord et de l'Est de la France (USINOR), France describe a process for treating a steel sheet for the purpose of enamelling, and particularly for direct one-coat enamelling. This comprises depositing on the sheet, after rolling, a coat of a material selected from nickel and cobalt and the acetates and nitrates of these metals, in an amount of 0.45 to 20 g/m^2 calculated as the elemental metal, annealing the sheet in a decarburizing atmosphere consisting of hydrogen, water vapor and nitrogen, the hydrogen content being 10 to 75% and the H_2/H_2O ratio being within the range from 3 to 6, to substantially totally decarburize it and, during the cooling of the annealing cycle, subjecting the coated sheet to the action of an oxidizing atmosphere of hydrogen, water vapor and nitrogen containing 2 to 7% water vapor and having a H_2/H_2O ratio below 6.

The deposition may be carried out after cold rolling or after pickling in the case of sheets which have been merely hot rolled.

Example: *Nickel coating by reduction* – Sheets cold rolled with a continuous band rolling mill were used for the tests. The specimens taken after rolling were degreased and pickled as follows:

An alkaline degreasing step is effected at a temperature of 80°C by dipping or spraying. This treatment is followed by a hot rinse and then by a cold rinse with softened water. This treatment may be replaced by a continuous electrolytic degreasing step.

The degreasing step is followed by a pickling step with dilute sulfuric acid at a concentration of 8%, by dipping or spraying. This treatment is carried out at a temperature of 75°C and is followed by a rinse with softened water made acidic with sulfuric acid at a pH of about 2.8.

Both the degreasing and pickling treatments are repeated twice, in that order. The first pickling step is effected at a lower H_2SO_4 concentration and is followed by a cold softened water rinse instead of the previously described acidic rinse prior to effecting the second alkaline degreasing and the second pickling steps.

The nickel plating bath used has the following composition: 32 g/ℓ nickel sulfate ($NiSO_4 \cdot 7H_2O$); 12 g/ℓ sodium acetate; and 7 g/ℓ sodium hypophosphite.

The pH of the solution is adjusted between 4.5 and 5.8 and the nickel is coated at a temperature of 26°C.

The treatment may be carried out by dipping or spraying and is followed by a cold acidic rinse, a cold rinse and a hot rinse with softened water with added sodium hydroxide (0.5 to 4 g/ℓ) at 70°C, and then by drying by evaporation.

The nickel plated specimens are decarburized at 700°C by expanded annealing under a nitrogen-hydrogen atmosphere containing 20% hydrogen with a dew point adjusted at +30°C by addition of water vapor.

After decarburizing, the cooling is carried out under a wet oxidizing atmosphere containing 10% hydrogen with a dew point adjusted at +25°C.

The table below gives the results of the adhesion of the enamel to sheet specimens having thicknesses of 0.80 mm and 1.30 mm, nickel plated by reduction during 3 to 8 minutes and the amount of nickel deposited in g/m^2.

Sheet Thickness (mm)	Nickel Plating Time (min)	Adhesion of Enamel	Nickel Deposited (g/m^2)
0.80	3	nil	0.32
	4	good	0.45
	5	good	0.53

(continued)

Sheet Thickness (mm)	Nickel Plating Time (min)	Adhesion of Enamel	Nickel Deposited (g/m^2)
0.80	6	good	0.66
	7	good	0.80
	8	good	0.93
1.30	3	bad	0.26
	4	bad	0.40
	5	good	0.51
	6	good	0.69
	7	very good	0.87
	8	very good	1.07

The great flexibility of the process is apparent from the above results: thus, there is good adhesion for nickel plating times of 5 to 8 minutes, corresponding to a nickel plating in excess of 0.45 g/m^2.

For the tests tabulated in the above table, the weight losses due to the prepickling step were between 20 and 30 g/m^2.

Steel Sheet Treated with Corrosion-Resistant Base Film Then with Organic Silicon Compound

The work of *K. Yoshida, Z.-I. Morita and K. Koyama; U.S. Patent 4,026,728; May 31, 1977; assigned to Nippon Steel Corporation, Japan* relates to a steel sheet and strip (hereinafter called a sheet) having good paint adhesion.

The steel sheet is obtained by subjecting the base steel sheet to the first step of process where a corrosion-resistant base film is formed on the surface of the steel sheet by an electrolysis or dipping treatment carried out in a solution containing one or more members selected from the group consisting of chromic acid, phosphoric acid, salts of chromium, molybdenum, silicon, cobalt, manganese, copper, nickel, aluminum and titanium and then subjecting the steel sheet to the second step of process where the steel sheet is treated in a solution containing one or more members of the organic silicon compound which are expressed by the general formula $RSiX_3$, where R is a group of atoms having 2 to 10 carbon atoms and contains one or more functional groups selected from the group consisting of vinyl, epoxy, acryl and amino groups, and X is a halogen atom or an alkoxyl group.

The organic silicon compounds above include:

$$CH_2{=}CHSi(OCH_2CH_2OCH_3)_3$$

$$\underbrace{CH_2{-}CH}_{O}CH_2O(CH_2)_3Si(OCH_3)_3$$

$$CH_2{=}C(CH_3){-}C({=}O)O(CH_2)_3Si(OCH_3)_3$$

$$H_2N(CH_2)_2NH(CH_2)_3Si(OCH_3)_3$$

Commercial products thereof are for example KBC 1003, KBM 403, KBM 503 and KBM 603 (Shinetsu Chemical Industry). The organic silicon compounds may be applied in the amount ranging from 5 to 50 mg/m^2 (as Si) to produce full effects.

Steel sheets, including plated steel sheets, to which the treatment is applied exhibit much better paintability in comparison with those steel sheets which are not treated by the process.

This effect has been caused by the action of the organic silicon compound $RSiX_3$ as binder of the chromium oxide, phosphate metal salt or a composite base film to the upper organic resinous coating, because X in the organic silicon compound has a strong affinity to inorganic substances while R shows affinity to organic matters.

More particularly, the corrosion-resistant base film formed in the first step of the process contains on the uppermost layer thereof reactive groups of atoms such as MOH, where M is a metal atom selected from the group consisting of Cr, P, Si, Co, Mn, Cu, Ni, Al and Ti, which are in some way oriented. This group of atoms reacts with one or more of the organic silicon compounds expressed by the general formula $RSiX_3$ on the uppermost surface of the formed film, when a molecular segregation reaction occurs to separate group of –HX so that a layer of the form: steel sheet–(Cr, P, Si, Co, Mn, Cu, Ni, Al, Ti)–Si–R is formed on the surface of the steel sheet after the organic silicon compounds have been applied. The excellent paintability realized on the steel sheets may reasonably be attributed to the presence of R which is highly reactive with and has strong affinity to organic matters, that is resinous materials in the paint.

Example 1: A cold rolled and electrolytically degreased low carbon steel sheet (0.3 mm thick) was immersed for 3 seconds at 50°C in a solution containing 330 g/ℓ CrO_3, 280 g/ℓ SiO_2 and 10 g/ℓ $Ni(NO_3)_2$, and after being roll-squeezed the sheet was box-annealed in a tight coil shape for 3 hours in a mixed gas at 650°C consisting of 4 to 6% H_2 and 94 to 96% N_2 (designated hereinafter as HNX gas). Then the sheet was immersed for 2 seconds in a 30°C aqueous 1% solution of KBM 403, roll-squeezed and dried with 100°C hot air for 30 seconds. The film obtained in the first step amounted to 58 mg/m² (as metallic chromium) and that in the second step to 15 mg/m² (as Si). This product proved excellent in the evaluation test.

Example 2: A cold rolled and an electrolytically cleaned low carbon steel sheet (0.17 mm thick) were immersed for 3 seconds in a 50°C solution containing 50 g/ℓ CrO_3, 100 g/ℓ aluminum sol and 30 g/ℓ $Cu(NO_3)_2$, roll-squeezed and then thermally treated for 30 seconds in HNX at 500°C. This was then immersed for 2 seconds in a 1% aqueous solution of KBC 1003 at 30°C, roll-squeezed and dried by blowing it with 100°C hot air for 5 seconds. The product was subjected to adjustment rolling with the draft of 2.2%. The coating thus obtained in the first step amounted to 27 mg/m² (as metallic chromium) and that in the second step to 15 mg/m² (as Si). This product proved excellent in the evaluation test.

Evaluation: In Examples 1 and 2, no change was observed in the first adhesion, the second adhesion, the corrosion resistance of the painted layer, or the undercutting corrosion test.

Except for the undercutting corrosion test, a melamine alkyd paint (Orga 100-II Light Vermillion) was applied in an amount of 250 mg/dm² and baked for 20 minutes at 120°C.

For the undercutting corrosion test, an epoxyphenyl paint (83-088, Dainippon

Ink Chemical Industry) was applied in an amount of 55 mg/dm^2 and baked at 210°C for 10 minutes.

The first adhesion was tested by an extrusion working (5 mm Erichsen extrusion), cutting with a knife in an asterisk, followed by peeling off of the cellophane tape.

The second adhesion was tested, after the same extrusion working as above, by immersing for 24 hours in 40°C pure water, immediately cutting with a knife in an asterisk, followed by peeling with cellophane tape.

The corrosion resistance of the paint layer was tested, after the working as above, by applying a knife cut in an asterisk, followed by a salt spraying test (JIS Z2371) for 48 hours and peeling off with cellophane tape.

The undercutting corrosion test was carried out by applying a knife cut in an x, immersing for 4 days in a 25°C aqueous solution containing 1.5% citric acid and 1.5% sodium chloride, followed by peeling with cellophane tape.

Lead Coated Steel Surface Treated with Hydrochloric Acid

R.J. Shaffer and V.J. Schwering; U.S. Patent 4,089,707; May 16, 1978; assigned to Republic Steel Corporation found that the corrosion resistance of lead and lead alloy coated metal, such as terne and the like, can be unexpectedly improved by treating the lead surface with acid. Excellent results are obtained using hydrochloric acid. The method comprises the steps of applying hydrochloric acid to the surface of the coated metal and thereafter rinsing the surface.

The method of application can be similar to that of metal cleaners or metal treatment chemicals, Spray, dip immersion, roll immersion, brushing, flowing and roll application are suitable techniques for both continuous processing and batch application of the acid.

Examples 1 through 10: Table 1 sets forth the results of subjecting steel specimens coated with various lead alloy coating compositions to salt spray corrosion and porosity tests. Specimens were provided with an acid surface treatment in accordance with the process and were tested in comparison to identical specimens which were not given an acid treatment. The specific coating compositions for the specimens in each example are given in Table 2.

The salt spray tests were conducted in accordance with the requirements specified in ASTM B117-64 except that testing was limited to 7 hours of exposure during each 24 hour period. The measurements of pores per square inch were obtained by a porosity test in which the specimens were immersed for 6 hours in distilled water maintained at 200°±5°F. The pore count was then determined by visually counting the red rust spots within the one square inch of surface having the maximum density of pores.

Table 1

Ex. No.	Basic Coating Material	Salt Spray Corrosion Resistance, Av hr to 10% Red Rust		Porosity, Av Pores/sq in	
		Regular	Acid Treatment	Regular	Acid Treatment
1	Pb-Sn	21	500*	7	0

Table 1 (continued)

Ex. No.	Basic Coating Material	Salt Spray Corrosion Resistance Av hr to 10% Red Rust: Regular	Acid Treatment	Porosity Av Pores/sq in: Regular	Acid Treatment
2	Pb-Sn	21	500*	48	1
3	Pb-Sn	14	500*	32	5
4	Pb-Sn	14	449	40	0
5	Pb-Sn	7	16	28	0
6	Pb-Sn-Zn	7	266	100	0
7	Pb-Sn-Zn	78	137	5	0
8	Pb-Sn	24**	73	23	0
9	Pb-Sn	24**	115	4	0

*Test discontinued at 500 hours – coatings had not achieved 10% red rust.
**Significantly more than 10% red rust at 24 hours of exposure.

Table 2

Example	Coating Composition
1	12% tin*
2	10.5% tin*
3	7.8% tin*
4	5.9% tin*
5	4.0% tin*
6	2.0% tin, 0.01% zinc*
7	6.0% tin, 0.22% zinc*
8	Nominal 12% tin, $<$0.05% antimony*
9	Nominal 12% tin, $<$0.05% antimony*

*Balance lead plus incidental impurities

It will be seen from Table 1 that in each instance the application of an acid treatment to the lead alloy coating resulted in a significant improvement in corrosion resistance. At the same time, the acid treatment substantially eliminated any porosity of the coatings.

Table 3 shows the results of salt spray corrosion tests in which lead alloy coated steel specimens were treated with different concentrations of hydrochloric acid.

Table 3

Example No.	Treatment	Salt Spray Hours to 10% Red Rust ± Confidence Limits	Comments
10 A	Untreated	7 ± 0	Production of terne sample
10 B	1% HCl	53 ± 7	Vapor degreased
10 C	5% HCl	49 ± 0	Acetone scrubbed
10 D	10% HCl	52 ± 4	Acetone dip
10 E	20% HCl	50 ± 4	Xylol/alcohol dip
10 F	30% HCl	52 ± 4	Six-second immersion treatment
10 G	40% HCl	53 ± 11	Cold water rinse
10 H	50% HCl	55 ± 7	

Sulfurizing Cast Iron

M. Obayashi and N. Watanabe; U.S. Patent 4,230,507; October 28, 1980; assigned to KK Toyota Chuo Kenkyusho, Japan describes a method for treating the surface of cast iron. It comprises (a) an optional initial step of degreasing the surface of the cast iron; (b) thereafter, pretreatment by immersion in an aqueous solution of nitric acid; and (c) a sulfiding treatment step during which the material, which has been subjected to the pretreatment step, is held in molten sulfur to sulfide the surface layer of the material and thus to form a sulfurized layer. The surface-treating method permits the use of an aqueous solution of nitric acid, which is relatively easy to handle. The resulting sulfurized layer is very thick and stable; it has superior corrosion resistance to molten aluminum.

The surface treatment method is especially effective when applied to cast iron, but is similarly applicable to other materials, such as steel and alloy steels, to which conventional sulfurizing treatment methods are applied. When the surface treatment method is applied to cast iron, to steel and to alloy steels, far superior corrosion resistance to molten aluminum is imparted to the cast iron. It is not clear, however, why the method is more effective for cast iron than for the steel or alloy steels.

The aqueous solution of nitric acid used in the pretreatment step has a concentration of from 1 to 10%, preferably from 2 to 7%. (A 1% aqueous solution of nitric acid means a solution consisting of 1 volume of concentrated nitric acid and 99 volumes of water.)

The material is immersed in the aqueous solution of nitric acid in the pretreatment step for a period of from 20 minutes to 3 hours, at room temperature.

The pretreated material is then immersed and held in molten sulfur (a sulfiding treatment step). The purpose of the sulfiding treatment step is to convert the surface layer (composed mainly of a hydrate of iron and obtained by the pretreatment) into a sulfide layer composed mainly of FeS or FeS_2. The treatment temperature in this step is preferably within the range of from 100° to 145°C. The immersion time is for 1 to 5 hours.

In practice, it is beneficial to apply a diffusion treatment, a generally-known technique, to the material which has been subjected to the pretreatment step and the subsequent sulfiding treatment step. The diffusion treatment is generally performed by heating the material to a temperature ranging from 150° to 300°C in a nonoxidizing atmosphere (e.g., N_2 or Ar) or under reduced pressure (e.g., 10^{-1} mm Hg). It serves to remove excess sulfur adhering to the surface of the material and to render the sulfide layer (formed in the previous step) uniform and dense. It is preferred to perform this step in a nonoxidizing atmosphere or under reduced pressure with little oxygen because it thus prevents corrosion (ascribable to the copresence of sulfur and oxygen) of the material. Heating the material is effected at a temperature ranging from 150° to 300°C. Heating of the material is carried out for a period of from 2 to 5 hours.

Although such a diffusion treatment step is not always necessary, it is very effective for increasing the corrosion resistance of the material to molten aluminum because it has an effect of increasing the uniformity and denseness of the sulfurized layer on the surface of the material.

Example: This illustrates the practice of the method on a cast iron ladle for an aluminum die casting machine. Conventionally, this ladle has been produced from cast iron and coated with a commercially available facing material. The average life of this ladle is about 2 months.

(a) A ladle made of the same conventional cast iron (JIS FC-25, capacity about 10 ℓ) was used. The scale was simply shaved off of the ladle, and the ladle was then washed and degreased with trichloroethylene. It was then immersed for 2 hours in a 5% aqueous solution of nitric acid to form a surface layer on the surface of the ladle. It was then immersed for 4 hours in molten sulfur at 130°C to sulfide the surface layer and thus form a sulfide layer. The ladle was further subjected to diffusion treatment under reduced pressure at 170°C for 5 hours. The same conventional facing material was coated on the surface of the resulting ladle, and it was used in aluminum die casting. The aluminum was JIS ADC10 (9% Al-3% Si, Cu type aluminum alloy die casting). The ladle treated by the method of this process could be used for about 6 months. Thus, it became clear that it has about three times as long a life as the conventional one.

(b) The same ladle for an aluminum die casting machine as used in (a) was subjected to the pretreatment and sulfiding treatment under the same conditions as in (a). Without subjecting the treated ladle to diffusion treatment, its surface was coated with the same conventional facing material, and the coated ladle was used for aluminum die casting. As a result, the ladle could be used for about 4 months, and thus it became clear that it had two times as long a life as the conventional one.

Producing Blackening Oxide Layer with Controlled Combustion of Fuel Gas

In the so-called blackening process, also termed inoxidation, iron is coated with an oxide layer which is formed in an oxidizing atmosphere by annealing of the iron; this oxide layer is contemplated to consist predominantly of the intermediate oxide Fe_3O_4 which forms a blackish protective coating. The blackening is performed at a temperature of between 500° and 650°C, whereby the oxidizing atmosphere is produced by gas burnt in understoichiometric quantities (air ratio n between 0.9 and 0.99; n = 1 in a stoichiometric combustion). The composition of the furnace gases has a great bearing on the formation of the oxide layer.

In a process developed by *J.H. Valentijn; U.S. Patent 4,035,200; July 12, 1977; assigned to Smit Oven Nijmegen BV, Netherlands* the fuel gas is burnt at an apparent reaction temperature of from 500° to 800°C and the combustion products are thereafter passed to a heated blackening chamber at a temperature substantially above the dew point.

Thus, the process steps include the requirements that no water is separated from the combustion products on their way from the combustion chamber to the reaction chamber. The gas mixture is maintained in its entirety whereby, as known, reactions take place between the gases in correspondence with their temperature. In particular, no H_2O is separated.

The blackening step per se should be performed in an atmosphere the CO content of which is maintained constant. According to experience, the combustion in the combustion chamber may be conducted in optimum manner if the apparent reaction temperature is kept at about 750°C. The apparent reaction temperature is that temperature which results for a given $CO_2/CO{-}H_2O/H_2$ ratio from a diagram as seen in Figure 5.4a. The actually measured temperature at separate points of the combustion chamber may differ from this apparent reaction temperature.

Figure 5.4: Producing Blackening Iron Oxide Layer

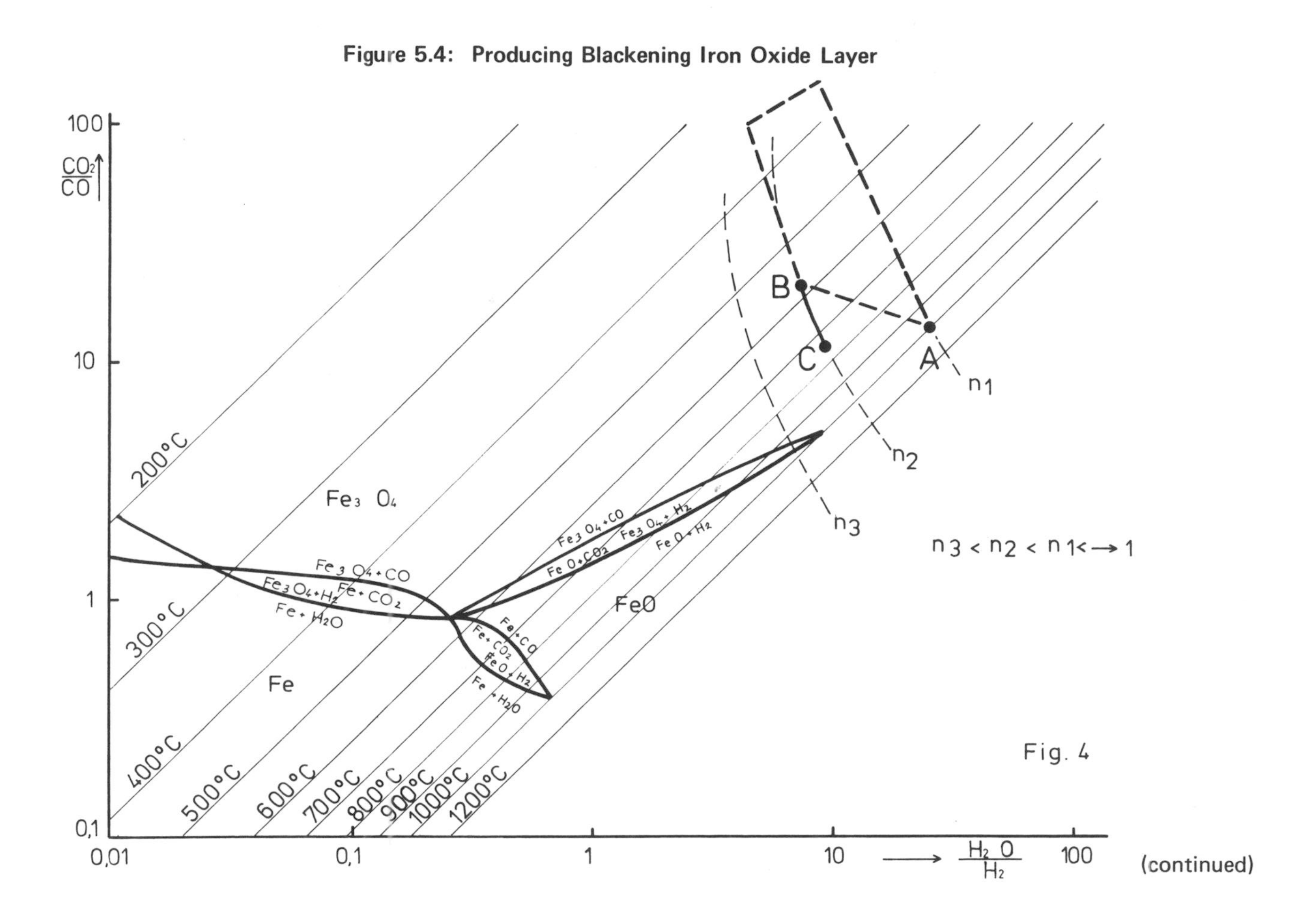

(continued)

Figure 5.4: (continued)

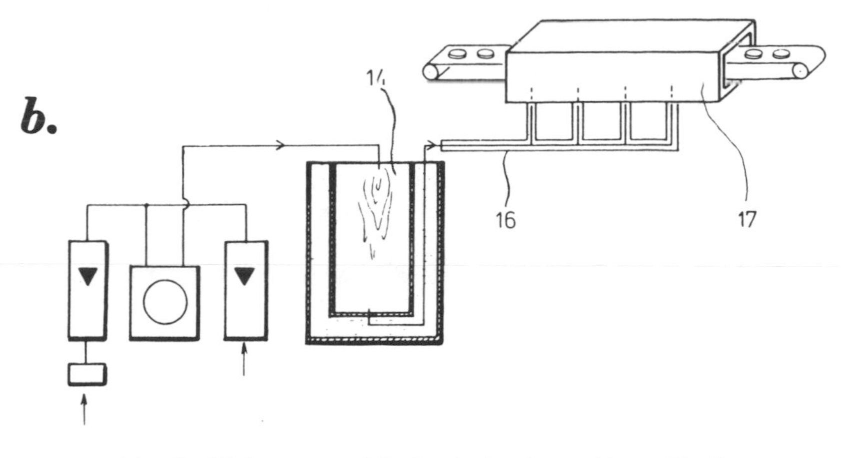

(a) Equilibrium states of the iron/various iron oxide modifications
(b) Combustion apparatus

Source: U.S. Patent 4,035,200

The desired volume and temperature conditions may be controlled or kept constant quite easily when the composition of the combustion product can be controlled through a control circuit or loop which comprises essentially a CO sensor and a device controlling the gas supply and operated by the measured value of the CO sensor. As known, for the gas atmosphere in question it is substantially easier to measure or control, respectively, the CO content, than the dew point. Besides, it is possible to control the system by means of measuring the total of the still combustible components present, i.e., by means of measuring the total proportion of $CO + H_2$.

In this way, the process results in a great simplification of the overall assembly for the blackening of metallic parts. The assembly merely comprises a combustion chamber adapted to be cooled and which communicates with a reaction chamber via a lightly heat-insulated, unheated conduit. Upon exit from the combustion chamber, the gases have a temperature of, for instance, from about 200° to 300°C, which temperature is well above the dew point and does not fall below this value even during the conveying of the gas to the blackening chamber when conduit is just lightly insulated. Accordingly, highly insulated or heatable conduits can be dispensed with, which fact likewise means a substantial simplification and lowering of the cost of the system.

The apparatus for carrying out the method may have one (or more) combustion chamber (chambers) disposed directly adjacent the reaction chamber, because cooling installation and accumulating means for the condensed water can be omitted. Furthermore, the combustion chambers may be arranged immediately within the reaction chamber such that the reaction gas is produced in the immediate vicinity of the products to be blackened.

Figure 5.4b shows an arrangement for carrying out the process. A gas-air mixture is fed by a mixing pump into a combustion chamber **14** and continuously burnt within the latter. The combustion chamber is provided with a cooling jacket (water cooling). Such cooling substantially reduces the apparent reaction temperature, e.g., to a value of about 750°C. Thereupon, the gases are further passed through the cooling jacket (line **16**) whereby these gases are cooled down to a temperature of, e.g., from 200° to 300°C. The gases produced within the combustion chamber have been formed by combustion with the correct air ratio from the outset, such that a different operating point in the $CO_2/CO-H_2O/H_2$ diagram is obtained, as will be explained below. Due to the high outlet temperature of about 200°C, the relative humidity of these gases is substantially lower, such that the temperature cannot fall below the dew point. Even during transport of the gases, the temperature would not fall below the dew point if the conduits are only provided with light thermal insulation. Heating can be dispensed with completely. This fact provides for substantial savings in the process, too.

The cooled fuel or heating gases are passed to a reaction chamber **17** wherein these gases are heated back to the temperature required for oxidation. The oxidation can be performed either continuously or with individual batches each.

Figure 5.4a is a $CO_2/CO-H_2O/H_2$ diagram wherein the equilibrium iron-iron oxide modifications are shown. When using, for example, natural gas as the fuel gas, then in the known process the combustion starts at point **A** (temperature 1000°C). Hereby, the abovementioned temperature is an apparent reaction temperature; in fact, the gases pass through various temperature stages in the combustion. The understoichiometric reaction gas produced is partially dehydrated by cooling. When reheated, the gas reaches point **B** (580°C). The last-mentioned temperature corresponds, e.g., to the operating temperature of the blackening process. Actually, the distance from point **A** to point **B** is not passed through directly, but rather indirectly via the dew point temperature, as indicated in the diagram by the broken line.

On the other hand, in this process the combustion starts not before point **C** (apparent reaction temperature 750°C). Point **B** can be obtained by following a line with constant n factor. Hereby, the lines designated with $\mathbf{n_1}$, $\mathbf{n_2}$ and $\mathbf{n_3}$ represent air ratios of gas mixtures which remain constant with respect to their total mass. The temperature is always substantially above the dew point of the combustion gases. Again, when point **B** is reached, the desired conditions for the blackening process are obtained. Heating and cooling are effected without removal of a gas portion with a constant total mass.

The gas composition or mixing by means of the mixing pump is controlled by a control circuit which comprises a CO probe as a sensor. The measured value provided by this probe controls the ratio between gas and air in such a way that the admixed volume of air is reduced in the case of insufficient quantities of CO which indicate an excessive volume of air. As the measuring probe, an infrared analyzer may be used. Hereby, the circuit arrangement and the determination of the measuring location are left to the expert's discretion. Control can be effected manually, too.

Reference may be made to the fact that the hot combustion gases from the combustion chamber may be introduced directly into the reaction chamber. However, in case that the heating must be conveyed over a certain distance, then it is advisable for reasons of safety, for reasons of conduit wear and the like, to cool the hot combustion gases to a transport temperature of from 200° to 300°C.

Increasing Rate of Deposition of Nickel by Heating Metal Before Coating

The process of *P. Philippe, L. Vincent and H. Graas; U.S. Patent 4,221,832; September 9, 1980; assigned to Centre De Recherches Metallurgiques-Centrum Voor Research In De Metallurgie, Belgium* is based on the observation in the course of experiments that the rate of deposition of nickel increases substantially when, instead of introducing the cold strip into the bath, the strip is preheated to a temperature above 300°C. This unexpected effect enables the metal strip to be coated with a relatively thick film in a very short time.

The process for treating metal strip comprises the following steps: (a) The strip is heated to a temperature above 300°C, preferably under a protective atmosphere if the temperature exceeds 500°C; and (b) the strip is dipped into a bath having a temperature higher than or equal to 80°C, preferably higher than or equal to 90°C, and containing at least one metal salt intended for the coating of the strip.

The coating bath is either a solution containing:

A nickel salt such as $NiCl_2 \cdot 6H_2O$ or $NiSO_4 \cdot 6H_2O$ in a concentration within the range from 5 to 50 g/ℓ, preferably from 15 to 35 g/ℓ;

A reducing agent such as sodium hypophosphite, $NaH_2PO_2 \cdot H_2O$, in a concentration greater than 0.1 mol/ℓ;

A buffer component, generally constituted by an alkali salt of a dibasic organic acid having the general formula $CO_2H-(CH_2)_n-CO_2H$ in which n is from 1 to 4, and/or sodium hydroxyacetate ($NaC_2H_3O_3$), sodium acetate ($CH_3CO_2Na \cdot 3H_2O$), the alkali salts of glycolic acid ($HOCH_2COOH$), citric acid, lactic acid, propionic acid, etc.;

or a solution for nickeling by displacement, containing simply a nickel salt and and acid, with a pH of 2 to 5.

The ratio between the quantity of nickel salt and the quantity of hypophosphite is in the range from 0.2 to 1.6 mols/ℓ, preferably from 0.3 to 0.8 mol/ℓ. The buffer component is present in the solution in a concentration between 0.04 and 0.55 mol/ℓ. The pH of the solution is 3 to 6, and is adjusted with NaOH or NH_4OH.

In order to avoid the precipitation of iron and nickel phosphites, it is highly expedient to add substances such as lactic acid (a nickel-complexing agent) and tartaric acid (an iron-complexing agent), and also to filter continuously the solution, for example through activated charcoal, because the particles of the precipitates may constitute deposit seeds.

Example: Samples of cold rolled LD rimming steel strip 0.8 mm thick were subjected to the following different treatments:

(a) Conventional annealing in a closed furnace;

(b_1) Heating to 700°C in 40 seconds, keeping at 700°C for 60 seconds, quenching in a bath of boiling distilled water;

(b_2) As above plus aging at 450°C for 60 seconds, final cooling by gas jets;

(c) Heating to 700°C in 40 seconds, under a protective atmosphere, holding at 700°C for 60 seconds, dipping in a bath at 97°C containing 24 g/ℓ $NiCl_2 \cdot 6H_2O$, 24 g/ℓ $NaH_2PO_2 \cdot H_2O$, and 20 g/ℓ lactic acid; the pH being adjusted to 5 by addition of NaOH; the time in the bath being adjusted to ensure cooling to 150°C and a residence time of

- (c_1) 0.5 second
- (c_2) 5 seconds
- (c_3) 10 seconds
- (c_4) 30 seconds

below this temperature;

(d_1) to (d_4) As under (c_1) to (c_4) but with additional aging for 60 seconds at 450°C and final cooling in air.

Results

Treatment	Final Hardness (Vickers 50 g)	Oxidation Test*	Phosphatization**
a	87-97	highly pitted	average
b_1	107	oxide film unchanged	good
b_2	100	some pitting	good
c_1	107	very slight pitting	good
c_2	110	slightly dull	very good
c_3	116	brilliant	very good
c_4	123	brilliant	good
d_1	107	brilliant	average
d_2	118	brilliant	average
d_3	124	brilliant	average
d_4	137	brilliant	average

*Oxidation in a bath of distilled water at pH 5 for 3 hr at 100°C.
**Grip-secured corrosion test of phosphatic and paint protection after 240 hr exposure to saline mist (ASMB Standard 117).

For the purpose of increasing the corrosion resistance, it is possible to add to the treatments described in the foregoing a chromatizing treatment in a solution containing $Na_2Cr_2O_7$ or CrO_3 in a concentration of 10 to 100 g/ℓ, in the presence of SO_4 ions in a concentration approximately 100 times lower with a current density of the order of 10 A/dm^2, the strip being cathodic. This chromatizing treatment may be carried out during the first cooling and/or during the second cooling, or after the latter. The resulting advantage is one of covering the previously formed nickel layer with a layer of metallic chromium and/or with a layer of trivalent chromium oxide.

SURFACE TREATMENT OF ZINC AND ZINC PLATED METALS

CORROSION-RESISTANT COATINGS FOR ZINC SURFACES

Aqueous Alkaline Coating Solution

It is known to coat zinc surfaces with aqueous coating solutions that are effective in forming thereon corrosion-resistant coatings which protect the surface from degradation due to attack by materials which tend to corrode the surface. In general, the coatings formed from such coating solutions also should have properties so that overlying coatings applied thereto adhere tightly and strongly.

Two basic types of such compositions are acidic compositions, e.g., those that form phosphate or chromate coatings on the surface, and alkaline compositions. An aqueous alkaline coating solution having a pH no greater than 10.2 for forming on zinc surface a corrosion-resistant coating and to which overlying coatings adhere excellently is disclosed by *E.R. Reinhold; U.S. Patent 4,278,477; July 14, 1981; assigned to Amchem Products, Inc.* This contains in solution one or more of the following metals: cobalt, nickel, iron and tin; an inorganic complexing material for maintaining the metal in solution; and optionally, a reducing agent.

The use of a mixture of iron and cobalt, added as ferric nitrate and cobalt nitrate, is preferred. A preferred inorganic complexing material is pyrophosphate and a preferred organic complexing material is nitrilotriacetic acid or a salt thereof. With respect to optional ingredients, it has been observed that an increase in the rate of coating formation can be realized by including in the composition a reducing agent, such as sodium sulfite.

Unless stated otherwise, each of the Zn surfaces treated with the compositions identified in the examples was a zinc panel of hot-dipped galvanized steel 4 by 12 inches in size, which was subjected to the following sequence of steps:

(A) spray cleaned with an aqueous alkaline cleaning solution for 20 sec at 160°F (71°C);

(B) rinsed with a cold water spray for 2 to 3 sec at ambient temperature;

(C) treated with a composition of the examples at a temperature of 125°F (52°C) by immersing in a laboratory immersion cell for 15 sec;

(D) rinsed with a cold water spray for 2 to 3 sec at ambient temperature;

(E) treated with a 0.5 wt % Cr^{6+}/reduced Cr aqueous solution Deoxylyte 41 (Amchem Products, Inc.) by immersing for 5 sec, followed by squeegeeing through wringer rolls and air drying; and

(F) painted with a single coat of polyester paint, CWS 9039 (Hanna Chemical Coatings Corp.) to a paint film thickness of about 0.8 to 1 mil, followed by baking for 75 sec in an oven having a temperature of 500°F (260°C) to a peak metal temperature of 420°F (216°C) and then quenching in cold water.

The degree of adherence of the paint film to the underlying treated surface and its degree of resistance to corrosion were evaluated by subjecting panels to tests used in industry to evaluate such properties.

Corrosion-resistant properties were evaluated by subjecting painted panels to salt spray conditions in accordance with ASTM B 117. A test referred to herein as T-Bend was used to evaluate paint adhesion.

As the results of the test are reported at the first T-bend at which no paint loss occurs, the lower the T-bend rating, the better the paint adherence. In general, a rating of 1 or 2 is considered excellent and a rating of 4 or more is considered poor.

Example 1: The first group of examples shows the use of a treating composition which comprises an alkaline solution of 25 g/ℓ of $K_4P_2O_7$ and 2.5 g/ℓ of $Fe(NO_3)_3 \cdot 9H_2O$, and the use of modified forms of this composition. The modification encompassed including in the composition amounts of $Co(NO_3)_2 \cdot 6H_2O$ as indicated in the table. In this group of examples, the paint used was an acrylic paint Durocron 630 and the thickness of the dry paint film was about 0.5 mil.

No.	Composition Including $Co(NO_3)_2 \cdot 6H_2O$ (g/ℓ)	pH	Paint Adhesion, Number of T-Bends
1	—	10.2	2
2	0.1	10.0	2
3	0.2	10.0	2
4	0.5	9.9	2
5	1	9.8	3
6	2	9.8	2

It was observed that the salt spray corrosion resistance of the coated panels increased proportionately to the cobalt concentration up to 0.5 g/ℓ $Co(NO_3)_2 \cdot 6H_2O$. Beyond that concentration, no further increase in corrosion resistance was evident. The color did, however, increase as the cobalt concentration increased to the limit tested (1 to 2 g/ℓ).

Example 2: The next group of examples shows the preparation of a concentrate from which there can be made a treating composition, the use of a bath of the composition to treat zinc panels, and the replenishment of the bath with a replenishing composition.

The concentrate contained the following and had a pH of about 7.

Concentrate

Constituents	Grams per Liter
$K_4P_2O_7$	100
$Co(NO_3)_2 \cdot 6H_2O$	5
$Fe(NO_3)_3 \cdot 9H_2O$	10
Water	969

A 2 ℓ bath of treating composition containing 10% by volume of the concentrate was prepared by diluting the concentrate with water. Accordingly, the make-up bath contained 10 g/ℓ of $K_4P_2O_7$, 0.5 g/ℓ of $Co(NO_3)_2 \cdot 6H_2O$ and 1 g/ℓ of $Fe(NO_3)_3 \cdot 9H_2O$. To replenish this bath as it was used to coat zinc panels, the following replenisher was prepared.

Replenisher

Constituents	Grams per Liter
$Na_2H_2P_2O_7$	25
Nitrilotriacetic acid, disodium salt	5
$Co(NO_3)_2 \cdot 6H_2O$	6
$Fe(NO_3)_3 \cdot 9H_2O$	9
Aqueous solution of NaOH, 50 wt %	12.6
Water	974.4

The table below summarizes the manner in which the bath was used and replenishment thereof.

		. .Concentration (ppm). .		
	pH	Co	Fe	Zn
Bath makeup	9.80	110	153	0
Processed 25 panels	9.9	–	–	–
Added 16.7 ml replenisher	9.80	108	152	57
Processed additional 25 panels (total 50)	9.92	–	–	–
Added 16.7 ml replenisher	9.78	110	153	101
Processed additional 25 panels (total 75)	10.06	–	–	–
Added 16.7 ml replenisher	9.85	108	153	153
Processed additional 25 panels (total 100)	9.95	–	–	–
Added 16.7 ml replenisher	9.78	110	153	190
Processed additional 25 panels (total 125)	9.96	–	–	–
Added 16.7 ml replenisher*	9.78	113	153	217
Processed additional 25 panels (total 150)	9.98	–	–	–
Added 16.7 ml replenisher	9.77	115	155	247
Processed additional 25 panels (total 175)	10.00	–	–	–
Added 16.7 ml replenisher	9.80	117	160	273
Processed additional 25 panels (total 200)	10.00	110	150	300
Added 16.7 ml replenisher	9.80	120	160	300

*And also added 100 ml of the composition used to make up the bath to replace drag-out.

During the coating of 200 panels, the bath remained free of sludge and other precipitate.

Example 3: In these examples the compositions that were formulated included sodium sulfite as the reducing agent in the amounts of 1 to 50 g/ℓ. In addition to the reducing agent, each of the compositions contained 25 g/ℓ of $K_4P_2O_7$, 2.5 g/ℓ of $Fe(NO_3)_3 \cdot 9H_2O$ and 2.5 g/ℓ of $Ni(NO_3)_2 \cdot 6H_2O$.

In utilizing the composition containing 1 g/ℓ Na_2SO_3, it was observed that the coating that was formed on a hot-dipped galvanized steel panel was substantially darker in color than a coating formed on a like panel utilizing a composition alike in all respects to the composition except for the absence of sodium sulfite. The darker the color, the greater the amount of coating, and this is an indication of a higher rate of coating formation inasmuch as each of the panels was treated with the composition for the same amount of time (15 sec). It was observed also that the use of increased amounts of sodium sulfite resulted in darker colors up to a concentration of 10 g/ℓ of sodium sulfite. At this concentration, the coating was somewhat lighter than the coating that was formed from the composition which included 5 g/ℓ of sodium sulfite. The use of the compositions containing 20 to 50 g/ℓ produced coatings which were about the same in color as that of the coating formed from the composition containing 10 g/ℓ.

Reducing Sludge Formation in Phosphating Bath

K. Yashiro, S. Saida and Y. Sano; U.S. Patent 4,142,917; March 6, 1979; assigned to Oxy Metal Industries Corporation found that the rate of sludge formation in a phosphating bath for forming a zinc phosphate coating on a zinc surface may be reduced by employing an aqueous treating solution free of nitrate or ammonia ions and containing 0.1 to 5 g/ℓ zinc ion, 5 to 50 g/ℓ phosphate, 0.5 to 5 g/ℓ hydrogen peroxide and 1 to 10 pbw of nickel and/or cobalt ions per part of zinc. Preferably, the coating thus applied is dried in place and then contacted with an aqueous mixed chromium composition having a ratio of hexavalent to trivalent chromium of from 2 to 10 and a pH of from 2 to 5 and dried without rinsing.

Examples 1 through 10: The chromate coating was removed from the surface of hot-galvanized steel sheets having a thickness of 0.35 mm, a width of 200 mm and a length of 300 mm by polishing them by means of a wet buff wheel. The polished surface was then treated with a suspension of a colloidal titanium phosphate surface conditioner at 60°C in a concentration of 1 g/ℓ by spraying for 2 sec. The treated surface was passed through squeeze rolls immediately thereafter and then contacted with an aqueous acidic phosphoric acid solution at 68°C having the compositions as specified in Table 1, by flooding the solution at a rate of about 1,000 ml/sec. Immediately after contact with the solution, the sheets were passed through squeeze rolls and then dried by hot air for Examples 1 through 4 and 6 through 10 and comparative Example 1 or rinsed with water for Example 5 and comparative Example 2.

The sheets dried without rinsing were coated with an aqueous solution having a ratio of Cr^{6+}/Cr^{3+} of 3, a pH of 2.6 and a total chromium content of 5 g/ℓ obtained by reacting 130 parts of chromium trioxide with 8 parts of methanol in an aqueous solution to reduce the hexavalent chromium partially, the coating being carried out in an amount of about 2 ml/m^2 at room temperature.

For Example 5 and comparative Example 2, the phosphated surface was water rinsed and then treated with a chromium solution as above at a total chromium concentration of 2 g/ℓ at 60°C by spraying. After removing excess solution by means of squeeze rolls, the rinsed sheets were dried in hot air.

Table 2 shows results obtained by subjecting the surface-treated sheets to the salt spray test according to JIS Z-2371. For reference, the weight of the coatings and amount of chromium deposited are also shown in Table 2. Thus treated sheets were also coated with a paint of two-coat, two-bake type for colored galvanized steel sheet (KP Color 2105, Kansai Paints Co.) to a total film thickness of 18 μ. The painted sheets were subjected to the bending adhesion test about a diameter equal to the thickness of two sheets. In another group of the thus painted sheets, the paint film was scribed by means of an NT cutter and the cross-hatched surface was then subjected to the salt spray test according to JIS Z-2371. Results obtained are shown in Table 2.

Table 1

	...... Treating Solution Composition.......								Contact	Rinsing
Ex.	Zn^{2+}	Ni^{2+}	Co^{2+}	PO_4^{3-}	H_2O_2	BF_4^-	F^-	Acid	Time	After
No.	.... (g/ℓ)							Ratio	(sec)	Conversion
1	1.00	3.00	–	14.8	2.00	–	–	7.6	2	No
2	1.00	2.00	1.00	14.8	2.00	–	–	7.6	2	No
3	1.00	–	3.00	14.8	2.00	–	–	7.6	2	No
4	2.30	6.20	–	28.6	1.10	4.50	–	6.3	2	No
5	2.30	6.20	–	28.6	1.10	4.50	–	6.3	2	Yes
6	2.30	4.10	2.10	28.6	1.10	4.50	–	6.3	2	No
7	2.30	–	6.20	28.6	1.10	4.50	–	6.3	2	No
8	1.00	3.00	–	14.8	2.00	–	0.25	6.5	2	No
9	1.00	2.00	1.00	14.8	2.00	–	0.25	6.5	2	No
10	1.00	–	3.00	14.8	2.00	–	0.25	6.5	2	No
*1	–	–	–	–	–	–	–	–	8	No
*2	–	–	–	–	–	–	–	–	8	Yes

*Comparative—an acidic phosphating solution containing the following in g/ℓ: Zn^{2+}, 2.46; PO_4^{3-}, 9.00; NO_3^-, 2.70; SiF_6^{2-}, 2.00; F^-, 0.25 and starch phosphate, 2.00. The solution had a total acid/free acid ratio of 8.0.

Table 2

Ex. No.	Coating Weight (g/m²)	Weight of Deposited Chromium (mg/m²)	Salt-Spray Corrosion Resistance 48 Hours (% rusted area)	Bend Adhesion (% peeled)	Painted Sheet Salt Spray Corrosion Resistance 1,000 Hours (minimum creepage)
1	1.51	10.3	0	no change	0
2	1.46	13.8	0	no change	0
3	1.48	9.6	0	no change	0.5
4	1.62	14.2	0	no change	0.5
5	1.58	11.1	0	no change	0.5
6	1.64	10.5	0	no change	0
7	1.60	12.4	0	no change	0
8	1.37	10.8	0	no change	0.5
9	1.48	9.2	0	no change	0
10	1.38	13.0	0	no change	0

(continued)

Table 2 (continued)

Ex. No.	Coating Weight (g/m²)	Weight of Deposited Chromium (mg/m²)	Salt-Spray Corrosion Resistance 48 Hours (% rusted area)	Bend Adhesion (% peeled)	Painted Sheet Salt Spray Corrosion Resistance 1,000 Hours (minimum creepage)
*1	1.55	11.2	40	no change	2-5
*2	1.52	11.8	15	no change	1

*Comparative example.

Tannic Acid plus Water-Dispersible Polymer

S. Tsuda, E. Tarumi, H. Kawaskai and T. Watanabe; U.S. Patent 4,247,344; Jan. 27, 1981; assigned to Nippon Steel Corp. provide a method for surface treatment of metal-plated steel products, such as zinc-plated and zinc-alloy-plated steel products, which comprises treating the metal-plated steel products with an aqueous solution containing 0.01 to 10% by weight, preferably 0.5 to 5.0% by weight of tannic acid and 0.001 to 10% by weight, preferably 0.01 to 1.0% by weight of one or more of water-soluble or water-dispersible polymers which are stable when mixed with the tannic acid, and forming a corrosion-resistant film on the surface of the metal-plated steel product.

When an inorganic polymer is used, the inorganic polymer is converted by drying into a glassy substance to form a dense film, so that the film provides a high resistance against the permeation of air and water which promote the corrosion of the metal.

On the other hand, when the polymer used is an organic polymer, the film thus formed is less resistive against the permeation of air and water than the film formed with the inorganic polymer addition, but provides better workability because the film is softer. Therefore, depending on the final application of the metal-plated steel products treated by the process, the film may be formed by selecting the inorganic polymers or the organic polymers.

The water-soluble or water-dispersible inorganic polymer includes biphosphates of metals such as aluminum biphosphate, and magnesium biphosphate; metal silicates, such as water glass; partially hydrated compounds of alkyl silicate; and metal oxide sols, such as silica sol.

The water-soluble or water-dispersible organic polymer used which should be stable when mixed or in contact with the tannic acid includes starch, polyvinylpyrrolidone, polyacrylamides, polyacrylic acids, sodium alginate, polyacrylic acid esters and polyvinyl acetate.

Example 1: A steel wire of 3 mm diameter coated with 300 g/m² zinc by dipping was treated in various aqueous solutions containing tannic acid and a water-soluble inorganic polymer for 3 sec, dried, and subjected to the wet corrosion test (MIL STD 202 C) to determine resistance against white rust. The test was done after storing the test pieces in a room at 20°C with 75% humidity for 1 month. Treating conditions and test results are shown in Table 1.

It is clearly demonstrated by the test results that the steel products, treated with the solution containing 0.01 to 10% by weight of tannic acid and 0.001 to 10% by weight of a water-soluble inorganic polymer show excellent corrosion resistance.

Table 1

Treating Solution	Wet Corrosion Test (200 hr)*
Nontreated	(A)
Comparison	
0.005% tannic acid + 1% aluminum biphosphate	(A)
0.02% tannic acid	(B)
0.02% tannic acid + 0.1% aluminum biphosphate	(C)
3% tannic acid + 0.1% aluminum biphosphate	(D)
Process	
5% tannic acid + 0.05% potassium silicate	(E)-(D)
2% tannic acid + 0.01% potassium silicate	(E)-(D)

*(D) is no rust; (E) is not more than 3% white rust; (C) is 4 to 9% white rust; (B) is 10 to 50% white rust; (A) is 51% or more white rust; % is percent of the rusted dimension to the total dimension tested.

Example 1: An 0.4 mm thick steel sheet coated with 70 g/m^2 zinc was treated in various solutions containing tannic acid for 5 sec and dried, pieces thus prepared were left in a room at 20°C with 75% humidity for 1 month and subjected to wet corrosion test (MIL STD 202C) to determine resistance against white rust. Test pieces were extruded 4 mm by an Erichsen test machine and white rust formation on the extruded portion measured. Treating conditions and test results follow. It is clear that process treatments give better corrosion resistance than comparison treatments.

Table 2

Treating Solution	Wet Corrosion Test (200 hr)
Nontreated	(A)
Comparison	
5% tannic acid	(A)-(C)
2% tannic acid + 0.1% silica sol	(C)-(E)
2% tannic acid + 0.5% polyacrylic acid ester	(C)-(E)
Process	
2% tannic acid + 0.1% silica sol + 0.1% polyacrylic acid ester	(E)-(D)
4% tannic acid + 0.1% starch + 0.1% magnesium biphosphate	(D)
1% tannic acid + 0.1% polyacrylamide + 0.2% partially hydrated compound of ethyl silicate*	–

*Aqueous solution obtained by partially hydrating ethyl silicate from hydrochloric acid; percent as ethyl silicate before hydrolysis.

Galvanized Zinc Coating, Chromate Coating, plus Fluoroplastic Coating

The process of *W. Labenski, H.P. Schapitz and H.-P Wessel; U.S. Patent 4,003,760;*

January 18, 1977; assigned to Mecano-Bundy GmbH, Germany relates to the production of metal articles having a multiple-layer protective coating comprising successive layers of zinc, chromium and synthetic resin.

Zinc Coating: The first step in the coating process is to galvanically apply a finely crystalline, highly homogenous zinc coating over the metal article. This is accomplished in a manner similar to the process described in U.S. Patent 3,808,057 by introducing the metal article into a zinc plating bath using a sulfuric acid electrolyte, at extremely high current densities of 40 to 100 A/dm^2, with the electrolyte being at a temperature of 50° to 55°C and under vigorous agitation, and containing an organic compound which causes a fine crystalline structure of the zinc precipitate.

	Bath 1 (g)	Bath 2 (g)
$ZnSO_4+H_2O$*	850	850
$Al_2(SO_4)_3+14H_2O$**	26	26
H_3BO_3 (boric acid)	6	6
$ZnCl_2$ (zinc acid)	4	4
Saccharin	0.25	0.25
Thiocaramine	0.25	0.45
Polyethyleneimine	—	0.75

*Zinc sulfate and water of crystallization.
**Aluminum sulfate and water of crystallization.

Chromate Coating: The zinc-plated articles are advanced into chromating equipment wherein a chromating coating is applied over the zinc coating. A chromic acid solution is used for this step which has a pH value of higher than 2, most usefully within the range of from 2.1 to 2.6. The chromic acid solution may contain sodium dichromate, sodium nitrate, nitric acid and concentrated acetic acid. Also, the chromic acid solution is to contain a reducing agent such as formic acid, formaldehyde or similar substances. With such a solution it is possible to obtain an unusually heavy layer thickness of from 5 to 10 μ during the short treatment period which is necessary for a continuously progressive method for coating articles.

Suitable chromate solutions may have the following composition (per 100 ℓ of solution): (a) 8 to 15 ℓ of solution, containing 250 to 350 g/ℓ sodium bichromate and 200 to 300 g/ℓ sodium nitrate, which is standardized to a pH value of 0.3 by using nitric acid, (b) 3 to 8 ℓ of solution, containing 500 to 700 g/ℓ formic acid of 85% and (c) 0.5 to 3 ℓ glacial acetic acid by which the chromate solution is standardized to the appropriate pH value. The balance of the solution is water.

Rinsing and Drying the Chromate Coating: The chromate-coated article is rinsed with fresh water and then dried immediately after its formation to a spongy but rough condition. This produces an activating surface of the chromate coating that is particularly effective to produce intimate contact with the next layer of synthetic resin, and the dried chromate coating is hydrated and contains water of crystallization which is to be preserved therein upon application of the subsequent synthetic resin coating. The chromate coating is dried with hot air; it is made up largely of trivalent chromium and has an olive green color at this stage of the process. Care should be taken to insure drying is sufficient to prevent water or chromic acid solution remaining on the article in amounts that would deleteriously affect the plastic coating next applied.

Application of the Plastic Coating: After the chromate coating is dried, a coating of synthetic resin is applied over the chromate coating. The synthetic material consists of fluoroplastic materials by which term is meant, utilizing the ASTM definition, resins that are paraffinic hydrocarbons in which all or part of the hydrogen atoms have been replaced with fluorine atoms and which may also include chlorine atoms in their structure. The term fluoroplastic thus includes fluorocarbon resins such as polytetrafluorethylene (PTFE), fluorinated ethylene-propylene (FEP), and polyhexafluoropropylene; fluorohydrocarbon resins such as polyvinyl fluoride, polyvinylidene fluoride and polytrifluorostyrene; chlorofluorocarbon resins such as polychlorotrifluoroethylene (PCTFE); and chlorofluorohydrocarbon resins.

These fluoroplastic substances have the advantage that a homogenous, closely pored surface forms even when the plastic film is very thin. Due to the property that these fluoroplastic substances absorb no water (they cannot be conditioned) the water of crystallinity remains within the chromate plating under the plastic coating. The fluoroplastic coating is applied as a dispersion, preferably in a high-boiling solvent; a suitable example being dispersion of polyvinyl fluoride resin having 40% solids by weight with a grain size below 2 μ in a solvent mixture such as propylene carbonate (56%) and diethylene glycol (4%). The chromate coating should be thoroughly wetted with the fluoroplastics coating. The fluoroplastics coating should be applied evenly to the article, and suitable air nozzles may be employed to smooth out the coating.

Thermal Treatment: The coated article is next subjected to thermal treatment to solidify the fluoroplastic coating and cause it to interlock with and adhere to the chromate coating, and at the same time preserve the water of crystallization in the chromate coating. The coated article is heated for a short time at a temperature of about 100° to 200°C, but most usefully at a temperature near 250°C to evaporate the solvent and jell the fluoroplastic coating without impairing the chromate coating.

Thermal treatment is accomplished by passing the coated article through an oven, and the article is maintained therein for a short time to prevent sagging of the coating. Also, short drying time prevents loss of water of crystallization from the chromate layer.

High circulation of hot air at a temperature of about 380°C yields article surface temperatures of about 250°C when the article is in the drying oven for 8 to 10 sec and leads to rapid coagulation of the plastic coating. The actual time in the drying oven is increased by the time needed to heat the article, which varies with the running speed of the article and thickness of the plastic coating, being about 20 sec for an article with a 14 to 18 μmm thick plastic coating moving at 11 m/min. After the thermal treatment, the coated articles are cooled rapidly to room temperature to consolidate the plastic coating with the chromate layer, cold water being suitable to achieve the desired rapid chilling. One or more additional plastic coatings can thereafter be applied if so desired.

Example: Steel tubing suitable for brake conduits was coated to form coated articles having a zinc plating thickness of 25 μ, a chromate plating thickness of from 6 to 8 μ and a plastic coating film thickness of 22 to 25 μ. The coated tubing was subjected to the above tests, and the results are set forth in No. 2 of the following table. For comparison, No. 1 shows the test results for corrosion protection of the same tubing with a zinc plating thickness of 25 μ (applied as described above) and a typical prior art blue chromate coating of 2 to 5 μ thick. It can be seen from the table, that the coated articles made according to this process exhibit significantly greater corrosion resistance.

Shaping of the Test Specimens	Condensed Water Test (DIN 50 017) White Rust	Salt Spray Test . . . .(ASTM-B117/64) . . . . White Rust	Red Rust
No. 1			
Sawed	1. Round	24 hr	240-312 hr
Straightened	(100%)	(100%)	
Bent			
No. 2			
Sawed	261. Round	2,400 hr	10,000 hr
Straightened	(under 1%)	(10-30%)	
Bent			
Edged			

Note: 30 specimens were tested in each and results all fell within the indicated range.

NONCHROMATE CONVERSION COATINGS

Sulfuric Acid, Hydrogen Peroxide and Silicate Treatment

A.M. Bengali, R.F. Zuendt, J.L.H. Allan and P.D. Readio; U.S. Patent 4,222,779; September 16, 1980; assigned to Dart Industries Inc. provide a conversion coating solution which comprises an aqueous solution of from about 0.2 to 45 g/ℓ of free H_2SO_4, from about 1.5 to 58 g/ℓ of H_2O_2, and from about 3 to 33 g/ℓ of SiO_2. The last component is conveniently provided in the form of a soluble silicate, e.g., sodium silicate or potassium silicate, of predetermined contents of SiO_2 and Na_2O or K_2O. Ammonium and lithium silicates are also useful in providing the SiO_2 component.

The mol ratios of SiO_2 to either Na_2O or K_2O generally range between 1 and 4, and it is preferred to use those silicates wherein the mol ratio is at least about 1.8 and most preferably at least about 2.2. The solution is easily prepared, e.g., by first adding sufficient sulfuric acid to the water under agitation to provide the desired free H_2SO_4 content and taking into account that some of the free acid will be subsequently neutralized by the Na_2O or K_2O portions introduced with the silicate. The silicate is added under agitation to the cooled acidic solution until it is completely dispersed. The peroxide addition is made last, preferably just prior to use. The sequence of addition can be changed, however, without any detrimental effect, provided that the silicate is acidified with sulfuric acid prior to mixing with the hydrogen peroxide, or peroxide decomposition will occur.

The preferred concentrations of the components in the aqueous solution are from about 1.8 to 18 g/ℓ of free H_2SO_4, from about 7 to 29 g/ℓ of H_2O_2 and from about 8 to 18 g/ℓ of SiO_2.

The resulting conversion coatings have very good resistance to corrosion as determined by the accepted accelerated corrosion test ASTM B-117-64. By the use of one or more of certain organic promoters, either as additives to the solution of sulfuric acid-hydrogen peroxide-silicate or employed in a subsequent treatment, the corrosion resistance of the coatings can be further enhanced. The group 1 organophosphorus compounds and the group 2 organic nitrogen compounds hereinafter have been found to be especially useful in this respect.

The group 1 promoters are organophosphorus compounds including C_{1-4} alkyl phosphonic acids, C_{1-4} hydroxyalkylenephosphonic acids, amino tri-C_{1-4} alkylenephosphonic acids, etc., as well as the acid or neutral sodium or potassium salts of the phosphonic acids. 1-Hydroxyethylidene-1,1-diphosphonic acid is a preferred compound.

The organophosphorus compound or mixture of such compounds is added either to the conversion coating solution or to a subsequent aqueous bath to provide a concentration therein of from about 0.15 to 10 g/ℓ, preferably from about 0.5 to 2 g/ℓ.

The group 2 promoters are organonitrogen additives selected from thioacetamide, urea, thiourea, N,N'-alkyl-substituted ureas or thioureas and cyclic N,N'-alkylene-substituted ureas and thioureas, wherein the alkyl and alkylene groups each contain from 1 to 4 carbon atoms.

Particular examples of suitable promoters belonging to group 2 include tetramethylurea, tetramethylthiourea, dimethylthiourea, di-n-butylthiourea, di-tert-butylthiourea, ethylenethiourea, etc. Thiourea is one preferred group 2 compound.

The organonitrogen compound or mixture of such compounds is either added to the conversion coating solution or to a separate aqueous solution to provide a concentration in either case of from about 0.5 to 50 g/ℓ, preferably from about 1 to 10 g/ℓ.

The solution is useful for forming conversion coatings on various metallic surfaces, such as those of zinc, cadmium, silver, copper, aluminum, magnesium, and zinc alloys.

The most common application is, however, in the formation of conversion coatings on zinc-plated articles such as zinc-plated steel articles.

Example 1: The aqueous conversion coating solution was prepared to contain 2.4 g/ℓ free H_2SO_4, 16.2 g/ℓ SiO_2 and 11.7 g/ℓ H_2O_2. The SiO_2 ingredient was added in the form of sodium silicate (SiO_2 = 33.2% w/w; Na_2O = 13.85% w/w) and a sufficient excess of sulfuric acid was provided to result in the indicated free H_2SO_4 content after neutralization of the Na_2O in the sodium silicate.

Standard Hull cell steel panels (10 x 6.8 x 0.03 cm) were plated with zinc using a cyanide electrolyte. After thorough rinsing and drying, the samples were then immersed for 20 seconds in the conversion coating solution maintained at room temperature. The treated samples were then rinsed in water and then dried with a hot air gun.

The dried, coated test specimens, which had a bright luster, were then subjected to the accelerated salt spray corrosion test in accordance with the ASTM test B-117-64. The tests were carried out for 6 hours and 24 hours and showed only traces of any corrosion after 6 hours and medium corrosion after 24 hours on the following rating scale: No = no corrosion; Tr/s = trace (scattered); Tr = trace; Mi/S = mild (scattered); Mi = mild; Me/S = medium (scattered); Me = medium; H/S = heavy (scattered); H = heavy.

Examples 2 through 10: The additional beneficial effects of organophosphorus compound additives are demonstrated in these examples. The general procedures of Example 1 were followed except that the conversion coating solutions contained the organophosphorus additives in the amounts specified in the table below, which also includes the results of the corrosion tests performed on the bright, coated test samples.

Ex. No.	Additive (g/ℓ)		Extent of Corrosion After 6 hr	24 hr
1	None	–	Tr/S	Me/S
2	Aminotri(methylene phosphonic acid)	0.75*	Tr/S	Me/S
3	Aminotri(methylene phosphonic acid)	7.50*	~No	Me
4	1-Hydroxyethylidene 1,1-diphosphonic acid	0.75**	~No	Mi/S
5	1-Hydroxyethylidene 1,1-diphosphonic acid	1.50**	~No	Tr
6	1-Hydroxyethylidene 1,1-diphosphonic acid	7.50**	~No	Tr/S
7	Ethylenediamine tetra(methylene phosphonic acid)	0.50***	~No	Me/S
8	Hexamethylenediamine tetra(methylene-phosphonic acid)	0.50†	~No	Mi
9	Diethylenetriamine penta(methylene phosphonic acid)	0.75*	~No	Mi
10	Diethylenetriamine penta(methylene phosphonic acid)	7.50*	~No	Me/S

*Active content about 50%.
**Active content about 60%.
***Active content about 90%.
†Active content about 97%.

In accordance with a further process of *R.F. Zuendt, A.M. Bengali, P.D. Readio and J.L.H. Allan; U.S. Patent 4,225,351; September 30, 1980; assigned to Dart Industries Inc.* there is provided a conversion coating solution which comprises an aqueous solution of from about 0.2 to 45 g/ℓ of free H_2SO_4, from about 1.5 to 58 g/ℓ of H_2O_2, from about 3 to 33 g/ℓ of SiO_2, from about 0.15 to 10 g/ℓ of at least one of the organophosphorus compound promoters and from about 2 to 20 g/ℓ of at least one secondary promoter selected from the group consisting of ascorbic acid, boric acid, gluconic acid, glycolic acid, tartaric acid and salts of these acids. The organophosphorous compounds are described above in U.S. Patent 4,222,779.

The general procedures used in the examples for preparing the conversion coating solutions, test specimens and forming the conversion coatings are described below.

The aqueous conversion coating solutions were each prepared to contain 2.4 g/ℓ free H_2SO_4, 16.2 g/ℓ SiO_2, 11.7 g/ℓ H_2O_2 and 0.85 g/ℓ of 1-hydroxyethylidene 1,1-diphosphonic acid. The SiO_2 ingredient was added in the form of sodium silicate (SiO_2 = 33.2% w/w; Na_2O = 13.85% w/w) and a sufficient excess of sulfuric acid was provided to result in the indicated free H_2SO_4 content after neutralization of the Na_2O in the sodium silicate.

Standard Hull cell steel panels (10 x 6.8 x 0.03 cm) were plated with zinc using a cyanide electrolyte. After thorough rinsing and drying, the samples were then immersed for 40 seconds in the conversion coating solution maintained at room temperature. The treated samples were then rinsed in water and then dried with a hot air gun.

The dried coated test specimens were then subjected to the accelerated salt spray corrosion tests in accordance with ASTM test B-117-64. The tests were carried out for various periods of time, i.e., 6, 16, 24 and 30 hours. After each test the specimens were examined for evidence of corrosion on a rating scale from 1 (heavy corrosion) through 10 (no corrosion).

Examples 1 through 3: The beneficial effects of boric acid and zinc gluconate as secondary additives are demonstrated in these examples. The general procedures described above were followed except that the solutions of Examples 2 and 3 also contained the additives indicated in Table 1, which includes the results of the corrosion tests performed on the bright, coated test samples.

Table 1

Ex. No.	. . . Additive (g/ℓ)		Extent of Corrosion After 6 Hours	24 Hours
1 (Control)	None	–	9	7
2	Boric Acid	5	10	8
3	Zn gluconate	5	9	8

Examples 4 through 12: In this series of experiments all the conversion coating solutions contained triarylmethane dyes in addition to the secondary additives shown in Table 2. These dyes used were a mixture of E.I. Du Pont de Nemours' liquid dyes Victoria Pure Blue BOP solution (0.2 ml/ℓ Basic Blue 7, C.I. 42,595) and Paper Blue R Liquid (0.1 ml/ℓ Basic Violet 3, C.I. 42,555).

The results of corrosion tests on the bright, colored, coated test specimens are shown in Table 2.

Table 2

Ex. No.	. . . . . . Additive (g/ℓ)		. . . Extent of Corrosion After . . . 16 Hours	24 Hours	30 Hours
4 (Control)	None	–	7	6	6
5	Boric acid	5	9	8	7
6	Boric acid	20	9	8	7
7	Ascorbic acid	5	9	8	7
8	Potassium sodium tartrate	5	10	8	–
9	Glycolic acid	5	9	9	–
10	Zn gluconate	5	9	8	7
11	Na gluconate	5	9	7	–
12	Na gluconate + Zn sulfate*	3.2	9	8	–

*The amount of Zn in 3.2 g $ZnSO_4 \cdot 7H_2O$ is equivalent to that in 0.5% Zn gluconate.

J.L.H. Allan, P.D. Readio, A.M. Bengali and R.F. Zuendt; U.S. Patent 4,225,350; September 30, 1980; assigned to Dart Industries Inc. have discovered that certain dyes which heretofore predominantly have been used in the dyeing of natural fibers such as paper, cotton, wool, silk, etc., when incorporated into a sulfuric acid-hydrogen peroxide-silicate conversion coating solution (such as that described above in U.S. Patent 4,222,779), unexpectedly impart pleasing and lasting colors to the coated workpieces without detrimentally affecting the corrosion resistance quality of the coating or the stability of the coating solution. The dyes are cationic triarylmethane dyes. The triarylmethane dyes used are well known and are recognized as a separate generic group of dyes having a Color Index (C.I.) in the range from 42,000 to 44,999. They are commercially available in a wide variety of colors both in solid form or as aqueous solution concentrates with solids contents typically in the 40 to 50% range. The amount of dye to be added to the conversion coating solution depends obviously on the desired depth of color.

The solution is easily prepared, e.g., by first adding sufficient sulfuric acid to at least a major portion of the makeup water under agitation to provide the desired free H_2SO_4 content and taking into account that some of the free acid will be subsequently neutralized by the Na_2O or K_2O portions introduced with the silicate. The silicate is added under agitation to the cooled acidic solution until it is completely dispersed. The peroxide is added and then the dye, preferably in the form of a dilute solution in a minor portion of the water used in the preparation of the conversion coating solution. The sequence of addition can be changed, however, without any detrimental effect, provided that the silicate is acidified with sulfuric acid prior to mixing with the hydrogen peroxide, or peroxide decomposition will occur.

The preferred concentrations of the components in the aqueous solution are from about 1.8 to 18 g/ℓ of free H_2SO_4, from about 7 to 29 g/ℓ of H_2O_2, from about 8 to 18 g/ℓ of SiO_2 and from about 0.05 to 0.3 g/ℓ of the triarylmethane dye or mixture of dyes.

The most common application is, however, in the formation of conversion coatings on zinc-plated articles such as zinc-plated steel articles. The zinc plate provides the steel with cathodic protection against corrosion, and the conversion coating further improves the corrosion resistance, reduces the susceptibility to finger markings and enhances the appearance by chemical polishing of the article and by the color imparted by the dye. It is important that the zinc plate deposit is relatively smooth and fine-grained prior to coating, and that the thickness of the plate deposit is at least 0.005 mm since some metal removal occurs when the film is formed. The preferred plate thickness is between about 0.005 and 0.02 mm.

Usually the formation of the conversion coating follows immediately after the last rinse in the plating cycle.

The resulting conversion coatings have very good resistance to corrosion as determined by the accepted accelerated corrosion test ASTM B-117-64. By the use of one or more of certain organic promoters as additives to the solution of sulfuric acid-hydrogen peroxide-silicate the corrosion resistance of the coatings can be further enhanced.

Examples of these organophosphorus compounds are given above in U.S. Patent 4,222,779.

The general procedures used in the examples for preparing the conversion coating solutions, test specimens and forming the conversion coatings are described below.

The aqueous conversion coating solutions were each prepared to contain 2.4

g/ℓ free H_2SO_4, 16.2 g/ℓ SiO_2, 11.7 g/ℓ H_2O_2. The SiO_2 ingredient was added in the form of sodium silicate (SiO_2 = 33.2% w/w; Na_2O = 13.85% w/w) and a sufficient excess of sulfuric acid was provided to result in the indicated free H_2SO_4 content after neutralization of the Na_2O in the sodium silicate.

Standard Hull cell steel panels (10 x 6.8 x 0.03 cm) were plated with zinc using a cyanide electrolyte. After thorough rinsing and drying, the samples were then immersed for 20 seconds (unless otherwise noted) in the conversion coating solution maintained at room temperature. The treated samples were then rinsed in water and then dried with a hot air gun.

Examples 1 through 8: A number of conversion coating solutions containing various blue dyes were prepared and tested for color and hydrogen peroxide stability after 24 and 90 hours storage. The table below identifies the dyes, shows the dye concentrations and the results of the stability testing. Of the seven dyes tested in this series only those of Examples 4 and 8 did not appear to promote peroxide consumption of the bath nor undergo an undesired color change. These dyes were therefore used for conversion coating trials to determine if they would impart a desired blue color to zinc plates treated with the respective solutions. Results of 20-second immersion in each of the two baths were that no permanent color was imparted to the surface of the test panels.

Ex. No.	. . Dye, g/ℓ (ml/ℓ) . .		Initial Bath Color	. . . After 24 Hours. . . . Bath Color	% H_2O_2 Retention	. . . .After 90 Hours. . . . Bath Color	% H_2O_2 Retention
1	None	—	Pale yellow	Pale yellow	95	Pale yellow	94
2	Chromate Blue #1*	0.1	Blue	Purple	88	Yellow	81
3	Chromate Turquoise #5*	0.1	Turquoise	Green	86	Pale yellow	66
4	Blue #7**	0.1	Greenish blue	Greenish blue	94	Greenish blue	94
5	Merpacyl Blue SW***	(0.5)	Blue	Pink	94	Colorless	89
6	Pontamine Blue AB***	(0.5)	Dark blue	Dark blue	92	Pale violet	85
7	Brilliant Bond Blue A***	(0.5)	Blue	Blue	95	Blue	83
8	Sevron Blue 5G***	(0.5)	Dark blue	Dark blue	93	Dark blue	91

*Conversion coating dyes from Sandoz Colors and Chemicals.
**Conversion coating dye from Pavco Inc.
***E.I. Du Pont de Nemours and Co.

Examples 9 through 11: In Example 9 the procedures of the previous examples were followed exactly except that the dye was Basic Violet 3, which is a cationic triarylmethane dye having a Color Index of 42,555. The particular dye used in this example was Paper Blue R solution (E.I. Du Pont de Nemours) provided in the form of an aqueous acetic acid solution of about 1.115 SG and a solids content of about 50 wt %. When 0.5 ml/ℓ of the dye solution was added to the bath there resulted a dark blue color which after 90 hours of storage did not change. The peroxide concentration was not significantly affected after conclusion of the testing (92% retention vs 94% without any dye).

Results of 20-second immersion coating tests in the dye bath of zinc-plated test panels showed a permanent reddish blue color to the surface. Repeating the immersion coating tests with varying concentrations of the dye from 0.1 to 0.5 ml/ℓ showed that any desired depth of color could be imparted to the surface merely by changing the dye concentration.

Examples 10 and 11, in which the triarylmethane dyes were respectively a Basic Blue 7 (C.I. 42595) and Basic Green 4 (C.I. 42000), showed the same successful coloration in the concentration range used in Example 9.

Example 12: In this example the conversion coating solution contained 0.85 g/ℓ (dry basis) of 1-hydroxyethylidene 1,1-diphosphonic acid as a further promoter for corrosion resistance. The dyes used were a mixture of Basic Blue 7 and Basic Violet 3 (0.2 ml/ℓ Dupont Victoria Pure Blue BOP solution, and 0.1 ml/ℓ DuPont Paper Blue R Liquid). Hull cell panels plated in a small scale as well as commercially plated clamps and elbow brackets served as zinc-plated specimens for conversion coating, which was carried out for 20 seconds.

Visual examination of the coated specimens showed a desirable blue color of excellent uniformity and shade, closely matching those obtained with conventional blue chromate treatment.

Titanium, Aluminum, Vanadium Based Films

A need has arisen for a chromium-free or low-chromium passivation system that has the attributes of a traditional chromium conversion system which is not found in phosphating systems or in bright dipping, transparent lacquer systems. Such a chromium-free or low-chromium system should, as does a blue bright chromate conversion coating, have considerable wet strength when suitably dried and aged such that it becomes impervious and passive to moisture, handling stains, and mildly corrosive media during storage and use to the extent that the treated surface will pass a standard salt spray test for up to about 24 to 32 hr or more.

The attributes of chromate conversion coatings can be attained in aqueous bath formulations as described by *J.L. Greene; U.S. Patent 4,298,404; November 3, 1981; assigned to Richardson Chemical Co.* which combine certain chromium-free film-forming agents in combination with activating agents that are carboxylic acids or derivatives thereof.

With proper maintenance and pH control, baths according to this process have a useful life up to about twice as long as that of a conventional chromated blue-bright solution, thereby reducing the frequency of required bath dumpings. About 130 to 135 ft^2 of surface area can be passivated on a commercial scale for each gallon of bath prepared.

It has also been found that baths in accordance with this process will tolerate up to 1.5 g/ℓ of zinc metal removed from the plated surface of the item being coated during the passivation reaction, while a typical chromate conversion coating bath will begin to yellow and cause a yellowing of the finish being formed when this level of zinc concentration is approached.

Depending upon the particular bath, the metals being passivated, overall cost considerations, the additives, concentrations, and the desired appearance of the film to be produced, the film-forming agents that are typically preferred include one or more of titanium, aluminum, vanadium or silicon, preferably as bath-soluble alkali metal fluoride and/or oxalate salt complexes. Especially preferred are titanium salts. Concentrations of each film-forming agent when within a passivation bath can range between concentrations that are barely effective in forming a film, usually on the order of about 0.2 g/ℓ of bath and up to the solubility limit of the particular agent with the passivation bath, which is usually no higher than about 25 g/ℓ.

Typical exemplary solubility limits are about 25 g/ℓ for sodium silicofluoride, 12 g/ℓ for potassium titanium fluoride, 5 g/ℓ for sodium metavanadate, 20 g/ℓ for sodium orthovanadate, 2 g/ℓ for sodium aluminum fluoride, 3 g/ℓ for ceric sulfate, 4 g/ℓ for ferric nitrate, 5 g/ℓ for titanium sulfate, 5 g/ℓ for titanium fluoroborate and 22.5 g/ℓ for potassium titanium oxalate.

Concentrations above these would usually not be economically advantageous, and often substantially lower concentrations are preferred. Preferably, the concentration range for these agents in general is between about 0.4 and 6 g/ℓ of bath for each added film-forming agent.

Example 1: A powdered film-forming composition was prepared by blending together 3.7 g of potassium titanium fluoride (K_2TiF_6), 0.8 g of boric acid (H_3BO_3), 1.2 g of sodium sulfate (Na_2SO_4), 1.0 g of sodium nitrate ($NaNO_3$), and 1.0 g of sodium glucoheptonate ($C_7H_{13}O_8Na$). This composition was dissolved in 1 ℓ of water to which had been added 0.25 vol % of nitric acid, the resulting solution having a pH of 1.85.

A freshly zinc-plated steel panel was rinsed in water to remove adhering zinc plating solution and was then immersed in the passivation bath at room temperature for 25 seconds, after which it was removed from the solution, rinsed in cold water, and blown dry in a stream of warm air. The surface of the panel was covered by a uniform adherent blue bright film that showed definite hydrophobic characteristics and that was adequate to protect the underlying zinc surface for 24 hr in a 5% neutral salt spray in accordance with ASTM B-117.

Example 2: A powdered blended mixture of 1.5 g of sodium aluminum fluoride (Na_3AlF_6), 0.8 g of boric acid, 1.2 g of sodium nitrate, 1.0 g of sodium sulfate, and 1.5 g of sodium glucoheptonate was dissolved in 1 ℓ of water containing 0.25 vol % of nitric acid in order to form a bath having a pH of 1.95. Upon immersion for 25 seconds, a panel of zinc-plated steel developed a pale reddish-green film that protected the underlying zinc during a 16-hour ASTM B-117 salt spray exposure.

Example 3: The zinc surface of a panel formed a good, uniform blue bright film and was protected against an ASTM B-117 salt spray test for 16 hours upon being immersed for 25 seconds within 1 ℓ of water to which had been added 0.40 vol % of nitric acid, together with a powdered blended mixture of 3.5 g of sodium orthovanadate ($Na_3VO_4 \cdot 16H_2O$), 3.7 g of sodium silicofluoride (Na_2SiF_6), 1.0 g of sodium glucoheptonate, 1.2 g of sodium nitrate, and 0.8 g of boric acid, the solution having a pH of 1.7.

PROTECTING GALVANIZED SURFACES AGAINST WATER CORROSION

Zinc Phosphate and Zinc Silicate Coatings

Zinc, being more electronegative than steel, protects the latter against corrosion by preferentially dissolving itself. This dissolution is delayed and slowed down by a layer of insoluble salts formed by displacement of mineral salts in solution in water. However, the formation of these protective salts is slow and does not always ensure the complete protection of zinc, especially if the water is hot, if it has a low mineral salts content (degree of water hardness near zero) or if it contains metal such as copper which acts as a catalyst for the attack on zinc. It has been stated that corrosion is considerably accelerated by water having a temperature between 60° and 80°C and that traces of copper even at a concentration as low as 1 ppm have a pronounced catalytic corrosive effect. Moreover, the more and more frequent use of water softeners, which lowers the degree of hardness by ion exchange in the water distribution systems, increases the danger of premature corrosion of galvanized parts.

M. Longuepee and N. Dreulle; U.S. Patents 4,126,469; November 21, 1978; and 4,110,127; August 29, 1978; both assigned to International Lead Zinc Research Organization, Inc. have discovered that it is possible to bring about a continuous deposit, insoluble in hot or cold water which efficiently protects zinc against corrosion, even in hot water having a low degree of hardness and a significant copper content, by bringing the galvanized parts to be protected into contact with a solution containing silicates of metaphosphates in a phosphoric medium having a low pH value, in the presence of zinc ions; the deposit can be rapidly achieved and gives lasting protection.

There is provided a solution intended for rapidly depositing a coating which protects the surface of zinc-coated ferrous metal parts against corrosion in the presence of water, and is characterized by the fact that it is built-up by adding to water, per liter of final solution, from 1 to 40 g of sodium metasilicate, 14 to 40 ml of phosphoric acid of a density of 1.71 g/ml, from 1 to 40 g of sodium nitrate and 10 to 50 g of anhydrous zinc chloride; the pH value of the solution being adjusted to a value between 2.3 and 3.8 by the addition of calcium carbonate. Preferably nickel chloride can be added as a deposit accelerator.

These solutions are used in a process of depositing a protective coating on zinc-coated ferrous metal parts characterized by the fact that a solution of the type that precipitates at room temperature is made up, that this solution is filtered after adjustment of the pH value, heated to a working temperature ranging from 15° to 75°C, that this solution is brought into contact with the parts for a period ranging from 20 to 72 hours, at the working temperature, and that the part is then rinsed with water.

According to an advantageous form of the process of depositing a protective coating on the inner surface of zinc-coated steel piping, the solution is brought into contact with the inner surface by circulation, preferably intermittent, of the solution, and a filter is placed in the circulating system.

Example 1: A solution is made up according to the following formula for 1 liter of solution: 35 g of sodium hexametaphosphate, 5 g of sodium metasilicate, 15 ml of phosphoric acid (specific gravity = 1.71), 14 g of sodium nitrate,

5 g of crystallized nickel chloride ($NiCl_2 \cdot 6H_2O$), 20 g of anhydrous zinc chloride and calcium carbonate to bring a pH value of about 2.8.

Example 2: A solution is made up, according to the following formula for 1 liter of solution: 20 g of sodium metasilicate, 17 ml of phosphoric acid (specific gravity = 1.71), 20 g of sodium nitrate, 5 g of crystallized nickel chloride ($NiCl_2 \cdot 6H_2O$), 20 g of anhydrous zinc chloride and calcium carbonate to bring the pH value of the solution to 3.6.

Example 3: A solution is used according to Example 1. The working temperature of the solution is 65° to 75°C. The time necessary for the formation of the deposited coating is 20 hours or preferably 24 hours. The mode of formation is by continuous, or preferably intermittent circulation of the solution, after passing through a glass wool filter. The nature of the coating is zinc phosphate.

Example 4: A solution is used according to Example 2. The working temperature of the solution is between 15° and 25°C. The time of formation of the deposited coating is 72 hours. The mode of formation is by continuous, or preferably intermittent circulation of the solution. The nature of the coating is a mixture of phosphates and zinc silicates.

Results of Corrosion Testing: The tests have been carried out on test pieces which have been treated according to Example 3 as well as on untreated reference test pieces, in the following conditions: water temperature, 80°C; French degree of water hardness, 0; and copper content, 1 ppm. The duration of the tests are shown in the following table:

2 Months	
Test piece	Intact
Reference test piece	Strongly rusted as scattered corrosion pits
9 Months	
Test piece	Practically intact
Reference test piece	All over covered with corrosion pits

The deposits obtained are continuous, hard, and resistant to abrasion and shocks without flaking. The protection of zinc obtained is efficient in the most unfavorable corrosion conditions, i.e., in hot water having a low degree of hardness and a significant copper content. The formation of the protective deposit is rapid. When a water distribution system of a new building is put into service, the treatment of the pipings does not cause an appreciable delay to the completion of the assembly. The coating obtained remains hard and adherent in hot and cold water and does not contaminate these waters, which remain suitable for consumption by human beings.

Hydrated Zinc Pyrophosphate Coatings

Corrosion is often observed in hot and cold water distribution systems or appliances which are made of galvanized steel. This corrosion, which is particularly frequent in the first months of service, is undesirable whenever it occurs; it is particularly undesirable when the galvanized steel is in the plumbing of buildings.

The presence and degree of corrosion is closely connected with certain factors, including water temperature, degree of water hardness, and the presence of traces of copper in the water. There is a considerable acceleration of the corrosion in hot water between 60° and 80°C. Corrosion is also more severe with a low degree of water hardness, and with water which contains copper, even at very low concentrations.

P. Dreulle, M. Longuepee and D. Dhaussy; U.S. Patent 4,110,128; August 29, 1978; assigned to International Lead Zinc Research Organization, Inc. have discovered that a coating of hydrated zinc pyrophosphate, $Zn_2P_2O_7 \cdot 3H_2O$, has desirable properties. This coating may be obtained from an aqueous solution containing hexametaphosphate and metasilicate, using the sodium salts of each of these.

Example 1: The solution is made up containing the following ingredients with quantities of solute given per liter of final solution: 10 to 70 g of sodium hexametaphosphate, 1 to 40 g of sodium metasilicate, 15 to 40 ml of orthophosphoric acid (density = 1.71 g/ml), 10 to 50 g of anhydrous zinc chloride and calcium carbonate to bring the pH value to between 2.0 and 3.0.

Nickel ion in the coating solution serves as an accelerator for the depositing of the coating on the galvanized part. The amount of nickel ion added to the solution may be varied in accordance with the desired rate of deposition. For example, from 0.5 to 20 g of hexahydrated nickel chloride per liter of final solution may be added to the abovedescribed solution.

Example 2: A solution was made up containing the following ingredients with quantities of solute given per liter of final solution: 35 g of sodium hexametaphosphate, 5 g of sodium metasilicate, 15 ml of orthophosphoric acid (density = 1.71 g/ml), 20 g of anhydrous zinc chloride, 5 g of crystallized hexahydrated nickel chloride and calcium carbonate to bring the pH to about 2.8.

This solution, used at an average temperature of 65°C, leads to the formation of a protective coating on galvanized steel by circulation or immersion, the solution being filtered and stirred continuously.

The rate of formation of the protective coating, and the temperature at which the coating may be obtained, are affected by the amount of chlorate added to the initial solution. Up to 20 g of sodium chlorate per liter of final solution may be added to the solution described in Example 1. A deposit may be obtained in 3 days at 40°C with a small amount of sodium chlorate, or in 8 days at room temperature, at a more acid pH value, with a greater added amount of sodium chlorate. Thus, the treatment of galvanized parts where the circulation of a hot solution is impossible, such as through a cold water distribution system which is not heat-insulated, is feasible.

If the solution contains chlorate, the preferred pH range is from 2.5 to 3.0. A solution containing no chlorate is preferably used at a pH value of from 2.7 to 3.0. Solutions containing either nickel or chlorate, or both, may be used at a temperature of from 10° to 70°C. Solutions containing neither nickel nor chlorate should be used at from 40° to 70°C.

Since the formation of the pyrophosphate is necessary for the proper coating on the galvanized part, it is important to avoid hydrolysis of pyrophosphate to orthophosphate. Therefore, the presence of nitrate ions in the coating solution should be avoided.

The regeneration of used coating solution may be accomplished in the following series of steps. First, the amounts of phosphorus and zinc which must be replaced are determined by analysis of the used coating solution. The regenerating solution should contain an amount of metaphosphoric acid, HPO_3, corresponding to the amount of phosphorus which must be replaced in the coating solution. Metaphosphoric acid can be formed in the regenerating solution by reacting a corresponding amount of a metaphosphate salt with a strong acid. For example, the desired amount of sodium hexametaphosphate can be dissolved in water, and reacted with a sufficient amount of sulfuric acid to convert the hexametaphosphate to metaphosphoric acid, according to the following equation:

$$(NaPO_3)_6 + 3H_2SO_4 \rightarrow 6HPO_3 + 3Na_2SO_4$$

Zinc chloride and zinc oxide are next added to the regenerating solution, in a sufficient combined amount to supply the necessary amount of zinc to the used coating solution. Sufficient zinc oxide must be added with the zinc chloride so that the pH value of the regenerating solution is maintained at between 3.0 and 3.3. Excessive addition of zinc oxide may cause a neutralization of the regenerating solution. If the solution is allowed to have very high concentrations of both zinc and phosphorus, zinc phosphate may precipitate.

The regenerating solution thus formed is added to the spent coating solution, and the pH value of the resulting solution is then adjusted to between 2.0 and 3.0.

Tests have shown that regeneration of the same initial solution about 30 times led to the formation of deposits having the same general properties as those obtained after the first use of the initial solution. Corrosion tests made on galvanized pipes treated with different solutions regenerated according to this process were carried out at a water temperature of 80°C, a total water hardness of zero, a copper content of 1 ppm, and at a water renewal rate of one-third of the total volume per day. The pipes treated in accordance with this process showed an excellent corrosion resistance compared with untreated pipes; the properties of pipes treated with regenerated solutions were comparable to those of pipes treated with a fresh solution. After 14 months of testing under the above conditions, the treated pipes had only some rust pits whereas the reference pipes presented a general corrosion.

Inhibiting Wet Storage Staining

Surfaces of zinc and metallic materials coated with zinc, such as galvanized steel, are subject to the so-called "wet storage staining." This means that during storage and transportation in humid environments, the sheets or articles made of zinc or coated with zinc become oxidized and form powdery surface stains which are commonly known as "white rust." The presence of white rust greatly impairs the appearance of the articles and also the adhesion of paints or other coatings which one may wish to apply to the metal.

R.L. LeRoy; U.S. Patent 4,093,780; June 6, 1978; assigned to Noranda Mines

Limited, Canada found that certain organic compounds, e.g., esters and polyesters of thioglycolic acid, are eminently suitable for inhibiting wet storage staining of zinc and zinc coated materials because they form a water-insoluble protective coating with zinc on the metal surface.

Basically, all esters and polyesters of thioglycolic acid which form a water-insoluble complex with zinc atoms are suitable as inhibiting agents. By water-insoluble, it is meant that the complex should have a solubility in water below 0.1 g/ℓ at 20°C. Preferably, the water solubility of the complex should be below 1 mg/ℓ at 20°C.

The active organic compounds which are generally inexpensive and easy to produce can be applied onto the zinc surface in the form of a solution in any suitable solvent, such as water, alcohol, ketones, petroleum solvents, etc. by dipping, spraying, brushing, rubbing, etc. They can also be applied in the form of dispersions or even as compositions which contain up to 99% of the active compound, with a very small amount of water or other solvent being added thereto to promote formation of the complex.

Most commonly, solutions or dispersions containing between 0.15 and 3% by weight of the active compound are preferred because they provide satisfactory protection and at the same time are inexpensive due to the small concentration of the active compound therein. Particularly preferred are aqueous solutions or dispersions, again because of their low cost.

An example of a suitable active compound which has been found to be an effective wet stain inhibitor for zinc surfaces is $HSCH_2C(O)OC_nH_{2n+1}$ where n is between 3 and 18 inclusive.

Example 1: *Synthesis of 1,2,6-hexanetriol trithioglycolate* – A reaction mixture of 72 g of 1,2,6-hexanetriol, 150 g of thioglycolic acid and 40 g of xylene was refluxed in a nitrogen atmosphere for 2 hours at 125° to 155°C and 2 hours at 155° to 160°C. 29.0 g of water of condensation were collected, compared to an expected 29.5 g. The mixture was freed of solvent, water, and unreacted materials by stripping to a pot temperature of 155°C at 2 torrs pressure. The residual product was a clear liquid and weighed 217 g, compared to an expected 222 g.

Example 2: *Synthesis of glycerol dithioglycolate* – A reaction mixture of 20 g of glycerol, 40 g of thioglycolic acid, 0.3 g of p-toluenesulfonic acid, and 30 g of toluene was refluxed in a nitrogen atmosphere for 1 hour at 108° to 117°C and 1 hour at 117° to 120°C. 7.6 g of water of condensation were collected, compared to an expected 7.8 g. The reaction mixture was stripped to a pot temperature of 120°C at 5 torrs. The residual product was a clear and colorless liquid.

Example 3: Galvanized panels (4" x 8") were treated by dipping in a 1.5% methyl hydrate solution of 1,2,6-hexanetriol trithioglycolate. After exposure in a humidity cabinet (100°F, 100% RH) these panels suffered less damage than similar panels treated with the chromate based formulation known as Iridite.

Example 4: Galvanized panels (4" x 8") were treated by dipping in an 0.15%

aqueous dispersion of 1,2,6-hexanetriol trithioglycolate at 55°C. The panels were sprayed with distilled water and clamped together in a stack, which was exposed out-of-doors. Iridite-treated and untreated panels were included in the stack. After 10 day's exposure, untreated panels were heavily stained with white rust and Iridite-treated panels had 20 to 30% white rust damage on their surfaces and were dulled. Panels treated with 1,2,6-hexanetriol trithioglycolate solution had no evidence of white rust damage.

Example 5: Galvanized steel coupons (2" x 2") were treated by dipping in a 1.5%, 50°C, methyl hydrate solution of glycerol trithioglycolate, followed by rinsing with cold water. Treated coupons resisted visible corrosion damage for 36 hours on exposure by partial immersion in water.

CHROMIUM-FREE POSTTREATMENT OF CONVERSION COATINGS

Solution Containing Titanium plus Phosphoric Acid, Phytic Acid or Tannin and H_2O_2

Chromates are conventionally employed not only in the chromating process but as posttreatments in the phosphating and compound oxide coating processes. The anticorrosion effect of the chromating treatment on metals, especially on steel, zinc or aluminum is outstanding and consequently chromates have been widely employed in the field of metal surface treatment. Nevertheless, the environmental and health hazards of the toxic chromium compounds have come into question, and it has become an objective to develop safer surface treating agents which can be employed in place of chromium.

Y. Matsushima, K. Yashiro, H. Kaneko and M. Suzuki; U.S. Patent 4,110,129; August 29, 1978; assigned to Oxy Metal Industries Corporation found that the chromium-free posttreatment of conversion coated zinc or zinc alloy surfaces may be accomplished by contacting the surface with an aqueous solution containing titanium ion and at least one adjuvant compound selected from the group consisting of phosphoric acid, phytic acid, tannin, the salts and esters of the foregoing, and hydrogen peroxide. The aqueous solution contains at least 0.02 g/ℓ of titanium ion, from at least 0.1 g/ℓ of the adjuvant and is adjusted to a pH value from about 2 to 6.

Examples 1 through 10: *Sealing treatment for colored galvanized steel plates* – Specimens were of hot-galvanized steel plate [zero-spangled chromated plate (Shin Nittetsu Co.)] having a size of 100 x 200 x 0.27 mm and having been processed as a base for a colored galvanized steel plate. The specimens were polished by passing them through a wet buffing machine in 3 passes to scrape the chromate film, followed by rinsing with hot water for 3 seconds. The specimens were then contacted with aqueous cleaning solution at 40°C by spraying for 3 seconds. Immediately thereafter, the specimens were treated with a conventional phosphating bath for colored galvanized steel at 65°C for 10 seconds by the spraying process, followed by rinsing with hot water at 40°C for 6 seconds. The phosphated test pieces were sprayed with one of the water-soluble compositions according to the process as specified in Examples 1 through 10 and those of Comparative Examples 1 through 8 in Table 1 at 40°C for 3 seconds. The treated test pieces were then passed through electrically driven rubber rolls to remove the excess water-soluble composition, dried in hot air at

120°C and then allowed to stand until they were cooled to a temperature of lower than 40°C.

The thus-treated specimens were coated with paint (an alkyd resin type, Nippon Paint Co.) by means of a No. 12 coater and the coated paint was baked by heating the specimens to a temperature of 210°C to obtain a coating thickness of 7 μ. The edges of the specimens were seal-painted with a ship bottom coating (Silver VSI, Takada Toryo Co.). The specimens were crosshatched on the lower half thereof by means of a NT cutter and subjected to the salt spray test according to JIS-Z-2371 for 240 hours. The test pieces were then rinsed with water and dried.

Scotch tapes having a width of 50 mm were applied to the crosshatched portion and the flat portion (the upper half of test pieces without the crosshatch) and tightly adhered on the surfaces by means of a rubber roller. One minute thereafter, the tapes were stripped off rapidly and the peeled width in mm at both sides of the crosshatch and the number of blisters on the flat portion were observed. Table 4 shows the results evaluated in accordance with the evaluation criteria specified in Tables 2 and 3.

Table 1

Ex. No.	$(NH_4)_2TiF_6$	$K_2TiO(C_2O_4)\cdot 2H_2O$	H_2TiF_6	H_2O_2 (100%)	$NH_4H_2PO_4$	Phytic Acid (100%)	Tannic Acid	Chromate Solution	Remarks
1	3	–	–	1.5	–	–	–	–	–
2	3	–	–	1.5	1	–	–	–	–
3	3	–	–	1.5	3	–	–	–	–
4	3	–	–	–	–	0.5	–	–	*
5	1	–	–	–	–	1	–	–	*
6	–	6	–	1.5	–	0.5	–	–	*
7	–	2	–	0.5	–	1	–	–	*
8	–	–	3	–	–	0.5	–	–	–
9	3	–	–	–	–	–	0.5	–	**
10	3	–	–	1.5	–	–	1	–	**
1***	–	–	–	–	–	–	–	20	–
2***	–	–	–	–	–	–	–	–	†
3***	–	–	–	1.5	–	–	–	–	–
4***	3	–	–	–	–	–	–	–	–
5***	–	6	–	–	–	–	–	–	††
6***	–	–	–	–	3	–	–	–	–
7***	–	–	–	–	–	1	–	–	*
8***	–	–	–	–	–	–	1	–	–

*pH adjusted by NH_4OH.
**Available from Fuji Kagaku Co.
***Comparable.
†Treated with pure water.
††pH adjusted by $H_2C_2O_4$.

Table 2: Criterion for the Evaluation of Crosshatched Portion

Stripped Width from Both Sides of Crosshatch, mm	Rating
0	5
0.5–1	4
2–3	3
4–5	2
Wider than 6	1

Table 3: Criterion for the Evaluation of Flat Portion

Appearance	Rating
No peeling	5
Peeling of very few spots (less than 1% of the total area)	4
Peeling of spots (2 to 10% of the total area)	3
Partial peeling (11 to 50% of the total area)	2
Severe peeling (more than 51% of the total area)	1

Table 4: Result of the Salt Spray Test for 240 Hours for Painted Specimens

Ex. No.	pH	Cross-Hatched Portion			. . . Flat Portion. . . .		
1	3.0	4	4	4	4	4	5
2	3.2	5	5	5	5	5	5
3	3.3	4	4	4	5	5	5
4	3.3	3	3	3	5	4	4
5	3.2	5	5	4	5	5	5
6	4.8	4	4	4	4	4	5
7	4.0	4	5	5	4	5	5
8	2.4	5	5	5	5	5	5
9	4.0	4	4	4	5	5	4
10	4.1	4	5	5	5	5	5
1*	3.9	5	5	5	5	5	5
2*	6.5	2	2	2	2	2	2
3*	6.4	3	3	3	2	2	2
4*	3.6	2	2	2	1	2	2
5*	4.5	3	3	3	3	3	3
6*	4.1	2	2	2	2	3	3
7*	3.4	1	1	2	2	1	2
8*	5.0	3	3	3	3	3	3

*Comparative.

Aqueous Solution of Thiourea and Vegetable Tannin

Y. Miyazaki, M. Suzuki and H. Kaneko; U.S. Patent 4,180,406; December 25, 1979; assigned to Oxy Metal Industries Corporation found that when an aqueous solution of thiourea and a vegetable tannin is applied to conversion coatings on the surface of zinc or an alloy thereof, the corrosion resistance and adhesion of the subsequently painted film is comparable to that obtained via the post-treatment process with chromic acid.

Thiourea compounds useful in the process include thiourea itself and derivatives thereof such as alkylthiourea, e.g., dimethylthiourea, diethylthiourea, guanylthiourea, etc. in a concentration from 0.1 to 20 g/ℓ, preferably from 0.5 to 5 g/ℓ.

The tannin may be used in a concentration from 0.1 to 20 g/ℓ, preferably from 0.5 to 3 g/ℓ.

The weight ratio of thiourea to tannin may range from 10:1 to 1:10, preferably from 3:1 to 1:3.

The pH range of the treating solution depends on the type of tannin, method and conditions of the application and the like but normally ranges from 2 to 10, preferably from 2.5 to 6.5.

Example 1: Test panels of hot galvanized steel plate 100 x 300 x 0.3 mm and pretreated with chromic acid were polished 5 times by wet buffing to remove chromate adhered on the surface and then immersed in an aqueous solution of a surface conditioner containing titanium phosphate and passed between rubber rolls to remove excess solution. The test panels were sprayed with an aqueous zinc phosphatizing solution, washed with water, passed between rubber rolls to remove excess solution and then posttreated by immersion in an aqueous solution containing thiourea at a concentration of 2 g/ℓ and gallotannin (Tannic Acid AL, Fuji Kagaku Kogyo Co.) at a concentration of 1 g/ℓ at 50°C for 2 seconds, followed by passing between rubber rolls to remove excess solution and drying with hot air.

Thus treated, panels were then coated with an alkyd resin based paint via draw-down bar and then baked in a hot air recycling oven at 280°C for 50 seconds to obtain a coating thickness of about 6 μ.

Additional test panels were coated with a primer of epoxy resin and baked in a hot air recycling oven at 280°C for 50 seconds to form a coating of about 4 μ thickness. The primed test panels were then coated with a top coating paint of acrylic resin type and then baked in a hot air recycling oven at 280°C for 60 seconds to form a double coating having a total thickness of 14 μ.

The adhesion of the painted films was tested by the bending test in which two test pieces were bent to 180° and then folded completely by means of a vise and then applying and removing rapidly a cellophane tape and the results were then rated from 5 to 1.

Criterion for rating results of the bending test were as follows: 5, no stripping; 4, not more than 5% stripping; 3, not more than 25% stripping; 2, not more than 50% stripping; and 1, more than 50% stripping.

The salt spray corrosion test was performed by scribing the test pieces to the depth of the base metal by means of a knife and then subjecting the panels to the salt spray test according to JIS-Z-2371 for 240 hours for the test panels coated with the single layer paint and for 1,000 hours for the test pieces coated with two layers. After washing with water and drying, cellophane tape was applied and removed rapidly from the scribed portion of each test panel to measure the maximum width stripped off from the surface in mm and to observe the stripping-off of the painted film due to blisters in the film.

The aqueous corrosion test was performed by immersing the test panels into boiling water for 2 hours. Blisters were observed and the test panels rated by the bending test. The blisters were rated as follows: 10, none; 5, blisters on a portion of the panel surface; and 0, blisters on the entire panel surface.

Examples 2 through 5: For comparison purposes, panels were treated in an identical manner as above except for the posttreatment step.

Example 2 employed an aqueous solution of thiourea in a concentration of

2 g/ℓ, Example 3, an aqueous solution of gallotannin (Tannin Acid AL, Fuji Kagaku Kogyo Co.) in a concentration of 2 g/ℓ, Example 4, a conventional aqueous solution of chromic acid at a concentration of 18 g/ℓ, and Example 5 omitted the posttreatment entirely. Results are presented in the following table.

Ex. No.	Single Layer Paint . . . 240 hr Salt Spray. . . . Blisters	Corrosion Width (mm)	Bending	Two Layer Paint Aqueous Corrosion Blisters	Bending	1,000 hr Salt Spray (mm)
1	10	0-0.5	4	10	4	0-1
2	5	3	2	5	2	4
3	5	3	2	5	2	3
4	10	0-0.5	3-4	10	4	0-1
5	0	5	2	0	2	5

The results obtained by the process are comparable with those obtained by the conventional chromate process, except that the adhesion result obtained by the process is somewhat superior to that obtained by the conventional process.

SELECTIVE OR ONE SIDE PLATING

Use of Hydrated Compounds to Prevent Zinc from Adhering

Galvanized steel has been proposed to be used in the automotive industry to provide improved corrosion resistance for specific components. Unfortunately, the gloss of painted zinc surfaces and mild steel are difficult to match.

Therefore, it has been proposed to provide the zinc coating on the intended inner surface only of the steel strip, thereby allowing painting of the exterior steel metal surface in conventional manner.

In a method developed by *R.F. Hunter, S.R. Koprich, R.G. Baird-Kerr and D.S. Sakai; U.S. Patent 4,101,345; July 18, 1978; assigned to The Steel Company of Canada, Limited, Canada* mild steel strip is galvanized in selected areas only, typically on one side only of the strip, by pretreating the areas selected for absence of galvanizing to form thereon a readily chemically strippable coating of chemically hydrated compounds nonwetting in molten zinc. Preferably, the pretreatment is carried out using phosphoric acid solution to form thereon a thin coating of a composite of hydrated iron phosphates and iron hydroxides. After passing the thus-treated steel strip through the molten zinc bath and solidification of the zinc coating, the composite coating is removed by dilute mineral acid. Mild steel strip galvanized on one side using this procedure has particular utility in the automotive industry.

Owing to their ready availability and effectiveness, phosphoric acid-containing treating solutions are preferred, generally containing about 5 to 50 g/ℓ of total phosphate.

The formation of the preferred coating layer usually is achieved by contacting the surface of the steel on which the coating is to be formed with a phosphoric acid and sodium phosphate solution having an acid pH at a temperature above about 140°F.

The treatment solution used preferably is a diluted form of a commercially available phosphate solution concentrate such as Bonderite, Fosbond, Granodine or Enthox.

The coating usually has a weight of less than about 50 mg/ft^2 of surface, preferably in the range of about 20 to 40 mg/ft^2. Coating weights of this size have been found to be satisfactory in preventing wetting of the coated surface by molten zinc in the galvanizing step and hence, in preventing adherence of zinc to the coated surface. Coating weights of this size may be achieved by contact times of only 1 to 2 seconds at about 150°F between the treatment solution and the mild steel surface, using spray or dip application.

The coating layer is preferably a composite of a hydrated iron phosphate and iron hydroxide, such as a composite of $Fe_3(PO_4)_2 \cdot 8H_2O$ and $Fe(OH)_3$, particularly in the proportions of about 60% of the iron phosphate and 40% of the ferric hydroxide.

The coating layer may be removed from the galvanized strip after conventional adhesion of zinc to the uncoated surface and cooling to provide a clean steel surface indistinguishable from the initial mild steel surface and to which paint may be applied by conventional automobile painting techniques.

Removal of the preferred composite layer is achieved in rapid manner by mild acid treatment thereof. One preferred removal operation involves contacting the coating layer with dilute hydrochloric acid, preferably about 3 to 4% HCl, at an elevated temperature, preferably about 150° to 160°F. One beneficial side effect of the acid removal treatment is an apparent smoothing of the mild steel surface, thereby providing an improved surface for painting.

Example 1: Steel of commercial bottle top grade was batch annealed, temper rolled to a No. 5 finish without oiling and then slit into coils having a strip width of 12". This material was fed continuously to a pilot plant processing line at a rate of 60 fpm. The processing received by the steel strip in the line consisted of the following sequential steps:

(1) alkaline cleaning using Oakite 20 at 180°F and at a concentration of 36 g/ℓ for 12 seconds;

(2) cold water rinsing for about 4 seconds;

(3) pickling in hydrochloric acid at 70°F at a concentration of 12% by weight for about 6 seconds;

(4) cold water spray rinsing for about 4 seconds;

(5) hot water dipping at 200°F for 7 seconds;

(6) dipping in Bonderite 901 solution at 160°F and at a concentration of 11.8% by volume for about 1.5 seconds, with appropriate masking to treat one side only of the strip;

(7) air wiping of excess solution from the surface and drying with hot air at a temperature of about 250°F. The time between dipping in the solution and complete drying of the coating was about 6 seconds;

(8) zinc ammonium chloride fluxing of the nontreated side

by use of a low pressure spray at a concentration of 75 g/ℓ and a flux temperature of 185°F;

(9) spray rinsing of the nonfluxed side at a water flow rate of about 2 imp gal/min and a water temperature of 118°F;

(10) air wiping of both sides and drying with hot air at about 250°F for about 4 seconds;

(11) heating the steel to about 450°F in an electrically heated oven for about 18 seconds;

(12) dipping in molten zinc containing 0.12% aluminum at a temperature of 860°F for about 4.8 seconds;

(13) wiping both sides of the strip upon emerging from the zinc bath with steam superheated to a temperature of 500°F;

(14) cooling and solidifying the zinc coating using cold air impingement for 7 seconds;

(15) stripping the non-zinc-coated side by contacting the surface with 2% by weight hydrochloric acid at a temperature of 140°F for 2 seconds, with minimization of attack of the zinc-coated side by the hydrochloric acid;

(16) cold water spray rinsing of both sides before the acid wet strip dried for a total spray time of 5 seconds;

(17) air wiping and hot air drying of the strip at a temperature of about 250°F for 4 seconds; and

(18) recoiling of the strip.

The abovedescribed sequence of process steps resulted in a zinc-free surface of mild steel and zinc-coated side having conventional hot dip galvanize spangle characteristics. The zinc coating had excellent adhesion and coating weights were in the range of 0.26 to 0.46 oz/ft^2. The zinc-coated side had bare edges of up to 0.25".

The non-zinc-coated side was found to react identically to normal cold rolled steel with respect to painting pretreatments, such as zinc phosphating. The surface finish of the non-zinc-coated side was not found to have significantly altered during the process.

Therefore, in continuous pilot plant operation, the coating rapidly produced from Bonderite 901 on the mild steel surface effectively prevents the galvanizing of the so-treated side of the steel strip. The coating was readily removed from the one-side galvanized strip and the coating and subsequent removal operations did not adversely affect the mild steel strip surface.

Example 2: The procedure outlined in Example 1 was repeated, except that in this instance the pretreatment step (6) was omitted.

The nonfluxed side of the strip was found to have a generally spotty zinc pickup with a zinc coated edge of approximately 0.5 to 0.8" in width.

Thus, when the nonfluxed side of the sheet is left bare, sporadic pickup of zinc occurs, which is considered unsatisfactory where a wholly mild steel surface is required on the nongalvanized side.

Applying Silicone Resin to Nonplating Area for Partial Galvanizing

In recent years, in automobile and electric industries, for example, demands have been increasingly made for partially or one side coated steel sheets, galvanized partially or on one side, as a steel material having good corrosion resistance on one side and good weldability and paintability on the other side.

In the process of *K. Yoshida and Y. Kitajima; U.S. Patent 4,047,977; Sept. 13, 1977; assigned to Nippon Steel Corporation, Japan* silicone resin is applied partially or on one side of a steel strip. The silicone resin applied steel strip is baked at a temperature ranging from 300° to 800°C in an oxidizing atmosphere and is introduced to a galvanizing bath.

One or more metallic oxides, metallic hydroxides, metallic nitrides, metallic carbides, metallic carbonates, metallic phosphates, metallic silicates, etc., may be added to the silicone resin.

To improve heat resistance of the resin coating, metallic oxides such as SiO_2, Al_2O_3, MgO, TiO_2, metallic nitrides such as SiN_4, carbides, phosphates and silicates such as WC, $CaCO_3$ Na_2CO_3, $Ca_3(PO_4)_2$, $AlPO_4$, or $CaSiO_3$ can be added to the silicone resin.

The amount of these additives to be added to the silicone resin is less than 50% by weight.

Silicone resins which may be used are KF 96, KM 722, KS 66, KE 45 RTV and KR 255 (Shinetsu Chemical Industries Co., Ltd.) and SH 200 (Toray Silicone Co., Ltd.).

In some cases, a very small amount of metal coating attaches in fragments on the resin coated steel strip surface. Therefore, it is desirable to apply brushing to the resin coated surface of the steel strip coming out of the hot dipping bath to remove the coated metal as well as to remove the silicone resin coating.

One example of the apparatus for practicing the process is described with reference to Figure 6.1.

The steel strip **2** is uncoiled from the steel strip coil **1**. One side of the steel strip **2** is applied with the silicone resin by means of the coating roll **3** (a spray may also be used).

The resin coated steel strip **2** is baked at a temperature ranging from 300° to 800°C in the oxidizing furnace **7**, then annealed in the reducing pretreatment equipment **4**.

The resin coated steel strip **2** is introduced into the hot dipping bath **5** where it is coated with zinc on the other side, the amount of the zinc coating being controlled. Finally, the silicone resin coating is removed from the strip by means of the brushing roll **6** and the strip is coiled.

Figure 6.1: Apparatus for Selective Continuous Galvanizing

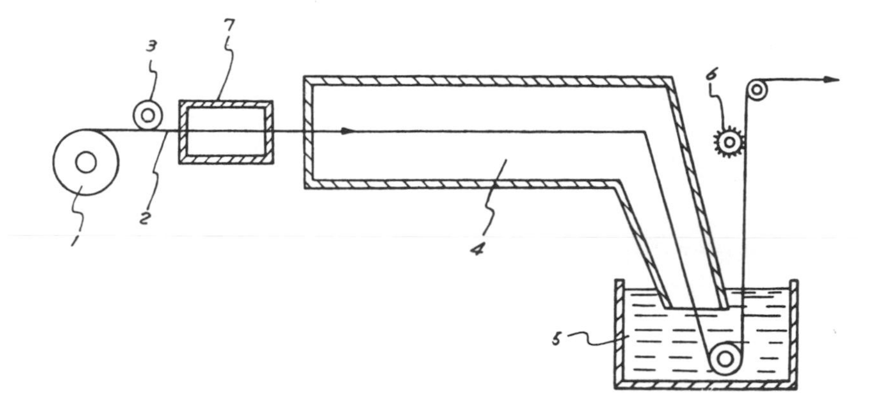

Source: U.S. Patent 4,047,977

Examples 1 through 15: These examples were conducted using Figure 6.1. The results are shown in the following tables.

Ex. No.	Silicone Resin	. . Coating, g/m². . .	Annealing Conditions (°C)	Hot Dipping Bath Temp. (°C)	Coated Metal (g/m²)	Speed of Strip Pass (m/min)
1	KM 722	6	740	450	183	50
2	KF 96	32	740	450	183	50
3	KE 45 RTV	47	740	450	183	50
4	KR 255	17	740	450	183	50
5	SH 200	0.7	740	450	183	50

Ex. No.	Material to Be Hot Dipped	Silicone Resin	Silicone Resin Additives, %		Resin Coating (g/m²)	Coating Position	Metal Coating	Bath Temp. (°C)	Dip Time (sec)
6	steel strip	KR255	$Ca(OH)_2$	20	8.0	one side wholly	Zn	470	30
7	steel strip	KR255	$Cr(OH)_2$	10	5.0	one side wholly	Zn	470	20
8	steel strip	SH 200	SiO_2	30	1.2	one side wholly	Zn	470	40
9	steel strip	SH 200	SiN_4	2	1.0	one side wholly	Zn	470	50
10	steel strip	SH 200	TiC	25	27	one side wholly	Al	700	40

Ex. No.	Oxidating Heating . . Conditions* . . Strip Temp. (°C)	Heating Time (sec)	Reducing Atmosphere Treating Conditions** Time (sec)	Zinc Bath Temp. (°C)	. . . . . Results*** . . . . Side to Be Nonplated	Side to Be Plated
11	300	20	160	460	A	C
12	400	10	80	460	A	D
13	500	8	64	460	A	D
14	700	6	64	460	A	D
15	800	6	64	460	A	C

(continued)

Ex. No.	Oxidating Heating Conditions* Strip Temp. (°C)	Heating Time (sec)	Reducing Atmosphere Treating Conditions** Time (sec)	Zinc Bath Temp. (°C)	Results*** Side to Be Nonplated	Side to Be Plated
†	250	20	160	460	B	E

*Burner heating in air.
**Reducing atmosphere: 75% H_2, balance N_2. Maximum strip temperature is 720°C.
***Code: A is completely nonplated; B is plated discontinuously; C is very small nonplated spots (no practical problem); D is completely satisfactory; and E is many button-like nonplated portions.
†Comparative.

Use of Oxidized Film for One Side Plating

K. Asakawa and M. Yoshida; U.S. Patent 4,107,357; August 15, 1978; assigned to Nippon Steel Corporation, Japan provide a method of producing a metal such as a steel sheet having molten metal plating on one side thereof which can be produced with low cost, high productivity and high quality. A metal strip such as a steel strip is subjected to activation treatment in a furnace having a reduction atmosphere of a Sendzimir system, Selas system or nonoxidation system. It is then dipped into a plating bath for molten metal coating, which comprises blowing an oxygen-containing gas on only one side thereof to produce an oxidized film, dipping the resulting material into a plating bath of molten metal, and maintaining a nonoxidizing atmosphere on the outlet side of the bath so as to prevent mechanical adherence of the plating metal to the oxidized surface whereby the plating can be effected on only the unoxidized surface.

The process may be practiced in the conventional continuous plating line by oxidizing one side of a strip at the terminal end of the furnace filled with a reducing atmosphere, for example, at the snout part, dipping the strip into the plating bath while the opposite or reverse side thereof is kept activated so that formation of an alloy layer is prevented, and providing a seal box at the outlet side of the bath whereby the repelling of the plating metal adhering to the oxide surface is accelerated and the plating on only the nonoxidized surface is carried out. In this way, a one side plated steel sheet of lower cost and higher quality than the conventional one can be obtained.

In Figure 6.2a the numeral **8** is a part which is directly connected to a reducing furnace and which is filled with a reducing atmosphere. The numeral **11** is a one side oxidizing treatment chamber. **12** is a space where a nonoxidizing gas is allowed to pass so as to prevent the oxidizing gas from bypassing thereinto and causing oxidation. **18** is a seal chamber for the surface of a bath connected to a gas supply source **17** where a nonoxidizing atmosphere is maintained to prevent oxidation of the bath surface. **15** and **16** are sealing elements adapted for preventing mutual interconnection of gases in each chamber, respectively.

Also in Figure 6.2a the numeral **10** is a nozzle for blowing an oxidizing gas which has an outlet for blowing gas of a slitlike shape extending widthwise of a strip and facing thereto. It is connected to an oxidizing gas supply means **9**. On the side of the strip opposite to the nozzle **10** is the space or chamber **12** as described above. In its center there is a slitlike outlet **14** for blowing a sealing gas extending along the travel of the strip. Chamber **12** is connected to a sealing gas supply means **13**.

Figure 6.2: Use of Oxidized Film for One Side Plating

a.

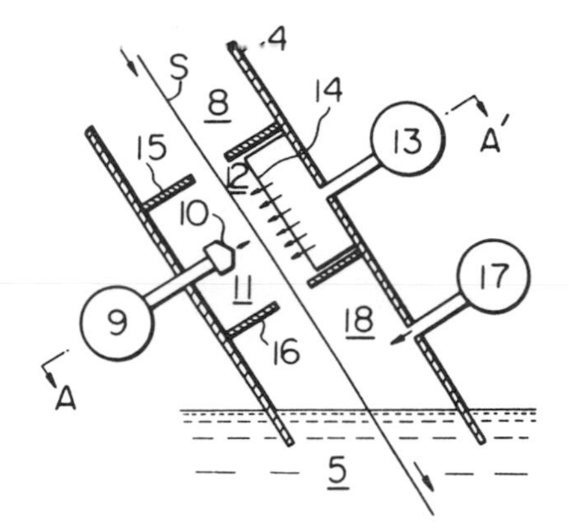

b.

(a) One side oxidizing apparatus
(b) Seal box at outlet side of bath

Source: U.S. Patent 4,107,357

In Figure 6.2b there is shown the seal box **6** for the outlet side of the plating path, the lower side of which is open and dipped into the bath **5**, and the upper side of which is sealed except for an outlet **23** for the strip. The numeral **21** indicates nozzles for blowing a gas adapted for use as wiping, which are provided on each side of the strip facing to each other. Each nozzle **21** has a slit-like gas opening extending along the width of the strip and is connected to the gas supply means **22**.

The strip **S** which has been oxidized on its one side is dipped into the bath **5** and then pulled upwardly as mentioned above, where a nonoxidizing atmosphere is kept inside the seal box and is also supplied from the wiping nozzles **21** after adjustment with respect to its temperature, pressure, and flow rate. Since a nonoxidizing atmosphere is maintained inside the seal box **6** to cause the repelling phenomenon and also gives the wiping or scraping action, the plating metal does not adhere to the oxide surface of the strip while it adheres to the nonoxidized surface in a suitable thickness by means of the wiping action.

Example 1: In a Selas process (furnace temperature 900°C, atmosphere 80% N_2 + 20% H_2) for galvanizing a preannealed steel strip (thickness 0.8 mm, width 914 mm), a one side oxidizing apparatus as shown in Figure 6.2a is provided at the snout part of the line. An oxidizing gas consisting of 0.1% oxygen and the rest nitrogen is blown to the strip from the oxidizing nozzle at the rate of 500 ℓ/min. Thereafter the strip is dipped into a zinc-plating or galvanizing bath for conducting plating at the rate of 30 m/min. At the outlet side of the plating bath a seal box as shown in Figure 6.2b is provided where a nitrogen gas heated at 300°C under pressure of 1 kg/cm^2 is blown to the strip for wiping. As a result of this, there is no adherence of zinc to the oxidized surface of the strip while there is no unplated portion on the reverse side, and thus a one side zinc plated steel strip showing excellent adherence of plating is produced.

Example 2: In a Sendzimir process for producing one side 10% lead-tin alloy plated steel, a one side oxidizing apparatus as shown in Figure 6.2a is provided at the snout part of the line, where an oxidizing gas containing 0.05% oxygen and the rest nitrogen is blown from the oxidizing nozzle to one side of the strip at the rate of 500 ℓ/min for effecting only one side oxidation thereof. The strip thus treated is dipped into a bath where the plating is conducted at the plating rate of 20 m/min. At the outlet side of the bath, a seal box is provided, into which a reduction gas consisting of 95% N + 5% H is introduced at the rate of 50 ℓ/min so as to maintain a nonoxidizing atmosphere therein. The lead-tin one side plated steel sheet thus obtained has no plating metal adhering to its oxide surface while having no unplated parts on its plated surface. Moreover, it has very few pinholes and displays excellent anticorrosion property.

Plating Inhibitor

The plating inhibitor of the process of *K. Doi, H. Kato and T. Kurauchi; U.S. Patent 4,201,592; May 6, 1980; assigned to Nisshin Steel Co., Ltd. and Nippon Paint Co., Ltd., Japan* is characterized by containing one or more phosphates and one or more inorganic inert materials which are heat-resistant and do not react with molten metal.

The phosphate ingredient used in the plating inhibitor must have a film-formability and high heat resistance as essential conditions. It is classified into the following three groups: the metal phosphate group, the metal condensed phosphate group, and their modified phosphate groups.

A part of the water-soluble phosphate type bases optionally can be replaced by one or more of the group of alkali metal silicates, quaternary ammonium silicate, silica sol and alumina sol.

Examples of inert inorganic materials of the plating inhibitor include titanium oxide, zinc white, chromium oxide, cobalt oxide, barium sulfate, talc, clay, mica, kaolin clay, asbestine, asbestos, calcium carbonate, alumina, siliceous sand, magnesium carbonate; and natural minerals such as feldspar, garnet, gypsum, quartz, olivine, chlorite, serpentine, lithium spodumene, alum, melilite, benitoite, wollastonite, analcite and the like; and synthetic mineral powders such as synthetic mica and the like.

The abovementioned inorganic inert materials are effective in preventing molten metal from adhering to the plating-inhibitor layer and also in preventing molten metal from contacting the steel plate where the plating inhibitor is coated. However, since these inorganic inert materials alone cannot adhere to a steel plate and cannot form a continuous film, it is impossible to prevent the occurrence of temper color and the separation of the coated film if they are used alone.

The interaction between the phosphate type base and the inert inorganic material varies depending on the ratio of the two. If the interaction is great, the effects appear in the prevention of adherence of molten metal and the removal of the plating inhibitor, and if the interaction is little or nonexistent, the effect appears in the prevention of temper color.

Inorganic inert materials having a particle size of at least 0.1 μ, preferably more than 1 μ are mixed and dispersed.

In order to fully achieve the effect of the process, the mixing ratio of phosphate type base:inorganic inert material in the plating-inhibitor must be 5-70:95-30 on the basis of the nonvolatile content volume ratio.

If desired, in addition to the above main ingredients, organic or inorganic thickeners, surface active agents, water-soluble resins or emulsions may be added in order to control viscosity or to improve wetting property on a steel plate, dispersibility, storage stability, coating efficiency, film-formability or the like.

Example 1: Plating inhibitor (1) (nonvolatile content = 8.7% by volume) was prepared by mixing 11.8 parts by volume of primary aluminum phosphate aqueous solution ($3Al_2O_3/P_2O_5$ mol ratio = 1.0, nonvolatile content = 28.6% by volume) and 4.3 parts by volume of titanium oxide in a mixer and then adding 70 parts by volume of water to the resultant dispersion. The plating inhibitor (1) thus prepared was coated by an air sprayer on one side of a clean steel plate, which had been previously degreased, water-washed and dried, in such a manner as to obtain a dry film thickness of 10 μ, at a rate of 115 cc/m^2.

The coated steel plate was then fully dried in a drying furnace at 400°C to remove free moisture, and the surface of the dried film was checked. The steel plate with the film was then passed through a preheated furnace (a slight oxidative or nonoxidative atmosphere of 700° to 880°C) and a reductive furnace (an atmosphere containing hydrogen gas at 840°C) for about 2 minutes. The steel plate was then dipped in a molten zinc bath at 460°C for 5 seconds. The steel plate was then subjected to rapid cooling, and was checked with regard to the adherence of zinc to the surface of the plating inhibitor, the occurrence of temper color and the removability of the plating inhibitor layer.

The adherence of zinc to the surface of the plating-inhibitor layer was checked with the naked eye. The occurrence of temper color was checked by bending the steel plate around a mandrel of a bending tester having a diameter of 10 mm, stripping the plating-inhibitor layer off with cellophane tape and observing the surface of the steel plate. The removability of the plating-inhibitor layer was checked by measuring reciprocation times of a brass wire brush of a Gardner washability tester loaded with 500 g until the naked surface of the steel plate was revealed. The surface of the coated steel plate was checked with the naked eye with regard to flowing, cracking, cissing, foaming, uniformity of film thickness and the like. The storage stability of the plating inhibitor (1) was checked by putting a sample of the plating-inhibitor (1) in a sealed glass bottle, placing the bottle at room temperature and measuring the degree of settling and redispersibility of the sample after one week.

Example 2: 20 parts by weight of aluminum hydroxide (chemically pure reagent) was added to 100 parts by weight of orthophosphoric acid (chemically pure reagent) and was heat-dissolved at 80°C. 18 parts by weight of magnesium oxide (chemically pure reagent) was heated with the resultant solution while mixing, and 50 parts by weight of water was added to the resultant mixture to obtain a modified aluminum phosphate aqueous solution having a nonvolatile content of 33% by volume. 9.5 parts by volume of the modified aluminum phosphate aqueous solution thus prepared was mixed with 17.2 parts by volume of siliceous sand powder and 40 parts by volume of water, and the resultant mixture was fully mixed in a pot mill for 16 hours to prepare plating inhibitor (2).

One side of a steel plate which had previously been degreased, water-washed, acid-washed, and water-washed again, was coated with a flux solution (20% aqueous solution of $3ZnCl_2/NH_4Cl$), and the other side of the steel plate was coated with the plating inhibitor (2) diluted with water in such a manner as to have a nonvolatile content of 28.5% by volume. The coating of the plating inhibitor was conducted by a roll coater at a rate of 140 cc/m^2 in such a manner as to obtain a dry film thickness of 40 μ. The coated steel plate was then dried at 300°C for 1 minute, and the surface of the coating film was checked. The steel plate was then dipped in a molten zinc bath at 460°C for 5 seconds. After subjecting the steel plate to rapid cooling, it was tested in the same manner as in Example 1.

Example 3: Plating inhibitor (3) (nonvolatile content = 25% by volume) was prepared by mixing 32.7 parts by volume of primary magnesium phosphate aqueous solution ($2MgO/P_2O_5$ mol ratio = 1.0, nonvolatile content = 28.6% by volume) with 10 parts by volume of clay in a colloid mill, adding 25.7 parts by volume of water to the resultant mixed dispersion with stirring and then adding 15.6 parts by volume of alumina sol (nonvolatile content = 3.2% by volume, pH = 4) to the resultant mixture.

The plating inhibitor thus prepared was coated by an airless sprayer on a steel plate in such a manner as to obtain a dry film thickness of 5 μ at a rate of 20 cc/m^2 according to the same procedure as in Example 2. The coated steel plate was then dried at 280°C for 1 minute, and the surface of the coating film was checked. The steel plate was then dipped in a molten zinc bath at 460°C for 5 seconds. After subjecting the steel plate to rapid cooling, it was tested in the same manner as in Example 1.

Example 4: 30 parts by weight of magnesium hydroxide (chemically pure reagent) was added to 100 parts by weight of orthophosphoric acid (chemically pure reagent), and was heat-dissolved at 100°C. 3 parts by weight of zinc white was mixed and reacted with the resultant solution and 100 parts by weight of water was added to the resultant mixture to prepare a modified magnesium phosphate aqueous solution having a nonvolatile content of 24% by volume. 18.9 parts by volume of a modified magnesium phosphate thus prepared, 15.9 parts by volume of siliceous sand powder and 37 parts by volume of water were intimately mixed in a pot mill for 8 hours to prepare plating inhibitor (4) (nonvolatile content = 28% by volume).

The plating inhibitor (4) thus prepared was coated by an air sprayer on a clean steel plate which had previously been degreased, water-washed, acid-washed and water-washed in such a manner as to obtain a dry film thickness of 20 μ, at a rate of 71 cc/m^2. The coated steel plate was then dried and the surface of the coating film was checked. After plating and cooling the steel plate, it was tested in the same manner as in Example 1.

Example 5: 13.5 parts by weight of orthophosphoric acid (chemically pure reagent), 3 parts by weight of aluminum hydroxide (chemically pure reagent), 1 part by weight of boric acid (chemically pure reagent) and 60 parts by weight of water were mixed and reacted at 60°C to prepare a modified aluminum phosphate aqueous solution having a nonvolatile content of 9% by volume. 30.8 parts by volume of the modified aluminum phosphate aqueous solution thus prepared was mixed with 8.8 parts by weight of clay in an SG mill to prepare plating inhibitor (5).

As in Example 2, the plating inhibitor (5) was diluted with 12 parts by volume of water in such a manner as to have a nonvolatile content of 22% by volume. The diluted plating inhibitor was coated on a steel plate by an air sprayer in such a manner as to obtain a dry film thickness of 15 μ at a rate of 68 cc/m^2. The coated steel plate was then dried at 200°C for 2 minutes and the surface of the coating film was checked. The steel plate was then dipped in a molten zinc bath at 460°C for 4 seconds. The steel plate was then subjected to rapid cooling. A part of the steel plate was tested in the same manner as in Example 1, and the remaining part of the steel plate was tested with regard to the removability of the plating inhibitor with Scotch Bright (abrasive).

Example 6: 30 parts by weight of primary magnesium phosphate aqueous solution ($2MgO/P_2O_5$ mol ratio = 1.0, nonvolatile content = 50% by weight), 0.6 part by weight of boric acid (chemically pure reagent), 0.5 part by weight of active aluminum hydroxide (Al_2O_3 = 49.8%) and 0.3 part by weight of heavy magnesium oxide were mixed and dissolved in a hot water bath to prepare a modified magnesium phosphate having a nonvolatile content of 30.4% by volume. 6.2 parts by volume of the modified magnesium phosphate thus prepared, 8.1 parts by volume of titanium oxide, 0.7 part by volume of mica, 2 parts by volume of 10% aqueous solution of methyl vinyl ether/maleic anhydride copolymer resin (Gantrez AN-119) and 27.6 parts by volume of water were mixed in an SG mill. 12.0 parts by volume of acid-stabilized colloidal silica (nonvolatile content = 10% by volume, pH = 3.5) was then added to the resultant dispersion to prepare plating inhibitor (6) having a nonvolatile content of 21% by volume.

According to the same procedure as in Example 2, the plating inhibitor (6) thus prepared was coated by an air sprayer on a steel plate in such a manner as to obtain a dry film thickness of 35 μ at a rate of 167 cc/m^2. The coated steel plate was then dried at 180°C for 1 minute, and the surface of the coating film was checked. The steel plate was then dipped in a molten zinc bath at 460°C for 5 seconds. After subjecting the steel plate to rapid cooling, a part of the steel plate was tested in the same manner as in Example 1, and the remaining part was checked by a leveller with regard to the removability of the plating inhibitor.

The results of the tests are shown in the table below. In the table in column (1), (A) means there are substantially no defects and (C) means there are no defects at all such as flowing, cracking, cissing, foaming, unevenness of film thickness and the like. In column (2), (A) means that molten metal does not adhere to the plating-inhibitor layer substantially, (B) means the molten metal adheres to plating-inhibitor layer a little (but practically usable), and (C) means that the molten metal does not adhere to plating-inhibitor layer at all. In column (3), (A) means the temper color does not occur substantially and (C) means the temper color does not occur at all.

In column (4) Method A: Reciprocation times of a brass wire brush until the naked surface of the steel plate is revealed, (A) is 50 to 200 times, (B) is 200 to 500 times (but practically usable), and (C) is 50 times or less. In column (5) Method B (by Scotch Bright) and column (6) Method C (by leveller), (A) means the plating inhibitor does not remain substantially. In column (7), (A) means the solid contents do not settle substantially and (C) means the solid contents in the plating inhibitor do not settle at all and the original dispersion state is maintained.

Performance of Plating Inhibitor

Ex. No.	(1) Surface State of Plating Inhibitor Film	(2) Adherence of Molten Metal to Plating Inhibitor Layer	(3) Occurrence of Temper Color on Steel Plate	(4) Removability of Plating Inhibitor Method A	(5) Removability of Plating Inhibitor Method B	(6) Removability of Plating Inhibitor Method C	(7) Storage Stability of Plating Inhibitor
1	(A)	(B)	(C)	(B)	–	–	(A)
2	(A)	(C)	(A)	(C)	–	–	(A)
3	(A)	(A)	(C)	(A)	–	–	(A)
4	(A)	(C)	(A)	(A)	–	–	(A)
5	(A)	(A)	(C)	(A)	(A)	–	(C)
6	(C)	(A)	(A)	(A)	–	(A)	(C)

SURFACE TREATMENT OF EQUIPMENT AND MANUFACTURED ARTICLES

METAL TIRE CORD

Promoting Cord to Rubber Adhesion with Benzotriazole

Pneumatic vehicle tires are often reinforced by means of cords prepared from brass coated steel filaments. This tire cord is frequently high carbon steel or high carbon steel cord with a thin layer of alpha brass. The cord may be a monofilament, but normally is prepared from several filaments which are stranded together. The filament is coated with brass, cold drawn and then stranded to form the cord. In most instances, normally depending upon the type of tire being reinforced, the strands of filaments are further cabled to form the final cord.

Brass plated steel wire tire cords are subject to corrosion of the steel structure and oxidation of the brass plating if improperly handled prior to incorporation into a tire. Corrosion and oxidation can result in poor adhesion between the cord and rubber and more importantly in a deterioration of the physical properties of the cord.

G.W. Rye; U.S. Patent 4,192,694; March 11, 1980; assigned to The Goodyear Tire & Rubber Company provides an efficient, low cost method of applying protective agents to brass coated steel wire, the method being capable of rapidly treating the wire and reaching even remote surfaces of the wire. The method does not require the use of drying equipment or other expensive and time consuming follow-up treatments. The treated brass coated steel tire cord possess effective corrosion resistance.

The objects of the process can be accomplished by treating the cord during the course of its preparation or thereafter with reagents capable of promoting and/or retaining adhesion between the metal cord and adjacent vulcanized rubber and/or capable of improving the resistance of the cord to corrosion prior to incorporation into the tire during and after incorporation into the tire, the reagent being in a solid or molten state. The reagents include compounds capable of preventing the oxidation of the steel substrate and/or capable of preventing the corrosion of the brass. The only limitation regarding the use of the reagent in solid form is that on coming in contact with the wire, that portion directly in contact with the wire becomes molten.

The process can be used to treat the filament after drawing but before stranding, after stranding to form the cord, in the form of woven fabric or as multiple ends such as may be used at a creel calendering operation. In fact, the method can be used at any point in the manufacture of the cord and even subsequent thereto, the only requirement being that the cord be treated at some point before it becomes a reinforcing element in the tire or other rubber product.

The reagents of the process include, but are not limited to, reagents selected from the group consisting of precipitation compounds, oxidizing compounds, and compounds having the following structural formula:

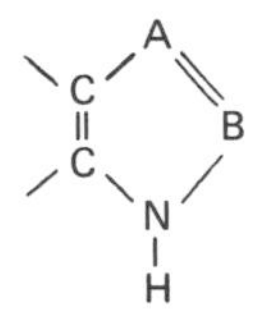

where the adjacent carbon atoms are joined to form a benzene or naphthylene ring, the ring being substituted (for example, with a single methyl group) or unsubstituted and where A and B are selected from the group consisting of –N– or –CH–, with the proviso that A and B are never both –CH–, the agent being in the form of a solid or a liquid. The precipitation compounds include compounds selected from the group consisting of organic borates, organic phosphate and organic metaphosphates. The oxidation compounds include organic nitrites.

The precipitation compounds offer their protection through an indirect oxidizing (buffering) mechanism. The oxidation compounds offer protection by directly oxidizing metallic ions in the substrate surface.

Examples of organic compounds which can be used include organic alkyl, cycloalkyl and aryl derivatives of metaboric acid, orthoboric acid and pyroboroic acid as well as meta-, ortho-, pyro- and hypo-phosphoric acid.

The agents can be used alone or in combination. Likewise a series of units can be used, each containing a different agent. It is preferred that one stage of the treatment involve the use of a (benzotriazole) BTA-type chemical agent.

If moisture is desired, it can be introduced, for example, by the introduction of steam into the reaction area, the addition of water to the molten agents, etc. The manner by which the water is introduced is not critical.

The brass coating of a typical brass coated steel cord is microscopically porous, thereby exposing small areas of steel surface to any surrounding environment. It is believed that BTA interacts with copper in a brass coating to form a polymeric complex of BTA plus copper. This polymeric complex is insoluble in most solvents and serves as a protective barrier to any environmental degradation of the underlying brass. On the other hand, anions from the precipitation and oxidation compounds, it is theorized, interact with iron and iron oxide from steel surfaces exposed through microscopic pores to form an adherent oxide film which protects the steel. It is not necessary that the barrier layers of polymeric complexes adsorbed be extremely thick. In fact, such layers should not be so thick as to interfere with the sulfur reaction required for bonding the wire to the rubber, the adhesion of rubber to metal cord requiring the formation of copper-sulfur bonds.

The process results in increased surface protection of brass coated steel prior to rubber encapsulation and improved aged adhesion of vulcanized brass coated steel/rubber composites. It also prevents cord failure due to excessive corrosion during the use of the product, e.g., a tire being reinforced with the cord.

Example 1: A drawn brass coated steel cable at room temperature was passed slowly (possibly at a speed of about 25 to 50 yd/min) through a tube containing molten benzotriazole (BTA). The cable was then subjected to an ultrahigh temperature to flash off the excess BTA. The initial adhesion of the treated cable to vulcanized carbon black loaded natural rubber was 37 kg.

The corrosion resistance of the cable in a high temperature, high humidity environment was rated excellent.

Example 2: A tube of molten BTA was mounted on a draw machine. A hot drawn brass coated steel filament as it exited the draw machine was passed through a ¾-inch thickness of molten BTA at a speed of 900 m/min. A cable prepared therefrom had an initial adhesion of 40.4 and a wet adhesion of 40.0. Its corrosion resistance was excellent. An untreated control had an initial adhesion of 59 and a wet adhesion (unvulcanized rubber soaked in water before the cable was embedded therein) of 34. The corrosion resistance was rated poor.

Example 3: A brass coated steel filament was passed through a 12-inch tube of powdered BTA at a rate of 900 m/min after exiting from a draw machine. The filament was at a temperature of about 200°C. The powder temperature was varied from room temperature to 120° to 180°C. A cable was prepared from the resultant filament and checked for adhesion.

Powder Temperature (°C)	. . . Adhesion . . . . Initial	Wet	Corrosion Resistance
Room temperature	49	43	Excellent
120	41	31	Excellent
180	45	38.5	Excellent

The untreated control for room temperature and 180°C had initial and wet adhesion values of 57 and 29 respectively. The untreated control for 120°C had initial and wet adhesion values of 44 and 24. Therefore the treatment resulted in wet adhesion improvements at each temperature.

In this process of *G.W. Rye and K. Marencak; U.S. Patent 4,189,332; Feb. 19, 1980; assigned to The Goodyear Tire & Rubber Company* the treating reagent described above in U.S. Patent 4,192,694 is in the form of a gaseous state, either saturated or unsaturated.

The vapor technique has the advantage of permitting the reagent the furthest recesses of the stranded cord even when wrapped upon a spool. It also permits the reagent to enter the microscopic pores of the brass coating.

Any manner of exposing the cord to the vapors will result in some improvement in corrosion or oxidation resistance.

Example 1: A chamber 30 cm long was mounted at the exit of a wire drawing machine operating at a speed of 900 m/min. BTA (benzotriazole) was vaporized in this chamber with the melt temperature being maintained at 120°C. Drawn filaments (brass coated steel) were passed through this chamber. Filaments treated in this manner were formed into a 5 x 0.25 cable.

The resulting cable was embedded in carbon black loaded polymer, and the polymer vulcanized. Corrosion resistance on the bare cable and adhesion data on the vulcanized composite were obtained. The results are listed below.

	Original Adhesion (kg)	Wet Compound Adhesion (kg)*	Corrosion Resistance
Untreated	57.0	29.0	Poor
Treated	45	38	Excellent

*Unvulcanized rubber soaked in water before the cable was embedded therein.

Example 2: A chamber 30 cm long and approximately 6 cm high was equipped with suitable heaters. BTA was melted in the bottom of this chamber and a wire cable passed through this vapor zone. Test results on cable treated in this manner were as follows.

Sample 1 Run	Speed (m/min)	Exposure Time	Melt Temperature (°C)	Original Adhesion (kg)	Wet Compound Adhesion* (kg)	Corrosion Resistance
1	30	0.6	180	53	32	Excellent
2	30	0.6	120	55	27	Excellent
3	90	0.2	180	54	28	Excellent
Untreated	–	–	–	57	25	Fair

*Unvulcanized rubber soaked in water before the cable was embedded therein.

Example 3: The chamber used in Example 2 was mounted on a strander running at 90 m/min. Test results on cables produced when exposing the cable with the melted BTA at a temperature of 180°C are as follows.

	Original Adhesion (kg)	Wet Compound Adhesion (kg)	Corrosion Resistance
Treated	52	30	Excellent
Untreated	48	21	Fair

In every one of the above examples the treated cord possessed a higher wet compound adhesion and better corrosion resistance than the untreated cord.

Improving Cord to Rubber Adhesion with Coating Combination of Vulcanization Accelerator and Phosphate Corrosion Inhibitor

A.J. Pignocco and M.E. Waitlevertch; U.S. Patent 4,182,639; January 8, 1980; assigned to United States Steel Corporation have discovered that significant improvements in adhesion of brass-coated steel cord to rubber may be achieved through the use of certain compositions and techniques for applying them to the cord. Inorganic or organic phosphates, preferably tricresylphosphate (TCP), and various sulfur-containing rubber vulcanization accelerating agents are employed. The accelerators are believed to interact with ZnO to form complexed zinc perthiomercaptides, and the perthiomercaptide is believed to be the active sulfurating agent that reacts with and bonds to the rubber hydrocarbon. The surface of a typical 70/30 brass (70% copper, 30% zinc) contains ZnO, and after the reaction of accelerator and ZnO is completed, the remaining active copper sites on the brass are covered or rendered inert by treatment with corrosion inhibitors or other such treatment. As a result of such treatments, the brass surface then contains the proper number of reactive polysulfide pendant groups, which are free to interact with and bond to the rubber compound. The number of reactive polysulfide pendant groups on the brass surface controls the extent of the interaction of the cord with the rubber because none or relatively few copper sites remain active on the cord.

The accelerator and the phosphate are used at ratios of from about 0.8:1 to 4:1 of phosphate to accelerator. While any commercial sulfur-containing accelerator would be satisfactory, the thiazyls, thiurams, and dithiocarbonates are preferred.

The phosphate/accelerator composition may be applied to the cord or the wire as a slurry or in a solvent such as trichloroethylene or chloroform. The slurry may be in an oil base such as, for example, corn oil.

The process is applicable to any brass-coated tire cord, but it is particularly useful with cord having a brass coating weight between about 3 and 8 g of brass per kilogram of wire, the brass having a copper content of about 60 to 70%.

Example: In the first series of experiments, a mixture consisting of 2 g of Thiotax and 2 ml of TCP in 100 ml of Trichlor was applied to a commercial cord with sponge contactors. This method of application was used for posttreating cord as it exited from a buncher which was operating at a speed of 170 ft/min. Results are shown in lines 3 and 5 of the table below.

The final series of experiments was conducted on a Morgan wire drawing machine which was operated at 3,500 ft/min.

To enhance the moderator-cord reaction kinetics, the contactor stage was located in close proximity to the final drawing die such that the hot wire was drawn directly through the reactants. To still further increase the interaction of the wire with the reactants in this series of tests, the dilute Trichlor solutions used in earlier experiments were replaced in turn with dry Thiotax powder, dry MBTS (benzothiazyldisulfide) powder, a slurry of Thiotax and TCP, and a slurry of MBTS and TCP.

The untreated cords, treated cords, and cords bunched from wires treated with slurries were fabricated into adhesion strip pads using a commercial rubber formulation believed to be natural rubber containing a significant amount, i.e., about 50% of polybutadiene and SBR synthetic rubber. The results of these strip adhesion tests are shown in the table. On the basis of a standard acceptable pullout load of at least 50 lb and a rubber coverage in excess of 75% for suitable adhesion, all the untreated cords (cords 1, 2 and 4) were unsatisfactory. The table further shows that the posttreated cords (cords 3 and 5) and the cords bunched from wires treated with slurries (cords 8 and 9) satisfied adhesion requirements.

Results of Strip Pad Adhesion Tests of Tire Cord in Rubber

Cord	Identification	Treatment*	Pullout Load (lb)	Coverage (%)
1	66.4% Cu-3.80 g/kg	None	42	61
2	65% Cu-5.1 g/kg**	None	42	59
3	65% Cu-5.1 g/kg**	(A)	53	92
4	65.4% Cu-3.0 g/kg***	None	52	71
5	65.4% Cu-3.0 g/kg***	(A)	56	91
6	65.6% Cu-3.5 g/kg	(B)	51	72
7	65.6% Cu-3.5 g/kg	(C)	58	74
8	65.6% Cu-3.5 g/kg	(D)	58	79
9	65.6% Cu-3.5 g/kg	(E)	64	88

Note: (A) is posttreated cord (Morgan Buncher) Thiotax + TCP in Trichlor; (B) is posttreated wire (Morgan wire drawing machine) dry Thiotax powder; (C) is posttreated wire (Morgan wire drawing machine) dry MBTS powder; (D) is posttreated wire (Morgan wire drawing machine) Thiotax + TCP slurry; and (E) is posttreated wire (Morgan wire drawing machine) MBTS + TCP slurry.

*Posttreated cord was bunched at 170 fpm and 3,500 fpm on the Morgan.
**Heavy coating weight.
***Light coating weight.

Use of Benzotriazole plus Precipitating Anion

Clean, untreated brass-coated steel wire will normally have sufficient good initial adhesion to the adjacent rubber. However, the adhesion usually will drop with time, i.e., with aging due to heat, stress and/or chemical degradation or corrosion effects.

R.M. Shemenski, Sr.; U.S. Patent 4,269,877; May 26, 1981; assigned to The Goodyear Tire & Rubber Company provides a technique to protect and to maintain a clean surface on brass-coated steel through inventory storage and factory processing. Further, the method improves initial or original adhesion between brass-coated steel and rubber, as well as aged adhesion between brass or brass-coated steel tire cord and the rubber immediately adjacent thereto.

The objects are accomplished by treating a clean brass-coated steel cord with benzotriazole (BTA) and/or an anion which aids in the formation of a protective film either through a precipitation and indirect oxidizing (buffering) mechanism, (hereinafter referred to as precipitation anion), and/or an anion which forms a film by directly oxidizing metallic ions in the substrate surface (hereinafter referred to as an oxidation anion). Normally the anions, whether precipitation or oxidation, will be donated by a salt. A salt donating precipitation anion will be referred to hereinafter as a precipitation compound and salt donating an oxidation anion will be referred to as an oxidation compound.

The precipitation anions include anions selected from the group consisting of borate, phosphate, metaphosphate, sulfate, silicate, carbonate, nitrate, arsenate and iodate.

The oxidation anions include anions selected from the group consisting of chromate, nitrite, tungstate, molybdate, dichromate, arsenite, and sulfite.

The specific cation of the salts containing these anions may be any cation, but preferred is cyclohexyl amine, iron, cobalt, nickel, tin or zinc. Potassium and sodium are illustrations of other cations which can also be used.

It is believed that protection of brass-coated steel cord is realized by formation of a complex polymeric film comprised of BTA plus precipitation anion and/or oxidation anion plus water molecules and/or hydroxyl ions supplied by the water. The precipitation anions require the presence of oxygen in the vicinity of the metal surface whereas the oxidation anions can operate in the absence of oxygen.

When BTA is applied to the metal in solution form, e.g., as an aqueous solution, the concentration of BTA in solution is from 0.001 wt % to 1.0 wt %, preferably 0.01 to 0.5 wt %. The precipitation and oxidation compounds can be applied to brass-coated steel surfaces by using aqueous solutions containing 50 to 10,000 ppm, preferably 200 to 2,000 ppm.

Example 1: Bare brass-coated steel cords (3 x 0.20 + 6 x 0.38) were dipped in various solutions and then subsequently aged for 7 days at 38°C and 100% relative humidity. After aging, the bare cords were vulcanized in standard test blocks and these blocks tested in the original condition and after additional pressurized aging of the cured blocks for 6 hours at 150°C. Adhesion ratings are summarized in Table 1.

Table 1: Effect of Surface Treatment on Initial and Aged-Adhesion for As-Dipped and Aged Cords

Wire Treatment*	
1	Control
2	0.2 wt % BTA in water
3	0.5 wt % BTA in water
4	0.5 wt % BTA + 300 ppm CHAB**
5	0.5 wt % BTA + 381 ppm sodium borate
6	0.5 wt % BTA + 380 ppm sodium phosphate
7	0.5 wt % BTA + 234 ppm sodium chromate
8	0.5 wt % BTA + 70 ppm sodium nitrite

*All in aqueous solutions.
**Cyclohexylamine borate.

Results: Adhesion Rating (All in Compound B)

	Vulcanized Original Adhesion		Vulcanized Aged Adhesion	
Wire Treatment	As Dipped Cord	Aged Cord	As Dipped Cord	Aged Cord
1	100	62	45	20
2	99	79	42	34
3	109	110	51	37
4	101	95	50	48
5	96	109	57	53
6	105	88	49	64
7	100	107	65	70
8	102	105	55	58

As indicated by the above data, dipping substantially improved the adhesion of the aged bare cord. In addition, vulcanized aged adhesion was improved for as-dipped cord.

Example 2: Vulcanized aged adhesion, both himidity and salt water aging, of cords dipped in several solutions is given on the following page in Table 2.

Table 2: Vulcanized Aged Adhesion

Wire Treatment*

1	Control
2	Sodium nitrite (1,000 ppm)
3	Sodium nitrite (1,000 ppm) + 0.5 wt % BTA
4	Sodium chromate (1,000 ppm)
5	Sodium borate (1,000 ppm)
6	Sodium phosphate (1,000 ppm)

*All aqueous solutions.

Results: Adhesion Rating (All in Compound B)

Wire Treatment	Original Adhesion	Humidity Aged Adhesion	Salt Water Aged Adhesion
1	100	73	53
2	111	27	49
3	136	112	71
4	103	69	74
5	110	101	87
6	86	79	71

These data show that dipping improves the level of vulcanized aged adhesion and that the combination of BTA plus anion is better than the anion viz, nitrite, used alone. It should be noted that the nitrite data (2) is atypically low.

Example 3: Brass-plated steel tire cords were dipped in an aqueous solution containing 0.5 wt % BTA, 300 ppm CHAB, and 234 ppm sodium chromate. Both as-received (control) and dipped cords were aged bare and in unvulcanized (green) test blocks for 2 and 5 days at 38°C and 100% relative humidity. Adhesion results are given in Table 3.

Table 3: Effect of Surface Treatment on Humidity Aged Bare Cords and Unvulcanized Rubber Blocks

Cord Condition

1	Initial condition (control or dipped)
2	Green block humidity-aged 2 days
3	Bare cords humidity-aged 2 days
4	Green block humidity-aged 5 days
5	Bare cords humidity-aged 5 days

Results: Vulcanized Adhesion Ratings (All in Compound C)

	Control Cord			Dipped Cord		
Cord Condition	Original Adhesion	Humidity Aged Adhesion	Salt Water Adhesion	Original Adhesion	Humidity Aged Adhesion	Salt Water Adhesion
1	100	82	42	71	63	45
2	65	116	28	69	78	46
3	39	38	29	75	76	48
4	24	47	19	67	66	34
5	21	28	17	72	79	28

These data show that dipping improves adhesion as the severity of aging increases, especially for bare cord.

Rubber compound A was prepared using the following formulation:

Ingredients	Parts by Weight
Natural rubber	100
Peptizer	0.1
Resorcinol	3
Processing oil	10
Stearic acid	2
Furnace black	55
Zinc oxide	10
Hexamethylenetetramine	2
Antioxidant	0.75
Accelerator	1
Retarder	5
Sulfur	5

Rubber B was the same as A with the exception that silica was substituted for the carbon black. Rubber C was the same as A with the following exceptions: Cobalt naphthenate (2.5 parts) replaced the resorcinol and hexamethylenetetramine, and the accelerator and sulfur levels were changed to 0.5 and 8 parts respectively.

Combination of Benzotriazole and Cyclohexylamine Borate

R.M. Shemenski, Sr.; U.S. Patent 4,269,645; May 26, 1981; assigned to The Goodyear Tire & Rubber Company has also found that the vulcanized, aged, adhesion of rubber to brass-coated steel cord is improved by the use of benzotriazole (BTA) and cyclohexylamine borate (CHAB). The BTA and CHAB are used alone or in combination. They are added directly to the surface of the cord or to the rubber immediately adjacent to the cord. BTA treatment will reduce surface contamination of the cord. CHAB will aid in improving aged adhesion.

Benzotriazole alone has been found to protect brass-coated steel tire cord during inventory storage and factory processing prior to rubber encapsulation. Cyclohexylamine borate increased the aged-adhesion of vulcanized brass-coated steel plus rubber composites. However, the use of BTA together with CHAB results in an optimum combination of surface protection of brass-coated steel prior to rubber encapsulation, plus adhesion promoting effects in the vulcanized brass-coated steel/rubber composite. This combined effect can be realized, for example, by dissolving BTA and CHAB together in a single alcoholic solution (e.g., methyl or ethyl alcohol) and dipping brass-coated steel into this solution, or by preparing separate solutions of BTA and CHAB and dipping the metal sequentially. The order of the sequential dipping process is not important. The metal surface must be adequately dried after dipping. The weight ratio of BTA to CHAB is normally from 1:10 to 25:1, preferably from 1:1 to 10:1. These ratios are guidelines only and in no way intended to be limitations.

The brass coating of a typical brass-coated steel cord is microscopically porous, thereby exposing small areas of steel surface to any surrounding environment. It is believed that BTA interacts with copper in a brass coating to form a polymeric complex of BTA plus copper. This polymeric complex is insoluble in most solvents and serves as a protective barrier to any environmental degradation of the underlying brass. On the other hand, CHAB, it is theorized, interacts with iron and iron oxide from steel surfaces exposed through microscopic pores to form an iron boro-ferrite compound which protects the steel. It is not necessary that the barrier layers of polymeric complexes adsorbed BTA/CHAB be extremely thick. In fact, such layers should not be so thick as to preclude all migration of copper from the brass through the protective coating to the rubber interface.

Improved Adhesion of Brass-Coated Steel Tire Cord Treated in BTA, CHAB, or a Combination of BTA and CHAB

Wire Treatment	Original Adhesion	Humidity Aged	Salt Water Corrosion Aged, hr 4	8	24
Control	100	34	35	30	25
Dipped in 0.5 wt % BTA in alcohol	126	36	60	32	29
Dipped in 2.0 wt % BTA in alcohol	107	41	73	32	30
Dipped in 0.01 M cyclohexylamine diborate in alcohol	122	42	88	71	33
Dipped in 0.01 M cyclohexylamine diborate + 2 wt % BTA in alcohol	102	39	67	31	38
Dipped in 0.01 M cyclohexylamine tetraborate in alcohol	118	41	103	56	39
Dipped in 0.01 M cyclohexylamine diborate + 0.01 M cyclohexylamine tetraborate + 2 wt % BTA in alcohol	112	43	56	35	27
Dipped in 0.01 M cyclohexylamine tetraborate + 2 wt % BTA in alcohol	101	40	83	39	30

Results: Adhesion rating (all in Compound A).

The treated wire in the above table exhibited superior overall adhesion when compared with untreated wire. Brass-coated steel cord was treated using sequential addition of the CHAB and BTA at different concentrations in alcohol (CHAB) and water (BTA), the cord being dipped first in the CHAB solution and then the BTA solution. Again the overall adhesion properties were improved. Compound A was the rubber used. The composition of rubber compound A is given above in U.S. Patent 4,269,877.

SEMICONDUCTORS

Immersion Plating Very Thin Films of Aluminum

Thin films of aluminum have become technologically very significant. For example, vacuum-deposited aluminum films in the range of 5000 Å to 2 μ in thickness are very commonly employed in semiconductor devices. This is because the aluminum is readily deposited, easily etched in fine configurations, has high electrical conductivity and is relatively inexpensive.

A major problem associated with any of these applications is the tendency for aluminum films to rapidly form a surface layer of aluminum oxide when exposed to common oxidizing ambients such as air or moisture. This oxide layer in turn interferes greatly with the establishment of metallurgical bonds between the aluminum film and almost all other metals deposited or bonded to it. This is because most other metals are incapable of chemically reducing the aluminum oxide film to aluminum, thus enabling the establishment of metallurgical bonds.

To eliminate the aluminum oxide film from thin aluminum films in order to enhance metallurgical bonding between the aluminum film and overlying metals,

L. Richardson; U.S. Patent 4,235,648; November 25, 1980; assigned to Motorola, Inc. has disclosed a two-step process: (1) chemical removal of the aluminum oxide, and (2) replacement of the outer layers of aluminum with a second metal which does not oxidize or which can readily be plated, or whose oxide is reduced by a solid state process by a third metal which is to be bonded to the aluminum film. Examples of metals whose oxides are easily reduced are silver, copper and nickel. Tin oxides and zinc oxides are readily reduced at elevated temperatures by aluminum so that either tin or zinc are suitable for the intermediate bond between a lower aluminum film and an upper aluminum film.

In the process, the removal of the surface aluminum oxide is achieved by etching at a relatively weak buffered fluoride solution followed by an immediate dip into hot water. For best results, this etch-dip process is repeated prior to the metal immersion plating. The immersion solution contains the fluoride ion which has a very high solvent capability for the aluminum surface oxide which normally precludes adhesion of coatings on thin aluminum layers. The use of the hot water treatment prior to the immersion plating appears to result in an aluminum oxide which is more susceptible to replacement than oxides formed by other preplating treatments.

An example of the use of zincation of a lower or first metal layer on an aluminum microcircuit in order to provide adherent and electrically conductive contacts to an upper conductor layer deposited through vias in an insulating layer will be given. Starting with an integrated circuit substrate having the patterned first layer aluminum metallization and an overlying insulating film which is primarily a silicon oxide and having via holes etched therethrough for the formation of contacts to the underlying layer of aluminum, the substrate was dipped for 1 second to a 50°C 4:1 solution of 40% NH_4F:49% HF followed immediately by a dip into 50°C water for 1 second. This two-fold sequence is repeated and the sample was then immersed into a zinc bath comprising 200 g $ZnSO_4 \cdot 7H_2O$, 10 ml HF and 1,000 ml H_2O. With the bath at 50°C, the plating occurred in approximately 4 seconds. A second layer of aluminum was then deposited without backsputtering and patterned to provide a contact-resistance test pattern. The contact resistance between the two layers of aluminum was measured and found to be approximately 30% higher than with the standard backsputtered surface preparation prior to the deposition of the aluminum. After sintering at about 500°C for 20 minutes, the contact resistance was reduced to a value substantially equal with the conventional backsputtered system.

The process of removing Al_2O_3 surface films from aluminum films and replacing them with metals that are readily reduced by an overlying metal to promote true metallurgical bonds are useful for many applications that require improved mechanical bonds and improved electrical contact to aluminum films.

Antioxidant Coating of Copper Parts

Heretofore, thermal compression gang bonding of semiconductive devices has been accomplished in this manner. A copper pattern of ribbon-shaped interconnect leads was coated with a nickel layer and the nickel layer was plated with gold to a thickness within the range of 30 to 60 microinches. The interconnect ribbon leads were then gang bonded to gold gang bonding bumps carried from and rising above the surface of the semiconductive device to be gang bonded. During the thermal compression gang bonding step, the nickel layer served as a barrier under the gold layer so that a gold-to-gold bond was obtained between the gold plated copper interconnect lead and the gold gang bonding bump. A similar gold-to-gold thermal compression bond was obtained at the outer end of the interconnect lead between the interconnect lead and the lead frame.

One of the problems with this gold-to-gold thermal compression bonding technique is that the expense is relatively high due to the cost of the gold employed in making the bonds.

In a process developed by *C.D. Burns; U.S. Patent 4,188,438; February 12, 1980; assigned to National Semiconductor Corporation* copper parts associated with thermal compression bonding of lead structures to a semiconductive device, such as the lead frame, the interconnect lead, and the gang bonding bumps, are coated with an antioxidant material. The antioxidant material and its thickness are chosen to be compatible with thermal compression bonding therethrough so that the completed thermal compression bond forms a bonding interface to the copper through the antioxidant coating. Suitable antioxidant coating materials include, gold, chromate, and copper phosphate.

The advantages of the process can be understood with reference to Figure 7.1.

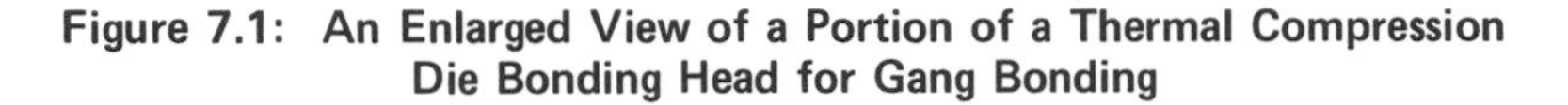

Figure 7.1: An Enlarged View of a Portion of a Thermal Compression Die Bonding Head for Gang Bonding

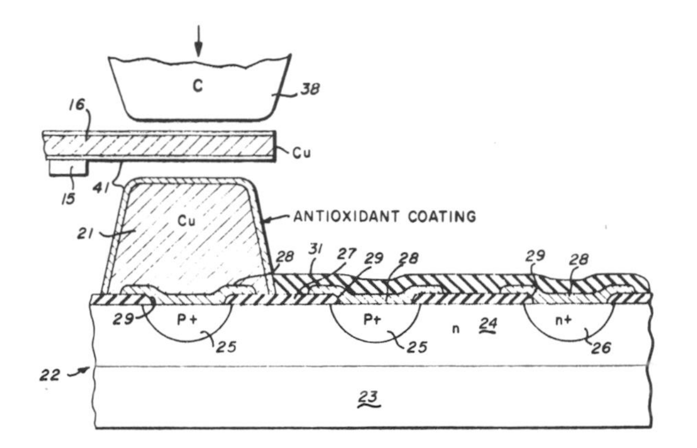

Source: U.S. Patent 4,188,438

In prior gang bonding attempts, it was found that when thermal compression bonds were attempted to be made between a copper interconnect lead **16** and a bare copper gang bonding bump **21** that relatively poor bonds were obtained due to the presence of a copper oxide layer formed on the surfaces of the copper. This copper oxide layer interferred with obtaining strong thermal compression bonds between the copper parts. However, it has been discovered that when the copper parts **16** and **21** are coated with an antioxidant coating to prevent oxidation thereof and when the antioxidant coating **41** is sufficiently thin so that a thermal compression bond is obtained through the antioxidant layer to the underlying copper, that strong thermal compression bonds are obtained to the copper. Typically, it is found that the copper-to-copper bonds obtained in this manner are stronger than the leads **16**.

Suitable antioxidant coatings **41** include gold, chromate, and copper phosphate. Such antioxidant coatings are deposited to thicknesses sufficiently thin to permit thermal compression gang bonding therethrough.

In the case of gold, the antioxidant coating **41** preferably has a thickness less than 6000 Å. The antioxidant coating permits the parts to be stored and handled without adversely affecting the thermal compression bond to be obtained thereto. The gold antioxidant coating is preferably applied, especially to the copper gang bonding bumps **21**, by the gold immerison coating process. Such a coating process is disclosed in U.S. Patent 4,005,472. Otherwise, the gold coating is applied by conventional electroplating processes.

Chromatic antioxidant coatings are applied to the copper portions by cleaning the copper portions with hydrochloric acid and then immersing the copper parts or devices having exposed copper parts in a plating solution of chromic acid mixed with H_2SO_4 acid such mixture being 2.0% chromic acid to 8.0% full strength H_2SO_4 acid to 90% deionized water by volume. The parts are immersed for 1 minute at room temperature then removed, rinsed in deionized water and dried. In this process the chromate is deposited on the copper surface to a thickness of between 10 and 100 Å.

The copper phosphate antioxidant coating is applied to the exposed copper portions by cleaning the copper portions with HCl followed by rinsing thereof in deionized water. The exposed copper portions are preferably roughened to an RMS surface roughness of about 20 microinches. The copper portions are then oxidized by heating thereof in air at 270°C for 10 minutes. The oxidized copper portions are then immersed for 15 seconds in a solution of phosphoric acid at 84% strength at room temperature. After immersion the parts are rinsed with deionized water and dried. The antioxidant copper phosphate coating applied by this process has a thickness falling within the range of 100 to 2000 Å.

ALUMINUM PRINTING PLATES

Hot Water Treatment of Aluminum Support Having Etched Anodized Surface

The work of *T. Mori and A. Ohashi; U.S. Patent 4,116,695; September 26, 1978; assigned to Fuji Photo Film Co., Ltd., Japan* relates to a method for producing an aluminum support for a printing plate in which an undesirable increase in adhesion between the support and a photosensitive material thereon with the passage of time is minimized. This is achieved by treating an anodized aluminum plate, which has been etched, with hot water or water vapor.

In general, suitable anodizing conditions are as follows: acid concentration, about 1 to 80% by wt; temperature of solution, about 5° to 70°C; current density, about 0.5 to 60 A/dm^2; voltage, about 1 to 100 V; period of electrolysis, about 30 sec to 50 min.

The resulting aluminum support having an anodized oxide coating thereon as a result of the anodic oxidation is subjected to an etching operation either by immersing the support in an acid or alkaline solution or by cathodic electrolysis in one of these solutions. The preferred range of etching should correspond to from about 5 to 75% by wt based on the total weight of the anodized oxide coating, and more suitably from about 10 to 65%.

The purpose of the process can be achieved by treating the thus provided aluminum support having an etched, anodized oxide surface coating with hot water or water vapor.

The types of water which can be used for the process include ion exchanged, distilled, natural or city water, with ion exchanged and distilled being preferred. One or more of the organic solvent, an amine compound, an organic acid or an oxyacid salt of phosphorus or boron can be added to the water in an amount preferably of 5% by wt or less, so as to promote the effect of the treatment and also to sequester the metal ions present in the water.

In the two treating methods, i.e., the treating method utilizing hot water and the other treating method using water vapor, the former not only requires simpler equipment but possesses a better operating efficiency. For the hot water treatment, a preferred temperature range lies preferably between about 85° and 110°C (for the hot water treatment, elevated pressure is not generally used; the boiling point of the water is increased to over 100°C by additives to the water) with a treating period preferably from about 15 sec to 10 min; and with a pH of the hot water preferably from about 4 to 9.5. In the case of water vapor treatment, a suitable temperature range is preferably from about 105° to 120°C. Suitable ranges for the treating period and the pH are the same as those for the hot water treatment.

Examples 1 through 8: An aluminum plate (material 2S) grained by brushing as is described in the examples of Japanese Patent Application OPI No. 33411/73 was immersed in a 10% sodium hydroxide aqueous solution kept at 55°C for 20 sec, and then rinsed with water. At this stage the aluminum surface looked grey. A clean, white aluminum surface appeared when this plate was immersed in a 70% nitric acid at room temperature (20°C) for 60 sec. Anodic oxidation of this aluminum plate was carried out by making the plate the anode and using a lead plate cathode in a 20% sulfuric acid aqueous solution (30°C) with a dc current density of 2 A/dm^2 for 3 min. The amount of the oxide coating was 3.0 g/m^2.

Then, the anodized aluminum plate was immersed in a 4% phosphoric acid aqueous solution at 70°C for 30 sec, subjected to the treatment of the process, treated with a potassium bichromate solution, a sodium silicate aqueous solution, a nickel acetate aqueous solution, a gelatin aqueous solution, or with a polyvinyl pyrrolidone, the conditions of each treatment being listed in Table 1.

Then the treated aluminum plate was coated with the following photosensitive coating composition using a whirler, and then heated to 100°C for 2 min. The coating weight was 1.7 g/m^2 on a dry basis.

	Grams
1,4-Di-β-hydroxyethoxycyclohexane-phenylene diacrylic acid condensate ($\overline{M}w$ = 8,000)	10.0
5-Nitroacenaphthene	1.0
Dibutyl phthalate	4.0
Phthalocyanine blue	2.0
Monochlorobenzene	300

Additional drying was carried out at 100°C for 5 or 30 minutes.

The photosensitive lithographic printing plate thus prepared was contact-exposed through an optically negative original using a 30 A carbon arc placed 70 cm from the plate for 30 sec, and then developed by wiping with a cotton pad soaked with the following developer.

Developer Composition	Milliliters
4-Butyrolacetone	1,000
Glycerol	100
Methyl abietate	10
Hydrogenated wood resin (Staybelite resin)	1
Wetting agent (Zonyl A)	10.2
Distilled water	100
85% phosphoric acid	25

The development performance, the time required for development, and the degree of staining at the nonimage area was evaluated at the 1,000th print obtained with a Multilith 1250 printer are shown in Table 1. Moreover, the printing plate having the support treated in accordance with the process proved capable of providing as many as 30,000 prints which were all of excellent quality, showing no quality degradation from the start of printing.

Table 1

		. . Dry at 100°C, 5 min . .			. . Dry at 100°C, 30 min . .		
		. Development .		Back-	. Development .		Back-
Ex. No.	Treatment Conditions	Perform-ance	Rate (sec)	ground Staining	Perform-ance	Rate (sec)	ground Staining
1	102°C, water vapor at 3 kg/cm² (60)*	Good	9	No	Good	15	No
2	95°C, distilled water (30)	Good	6	No	Good	12	No
3	95°C, distilled water (120)	Good	5	No	Good	11	No
4	85°C, deionized water (60)	Good	7	No	Good	13	No
5	85°C, deionized water (300)	Good	7	No	Good	14	No
6	1% triethanolamine aqueous solution, 85°C (30)	Good	9	No	Good	13	No
7	1% trisodium phosphate aqueous solution, 90°C (30)	Good	7	No	Good	10	No
8	1:1 mixture of solutions in Ex. 6 and 7, 90°C (30)	Good	8	No	Good	11	No

*Treatment time in seconds.

Antioxidant Gum Formulation

It is known to coat the surface of exposed and developed aluminum-based lithographic printing plates with a water-soluble gum composition, particularly aqueous gum arabic so as to prevent the formation of aluminum oxide due to the exposure of bare aluminum surfaces to the air. However, there are several disadvantages to the use of gum arabic for this purpose. A gum arabic solution entraps air and demonstrates excessive foaming especially in automatic developing machinery. This prevents the efficient application of gum to a plate surface.

The work of *C.D. Mukai; U.S. Patent 4,217,242; August 12, 1980; assigned to Polychrome Corporation* relates to a process of making a gum for protecting lithographic printing plates from atmospheric elements. The produced gum overcomes the high viscosity and consequential foaming problems encountered in employing such gums for lithographic purposes. The method comprises the steps of:

(1) Blending a solution of aqueous gum arabic with a mineral acid so as to adjust its pH to 2.0 or less;

(2) Heating the solution at a temperature which is less than its boiling point so as to obtain a viscosity of from about 1.5 to 10 cp;

(3) Adjusting the pH to from about 3 to 10.5 with sodium hydroxide;

(4) Adding from about 1 to 2% of formaldehyde based on a 37% formaldehyde solution;

(5) Adding from about 0.01 to 1.0% of a compatible surfactant capable of reducing the surface tension of the composition to about 20 to 35 dyn/cm at room temperature; and

(6) Filtering out suspended particles.

It is preferred that the starting material comprise an aqueous solution of gum arabic having a concentration of up to about 25% by wt. Most preferred is a concentration of from about 8 to 10%.

This aqueous gum arabic solution is then treated with sufficient mineral acid so that the pH of the resultant mixture is less than 2.0. The most preferred is phosphoric acid.

This acid treated gum arabic solution is then heated at a temperature just below its boiling point, preferably at 85° to 99°C at atmospheric pressure for a time sufficient to bring the mixture to a viscosity of preferably about 1.8 to 2.2 cp. This mixture is then adjusted to a pH of from preferably 4.5 to 4.7 by a compatible basic material, preferably sodium or potassium hydroxide.

The resultant composition is then treated with a preservative, preferably formaldehyde in an amount equivalent to from about 0.1 to 2% of a 37% formaldehyde solution. Added to this resultant product is a sufficient amount of a compatible surfactant so as to reduce the surface tension of the end material to from about 20 to 35 dyn/cm at room temperature. Typical amounts of such surfactants range preferably from 0.01 to 5% by wt of the end product material. This material is then separated of its suspended particles.

The thusly produced fluid has demonstrated superior printing plate protection properties with ease of application, drying and removal with a substantial reduction in foaming as compared to a similarly concentrated aqueous gum arabic solution without the aforementioned processing.

Example: 8.26 g of gum arabic, 90.25 g of water and 0.59% of phosphoric acid (85%) are mixed together and heated to 98°C with stirring for about 20 hours until its viscosity reaches 2.0 cp. The mixture is then neutralized to a pH of 4.5 with 0.27 g of 10 N sodium hydroxide. To this was added 0.59 g of formaldehyde (37%) and 0.05 g of surfactant FC128 (3 M). This mixture was then blended with sufficient Celite 535 to allow filtration of the suspended particles. The filtered fluid was applied to a typical lithographic printing plate which was thereby protected from oxidation for a 1-week period. At the end of this 1-week period, the gum was easily removed from the surface of the plate which produced clean, commercially acceptable reproductions.

IRON AND STEEL MATERIALS

Manufacturing a Corrosion Resistant Rotor Assembly

The work of *D.L. Henderson; U.S. Patent 4,214,921; July 29, 1980; assigned to Emerson Electric Co.* relates to a method of treating a squirrel cage rotor assembly for resisting corrosion on the exposed surfaces of the rotor assembly.

The method of manufacturing a rotor assembly involves heating a rotor core to a predetermined elevated temperature which is sufficient to permit a rotor shaft to be fitted in the bore of the core for being fixed or locked in place relative thereto by shrink-fitting, and further which is sufficient to cause at least some oxidation of the steel laminations, this oxidation being visually observable as "bluing" on the surface of the laminations. For example, the core may be heated in a gas fired heat treat furnace to a desired elevated temperature (e.g., to about 900°F or 482°C).

After being heated to its desired elevated temperature, the core is removed from the heat treat furnace and the shaft is inserted in the bore and accurately positioned with respect to the core. Then, the heated core together with the shaft inserted therein is immersed or quenched in an oil bath. Preferably, the oil in the oil bath is a water-dispersing, corrosion resistant, petroleum based, solventless oil which is maintained (i.e., cooled) to stay at or near normal room or ambient temperature. One such commercially available oil, Melcoat 4655, has been successfully used in carrying out the method.

The oil bath is maintained within a predetermined range (e.g., between about 60° and 100°F or about 15° and 30°C) so as to be approximately at ambient or room temperature. Of course, the repeated quenching of heated rotor assemblies in the oil will cause the temperature of the oil to rise beyond the above-noted preferred temperature range. Therefore, the oil bath is provided with a suitable expansion coil of a refrigeration system so as to carry away excess heat and to maintain the oil in the oil bath within its desired temperature range, and even more preferably to maintain the oil at a temperature of about 75°F (24°C). The rotor assemblies are maintained in the oil bath until they are quenched approximately to the temperature of the oil in the oil bath. During quenching, the oil in the oil bath "wicks" between laminations of the rotor core as well as covering all visible surfaces. The elevated temperature of the core seems to facilitate the wicking of the oil between the face-to-face laminations and thus the faces of the laminations are at least in part coated with the oil.

Following removal of the quenched rotor assemblies from the oil bath, a dry film will be observed to form on the exposed surfaces of the rotor assembly and on the face-to-face surfaces of the laminations. This dry film is capable of inhibiting the formation of corrosion or rust on the exposed surfaces of the rotor assembly for long periods of time even in high humidity environments.

Phosphatizing, Carburizing, Hardening and Tempering Steps

The work of *T. Bambuch, H. Wilusz, W. Englert, S. Czank and S. Surmiak; U.S. Patent 4,249,964; February 10, 1981; assigned to Huta Stalowa Wola-Kombinat Przemyslowy, Poland* provides a process for improving the strength, hardness and surface properties of a steel part by a chemical and thermal treatment of the parts which comprises the steps of phosphatizing, carburizing, hardening and tempering the parts carried out in the following sequential order: (1) phosphatizing the part by treatment with an acidic phosphatizing solution containing zinc ions and phosphate ions to produce on the part a coating ranging in thickness from 5 to 20 μ, the main ingredient of the coating being zinc phosphate; (2) carburizing the resulting phosphatized part in a gaseous carbon-containing atmosphere; and (3) hardening the part until the superficial layer is substantially all martensite, and finally tempering the part.

The gaseous atmosphere used for the carburizing step may contain some oxygen in combined form, for example, as carbon monoxide or carbon dioxide, without subsequent detriment to the surface of the part.

Example: Finished gear wheels with modulus 6 made of low alloy structural steel containing 0.2% of carbon, 0.8% of manganese, 0.3% of silicon, 0.6% of chromium, 0.5% of nickel, 0.2% of molybdenum, 0.03% of aluminum by weight, and the usual amounts of incidental ingredients such as phosphorus, sulfur and copper, were phosphatized, after previous cleaning and degreasing, by being dipped for 10 minutes, at a temperature of 97°C, in an aqueous solution having the following chemical composition: 35 g/ℓ of a salt mixture consisting of 35% by wt of manganese diacid phosphate, $Mn(H_2PO_4)_2$, and 65% by wt of ferric diacid phosphate, $Fe(H_2PO_4)_3$; 60 g/ℓ of zinc nitrate, $Zn(NO_3)_2$.

On the parts thus treated, a layer of phosphates, predominantly zinc phosphate with some additional iron and manganese phosphates, ranging in thickness from 9 to 12 μ was obtained.

After washing and drying, the gear wheels were carburized up to 0.9 mm thickness in a continuous process for 7.5 hours in an endothermal gaseous atmosphere in a zonal oven at a temperature of 920°C, the chemical composition of the gaseous carburizing atmosphere supplied from the generator being as follows: 20% of carbon monoxide, 38% of hydrogen, 0.1% of carbon dioxide and 40% of nitrogen.

The parts thus treated were next hardened by heating at a temperature of 820°C then quenched in oil and tempered for 1.5 hours at a temperature of 150°C.

After having checked the hardness and microstructure of the parts thus treated, and after having compared these results with those obtained for the carburized gear wheels which were not subjected to the preceding phosphatizing step, it was shown that the parts thus treated have a uniform high hardness throughout the whole area ranging within the limits from 740 to 780 HV, that is from 62 to 63 Rockwell hardness number. Furthermore, it also has been shown that the superficial layer of those parts contains only martensite. There were no traces of a thin defective superficial oxide layer in spite of the fact that the parts had been carburized in an atmosphere which contained some combined oxygen.

Plating with Nonferrous Metal plus Adherent Layer of Refractory Material

H.M. Maxwell; U.S. Patent 4,242,150; December 30, 1980 describes a method of forming steel reinforcing bars and other bar mill products with a protective coating of nickel to enhance the corrosion resistant characteristics of such bars.

Steel or ferrous billets are formed by any conventional process, such as a continuous casting process or forming the billet from an ingot by rolling, with each billet being formed with a predetermined length (e.g., 10 to 12 feet) and with a rectangular cross section (e.g., 4½ by 4½ inches). The particular dimensions of the billet are selected to provide a desired exterior surface having a predetermined ratio with the thickness of the final coating formed on the bar.

The entire exterior surface of each billet is then plated with a layer of protective metal, such as nickel, by any conventional plating process, preferably dull Watts electroplating, such layer having a thickness of about 28 mils, depending on the final product coating specified. It should be noted here that the protective nickel is a relatively expensive metal, and careful consideration must be given to all factors which might result in undue losses of this metal during the forming process.

The nickel plated billets are then individually covered with an adherent layer of a refractory material that will withstand high furnace temperatures, generally up to approximately 2350°F. A refractory material found to be particularly suitable for use in this process is a brick fire mortar composed of the following ingredients: silica, 82 to 87% by wt; aluminum, 3 to 6% by wt; ferrous oxide, 0.3 to 0.4% by wt; calcium oxide, 0.05 to 0.1% by wt; magnesium oxide, 0.05 to 0.1% by wt; titanium oxide, 0.5 to 1.5% by wt; and potassium sodium oxide, 2.3 to 2.8% by wt.

Before being applied to the billets, the refractory composition is preferably mixed with water at a ratio of about 1 part water to 5 parts refractory material (by volume) to form a slurry, and this slurry is then applied directly to all surfaces of the billet by any convenient method such as spraying, brushing or rolling, until the refractory material has a thickness of about 1/16 to 3/16 inch. The refractory material is suited to use in the method because it has a low volume loss (e.g., 5 to 6%) during heating of the billet at furnace temperatures up to 2350°F, and it is reasonably fast setting while also having adherence capabilities which assures that the integrity of the refractory covering is sustained during the movement of the billets through the furnaces. Further addition of charcoal granules may be added to provide a degree of local reducing carbon monoxide gas during part of heating cycle to further limit oxidation along with the hydrogen released from the slurry compound.

After the plated bars have a refractory covering applied to the exterior surfaces thereof, they are heated in a conventional hearth-pusher type furnace which is fired at both its top and bottom with zone heat control designed for three separate zones, namely a preheat zone having a preferred temperature of about 1600°F, a soak zone having a preferred temperature of about 1800° to 2000°F, and a final heat zone having a preferred temperature of 2000° to 2350°F. The billets are continuously pushed through the furnace on water-cooled rails in the first two furnace zones and on the bottom of the furnace in the last zone. To facilitate the pushing of the billets through the furnace, they are arranged with their lengthwise axes in parallel relation, and with each billet in direct contact or abutting relationship with the two adjacent billets on each side whereby when the billets at the leading or entry end of the furnace are pushed in a direction transversely to the lengthwise extent of the abutting line or billet, the entire line is moved serially through the furnace. Thus, a continuous heating process is established by constantly adding new billets at the entry end of the furnace and taking off fully heated billets at the exit end of the furnace after the billets have been pushed through all of the aforementioned stages of the furnace. Preferably, the cycle time for any billet moving through the furnace is approximately 1 hour, and when the billet is removed to a point that it is malleable. The cycle time of 1 hour, while generally desirable, may be varied substantially (e.g., extended to 2 hours) if desired. The degree of diffusion is primarily achieved in the 2000° to 2300°F zone at a rate approximately 10% per mil per half-hour.

The malleable billet is then hot rolled, using conventional rolling equipment, to reduce the thickness of the billet, and preferably to change its shape from one having a rectangular cross section of 4½ by 4½ inches to a bar having a typical circular cross section with a diameter of about ⅝ inch. During this rolling process, the nickel plating is distributed generally uniformly about the cylindrical exterior surface of the product and forms a protection layer of wrought nickel approximately 1 to 2 mils in thickness. Additionally, an iron-nickel alloy interface approximately 0.1 to 0.2 mil in thickness is formed between the ferrous core and the nickel to improve substantially the corrosion resistant characteristics of the finished bar. Normal handling of the billet, after it has been heated, will cause some, if not all of the refractory material to be separated from the billet, but conventional descaling equipment may be used, if necessary or desirable, to remove mechanically any lingering refractory material from the billets prior to the rolling thereof.

Composition for Priming Oxidized Steel Structures

The method of *G.T. Shutt; U.S. Patent 4,071,380; January 31, 1978* is adapted for treating, as by priming, oxidized steel substrates as customarily encountered in the form of underground pipes; underwater structures; aboveground structures, and the like. Fundamentally, the method resides in applying a partially prehydrolyzed alkyl silicate to an oxidized steel substrate, and to permit the alkyl silicate to cure or harden by becoming 100% hydrolyzed. The prehydrolysis would desirably be within the range of 40 to 98% and the solution may be applied to the substrate in any convenient manner, such as by troweling, brushing, spraying, or dipping. The solution is applied in sufficient quantity to saturate the existent oxidized film, with the state of saturation being readily determined visually. The hydrolysis may be completed as from ambient moisture with the more humid environment obviously promoting more rapid curing. Upon curing the alkyl silicate provides an inorganic film which causes the oxidized steel film to be stably maintained upon the substrate and to become an integral part of a rigid barrier protective of the substrate.

Among the alkyl silicates useful for the process are methyl silicate, ethyl silicate, propyl silicate, isopropyl silicate, butyl silicate, hexyl silicate, 2-ethylhexyl silicate, tetramethyl orthosilicate and tetraethyl orthosilicate. The alkyl silicates are of the character wherein each of the alkyl groups contains 1 to 8 carbon atoms but with the most common of these being the ethyl silicates.

Although this method may be practiced by utilization of the alkyl silicate solution alone, it has been found that the intermixing of dry particulate, inert matter with the prehydrolyzed alkyl silicate brings about certain additional desired results. Thus, such matter may have sufficient pigmentation so as to conduce to a more facile visual determination of the point of saturation as saturation will be evident when such matter constitutes a relatively solid surface coloration thereby signaling that the requisite quantity has been applied. Additionally, the usual oxidized metal surface is rough in texture so that such dry particulate matter serves as a filler to smooth such roughness. Such matter would be preferably within a mesh size of about, desirably, -325, as the finer the particle size the more adherent the developed coating.

The following are exemplary of dry particulate matter of this type: silica, talc, mica, kaolin, bentonite, asbestos, fireclay, aluminum oxide, zircon, ferric oxide, tin oxide, titanium oxide, chromium oxide, carbon, copper oxide, and metal pigments such as copper, aluminum, stainless steel, etc., as well as zinc in quantities less than required to cause galvanic reaction with the steel substrate.

To determine the amount of dry particulate matter for coloring and filling purposes, it has been found that 1 part of such matter to as much as 100 parts of the alkyl silicate solution, by volume, with the latter being prehydrolyzed within the range of 40 to 98% would be effective for an oxidized steel layer thicker than 0.003 inch, whereas 1 part particulate matter to as much as 25 parts of alkyl silicate solution, by volume, in the aforesaid prehydrolyzed range, is useful for oxidized metallic layers having thicknesses between 0.001 and 0.003 inch. With oxidized layers of less than 0.001 inch, the volume mixture may be as much as 1 part particulate matter to 1 part of the alkyl silicate solution. The ratios above set forth assure sufficient liquid solution to penetrate the oxidized metal layer without depriving the coating provided by the particulate matter of adequate solution, which deprivation could result in a chalking or dustlike texture of the mixture upon curing.

The method also comprehends the addition to the solution of certain corrosion-inhibiting compounds capable of producing ions for reacting with the metal substrate. To effect appropriate protection, such ions must be available under oxidizing conditions at the steel substrate and by reason of the cathodic metal oxide, such as rust, on the steel substrate, the appropriate oxidative environment is provided.

Corrosion controlling compounds useful for this purpose are from the class consisting of inorganic water soluble phosphates, such as sodium phosphate, calcium phosphate, potassium phosphate, ammonium phosphate, and magnesium phosphate.

Exemplary of the structures amenable to treatment by this process are the following: freshwater and saltwater barges (above and below water); decks, fittings, machinery; off-shore structures, cabins, bulkheads, masts, etc., on ships, steel stacks, boiler breaching, bridges, tank farms, boot tops on ocean freighters, steel piling in brackish water, chemical equipment, structural steel, food plant floors, walls, ceiling, galvanized structures, railroad cars, etc.

Simultaneous Neutralization and Passivation of Acid Treated Boiler Surfaces

W.P. Banks and L.D. Martin; U.S. Patent 4,045,253; August 30, 1977; assigned to Halliburton Company describe a method of removing deposits, including metallic oxide-containing deposits, from ferrous metal surfaces. More particularly, it relates to a method of passivating freshly cleaned ferrous metal surfaces, such as the internal surfaces of boilers, feed water heaters, heat exchangers and similar equipment.

The passivating solution is an aqueous solution containing a base, an oxygen containing gas, and a material selected from the group consisting of hydrazine, an iron complexing agent and mixtures thereof.

The alkaline solution which contains hydrazine and/or an iron complexing agent provides excellent passivating results at temperatures greater than about 220°F and up to temperatures wherein the chemicals utilized in the solution, specifically the organic chemicals, commence to degrade.

A passivating solution having high temperature operability is particularly useful in equipment, for example natural circulation boilers, wherein circulation of the solution in the equipment is caused by heating rather than by pumping. The passivating solution is therefore highly useful as a boil-out solution, that is, as a passivating solution, for natural circulation boilers.

The passivating solution of this process, containing a mixture of hydrazine and an iron complexing agent, provides not only satisfactory passivation at high temperatures, but in addition such a solution provides a tightly adherent film and also permits the elimination of the rinsing steps and the need for nitrogen blankets. The quantity of hydrazine useful herein is in the range preferably from about 0.025 to 0.05% hydrazine by total weight of solution.

Specific examples of iron complexing agents useful herein include 1-hydroxyethylidine-1,1-diphosphonic acid, the sodium salt of ethylenediaminetetraacetic acid, sodium gluconate and mixtures thereof. The quantity of iron complexing agent is in the range preferably from about 0.1 to 0.3% complexing agent by weight of solution.

In preparing the passivating solution, it is preferred that the base material be added to the water to form a base solution; thereafter, add the hydrazine to the base solution or the complexing agent to the base solution. Where the passivating solution contains both hydrazine and a complexing agent, it is desired, but not required, that the complexor be added to the base solution prior to the addition thereto of the hydrazine.

Experimental Procedure: (1) Remove mill scale, corrosion products and other deposits from all surfaces of a 1 x 2 x ⅛ inch mild steel coupon by striking the surfaces of the coupon with a high velocity stream of grade 625F glass beads and thereafter soaking the bead-blasted mild steel coupon for 10 minutes in a bath containing 5% hydrochloric acid solution which is maintained at a temperature of of 75°F.

(2) Remove the acid-cleaned coupon from the acid bath and permit the acid to drain from the coupon under room atmosphere for 5 minutes.

(3) Rinse the coupon with approximately 400 ml of deionized water.

(4) Repeat step (3) using an additional 400 ml of deionized water.

(5) Permit the rinse water to drain from the coupon under room atmosphere for about 5 minutes.

(6) Place the acid-cleaned coupon in a glass container. One edge of the coupon rests on the bottom of the container and one edge of the coupon leans against the side of the container. Introduce a quantity (about 80 ml) of the passivating solution into the glass container sufficient to immerse and fully cover the coupon with the solution. Place a cover on the glass container which does not seal the

container. Place the covered container which holds the coupon and test solution into a second container which is capable of withstanding high internal pressure. Introduce a sufficient additional quantity (about 25 ml) of the passivating solution utilized in the glass container into the annulus space between the first container and the second container such that the additional solution surrounds the exterior of the glass container, but does not enter the covered glass container. Seal the second container and apply heat thereto until the temperature on the interior thereof is about 350°F. The 350°F temperature is then maintained for a period of 6 hours.

(7) At the end of the 6 hour time period, place the second container into a water bath which is maintained at a temperature of approximately 75°F. When the interior temperature of the second container is approximately 200°F, unseal the second container and remove the coupon from the first container. Permit the passivating solution to drain from the coupon under room atmosphere for 5 min.

(8) Place the coupon into an air chamber in which the air is approximately saturated with water vapor and which is maintained at a temperature of 150°F. Permit the coupon to remain in the high humidity chamber for 4 days. Remove the coupon and observe the extent of rusting of the coupon surfaces.

The passivating solutions utilized in the tests all consist of chemicals dissolved in water. The specific chemicals utilized in each test are those identified in the table below having numbers in a column headed by a chemical. The numbers indicate percent chemical by total weight of solution. For example, in the table, Run No. 1, the solution utilized consists of 1% sodium hydroxide by weight of solution and 0.05% hydrazine by weight of solution. The remainder of the solution is water.

Because no attempt is made to exclude the presence of air from the test solutions or from the atmosphere within the first and second containers, the solutions contain air under the test pressures and temperatures.

For purposes of comparison, a mild steel coupon is prepared, as set out above in Steps 1, 2, 3, 4, 5 and 8. The extent of rusting of the thus treated coupon is observed to be heavy, which carries the numerical value of 8 as is further explained below.

In the table which follows, the observed degrees of rusting are identified by number according to the following schedule: 0 = none; 1 = none to trace; 2 = trace; 3 = trace to light; 4 = light; 5 = light to moderate; 6 = moderate; 7 = moderate to heavy; 8 = heavy; 9 = heavy to very heavy; and 10 = very heavy.

Run No.	NaOH	Na_2CO_3	$Na_2B_4O_7$	TEA*	Chemical A**	Na_4 EDTA***	Na Gluconate†	N_2H_4	$NaNO_2$	Extent of Rusting
1	1.0	–	–	–	–	–	–	0.05	–	1
2	–	1.0	–	–	–	–	–	0.05	–	2
3	0.5	0.5	–	–	–	–	–	0.05	–	2
4	–	–	–	0.3	–	–	–	0.05	–	2
5	–	–	–	0.1	–	–	–	0.05	–	2
6	0.5	0.5	–	–	–	0.25	0.25	–	–	2
7	–	–	–	–	0.25	–	0.25	0.05	–	2

(continued)

Run No.	NaOH	Na_2CO_3	$Na_2B_4O_7$	TEA*	Chemical A**	Na_4 EDTA***	Na Gluconate†	N_2H_4	$NaNO_2$	Extent of Rusting
8	0.1	0.5	–	–	0.1	–	–		–	3
9	–	0.5	–	0.3	–	–	–	0.05	–	4
10	0.5	–	–	0.3	–	–	–	0.05	–	4
11	1.0	–	–	–	0.25	–	–	0.05	–	4
12	–	–	1.0	–	–	–	–	0.05	–	4
13	–	0.5	0.5	–	–	–	–	0.05	–	4
14	–	–	0.5	0.3	–	–	–	0.05	–	4
15	–	–	–	–	–	0.25	0.25	0.05	–	4
16	–	1.0	–	–	–	–	–	–	–	4
17	0.5	0.5	–	–	0.25	–	0.25	–	–	4
18	1.0	–	–	–	0.25	–	–	–	–	4
19	–	–	–	0.1	–	–	–	–	–	5
20	–	–	–	–	–	0.25	0.25	–	0.5	5
21	–	–	–	–	0.25	–	0.25	–	0.5	5
22	–	1.0	–	–	–	–	–	–	0.5	5

*Triethanolamine.
**1-Hydroxyethylidene-1,1-diphosphonic acid.
***Tetrasodium salt of ethylenediaminetetracetic acid.
†Sodium salt of gluconic acid.

The above table clearly shows the contribution of hydrazine in obtaining good passivating results and the improvement of this process over those passivating solutions containing nitrite. Runs 1 through 15 and 17 and 18 demonstrate the use of solutions within the scope of this process. Runs 16 and 19 through 22 demonstrate passivating solutions utilized in the prior art when run under elevated temperatures. The table shows that the use of a passivating solution containing a base, plus hydrazine, in the presence of dissolved air with or without the presence of a complexing agent is effective in the simultaneous passivation and neutralization of mild steel at elevated temperatures.

The table also shows that a passivating solution containing a base and a complexing agent in the presence of dissolved air is effective in simultaneously neutralizing and passivating mild steel at elevated temperatures.

It is also shown that a passivating solution containing a base and a nitrite is not effective in passivating mild steel at elevated temperatures.

Comparing Runs No. 1, 2, 3, 4 and 5 clearly indicates the equivalency of the inorganic base, sodium hydroxide, the salt, sodium carbonate, and the organic base, triethanolamine, when combined with hydrazine and air for the purpose of passivating acid-cleaned steel. Such equivalency is unexpected and surprising.

Cleaning Tin-Plated Steel Cans

As the result of an extensive study on a cleaning process for tin-plated steel cans after molding, *M. Kimura, T. Sobata and H. Wada; U.S. Patent 4,265,780; May 5, 1981; assigned to Nippon Paint Co., Ltd., Japan* found that when an alkaline solution containing phytic acid or its derivative and having a specific pH value is used, the surfaces of such cans can be well cleaned with prevention of dissolution of tin.

The process for cleaning tin-plated steel cans comprises washing the surface of a tin-plated steel can with an alkaline solution comprising at least one of the esters of myoinositol with 2 to 6 molecules of phosphoric acid and their alkali metal salts, alkaline earth metal salts, ammonium salts and amine salts in a concentration of at least 0.05 g/ℓ and having a pH of 8 to 13.

The esters of myoinositol with 2 to 6 molecules of phosphoric acid may be, for instance, myoinositol diphosphate, myoinositol triphosphate, myoinositol tetraphosphate, myoinositol pentaphosphate or myoinositol hexaphosphate. Among them, the ester with 6 molecules of phosphoric acid corresponds to phytic acid, which is contained in various kinds of plants including grains such as rice, barley, soybean and corn. This is a quite harmless substance and has been employed in the canning industry as an additive for prevention of discoloration of canned foods and of smelling and corrosion of cans. The esters with 2 to 5 molecules of phosphoric acid are partially hydrolyzed products of phytic acid. The phytic acid and its hydrolyzed products may be employed in the form of various salts such as alkali metal salts (e.g., sodium salt, potassium salt), alkaline earth metal salts (e.g., calcium salt, magnesium salt, barium salt), ammonium salt and amine salts. The phytic acid, its hydrolyzed products and their salts are hereinafter referred to as phytic acid compounds.

The concentration of the phytic acid compound is preferably from 0.05 to 50 g/ℓ of the alkaline solution.

Examples 1 to 6 and Comparative Examples 1 to 4: A tin-plated steel can body (amount of plated tin, 11.2 g/m^2 on one surface) obtained by the DI molding process was washed by spraying thereto a base cleaning solution as shown below (No. 1 to 4) incorporated or not with phytic acid under the following conditions: temperature, 50° to 70°C; spraying time, 50 seconds; spraying pressure, 3 kg/cm^2. The resultant can was subjected to the test for cleaning efficiency and to the aqueous sodium chloride-immersion test.

Using the same can and the same washing solution as above, cleaning was effected under the same conditions as above but adopting a spraying time of 20 minutes instead of 50 seconds, and the amount of corrosion per each can (surface area: about 470 cm^2) was calculated from the difference of the weights of the can (mg/can) before and after spraying.

	Grams per Liter
Base cleaning solution No. 1	
Soda ash	4.5
Sodium tertiary phosphate	4.5
Emulgen 910	0.5
Purlonick L-61	0.5
Base cleaning solution No. 2	
Sodium metasilicate	1.5
Soda ash	0.5
Sodium bicarbonate	0.5
Sodium tripolyphosphate	1.0
Surfonic LF 17	0.2
Base cleaning solution No. 3	
Sodium metasilicate	4.0
Soda ash	2.0
Sodium bicarbonate	2.0

(continued)

	Grams per Liter
Sodium tripolyphosphate	3.0
Surfonic LF 17	0.5
Base cleaning solution No. 4	
Sodium bicarbonate	6.0
Sodium tertiary phosphate	1.0
Sodium secondary phosphate	1.0
Emulgen 910	0.4
Purlonick L-61	0.2

Test for cleaning efficiency – The can subjected to cleaning was immediately washed with water and allowed to stand at room temperature for 90 seconds, and the cleaning efficiency was evaluated by the wettability (%) on the can surface calculated according to the following equation:

$$\text{Wettability (\%)} = \frac{\text{Surface area wetted with water}}{\text{Total surface area}} \times 100$$

Aqueous sodium chloride-immersion test – The can subjected to cleaning was immersed into a 5% w/v aqueous sodium chloride solution at 25°C for 30 minutes, and the appearance of the can surface was observed: (a) no abnormality observed; (b) partial rusting observed; (c) considerable amount of rust observed.

The results of these tests are shown in the following table.

Ex. No.	Alkaline Washing Solution: Base Cleaning Solution	Alkaline Washing Solution: Phytic Acid Added (g/ℓ)	pH	Washing Temp. (°C)	Test for Cleaning Efficiency Wettability (%)	Aqueous Sodium Chloride-Immersion Test	Corrosion After Spraying 20 min (mg/can)
1	No. 1	1	10.1	50	98-100	b	52
2		2	10.1	50	98-100	b	24
1*		0	10.1	50	98-100	c	158
3	No. 2	1	10.0	70	98-100	a	2
2*		0	10.0	70	98-100	c	85
4	No. 3	0.5	11.0	50	98-100	–	3
5		1	11.0	50	98-100	a	2
3*		0	11.0	50	98-100	a	27
6	No. 4	1	9.7	60	95-98	a	10
4*		0	9.7	60	95-98	b	34

*Comparative example.

ADDITIONAL PROCESSES

Ground Cutting Surface Treatment with Chlorhexidine

T. Morgans; U.S. Patent 4,201,599; May 6, 1980 provides a method of restoring and/or protecting a used cutting surface which comprises bringing the cutting surface into contact with a solution or dispersion of a chlorhexidine compound, which solution or dispersion has dissolved therein a surfactant.

The chlorhexidine compound can be used in extremely low concentrations, and will usually be present in 0.001 to 3% by wt concentration.

Chlorhexidine, which is the common name for 1,6-di(4-chlorophenyldiguanido) hexane has been employed to useful effect in the treatment of metal cutting surfaces as a wide variety of its salts. The digluconate, diacetate and dihydrochloride compounds are particularly readily available commercially and of these, the digluconate, usually termed simply chlorhexidine gluconate, has been found to be particularly effective because of its good solubility in water. Aqueous chlorhexidine gluconate solutions having a concentration of from 0.5 to 1.3% by wt are suitable, a 1% solution being particularly preferable.

Although both anionic and cationic surfactants may be employed, the use of cationic surfactants is preferred, especially when the cutting surface being treated belongs to an implement for surgical use or a razor blade. Cationic surfactants dissociate in solution into a relatively large and complex cation which is responsible for the surface activity and a smaller inactive anion. The cation usually contains a pentavalent nitrogen atom which is often present as a quaternary ammonium group. In addition to possessing the emulsifying and detergent properties usually associated with surface-active agents, cationic surfactants have marked bactericidal activity against both gram-positive and gram-negative organisms. Quaternary ammonium compounds are preferred for use with surgical instruments, in particular because, inter alia, they combine readily with proteins.

Cetrimide is the preferred cationic surfactant for use in this process. The cationic surfactant may be used in a concentration of from 0.5 to 4% by wt, preferably about 3% by wt.

Example 1: An aqueous solution was prepared containing 1% by wt of cetrimide, 0.02% by wt chlorhexidine gluconate, 7% by wt of isopropyl alcohol, and small amounts of tartrazine (as coloring agent) and perfume. This solution was suitable for use in the restoration of blunted razor blades. Used razor blades which had been used, untreated, until they had become too uncomfortable to use were immersed in this solution for 20 minutes. After such treatment, they could be used in comfort for up to 65% of their original useful life.

Example 2: An aqueous solution was prepared containing 3% by wt of cetrimide, 1% by wt of chlorhexidine gluconate, 7% by wt of isopropyl alcohol and small amounts of tartrazine (as coloring agent) and perfume.

Tests were carried out using mild steel cutting blades normally employed in the cutting of Idem noncarbon copying paper. Each sheet of this paper is formed of the following layers: paper; China clay; microcapsules encapsulating a coloring dye; China clay; and paper.

In view of the composition of the paper, it is not surprising that this paper will blunt the edge of a cutting blade very quickly. Two cutting blades are employed in machines used for cutting Idem. These have to be sharpened in normal use when employing four layers of Idem at a time, every five days.

Two of these blades which had ceased to cut effectively were treated with the chlorhexidine gluconate solution employed in Example 2 and replaced in the machine. The blades were 7 feet 6 inches in length, 1 inch thick tapering to a knife edge up to 2 inches and weighed about 40 pounds each.

It was found that, after this treatment, the blades could be employed to good effect in cutting eight thicknesses of Idem instead of four for five weeks.

An apparatus for use in the testing of cutting surfaces for their cutting efficiency is also described in this patent.

Composition for Coating Metal Working Tools

Tools used in machining, drawing and general metal forming, cutting and other working operations are customarily lubricated with liquid lubricants in the form of straight oils or emulsions. Tool life however is not always satisfactory.

According to *D. Hartley, M.D. Barrett and P. Wainwright; U.S. Patent 4,243,434; January 6, 1981; assigned to Rocol Limited, England* a metal working tool, particularly a metallic tool, is provided with an adherent solid film comprising a molybdenum disulfide, graphite or other solid lubricant and a phosphate salt binder preferably containing a chromate also. Preferably the coating is 0.0001 to 0.001 inch thick.

The total composition may be as follows (parts by weight):

Orthophosphoric acid (calculated as the 88 to 93% acid)	3 to 25
Aluminum phosphate (calculated as aluminum hydroxide) or molar equivalent amount of other metal phosphate	1 to 5
Chromate calculated as chromium trioxide	0 to 5
Molybdenum disulfide, graphite or other solid lubricant	5 to 60

The composition as used preferably contains 50 to 80 parts of water total but may be sold with or without all or part of the water. A surface active agent, for example Texafor 85FP, is preferably present, for example 0.1 to 2 parts on the same basis.

Application may be by spray, dip, brush or other method, the dispersion desirably being kept thoroughly agitated. Application is followed by heat curing to improve adhesion, for example an hour or more at 200° to 250°C, or 3 hours or more at 180°C. A typical final coating thickness is 0.0001 or 0.0002 inch, but thicker coatings, for example up to 0.001 inch, can be obtained if required by multiple applications with intermediate drying. Air drying, for example 10 minutes at 20° to 25°C, or 2 to 3 minutes at 60° to 80°C, is suitable. The dispersion itself is preferably kept at 15° to 30°C during application and tools to be coated may be warmed if desired, for example to 60° to 80°C to speed drying. The final coating is heat cured as before. Neither drying nor curing temperatures are critical.

The tool should be clean before coating, but simple solvent washing or vapor degreasing, for example in Genklene (1,1,1-trichlorethane), is sufficient.

The following user tests illustrate the process. To carry them out the following aluminum phosphate solution was made up: 55.80 pbw orthophosphoric acid (88 to 93%); 12.31 pbw aluminum hydroxide; and 31.89 pbw distilled water. This solution was used to make up the following specific composition, again expressed in parts by weight.

Specific Composition	Parts by Weight
Monoaluminum phosphate solution	24.7
Chromium trioxide	1.3
Molybdenum disulfide powder	20.0
Texafor 85FP dispersing agent	0.4
Distilled water	53.6

The composition was then used as follows.

Example: Aluminum L93 and L94 forgings are milled with a double-flute router or slotter 2 inches in diameter and 3 inches long, made of high speed steel, at a cutting speed of 1,800 ft/min (3,500 rpm) taking a cut 2 inches wide and ½ inch deep. The forgings require approximately 24 hours machining time on an NC milling machine such as a Marwin Maximill Vertical 2 Spindle Head. With flood lubrication using a mineral oil emulsion such as Castrol Almasol A at 20:1 dilution, four components are normally milled before regrinding of the tool.

After coating the tool with the specific composition given earlier and with one spray coat dried 1 hour each at room temperature, 80° and 250°C, 15 or 16 such components can regularly be milled before regrinding is required.

Solder Cleaning and Coating Composition

J.M. Preston; U.S. Patent 4,014,715; March 29, 1977; assigned to General Electric Company provides a cleaner for the removal of oxidation and foreign contaminants from plated metal such as solder plated printed circuits, as well as for the deposition of a nonoxidizable film on the metal surface.

Excellent results were obtained in the cleaning of solder plate. It was found that the cleaning of solder plated printed circuits after etching makes possible reflowing the solder plating without a change in composition due to oxidation of the solder after etching, or from residues left on the solder from the copper etch. If cleaning is not done prior to reflowing, the solder may not melt or if fused a frosted coating may appear on the solder due to impurities. This composition inhibits the redeposition of metals on solder plate. This redeposition is further prevented in a heated solution by depositing or reacting with the cleaned solder forming a phosphate coating on the solder surface. The phosphate coating consists of complex polyphosphates. The exact composition has not been determined because of the micro amounts present.

The treating composition contains a mineral acid, water, thiourea, a wetting agent, superphosphoric acid concentrate having a P_2O_5 content of 72 to 80% and a dibasic acid such as succinic and others having the general formula $C_nH_{2n-2}O_4$.

In a gallon of the composition the quantity of the mineral acid used may vary as follows: sulfuric, 200 to 500 ml; fluoboric, 200 to 400 ml; and hydrochloric, 200 to 375 ml.

The quantity of the superphosphoric acid concentrate added to the cleaning composition may vary from 38 to 567 ml. Thiourea or its derivatives may vary from 55 to 76 g, and the wetting agent from 3.8 to 7.5 ml.

The additional succinic acid (0.1 to 1.0 g) may be added to the original superphosphoric acid solution or may be added separately to the composition batch.

The superphosphoric acid is prepared by heating a phosphoric acid containing 54.5% P_2O_5 with 2 to 5% of a monosaccharide or polysaccharide such as sucrose, maltose, glucose, fructose, lactose and mannose at a temperature of 300°F to concentrate the mixture to 72 to 80% P_2O_5. This concentrate contains about 0.2 g of succinic acid and small amounts of organic phosphates and succinates. Any phosphoric acid may be used that has a P_2O_5 content between 72 and 80%. It is preferred to use distilled water or deionized water in the composition so as not to alter the phosphorus content.

The wetting agent or surfactant may comprise Triton X 100, which is the octyl phenyl ether of a polyethylene glycol containing 9 to 10 ethoxy groups per molecule. Other types of surfactants may be used.

Example: The following ingredients were used to make up one gallon of the cleaning composition:

	Minimum	Maximum
Sulfuric acid	200 ml	500 ml or
Fluoboric acid	200 ml	400 ml or
Hydrochloric acid	200 ml	375 ml
Superphosphoric acid	38 ml	567 ml
Succinic acid	0.1 g	1.0 g
Thiourea	55 g	76 g
Wetting agent or surfactant	3.8 ml	7.5 ml
Water	. . . . . . .remainder	. . . .

All of the above ingredients were mixed while agitating and applied to the surface to be cleaned.

Preventing Crevice Corrosion in Chemical Apparatus

Titanium and titanium alloys find extensive use as the anticorrosion material used in chemical plants or like apparatus. It is used in severely corrosive environments or of component parts of such apparatus. However, where nonoxidized solutions such as hydrochloric acid solutions are handled, active dissolution of titanium occurs. Also, where chloride solutions at high temperatures are handled, the problem of abnormal corrosion of inner interstitial parts of apparatus or device corrosion has not yet been solved.

K. Shimogori, H. Sato and H. Tomari; U.S. Patents 4,154,897; May 15, 1979; and 4,082,900; April 4, 1978; both assigned to Kobe Steel, Ltd., Japan provide a chemical apparatus, in which sufficient prevention of crevice corrosion can be expected even under very severe conditions. This is achieved by providing a mixed oxide layer composed of an oxide of a platinum group element and an oxide of an anticorrosion metal on the surface of the titanium material of the chemical apparatus at least over those areas constituting the interstitial portions. The oxide layer is over an area no less than 0.001, more preferably no less than 0.002, of the titanium material surface. The oxide layer comprises a thickness no less than 0.01 μ, more preferably no less than 0.1 μ. The molar ratio of the platinum group element oxide to the anticorrosion metal oxide in the mixture is within the range of from 1:99 to 95:5, more preferably from 10:90 to 95:5. The oxide layer is provided through thermal treatment in an oxidizing atmosphere at a temperature ranging from 500° to 700°C and for a period ranging from 10 to 30 minutes.

Example 1: Pure titanium pieces 2-mm in thickness were subjected to sand blast treatment and then washed with hydrochloric acid, and then they were covered with respective PdO/TiO_2 mixture layers of compositions listed in Table 1. The resultant wafers were then individually coupled to pure titanium to prepare samples A. Also, there were prepared sample B by coupling PdO coated Ti to Ti; sample C by coupling Pd coated Ti to Ti; sample D by coupling Pd to Ti; sample E of the sole Ti-0.15% Pd alloy; and sample F of the sole Ti. Table 1 shows the results of measurements of the corrosion weight loss and hydrogen absorption of these samples, as measured after immersing them in boiling liquid containing 10% sulfuric acid for 20 hours.

Table 1

Sample Structure	Corrosion Weight Loss ($mg/15\ cm^2$-20 hr)	Hydrogen Absorption (ppm)
(A) Ti coupled with PdO/TiO_2 (1/99) coated Ti	4.1	6
Ti coupled with PdO/TiO_2 (10/90) coated Ti	4.1	0-3
Ti coupled with PdO/TiO_2 (30/70) coated Ti	4.0	0-3
Ti coupled with PdO/TiO_2 (95/5) coated Ti	4.0	0-3
(B) Ti coupled with PdO coated Ti	25.0	20
(C) Ti coupled with Pd coated Ti	27.0	28
(D) Ti coupled with Pd	25.0	10
(E) Ti-0.15% Pd alloy (alone)	32.5	36
(F) Ti (alone)	1,120	640

Note: In the coupled samples the area ratio of Ti to coupled material is 10:1. The proportions of PdO and TiO_2 in the samples (A) are in mol %. Figures of the hydrogen absorption in the samples (A), (B) and (C) represent the hydrogen absorption in noncoated Ti.

It will be seen from Table 1 that the corrosion weight loss and hydrogen absorption are least with the samples A according to the process.

Example 2: Mixture oxide coated titanium samples were prepared by using platinum group element oxides other than PdO, and corrosion resisting metal oxides other than TiO_2, and their anticorrosion and hydrogen absorption preventive property were measured after immersing the samples in various boiling liquids containing 5 to 10% sulfuric acid for 20 hours to obtain results as shown in Table 2. (The molar ratio between the platinum group element oxide and corrosion resisting metal oxide was set to 1:1, and the area ratio between coated portion and noncoated portion was also set to 1:1.)

Table 2

Sample	Corrosion Weight Loss ($mg/15\ cm^2$)	Hydrogen Absorption (ppm)
PtO/TiO_2	4.1	0-3
RuO_2/TiO_2	4.2	0-3
IrO_2/TiO_2	4.5	0-3
RhO_2/TiO_2	4.0	0-3
OsO_2/TiO_2	6.4	0-5
PdO/Ta_2O_5	4.0	0-3
PdO/ZrO_2	4.1	0-3
PdO/Nb_2O_5	4.1	0-3
Contrast PdO/TiO_2	4.0	0-3

It will be seen from Table 2 that both corrosion resistance and hydrogen absorption resistance were pronounced in all samples except for the sample of OsO_2/TiO_2, in which slightly high values resulted.

Example 3: Square pieces of titanium material, 25 mm long on each side and 1 mm in thickness, were covered over the entire surface with a mixture oxide layer which is shown in Table 3 and then coupled by galvanic coupling to nontreated Ti plates of the same size.

These samples were immersed in boiling 10% sulfuric acid solution for 20 hours, and then the corrosion weight loss and hydrogen absorption of their nontreated Ti were measured to obtain results as shown.

Table 3

Sample	Corrosion Weight Loss ($mg/15\ cm^2$-20 hr)	Hydrogen Absorption (ppm)
PdO 30/PtO 20/TiO_2 50	4.3	0-3
PdO 30/RuO_2 20/TiO_2 50	4.2	0-4
PdO 70/RuO_2 10/TiO_2 20	4.3	0-3
PtO 40/IrO_2 20/Ta_2O_5 40	4.3	0-5
RhO_2 30/RuO_2 10/ZrO_2 60	4.1	0-4
RhO_2 70/IrO_2 10/TiO_2 20	4.3	0-3
PdO 40/RuO_2 20/IrO_2 10/TiO_2 30	4.2	0-3
PdO 40/TiO_2 20/Ta_2O_5 40	4.3	0-3
PdO 70/TiO_2 30	4.0	0-3

It will be seen from Table 3 that the prevention of corrosion and hydrogen absorption can be effectively achieved by covering the Ti plate with the mixture oxide composed of at least two platinum group elements and an anticorrosion metal or at least two anticorrosion metals and a platinum group element.

Example 4: Crevice corrosion test pieces were prepared by forming PdO/TiO_2 mixture layers (3 μ thick) of various PdO contents on respective inch square piece assemblies consisting of two overlapping thin titanium plates having a central aperture as shown in Figure 7.2. In the figure, designated at **1** are the thin titanium plates, at **2** Teflon insulators, at **3** a titanium bolt, and at **4** a titanium nut. The PdO/TiO_2 mixture layer was formed by applying a solution containing palladium chloride and titanium chloride dissolved therein over the surface of each assembly, followed by thermal oxidation in an atmosphere at 550°C for 10 minutes.

The crevice corrosion of the samples prepared in this way was then observed after immersing them in a boiling aqueous solution containing 44% of ammonium chloride for 240 hours, and Table 4 shows the results. A noncoated piece assembly was also tested as contrast in the same manner.

Table 4

Test Specimen	Contrast	.... Coated Specimens.					....
PdO (mol %) in coating layer	–	0.5	1	30	70	95	97
Crevice corrosion	present	slight	none	none	none	none	slight

Figure 7.2: Test Piece to Be Provided with Crevice Corrosion Preventative Treatment

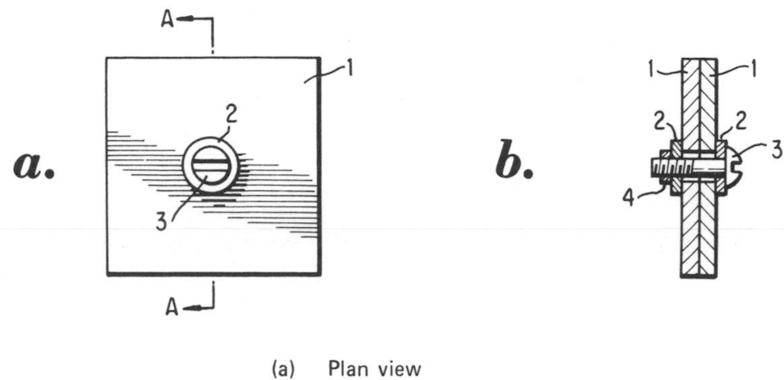

(a) Plan view
(b) View along A-A of (a)

Source: U.S. Patent 4,082,900

As is seen from Table 4, the crevice corrosion was reduced by the provision of the mixture coating layer, and particularly it was suppressed substantially perfectly when the PdO content was 1 to 95%.

Removing Sulfide-Containing Scale from Refinery Equipment

Many sources of crude oil and natural gas contain high amounts of hydrogen sulfide. Refineries processing such crude oil or natural gas commonly end up with substantial amounts of sulfide-containing scale on the metal surfaces in contact with the crude oil or gas. This scale is detrimental to the efficient operation of heat exchangers, cooling towers, reaction vessels, transmission pipelines, furnaces, etc. Removal of this sulfide-containing scale has been a substantial problem because conventional acid-cleaning solutions react with the scale and produce gaseous hydrogen sulfide.

In U.S. Patent 4,220,550 it was disclosed that acid-soluble, sulfide-containing scale could be effectively removed from metal surfaces without the release of gaseous hydrogen sulfide by use of an aqueous acid cleaning composition comprising an aqueous nonoxidizing acid having at least one aldehyde dissolved or dispersed therein, with the aldehyde being present in such compositions in an amount at least sufficient to prevent or substantially prevent the evolution of hydrogen sulfide gas. This required at least a stoichiometric amount of aldehyde in the cleaning solution (i.e., at least one mol of aldehyde per mol of hydrogen sulfide produced during the cleaning) and an excess of aldehyde was preferred. The best system was an aqueous sulfuric acid cleaning solution containing excess formaldehyde.

U.S. Patent 4,220,550 represented a technical breakthrough in the chemical cleaning industry. The discovery has been commercialized and widely accepted.

A chemical cleaning solution has been discovered by *G.R. Buske; U.S. Patent 4,289,639; September 15, 1981; assigned to The Dow Chemical Company* which comprises an aqueous nonoxidizing acid having dissolved therein, glyoxylic acid.

This chemical cleaning solution is a unique species within the generic disclosure of U.S. Patent 4,220,550 in that the reaction of products of the chemical cleaning solution and the sulfide-containing scale (e.g., iron sulfide) are soluble in the acidic cleaning medium and do not form solid precipitates. Under treatment conditions, the process cleaning composition removes the acid-soluble sulfide-containing scale from metal surfaces at a rate substantially equal to or greater than the preferred cleaning composition described in U.S. Patent 4,220,550 (i.e., solutions comprising aqueous sulfuric acid with excess formaldehyde).

U.S. Patent 4,220,550 describes generically the chemical cleaning compositions, the relative ratio of ingredients, and methods of use. This has been incorporated herein by reference and so it would be redundant to repeat the information here. The primary distinction between this process and U.S. Patent 4,220,550 resides in the use of the glyoxylic acid as the aldehyde in the cleaning solutions. Glyoxylic acid appears to be a unique species within the genus of aldehydes described in U.S. Patent 4,220,550.

As stated in U.S. Patent 4,220,550 the acidic cleaning solutions can utilize a variety of acids usually at concentrations ranging from about 5 to 15%. Sulfuric acid and hydrochloric acids are preferred, and hydrochloric acid is most preferred in this instance. U.S. Patent 4,220,550 also teaches the wisdom of adding a compatible acid corrosion inhibitor to the acid cleaning solution (preferably an amine-based corrosion inhibitor) and emphasizes the advantage of using the aldehyde in excess in the cleaning solution. The same teaching similarly applies here. Glyoxylic acid is added in an amount sufficient to prevent or substantially prevent the evolution of gaseous hydrogen sulfide when the cleaning solution is brought in contact with an acid-soluble, sulfide-containing scale.

Example: A solution of glyoxylic acid (7.5 g) and water (102 ml) was charged to a reaction vessel equipped with a gas scrubber containing 25% aqueous sodium hydroxide. The temperature of the glyoxylic acid/water solution was raised to 150°F in a water bath and iron sulfide (FeS, 7.5 g) was then added. After the temperature of this mixture reached 150°F, 35 ml of concentrated (36%) hydrochloric acid was introduced and the vessel was quickly sealed. When the acid was first added, there was a brief initial smell of hydrogen sulfide but no detectable amount of hydrogen sulfide after that. Analysis of the sodium hydroxide scrubbing system using an Orion S^{2-} electrode gave a zero reading for sulfide. The cleaning solution dissolved all of the iron sulfide and the spent cleaning solution was a clean liquid without any noticeable amounts of solid precipitate. No evolution of hydrogen sulfide gas was observed during the three hour test.

Protecting Aluminum Engine Element Against Galling and Corrosion

A cost saving method for protecting an engine element against operation galling and corrosion is disclosed by *W.A. Donakowski and J.R. Morgan; U.S. Patent 4,018,949; April 19, 1977; assigned to Ford Motor Company*. The element, particularly a piston, is cast of aluminum and cleaned free of dirt and organic matter. A warm stream of an aqueous solution containing a protective metal agent is directed onto a selected zone of the element. The agent consists essentially of potassium stannate and the zone is preferably the middle region of a piston skirt extending from the upper skirt periphery to lower skirt extremity (tail). The stream is maintained as a laminar flow as it traverses the selected zone. An ultrathin protective coating is adherently deposited on the zone of the piston.

Tin Plating: A tin plating immersion is prepared and may typically comprise an aqueous solution of potassium stannate ($K_2SnO_3 \cdot 8H_2O$). Potassium stannate should be 31 to 38% by wt of the chemical compound and have a 2.5% minimum of sodium gluconate or 12 to 16% sodium tripolyphosphate, either of the latter acting as a sequestering agent. The solution may have 1% maximum free alkali as potassium hydroxide. The moisture content of such tin plating compound should be no greater than 3.5% when sodium gluconate is utilized or 2.5% maximum when sodium tripolyphosphate is utilized.

The process as applied to the selective coating of a piston is shown in Figure 7.3.

Figure 7.3: Selective Coating of Piston

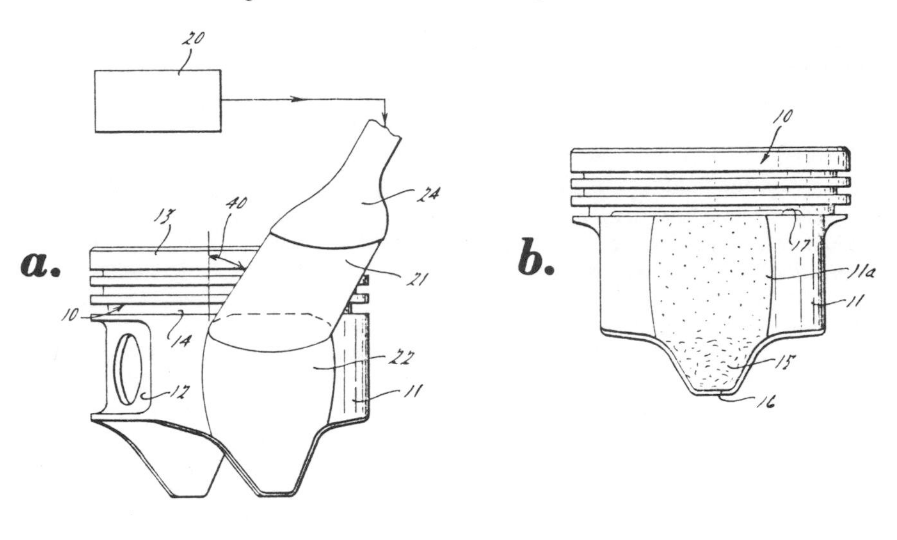

(a) Piston being coated
(b) Selected zone of a coated piston

Source: U.S. Patent 4,018,949

Stream Plating: The tin plating solution **20** is warmed and forced into a stream **21** directed onto each selected zone **11a** of the piston skirts, the stream being particularly directed to contact near the upper extremity **17** of the zone and follow along the surface **11** as a laminar flow **22** to the lower extremity **16** where the flow leaves the surface. It is preferable that a spread nozzle be employed to lay down a flat stream layer **21** which is commensurate in width to the width of a zone to be coated. The stream should be maintained at a temperature between 130° and 150°F. It is important that the angle of incidence **40** of the stream as directed upon the selected zone be within the angular range of 5 to 40° taken with respect to the plane of the zone. During such intimate contact with the cleaned aluminum zone **11**, the aluminum will be coated with a uniform thickness of tin in the preferably thickness range of 35 to 65 millionths of an inch (however, the operable range can be 10 to 150 millionths of an inch), and most typically the thickness will be about 50 millionths of an inch, assuming the stream is maintained flowing against such zone for a period of about 3 minutes. The tin plating aqueous solution is maintained in the heated condition of 130° to 150°F. The solution should not be allowed to drop below 120°F under any conditions since the plating rate of tin ceases at this temperature level. Also at temperatures higher than 160°F, the tin plating compound tends to break down chemically.

Regarding the rate at which tin is deposited for the rate at which the tin thickness builds up, it was found that after about 10 seconds, the peak rate is achieved even though the ultimate thickness has not yet been obtained. It has been determined by experiment, that tin deposition by a directed stream provides a 10% faster deposition thickness as compared to tin immersion characteristic of the prior art.

The product resulting from practicing the above method will have a tin coating of considerable strength and adhesion; the coating will require only 80% of the tin composition required according to total immersion methods.

Small Metal Parts Treatment

A method for treating metal parts either singly or in bulk barrel processing to provide durable and rust-inhibiting coatings is described by *R.A. Kelly and H.G. Pekar; U.S. Patent 4,165,242; August 21, 1979; assigned to R.O. Hull & Company, Inc.* This comprises the steps of (a) treating the metal parts with an aqueous phosphating solution to deposit a phosphate coating thereon, (b) electrophoretically depositing a siccative organic coating on the phosphate coated metal parts, and (c) treating the siccative organic coated part with an oil to deposit a corrosion-inhibiting top coat.

The oil which is applied as the top seal coat also may contain other compositions which improve the rust-inhibiting properties of the oil in amounts up to about 20 to 25% or higher. One example of a preferred type of additive composition is metal-containing phosphate complexes such as can be prepared by the reaction of (a) a polyvalent metal salt of the acid phosphate esters derived from the reaction of phosphorus pentoxide with a mixture of monohydric alcohol and from about 0.25 to 4.0 eq of a polyhydric alcohol, and (b) at least about 0.1 eq of an organic epoxide. Thin films of these complexes in oil over the phosphated and painted metal parts are effective in inhibiting the corrosion of the metal surfaces.

Example 1: *Zinc containing organic phosphate complex* – 49 parts (0.73 eq) of dipropylene glycol, 95 parts (0.73 eq) of isooctyl alcohol, and 133 parts of aromatic petroleum spirits boiling in the range 316° to 349°F are introduced into a reaction vessel. The whole is stirred at room temperature and 60 parts (0.42 mol) of phosphorus pentoxide is introduced portionwise over a period of about 0.5 hour. The heat of reaction causes the temperature to rise to about 80°C. After all of the phosphorus pentoxide has been added, the whole is stirred for an additional 0.5 hour at 93°C. The resulting acid phosphate esters show an acid number of 91 with bromphenol blue as an indicator.

The mixture of acid phosphate esters is converted to the corresponding zinc salt by reacting it with 34.5 parts of zinc oxide for 2.5 hours at 93°C. Thereafter 356 parts (one equivalent per equivalent of zinc salt) of butyl epoxystearate is added to the zinc salt at 88°C over a period of about 1 hour and the whole is stirred for 4 hours at 90°C. Filtration of the mass yields 684 parts of a zinc-containing organic phosphate complex having the following analysis: phosphorus, 3.55%; zinc, 3.78%; and specific gravity, 1.009.

Example 2: (a) An oil mixture is prepared containing 60 parts of mineral oil, 2 parts of triethanolamine, 3 parts of oleic acid, 15 parts of a sodium sulfonate wetting agent and 20 parts of the product of Example 1.

(b) An emulsion is prepared by vigorously mixing 20 parts of the oil of (a) with 80 parts of water.

Example 3: (a) A typical solution is prepared by dissolving 33.91 g of 75% phosphoric acid, 18.03 g of 42° Baume nitric acid, 14.11 g of zinc oxide and 8.81 g of zinc chloride in 25.06 g of water. These typical solutions would be dissolved in water at 2 to 5% by vol to produce a workable phosphate bath.

(b) Heat treated, 1050 steel spring fastener pieces are placed in a stainless steel barrel which is immersed in the zinc phosphate solution of 3a for a period of about 15 minutes at about 80°C, rinsed in water at room temperature and immersed in a chromic acid solution containing hexavalent chromium for about 1 minute at about 82°C. After drying the chrome rinsed parts, the barrel is immersed in a stainless steel paint tank. The barrel has a power connection to the positive side of a rectifier and insulated cathodes are submerged in the tank around the barrel. In this example the paint in the tank is a commercially available (Parr Inc.) water reducible epoxy ester resin pigmented black. The barrel is rotated intermittently and current is applied as indicated below. The paint temperature is about 28°C. The parts are removed from the paint tank, water rinsed, and heated in an oven for about 15 minutes at about 175°C to cure the paint. As the parts are removed from the oven, one of each type is quenched by immersion in the oil:water mixture of Example 2b, and thereafter air dried.

It has been found that when small metal parts such as steel U-bolts and spring clamps are treated in accordance with the procedure, improved rust inhibition is observed.

The results of the Salt Fog Corrosion test conducted on parts treated in accordance with the process compared to parts without the oil topcoat are summarized below.

Test	Paint Application* Time	Oil of Ex. 2**	Test Results (10 pieces) 168 hr (degree of rust)	264 hr
SF-1	90 sec	No	Pinpoints	5-1% 2-2% 3-5%
SF-1A	90 sec	Yes	OK	OK
SF-2	5 min	No	Pinpoints	8-2% 1-5% 1-10%
SF-2A	5 min	Yes	OK	OK

*Parr Epoxy at 225 V.
**Parts immersed at 65°C for 1 min.

Smoothing Irregular Metal Surface Without Precision Grinding

Many devices incorporating moving precision metal parts require that at least some surfaces of these parts, or the surfaces of stationary parts in proximity thereto, be true and as free of surface irregularities as possible. For example, the flat interior faces of front and rear heads for rotary compressors are in proximity to high speed rotors and must be free of even small irregularities.

R.E. Ahlf; U.S. Patent 4,181,540; January 1, 1980; assigned to Whirlpool Corporation provides a method of smoothing a metal surface whereby surface irregularities are removed without the use of a precision grinding operation. The method is advantageously used in smoothing the surfaces of metal parts to conform to each other or to a conforming surface, and provides a conversion coating on the surface of the part concurrently with the smoothing.

The metal surface is wetted with a solution of a compound which is known to rapidly oxidize the surface metal and to form a friable, relatively insoluble conversion coating thereon which is relatively impervious to the solution. A conforming surface is rubbed against the wetted metal surface so as to abrade and thereby remove the friable coating, thereby exposing freshly bared surface metal to the solution.

Continued rubbing of the surfaces, accompanied by rewetting of the metal surface, if necessary, results in repetitive formation and removal of a conversion coating on the metal surface and, therefore, gradual wearing away of surface irregularities. Upon removal of all irregularities and washing of the surface, a smooth conversion coating remains on the surface. Such a coating is highly desirable in many instances where the surface is to be used in a bearing application.

Typical of suitable conversion coating forming solutions are those normally used for coloring metals, such as those utilized in blackening or bluing metals. Solutions commonly used as a paint primer or those used for forming a lubricant-receptive coating on metal are also suitable. Specific examples of solutions which form conversion coatings are a manganese-iron phosphate complex (used on ferrous materials) and an aqueous alkali dichromate solution such as potassium or sodium dichromate (used on aluminum or zinc). Zinc or iron phosphates, known to be good paint primers, are further examples of conversion coating forming materials.

It has been found that surface irregularities of at least 0.05 mm may be easily removed by the procedure in a relatively short time as compared to conventional grinding times, but larger irregularities can also be removed by the process. The procedure generates very little heat and may be performed at substantially room temperature. Also, negligible surface distortion results from the use of this method. The method can be used on either hardened metal or soft metal with a negligible effect on the hardness.

Example 1: The cloth cover of a rotary metallurgical polishing lap was saturated with an aqueous solution of a manganese-iron phosphate complex. An iron casting with a flat face previously ground by an abrasive wheel of 180 grit was held stationary against the rotating cloth. Additional solution was periodically added to the cloth to maintain it in a saturated state.

After approximately 10 seconds of relative rotation, the casting was removed from the cloth and was observed to be highly polished. Unaided visual inspection revealed metallic crystals in the casting's surface, and microscopic examination of the surface revealed clearly defined graphite flakes. Conventional precision grinding of a surface to a comparable smoothness requires at least one minute, or six times the polishing time of this process.

Example 2: Aluminum surfaces, when in mutual contact and in relative rotation, exhibit a tendency to seize or gall. However, smooth surfaces of aluminum oxide do not experience such seizing under light loads.

Two flat surfaces of aluminum rotary compressor end housings having minor surface irregularities were wetted with an aqueous potassium dichromate solution (20 g/ℓ) maintained at a pH of between 8 and 10 by addition of sodium hydroxide. The dichromate solution was continuously applied between the aluminum surfaces as they were maintained in relative rotational contact. No galling or seizing was experienced. It is believed that aluminum oxide formed between the surface irregularities and was not abraded by the rubbing motion, but that aluminum oxide which formed on high spots was broken away by abrasion, exposing bare aluminum to the dichromate solution for further oxidation. Continued rubbing and wetting in this manner resulted in smooth, oxide-coated surfaces on each piece.

Each of the above examples was conducted at room temperature, and negligible temperature effects were noted.

Diffusion Coating an Article Having Fine Bores or Narrow Cavities

The work of *C. Hayman and J.E. Restall; U.S. Patent 4,156,042; May 22, 1979; assigned to U.K. Secretary of State for Defence, England* relates to processes for coating articles with diffusion coatings and particularly, though not exclusively, relates to coatings for gas turbine engine components, e.g., turbine blades, for increasing their high temperature corrosion resistance.

The process for coating an article with a diffusion comprises enclosing the article in a chamber together with a particulate pack including coating material in elemental or chemically combined form, the coating material selected from the group consisting of aluminum, chromium, titanium, zirconium, tantalum, niobium, yttrium, rare earth metals, boron and silicon, together with a halide activator of low volatility and cyclically varying the pressure of an inert gas or a reducing gas or a mixture of gases within the chamber while maintaining the contents of the chamber at a temperature sufficient to transfer coating metal onto the surface of the article and to form a diffusion coating thereon.

The halide activator is preferably selected from a group of inorganic halides wherein the equilibrium vapor, sublimation or dissociation pressure is equal to not more than atmospheric pressure at the coating process temperature, and advantageously from those in which this pressure is appreciably less than atmospheric.

Preferably the method is effectively carried out at a maximum pressure substantially below atmospheric pressure and most advantageously below about 100 torrs.

Preferably the cycle frequency is as high as is compatible with the transport of a sufficient quantity of the gas through the particular pack per cycle. The ratio of upper pressure limit to lower pressure limit is also preferably as high as is practicable and consistent with cycle frequency. Convenient pressure ranges are from about 50 to 10 torrs, preferably with cycle frequencies of at least 2 cpm. In general, higher frequencies are beneficial in increasing the ratio of coating thickness applied internally to that applied externally.

The particulate pack may include a filler such as a refractory oxide for support of the coating material or for dilution of the pack. A filler comprising a refractory oxide may support a coating material comprising liquid aluminum.

The articles coated by the process may be composed of any material that can be coated by pack cementation. Materials commonly coated by pack cementation are nickel-base, cobalt-base and iron-base alloys, and the refractory metals of Groups IV, V, and VI of the Periodic Table. In addition to these materials, carbon and carbon-containing materials, e.g., tungsten carbide, may be advantageously coated by the process. In particular, titanium carbide coatings may be produced on a cemented carbide article.

In one manner of operating the process, the article to be coated is kept out of physical contact with the particulate bed. This could be by placing the article inside a cage which is itself embedded within the particulate pack. A preferred construction for a cage is one that will permit vapors to pass from the particulate pack to the inside of the cage but which prevents or retards flow to the outside of the cage. One cage has sides and an upper face of imperforate material, e.g., nickel sheet or plate, and the base of a mesh or gauze through which vapor can pass.

Alternatively, the article to be coated may be suspended in the reaction chamber over a tray containing the particulate pack material.

Example 1: A gas turbine blade section in IN 100 alloy, bearing a hole of diameter about 1.5 mm and of length about 110 mm, was aluminized according to the method in a chamber. The method included embedding the blade section in a powder mix of 14 g AlF_3, 14 g Al and 388 g Al_2O_3, pumping out the chamber and admitting argon to displace any air, raising the temperature of the chamber and its contents to 900°C and setting the time-controlled valves to give a flow of argon into the chamber for 3 seconds to give a pressure of 28 torrs, maintain this pressure substantially constant for 20 seconds and then exhaust for 7 seconds to reduce the pressure to about 6 torrs, after which the cycle was repeated automatically. After 10 hours at the same temperature, the chamber was cooled and the blade section removed. On examination, the surface of the hole was found to be uniformly coated with an aluminized layer of mean thickness about 35 μm. The thickness distribution of the coating along the length of the hole can be seen from the following figures: Distance from one end of hole (mm): 10, 20, 30, 40 and 50 respectively. Coating thickness (μm): 40, 40, 35, 30 and 30 respectively.

Example 2: In a further example, a turbine blade in IN 100 alloy bearing holes of diameter about 1.5 mm and of length about 70 mm, was aluminized for 5 hours at 900°C inside a nickel gauze cage which was itself embedded in a powder pack mix of 6.5 g AlF_3, 10.6 g Al and 330 g Al_2O_3. The pressure range of argon was from 14 to 58 torrs and the pressure cycle frequency was 6 cpm. Bright metallic-looking and particularly smooth textured aluminized layers were produced on both the internal and external surfaces of the blade. The layer thickness within the hole measured close to the top, midspan and bottom was respectively 12, 8 and 12 μm. The layer thickness measured over the external surface at comparable positions was 25, 25 and 30 μm respectively.

Example 3: In an example of aluminizing with NaCl as halide activator, a turbine blade in IN 100 alloy was aluminized for 5 hours at 900°C within a nickel gauze cage which was itself embedded in a powder pack mix of 20 g NaCl, 14 g Al and 300 g Al_2O_3. The pressure range of argon was from 8 to 42 torrs and the pressure cycle frequency was 6 cpm. A layer of thickness 2 μm was produced within a hole of diameter 1.8 mm and a length 40 mm. The thickness of the external layer was about 16 μm.

Surface Treatment in a Rotary Treating Apparatus

The process of *A. Kirisawa; U.S. Patent 4,294,626; October 13, 1981; assigned to Japan Envirotic Industry Co., Ltd., Japan* relates to surface treatment such as washing, etching, plating, chromate treatment, and zinc-chromate treatment employing a rotary treating apparatus which is compact and does not require a wide space.

A surface treating system shown in Figure 7.4 can be employed for the chromate treatment and comprises a surface treating apparatus **A**, a nitric acid tank **B** and a chromate tank **C**.

Referring to the drawing, the surface treating apparatus **A** comprises a vessel **10** which is supported, springs **22** being between vessel **10** and legs **21** standing on a base **20**, a lid **37** pivotally attached on the upper edge of vessel **10** by a hinge **38** and connecting with a returning pipe **33** having a magnetic valve MV_1, a returning pipe **34** having a magnetic valve MV_2, an exhausting pipe **35** having a magnetic valve MV_4 and an exhausting pipe **36** having a magnetic valve MV_5 at its bottom. A supplying tank **18** which is fitted upon the outside of vessel **10** connects with the inside of vessel **10** through a connecting opening **19**, and has inserted a level gauge **L**. A holed container **15** having numberless holes **30** is rotatably attached on an axis **12** by a nut **17**, piercing through the bottom of vessel **10**, supported by a pair of bearings **13**, sealed by a mechanical seal **14** in a bearing box **11**. A pulley **24** is fixed at its under part, connecting with a pulley **23** fixed on an axis of a brake motor **M** through a belt **25**. One or more buffer blades may be fixed to the base of the holed container if desired, because the buffer blade gives a turbulence to treating liquids or washing liquids to result in effective contact of materials with the liquids. A washing nozzle **28** is inserted in vessel **10** at its upper part and connected with a washing liquid supplying pipe **29** having a magnetic valve MV_3.

The nitric acid tank **B** connects with a supplying pipe **31** having a pump P_1 and connecting with supplying tank **18** of the surface treating apparatus **A** at its upper part, and returning pipe **33** having a magnetic valve MV_1 and connecting with supplying tank **18** at its bottom. The chromate tank **C** connects with a supplying pipe **32** having a pump P_2 and connecting with supplying tank **18** at its upper part and a returning pipe **34** having a magnetic valve MV_2 and connecting with supplying tank **18** at its bottom.

A chromate treatment employing this surface treating system is as follows: Material **26** to be treated such as galvanized bolts, galvanized nuts, etc. are filled into a cage **16** which is made of wire net and fixed a pair of catches **27** at its upper edge.

Figure 7.4: Surface Treatment Using a Rotary Treating Apparatus

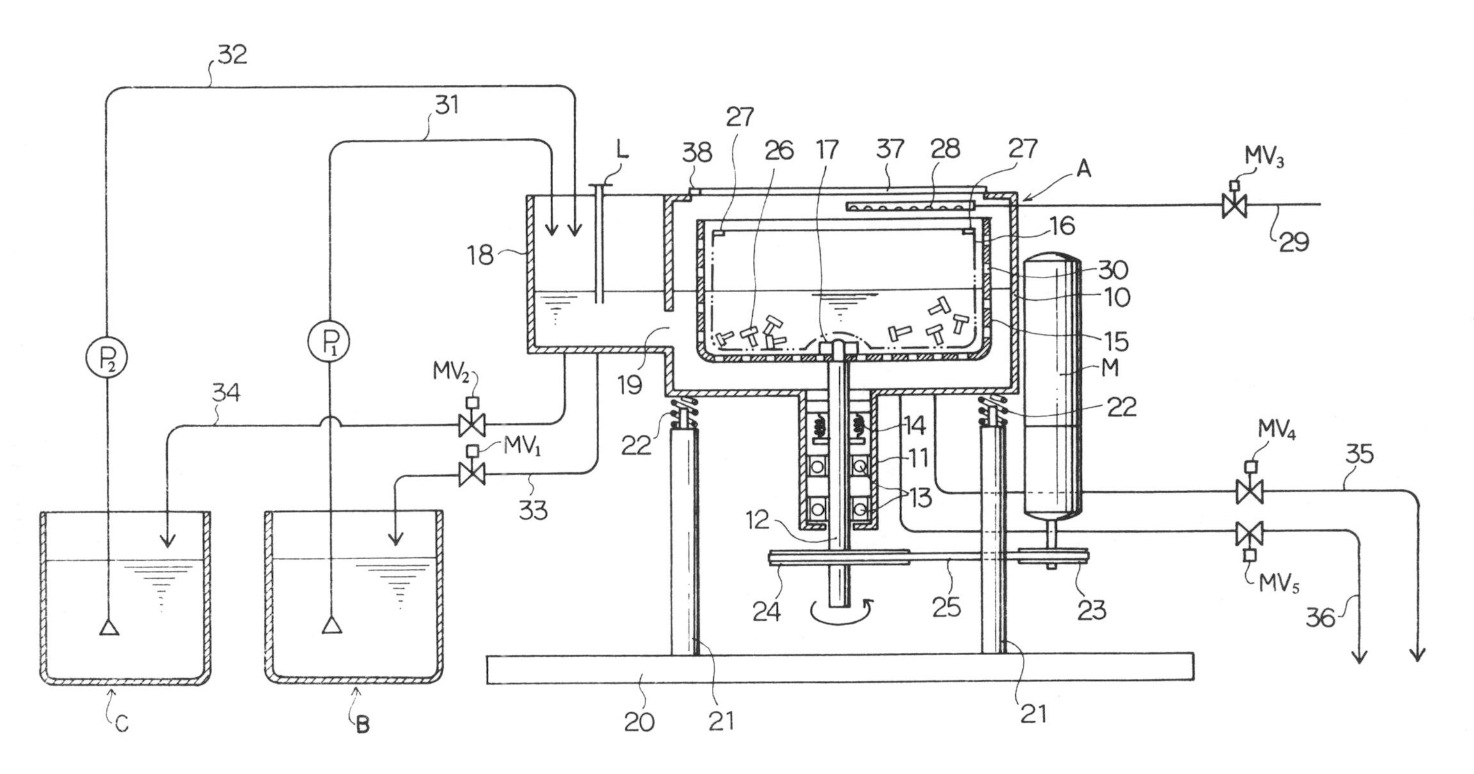

Source: U.S. Patent 4,294,626

One or more buffer blades may be fixed to the base of the cage if desired, because the buffer blade gives a turbulence to treating liquids and washing liquids to result in effective contact of materials with the liquids. The lid **37** of the vessel **10** is opened; then the cage **16** filling materials **26** is put into the holed container **15** of the vessel. The lid of the vessel is shut.

A chromate treatment process comprises, washing, nitric acid dipping, second washing, chromate solution dipping, third washing, dehydrating, and drying.

In the first washing process, the holed container **15** is driven to rotate by the brake motor **M** and washing water is sprayed from the nozzle **28** by opening of the magnetic valve MV_3. In this process, the holed container keeps a fixed rotation or the holed container is rotated intermittently or alternatively and washing water is successively exhausted through the exhaust pipe **35** by opening of the magnetic valve MV_4 or washing water is accumulated in the vessel and then exhausted by opening the magnetic valve MV_4 after washing has been completed. Dehydration after washing is carried out by a fixed rotation of the holed container with opening of the magnetic valve MV_4, and the wastewater is exhausted through exhaust pipe **35**. During the washing process, washing water is effectively contacted with the surface of materials **26** and removes effectively the galvanizing solution remaining on the surface of the materials.

In the nitric acid dipping process, nitric acid solution in tank **B** is poured into vessel **10** through the supply pipe **31** and supply tank **18** by pump P_1. The solution level is detected by level gauge **L** and the pump P_1 is stopped when the solution level reached a fixed level where materials **26** dip entirely in the solution. During the supplying of solution, the holed container keeps a fixed rotation or the holed container is rotated intermittently or alternatively, or the rotation is stopped. After the supplying of solution, the magnetic valve MV_1 of the return pipe **33** is opened with a fixed rotation of the holed container and nitric acid solution in the vessel is withdrawn to the nitric acid tank.

In the second washing process, magnetic valve MV_1 of return pipe **33** is shut and magnetic valve MV_5 of exhaust pipe **36** is opened; then the same process as the first washing process is carried out.

In chromate solution dipping process, magnetic valve MV_5 of exhaust pipe **36** is shut and chromate solution in chromate tank **C** is poured into the vessel through supply pipes **32** and **18** by driving of pump P_2. The solution level is also detected by level gauge **L** and pump P_2 is stopped to drive when the solution level reached a fixed level on which materials **26** dip entirely in the solution. During supplying of solution, holed container **15** keeps a fixed rotation or is rotated intermittently or alternatively, or rotation is stopped. After supplying of solution, magnetic valve MV_2 of return pipe **34** is opened with a fixed rotation of the holed container and chromate solution in the vessel is withdrawn to chromate tank **C**.

In the third washing process, magnetic valve MV_2 of pipe **34** is shut and magnetic valve MV_5 of exhaust pipe **36** is opened; then the same process as the first washing process is carried out. After the chromate treatment process has been completed, lid **37** of vessel **10** is opened and cage **16** is taken out. Materials in the cage are dried by heating and/or ventilating with or without centrifugal separating. Dehydrating and drying can be carried out in vessel **10** if the rotational speed of holed container **15** can be changed by connecting axis **12** of this container with either a low or a high speed motor or by equipping the vessel with a dryer.

METAL COLORING

COLORING STAINLESS STEEL

Controlling Potential Differences in the Coloring Solution

T. Takahari, S. Kondo, N. Sone, K. Hashimoto and T. Ishiguro; U.S. Patent 4,026,737; May 31, 1977; assigned to Nippon Steel Corporation, Japan have found that it is possible to obtain a desired color without color variation when the coloring potential difference of a standard material is compensated by the amount of the variation of the inflexion potential. This compensated value is used as a coloring potential difference for obtaining a desired color instead of controlling the variation of the inflexion point potential. For determining the compensation coefficient for the inflexion point potential, various experiments were made repeatedly as follows.

SUS 304 stainless steel sheets of BA finish as standard material were immersed in an aqueous solution containing 300 g/ℓ of chromic anhydride and 500 g/ℓ of sulfuric acid at 75°C, using a saturated calomel electrode on a platinum electrode with the coloring potential difference between the inflexion point potential and the coloring potential of the desired color.

This potential difference was used as a standard value, and the coloring potential difference of subsequent coloring treatments was sought by varying the coefficient (α) in the following formula. Figure 8.1a shows the relation between the compensation coefficient (α) and the color differences which were between the colors thus obtained and the color obtained with coloring potential difference of standard material. α_1 means that the individual inflexion point potential was noble and α_2 means that the potential was base in respect to the standard potentials.

Therefore, when the inflexion point potential of a material to be colored according to the process is (A') while the coloring potential is (B'), the coloring potential difference (A' - B') of the material to be colored is given by the following formula, and by completing the coloring treatment when the potential difference reaches the difference (A' - B'), it is possible to eliminate the color deviation and assure the reproductivity of a desired color.

Individual coloring potential difference (A' - B') =
Standard coloring potential difference (A - B) ± α[standard inflexion point potential (A) - individual inflexion point potential (A')]

(A' - B') is the potential difference between the inflexion point potential (A') and the coloring potential of an individual material;

(A - B) is the potential difference between the inflexion point potential (A) and the coloring potential (B) of a standard material;

(A') is an inflexion point potential of an individual material;

(A) is an inflexion point potential of a standard material;

(B') is a potential at which the coloring of an individual material is stopped;

(B) is a potential corresponding to a potential for a desired color on a standard material;

(±) varies depending on the kind of the reference electrode at the measurement of potentials, and in the case of a saturated calomel electrode it is (±), and in case of a platinum electrode it is (-); and

coefficient (α): when the individual inflexion point potential is nobler than the standard inflexion point potential, α_1 is used, and when the individual inflexion point potential is base compared to the standard inflexion point potential, α_2 is used. α_1 = 0.23 to 0.36. α_2 = 0.44 to 0.61.

The above control system is illustrated in Figure 8.1b.

Figure 8.1: Controlling Potential Differences in Coloring Solution

a.

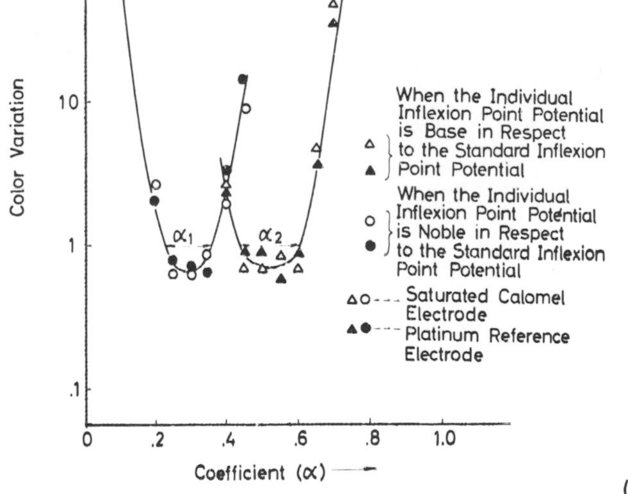

(continued)

Figure 8.1: (continued)

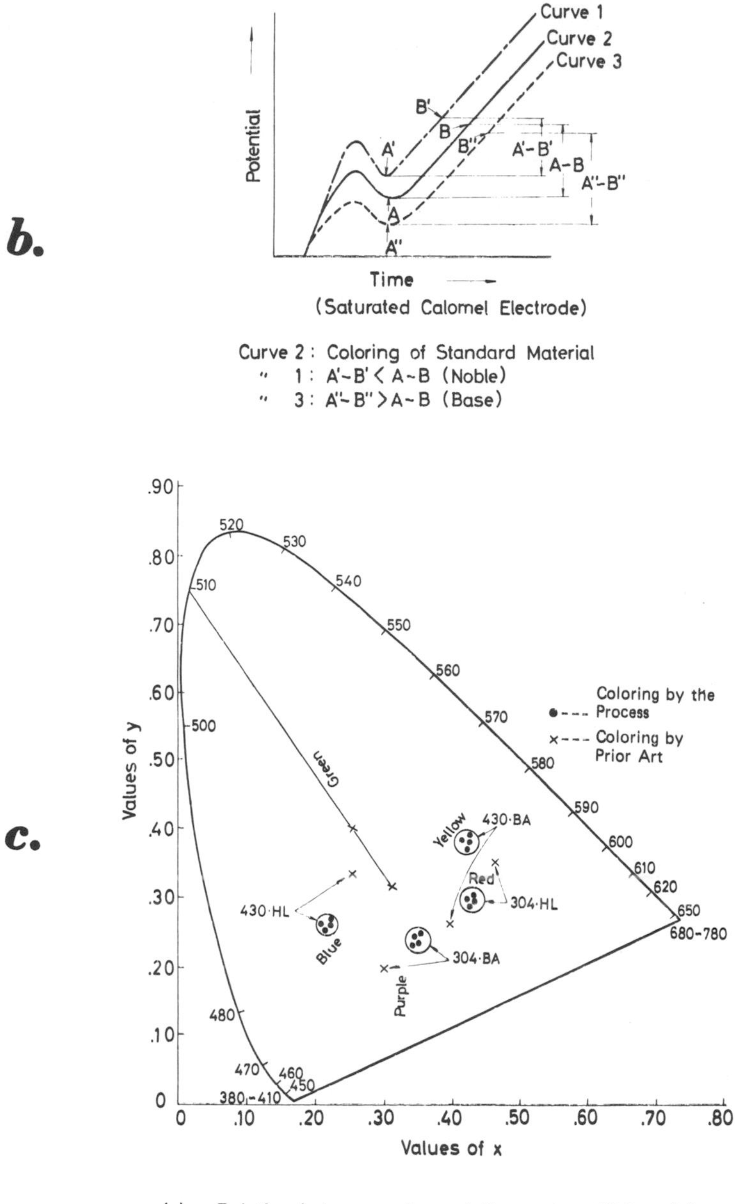

(a) Relation between color variation and coefficient (α)
(b) Relation between potential and time
(c) Color measurements of steel according to the process

Source: U.S. Patent 4,026,737

In the art the coloring potential difference is fixed as A' - B' = A - B so that good reproductivity of a desired color cannot be obtained, whereas according to the process, the standard coloring potential difference (A - B) is compensated as shown in Figure 8.1b, in which (A' - B') is smaller than (A - B) in the curve **1** and (A' - B') is larger than (A - B) in the curve **2**, so that good reproductivity of a desired color without variation can be assured.

Example: In this example, BA (bright annealed) finished or HL (hairline) finished SUS 304 and 430 stainless steels were used.

The BA finished SUS 430 stainless steel was subjected to an anodic electrolytic treatment as the precoloring treatment in a solution containing 30 g/ℓ sulfuric acid and 200 g/ℓ nickel sulfate at 1.0 A/dm^2, while the HL finished SUS 430 stainless steel was immersed in a precoloring treatment solution containing 50 g/ℓ sulfuric acid and 50 g/ℓ chromic acid at 70°C for 15 minutes. Next the above BA finished SUS 430 and the HL finished SUS 430 stainless steels as well as a BA finished SUS 304 and a HL finished SUS 304 stainless steel which were not subjected to the precoloring treatment were immersed in the coloring treatment bath containing 300 g/ℓ chromic anhydride and 500 g/ℓ sulfuric acid at 75°C.

The variation of the potential differences between the steel samples immersed in the bath and the platinum reference electrode was measured on a digital voltmeter and recorded continuously on a recorder. In this example, purple was intended for the BA finished SUS 304, red was intended for the HL finished SUS 304, gold for the BA finished SUS 430 and blue for the HL finished SUS 430, and the coloring was performed by calculating the individual coloring potential difference for each sample on the basis of the formula according to the process. For comparison, SUS 304 and SUS 430 stainless steels were used. The stainless steels were not subjected to the precoloring treatment and were colored by a conventional method under the following condition:

Individual coloring potential difference (A' - B') =
Standard coloring potential difference (A - B).

Method	Samples	Sample No.	Inflexion Point Potential (mV)	Coloring Potential Difference (mV)	Coefficient
		0	185.54*	14.40*	
Process	SUS 304 BA finish	1	186.65	15.07	0.64
		2	184.28	14.08	0.25
		3	185.90	14.76	0.55
		4	186.47	14.82	0.45
Conventional	SUS 304 BA finish	5	183.32	14.40	–
		0	187.28*	13.50*	
Process	SUS 304 HL finish	1	186.42	13.24	0.30
		2	178.36	10.38	0.35
		3	189.82	14.77	0.50
Conventional	SUS 304 HL finish	4	190.02	13.50	–

(continued)

Method	Samples	Sample No.	Inflexion Point Potential (mV)	Coloring Potential Difference (mV)	Coefficient
		0	167.34*	11.20*	
Process	SUS 430 BA finish	1	168.54	11.74	0.45
		2	167.50	11.29	0.55
		3	167.93	11.55	0.60
		4	166.48	10.94	0.30
Conventional	SUS 430 BA finish	5	165.32	11.20	–
		0	166.82*	8.00*	
Process	SUS 430 HL finish	1	165.42	7.58	0.30
		2	166.90	8.04	0.50
		3	166.23	7.79	0.35
Conventional	SUS 430 HL finish	4	167.29	8.00	–

*Standard values.

In the table, the inflexion point potential and the coloring potential difference for each sample are shown, and Figure 8.1c shows the results of the color measurements. It is clear from the figure that discrete colors with excellent reproductivity without color variation can be obtained by the process, while the colors produced by the conventional methods were completely different from those intended as shown below.

BA finished SUS 304, bluish purple;
HL finished SUS 304, reddish gold;
BA finished SUS 430, brown;
HL finished SUS 430, light brown.

Use of a Differentiation Curve to Obtain Coloring Starting Point

The method of *T. Takeuchi, H. Takamura, K. Takatsu and H. Shibata; U.S. Patent 4,269,633; May 26, 1981; assigned to Nisshin Steel Co., Ltd., Japan* for coloring stainless steel comprises dipping the stainless steel in a coloring liquor comprising a mixed aqueous solution of chromic acid and sulfuric acid to form an oxide film on the surface, the coloring to a desired color tone being controlled on the basis of the potential difference between the surface of the stainless steel and a reference electrode such as a platinum electrode dipped in the coloring liquor, wherein the coloring can be accurately controlled with a satisfactory reproducibility by checking a differentiation curve derived from a potential-time curve showing the variation with time of the potential difference between the surface of the stainless steel and the reference electrode by differentiating the variation amount of the potential difference per unit time (i.e., the variation amount with time of the potential difference) by time.

If a stainless steel to be colored has a relatively uneven surface, an inflexion point does not appear on a potential-time curve showing a variation with time of a potential difference between the surface of the stainless steel and a reference electrode, but the coloring starting point which corresponds to the inflexion point appears on a differentiation curve obtained by differentiating the variation with time of the potential difference by time.

Example: A SUS 304 stainless steel abraded with No. 150 abrasive material was used as a sample. This sample was immersed in a coloring liquor comprising a mixed aqueous solution of chromic acid (250 g/ℓ) and sulfuric acid (500 g/ℓ) to color the surface. The potential difference between the surface of the sample and a platinum reference electrode in the coloring liquor was measured to prepare a potential-time curve showing a variation with time of the potential difference and a differentiation curve obtained by differentiating the variation with time of the potential difference by time. As a result, the potential-time curve and the differentiation curve were obtained. In the potential-time curve, the variation of the potential difference slowly rose with time, and an inflexion point did not appear. Therefore, it was impossible to control the coloring on the basis of this curve.

On the other hand, in the differentiation curve obtained by differentiating the potential-time curve by time, the coloring starting point was clearly indicated by an inflexion point at which the variation amount of the potential difference per unit time converts from a falling tendency to a rising tendency, the coloring starting point thus being distinctly determined.

The change in potential from the coloring starting point to the finish potential at which the formation of the desired color is complete, i.e., the preferable potential range which gives the desired color was determined. When the finish potential was reached, the stainless steel was removed from the coloring liquor. The reproducibility of the color tone thus controlled on the basis of the differentiation curve was checked in accordance with Method of Measurement for Color of Materials Based on the CIE 1931 Standard Colorimetric System.

In the same manner as above, the reproducibility of the color tone of the stainless steel colored by controlling in accordance with the conventional method was checked.

The results are shown in the following table. The indication of the color was made according to The Color Difference Indication Method as defined in JIS Z8730.

Method and Target Color	Color Indicated by JIS Z8730			Color Difference	Color Evaluated by Naked Eye
	L	a	b	ΔE	
Process					
Green	31.92	-2.64	-1.31	–	Green
Green	31.80	-2.60	-1.33	0.13	Green
Green	31.90	-2.58	-1.27	0.07	Green
Green	31.60	-2.57	-1.29	0.33	Green
Gold	38.57	0.94	8.58	–	Gold
Gold	38.60	0.90	8.50	0.09	Gold
Gold	38.71	0.93	8.71	0.19	Gold
Gold	38.40	0.88	8.62	0.18	Gold
Conventional					
Green	31.92	-2.64	1.31	–	Green
Green	31.73	-2.01	-0.01	1.47	Blue
Green	31.60	-2.23	-1.51	2.87	Blue-green
Green	31.85	-3.05	-2.40	3.73	Yellow-green
Gold	38.57	0.94	8.58	–	Gold

(continued)

Method and Target Color	Color Indicated by JIS Z8730 L	a	b	Color Difference ΔE	Color Evaluated by Naked Eye
Conventional					
Gold	38.01	1.20	9.03	0.76	Gold
Gold	37.98	0.21	7.34	1.56	Yellow-green
Gold	38.98	-0.35	7.25	1.89	Yellow-green

Note: Color was measured by 307 type color analyzer (Hitachi Seisaku-sho KK).

The samples were respectively colored four times to the target color, green or gold. As can be seen from the above table, in the case of the coloring controlled by the method, the color difference ΔE was very small and the evaluation by the naked eye was also excellent, thus proving that the reproducibility of the color tone was quite satisfactory. On the other hand, in the case of the coloring controlled by the conventional method, the color difference ΔE was large, and the color evaluated by the naked eye included variously blue, blue-green, or yellow-green when the target color was green, and included yellow-green when the target color was gold. Thus, the reproducibility of the color tone was very bad.

DYEING ZINC AND ZINC ALLOYS

Multicolor

According to *N. Kasahara and K. Nonomura; U.S. Patent 4,238,250; Dec. 9, 1980; assigned to Mitsui Mining & Smelting Co., Ltd., Japan* there is provided a process for chemically dyeing zinc or zinc alloy in multicolor, which comprises the steps of (1) dyeing with a dyestuff zinc or zinc alloy having a dyeable chromate film bonded thereto, (2) polishing locally the dyed surface to expose the surface of zinc or zinc alloy, (3) making the chromate film on the exposed surface, (4) dyeing with another dyestuff having a different color the chromate film on the exposed surface, and then the steps (2), (3) and (4) may successively be repeated.

Example: After immersing the uneven surface of a zinc alloy die casting containing 3.5 to 4.3% aluminum (JIS H5301, ZDC 2) in an aqueous solution containing 100 g/ℓ of chromic acid, 10 g/ℓ of NH_4Cl and 30 g/ℓ of $ZnSO_4$ at 35°C for 10 seconds, the surface was washed and immersed in an aqueous solution containing 5 g/ℓ of acid Anthracene Brown RH at 50°C to dye the surface in brown.

After drying the surface, the surface was coated with a clear lacquer of melamine origin by about 2 μ thickness, baked and hardened at a definite temperature, and the surface polished slightly by a buff, whereby the convex surface parts of the zinc alloy were exposed.

Next, the exposed parts were immersed in an aqueous solution containing 20 g/ℓ of NaOH and 5 g/ℓ of NaF at 40°C for 3 minutes, washed with water, repeatedly immersed in the aforementioned chromate treating solution for about 10 seconds, and then dipped in an aqueous dye solution containing 1 g/ℓ of Alizarin Yellow GG at 50°C for 3 minutes. The resulting surface was washed with water and dried to yield a colored product rich in decorative feeling, of which concave parts become brown and convex parts were dyed in pale yellow having metallic gloss.

The product was further coated with a melamine clear lacquer of the same origin to be about 15 μ thick, baked and hardened. Abrasion resistance of the surface was remarkably increased and the surface showed such excellent characteristics as indicated in the following table, the properties of normal copper electroplated zinc die cast being shown therein as comparison.

Samples	Salt Spray Test by JIS Z 2371
Product treated by this method	No generation of rust until 480 hours later
Copper electroplate (zinc die cast ground)	48 hours later

Both samples in the table have antique decorative appearances which considerably resemble each other, but they are remarkably different in the point of corrosion resistance.

Dyeing Aluminum-Containing Zinc-Based Alloy

N. Kasahara and K. Nonumura; U.S. Patent 4,200,475; April 29, 1980; assigned to Mitsui Mining & Smelting Co., Ltd., Japan describe a process for dyeing an aluminum-containing zinc-based alloy. This process comprises immersing an aluminum-containing zinc-based alloy in an alkaline solution containing hexavalent chromium ion, further immersing the thus-treated alloy in an acidic aqueous solution and then dyeing the thus-further-treated alloy in an aqueous solution of an organic dyestuff thereby to obtain a dyed aluminum-containing zinc-based alloy having decorative appearance and corrosion resistance.

The pretreatment solution is required to have a pH value of 8 or higher, preferably 11 or higher, and contains as the main alkalizing sources alkali metal compounds such as sodium hydroxide, alkaline earth metal compounds or mixtures thereof. The hexavalent chromium compound used in the pretreatment solution includes sodium bichromate or chromic acid anhydride and is employed in a concentration of 1 to 200 g/ℓ, preferably 5 to 100 g/ℓ, calculated as Cr^{6+}. The temperature for the pretreatment is preferably 40°C to boiling temperature, and the feasible treatment time is from 10 seconds to 5 minutes.

The pretreated alloy is water rinsed and then (within 60 minutes, preferably 5 minutes, including the water rinsing time after the pretreatment) immersed in the preparatory solution in preparation for dyeing the alloy. The preparatory solution may be any one which has a pH value of preferably 0.5 to 3.5 and contains hexavalent chromium ion and suitable additives mainly comprising as the anion source sulfuric acid, nitric acid, hydrochloric acid or salts thereof, as disclosed by Japanese Patents 64/62, 9558/61, 4323/65, 18728/67, etc.

The pretreatment or immersion may be continued at about 10° to 40°C for 5 seconds to 3 minutes preferably with air agitation.

After treatment with chromic acid, the alloy is thoroughly rinsed with water and then immersed in a solution of an organic dye such as an acid dye or a mordant dye for 0.5 to 10 minutes thereby realizing uniform dyeing over the entire surface of the alloy. This dyeing operation can be conducted under conditions similar to those for ordinary zinc dyeing.

Example: A molded article of superplastic zinc-based alloy (a zinc-based alloy composed principally of 22% Al and 78% zinc) was immersed, after ordinary degreasing, in a mixed aqueous solution containing 30 g/ℓ of sodium carbonate, 20 g/ℓ of sodium hydroxide and 50 g/ℓ of chromic anhydride at 80°C for 1 minute as the pretreatment, rinsed with water, immersed in a chromate treatment solution in preparation for dyeing (disclosed in Japanese Patent 9558/61). containing 120 g/ℓ of chromic anhydride, 10 g/ℓ of potassium alum and 5 g/ℓ of zinc oxide, for 20 seconds at 18°C, thoroughly rinsed with water, immersed in an aqueous solution containing 5 g/ℓ of alizarin yellow GG for 3 minutes at 50°C, water rinsed and then dried, whereby the article was uniformly dyed yellow on the entire surface.

The article so dyed was subjected to a salt spray test with the result that it did not corrode until 60 hours passed after the start of the test. On the other hand, for comparison, the same procedure as above was followed except that the pretreatment was omitted, with the result that the article was hardly dyed and it did not rust until 16 hours passed after the start of a salt spray test when subjected thereto.

COLORING ALUMINUM

Transfer of Sublimable Organic Coloring Material to Absorbent Oxide Layer

There are many processes which can be used to produce color effects in the anodic oxide layer on aluminum. One of these involves the deposition of organic coloring agents in the pores in the fiber bundle making up the top layer of a transparent, colorless oxide layer by bringing the oxide into direct contact with a hydrolysis-resistant, coloring agent which can sublimate and which is printed on a substrate, e.g., a paper substrate, with the result that the anodic oxide layer sucks up the coloring agent into the pores in the fiber bundle under the influence of heat. The coloring agents which are suitable for this process are dispersion coloring substances with anthraquinone as the basis with at least one of the positions, 1, 4, 5 or 8 occupied by either H, OH, amino or amido groups and at least one active hydrogen, or azo coloring agents with an OH group in the ortho position of the azo group, or coloring agents with a 1,3-indandione group.

After the coloring substance has been deposited in the oxide, the pores in the anodic oxide layer containing the coloring substance are closed or sealed, as by a treatment in hot, deionized water. As a result of the hot water treatment, at least a part of the Al_2O_3 of the newly produced oxide layer is converted to AlOOH, so called pseudoboehmite.

The anodic oxide layers which are multicolored, patterned or carrying a picture can be produced commercially using the transfer of coloring material which can be sublimated from a paper substrate under the influence of contact pressure and heat by so-called heat-transfer printing.

This process had not been able to develop into a usable technology as it suffered from the serious disadvantage that on transferring the hydrolysis-resistant, sublimable organic coloring substance from the substrate to the absorbent 5 to 20 μm thick anodic oxide layer by heating to the temperature of 120° to 220°C necessary for that process, fine hairline cracks occurred and these were disturbing to the eye especially when viewed at acute angles of incident light.

H. Severus and H. Birkmaier; U.S. Patent 4,177,299; December 4, 1979; assigned to Swiss Aluminium Ltd., Switzerland found that an oxide produced by anodic oxidation can be given a colored image using sublimable, hydrolysis-resistant coloring substances by means of heat-transfer printing without the oxide layer afterwards exhibiting disturbing hairline cracks due to the effect of the temperature required for the sublimation process.

The oxide layer is 5 to 25 μm thick and exhibits a crack-free elongation of at least 0.65 part per thousand in the nonsealed condition. The ratio of crack-free elongation of the oxide layer in the nonsealed condition to crack-free elongation in the colored and nonsealed condition lies between 1:1.2 and 1:5.5.

The thickness of the anodic oxide layer is preferably between 10 and 22 μm, the crack-free elongation in noncolored, nonsealed condition between 0.7 and 4 parts per thousand, and the ratio of the crack-free elongation in the nonsealed, noncolored condition to the colored, nonsealed condition between 1:1.7 and 1:5.

The properties of the absorbent oxide layer required by the process are obtained by controlled interaction of the following parameters: (1) alloy composition and condition of the product or semifinished product to be anodized, in particular sheet and extruded section; (2) composition and concentration of the electrolyte; (3) electrolyte temperature; and (4) current density.

Oxide layers which have been found to be particularly suitable are those on Al-Mg alloys containing 0.5 to 4% magnesium, preferably 1 to 3% magnesium. These alloys are used preferably in the half-hard condition, as specified by the German Specification DIN 17007 sheet 4 (corresponding to the H-14 temper), in the rolled or recovered condition.

Example: A 19 μm thick oxide layer which was produced on a half-hard, rolled, 0.8 mm thick sheet of aluminum alloy Al-Mg 1.5 (anodizing grade—aluminum alloy 5050) exhibited a crack-free elongation of 0.8 part per thousand in the nonsealed condition. This material was obtained by anodic oxidation for a period of 45 minutes with a current density of 1.3 A/dm^2. The voltage was 14 volts and the temperature of the electrolyte was 23°C. The electrolyte contained 196 g H_2SO_4/ℓ and the aluminum content was 11 g/ℓ.

A substrate was provided made of paper suitable for low-pressure printing containing various hydrolysis-resistant, sublimable dispersion coloring substances such as are used in the low-pressure printing process in the form of a mirror image pattern. The substrate was then laid on the anodic oxide layer of the aforementioned Al-Mg sheet and held for 1 minute under a pressure of 0.1 kp/cm^2 at a temperature of 180°C, during which time the colored image was transferred to the anodic oxide layer which then bore the colored image in the correct, reversed manner. There were no hairline cracks in the anodic oxide layer which exhibited a crack-free elongation of 3.7 parts per thousand in this colored, nonsealed condition. The ratio of crack-free elongation in the noncolored condition to that in the colored condition was then 1:4.6.

Next, the colored anodic oxide layer was sealed by immersion for 45 minutes in a bath of deionized, boiling water containing additions of commercially available sealing salts, i.e., the pores of the anodic oxide layer which now contained the hydrolysis-resistant, sublimable coloring substance at their base were closed by forming aluminum hydrates.

After this sealing treatment the anodic oxide layer exhibited a crack-free elongation of 3.5 parts per thousand and a Haueisen abrasion hardness of 8.3 seconds per μm of oxide layer thickness.

Dark Grey to Black Aluminum Coating

The process of *J.F. Paulet; U.S. Patent 4,018,628; April 19, 1977; assigned to Swiss Aluminium Ltd., Switzerland* allows one to obtain a strongly adherent dark grey to black coating, with good corrosion resistance, on aluminum and its alloys. The degreased aluminum is pretreated in an aqueous solution containing at least one salt of a metal which is less electronegative than aluminum. It is then colored in another aqueous solution containing molybdate ions and at least one fluorine compound and/or a heavy metal chloride, and/or an organic chlorine compound with active chlorine atoms.

The following compositions (wt %) have been found to be particularly advantageous for the chemical pretreatment of metallic surfaces:

> An aqueous solution at 50° to 95°C, preferably 85° to 95°C, containing 2 to 15%, preferably 9 to 11% iron(III) chloride ($FeCl_3 \cdot 6H_2O$) and 0.1 to 5%, preferably 0.4 to 1% sodium fluoride (NaF). The duration of the pretreatment is 1 to 2 minutes.
>
> An aqueous solution at 18° to 25°C, preferably 20° to 23°C, which contains 0.5 to 10%, preferably 1.5 to 2.5% zinc oxide (ZnO), 2.5 to 50%, preferably 8 to 12% sodium hydroxide (NaOH), 1.5 to 30%, preferably 4 to 6% sodium-potassium tartrate ($KNaC_4H_4O_6 \cdot 4H_2O$), 0.1 to 1%, preferably 0.2% iron(III) chloride, and 0.002 to 2%, preferably 0.1% sodium nitrate ($NaNO_3$). The etching time is 0.25 to 3 minutes, in particular 0.5 to 2 minutes.
>
> An aqueous solution at 40° to 70°C, which contains 2 to 10% tin chloride ($SnCl_2 \cdot 2H_2O$), 5 to 20% sodium-potassium tartrate and 0.5 to 3% sodium hydroxide, and with which the metal surface is etched for 1 to 5 minutes.

The aqueous coloring solution is normally used warm and is preferably at a temperature of 70° to 95°C. Preferably it contains 1 to 10% ammonium heptamolybdate and 0.2 to 10% of an ammonium salt containing fluorine, in particular 4 to 6% ammonium heptamolybdate and 1 to 3% ammonium fluoroborate.

The aqueous coloring solution is stabilized in a pH range between 4 and 8.5, preferably between 6.5 and 7.5. The stabilizing compounds which have given the best results in this respect are the alkaline reacting amines and their derivatives, in particular triethanolamine.

Example 1: A sheet of aluminum alloy containing 1.2% Mn, 0.8% Fe, 0.4% Si and 0.1% Zn was degreased and then immersed for 2 minutes in an aqueous solution at 90°C, containing 10% iron chloride and 0.5% sodium fluoride. With this treatment the sheet was covered with a uniform, medium grey coating.

After rinsing in cold tap water the sheet was treated for 4 minutes in an aqueous solution at 90°C, containing 5% ammonium heptamolybdate and 1% ammonium fluoroborate (NH_4BF_4). The sheet was subsequently rinsed again in tap water. It then exhibited a matte dark grey, almost black, surface with uniform color distribution.

The same result was obtained using sheets of many different alloy compositions, and with pure aluminum.

Example 2: An aluminum sheet of the same composition given in Example 1 was degreased and then immersed for 30 seconds in an aqueous zincate solution at room temperature, having the following composition: 2% zinc oxide; 10% sodium hydroxide; 5% sodium-potassium tartrate; 0.2% iron chloride; and 0.1% sodium nitrate.

The sheet was coated with a uniform, strongly bonding light grey layer. The sheet was then rinsed in tap water and colored using the conditions given in Example 1. After rinsing with cold tap water once more, the sheet exhibited a shiny, uniformly black surface.

The same results were obtained using sheets of a large variety of aluminum alloys or pure aluminum.

Example 3: A sheet of aluminum of the same composition as in Example 1 was degreased and then immersed for 2 minutes in an aqueous pretreatment solution of the same composition as in Example 2. After rinsing in cold tap water, the sheet was treated for 3 minutes in an aqueous coloring solution at 90°C, which had been stabilized by an addition of triethanolamine and had the following composition: 5% ammonium heptamolybdate; 1.5% ammonium fluoroborate; and 8% triethanolamine. The pH of the solution was stabilized at a value of 7.

The surface of the sheet was a homogeneous, shiny black color. The lifetime of this stabilized coloring solution was about 8 m^2/ℓ, in contrast to 2 m^2/ℓ of a solution of the same composition but without a stabilizer.

At this point the coloring solution was exhausted of active ingredients, but was not contaminated by precipitating polymolybdates, so that it was possible without any difficulty to regenerate the solution by a further addition of active ingredients.

Blackening Aluminum to Obtain Light-Absorbing Surface

The sun's energy can be utilized in a relatively simple, ecologically beneficial manner by intercepting its rays with the aid of collectors having a light-absorbing surface, the accumulating heat being carried off to a load by suitable means such as a fluid-circulating system. Solar energy is mainly concentrated in the near-infrared and visible ranges of the spectrum having wavelengths of about

0.2 to 2.5 μ. An efficient absorber, whose absorption coefficient α approaches unity, must act as a black body for this radiation.

Unfortunately, such black bodies usually are also effective emitters of the same radiant energy, with an emission coefficient ϵ also approaching unity, i.e., with an absorption/emission ratio $\alpha/\epsilon \approx 1$. In order to reduce the resulting heat loss, this absorption ratio must be significantly increased. Body surfaces with a ratio $\alpha/\epsilon \geqslant 1$ are termed selective absorbers; with $\alpha/\epsilon \geqslant 5$ they are considered highly selective.

H. Meissner; U.S. Patent 4,145,234; March 20, 1979; assigned to Vereinigte Metallwerke Ranshofen-Berndorf AG, Austria provides a process for blackening aluminum substrates in a manner resulting in a high absorption/emission ratio. This can be attained by forming an oxide coating with a maximum thickness of about 2.5 μ on a surface of the aluminum substrate and dyeing this coating black in a hot acidic aqueous solution of potassium permanganate and a nitrate of cobalt and/or copper.

The use of a solution containing potassium permanganate facilitates the dyeing of this thin oxide coating whose initial hue ranges from colorless to light gray. The proportion of $KMnO_4$ may vary widely, upwardly of about 1 g/ℓ, with about 200 g/ℓ representing a practical upper limit. A preferred range is 5 to 30 g/ℓ.

The copper and/or cobalt nitrates may be present in a range between substantially 1 and 100 g/ℓ, preferably 5 to 25 g/ℓ.

The pH of the solution may vary between about 0.5 and 5, a preferred range being 2 to 3. The adjustment of the pH can be accomplished by the addition of nitric or acetic acid. The temperature of the solution should be close to the boiling point, advantageously between 90° and 100°C. The treatment time in the solution may range between 1 and 5 minutes.

Example 1: An aluminum substrate, degreased and briefly pickled, is treated for 2 minutes in an aqueous bath of 95°C containing 50 g/ℓ sodium carbonate and 15 g/ℓ sodium chromate. This treatment results in the formation of a light-gray oxide coating of 0.5 μ thickness. The coated substrate, upon thorough rinsing, is dyed in an aqueous solution of 90°C containing 10 g/ℓ $KMnO_4$ and 20 g/ℓ $Co(NO_3)_2$, with admixture of sufficient nitric acid to produce a pH of 2. After a 5-minute immersion, the oxide coating has turned black with an absorption coefficient $\alpha = 0.90$ and a ratio $\alpha/\epsilon = 7.4$.

Example 2: An aluminum substrate, cleansed as in Example 1, is oxidized for 1 minute in a bath of 100°C containing 45 g/ℓ Na_2CO_3 and 15 g/ℓ Na_2CrO_4. The subsequent dyeing treatment, after rinsing, is carried out in a solution of 90°C containing 100 g/ℓ $KMnO_4$, 10 g/ℓ $Co(NO_3)_2$ and 4 ml/ℓ nitric acid. The resulting surface layer has an absorption coefficient $\alpha = 0.85$ and a ratio $\alpha/\epsilon = 7.4$.

Chemical Process Using Aqueous Alkaline Ferric Citrate

J.N. Tuttle; U.S. Patent 4,212,685; July 15, 1980; assigned to Lea Manufacturing Company has developed a process for providing a protective, color-receptive coating over aluminum using chemical means without a requirement for the use of an electric current as in electrical anodizing. Absent conventional steps such as water rinses, the process comprises cleaning, including desmutting of the surface of an aluminum part as necessary, and treatment with an aqueous alkaline solution of ferric citrate for a time sufficient to provide the coating over the part. The treated aluminum part has enhanced corrosion resistance and the coating over the part acts as a base for other coatings such as paint, and is readily dyed.

Example 1: *(A)* – An aluminum panel measuring 2" x 4" x 0.016" may be prepared by soaking for 5 minutes in a conventional nonetching aluminum soak cleaner made up at 60 g/ℓ and maintained at 150°F. The panel would then be removed, water rinsed and immersed in a conventional mild alkaline etching cleaner consisting of 55 g of cleaner (Clepo No. 30R) dissolved in 1 liter of water. The cleaning bath should be maintained at about 150°F. The panel may then be removed after about 1 minute treatment and rinsed in cold water. The clean panel would then be immersed in a 10% nitric acid solution to desmut the same and provide a clean surface. A treatment time of one-half minute should be used. The clean panel would then be rinsed with cold water and would be ready for treatment in accordance with the process.

(B) – A clean aluminum panel, such as prepared by the above procedure, is immersed in an aqueous solution comprising 50 g of ferric citrate dissolved in 1 liter of water to which a 1 to 1 (molar ratio) mixture of trisodium phosphate and potassium carbonate is added in an amount sufficient to provide solution of pH of about 12.0. The temperature of the solution is held at about 72°F and an aluminum part, such as the panel treated as above, is immersed in the solution for about 30 minutes. Thereafter, it is removed and rinsed with water.

The panel, having coating, is dyed by immersion for about 2 minutes in a dye bath of Mordant Orange 6 maintained at about 120°F. The pH of the dye is adjusted to between 6.0 and 7.0. The dyed panel is rinsed with water and then may be sealed in a solution containing 50 g of sodium dichromate dissolved in 1 liter of water with the pH maintained at about 5.9. The time of sealing may be conveniently set at 15 minutes at a temperature of about 210°F. The panel may then be rinsed with water, dried in air and buffed by hand. It has a uniform brass coloration and good wear and corrosion-resisting properties.

Example 2: Following the procedure of Example 1, the formulation is altered by substituting sodium hydroxide to pH 12.0 for the carbonate/tribasic phosphate mixture. The results are not as good as color is less uniform and is dark. There is also a rapid build-up of sludge.

DYEING WOOD GRAIN PATTERN ON ALUMINUM

Anodic Oxide Film Contacted with Coloring Substance on Water Surface

The work of *H. Hidan; U.S. Patent 4,091,126; May 23, 1978; assigned to KK Hidan Seisakusho, Japan* relates to a method of dyeing a pattern analogous to the grain of wood onto the surface of a blank consisting of aluminum or alloys thereof having a film of anodic oxidation thereon.

It is known that the surface of metal in the group of aluminum treated with anodic oxidation is coated with a film oxide formed with a number of minute pinholes, the oxide film readily receiving any oil dye.

A coloring composition containing oil dye is dropped on a water surface to diffuse, distribute or spread the coloring composition on the water surface in a patterned manner. The surface to be dyed of the blank is brought into contact in flat or inclined fashion with the patterned composition on the water surface to thereby adsorb coloring components into pinholes in the surface. The water surface is a stationary water surface or a flowing water surface.

A coloring composition containing an oil dye is deposited by dropping or pouring onto the water surface in a water tank and spreads thereon, and an aluminum blank having a film of anodic oxidation is impregnated or brought into contact with the water surface upon which the coloring composition is spread or diffused in a pattern resembling wood grain.

The coloring composition used in this method includes a vehicle and a solvent as well as a known oil dye for dyeing aluminum. The solvent itself may be known matter, but should have, when the coloring composition is dropped into the water surface, the function to control the operation in which the oil dye is diffused or spread on the water surface. That is, the kind and quantity of the solvent are related to situation of density and distribution of patterns like the grain of wood obtained.

Although not in a sense of limitation, for example, used jointly with a water-soluble organic solvent are oil groups such as turpentine oil, long chain fatty acids such as oleic acid, or long chain alcohols such as hexyl alcohol.

The aluminum blank after being dyed causes the dyed coloring component to seal the pinholes in a known manner. In this manner, patterns like the grain of wood, which are beautiful and not skived, may substantially permanently be dyed on the surface of the aluminum blank.

In the following, several examples of coloring compositions used in the dyeing method and the results obtained by the use of these coloring compositions will be given. It is to be noted that components of respective coloring compositions are represented by capacity ratio.

Example 1:

	Capacity Ratio
Oil dye (red)	Suitable amount
Vehicle (printing ink vehicle)	1
Aluminum acetate	1-2
Solvent naphtha	1
Slowly drying solvent	3

Light and shade thick and thin lines are mixed and spread over the water surface, from the top of which a blank is immersed, and then, graceful patterns like the grain of wood are dyed on the surface thereof.

Example 2:

	Capacity Ratio
Oil dye (red)	Suitable amount
Vehicle	1
Oleic acid	2-5
Ethyl acetate	1-3

The dye is spread in netted fashion on the water surface, from which a blank is immersed, and then, patterns like the grain of wood analogous to the meshes of a net are dyed on the surface thereof. These patterns may be varied into thin meshes, thick and coarse meshed or the like by variously changing the aforementioned combination ratio or by further adding alcohol thereto.

Pattern Coating Carried Out in Dip Coating Process

In a process developed by *H. Hirono, K. Nagata and N. Doguchi; U.S. Patent 4,210,499; July 1, 1980; assigned to Yoshida Kogyo KK, Japan* an article formed of aluminum or its alloys, hereinafter referred to as "workpiece," is subjected to a formation of an oxide film such as, e.g., a chemically oxidized film, an anodic oxide film and a colored oxide film by known methods after degreasing, washing, drying and, optionally, etching or desmutting.

The chemically oxidized film is formed by dipping the workpiece into a solution containing chromate, phosphate, acetate, sulfate, nitrate, fluoride, etc. The anodic oxide film is formed by electrolytically oxidizing the workpiece in an acid electrolyte, such as sulfuric acid, oxalic acid, chromic acid, etc., and the colored anodic oxide film is formed by using an electrolyte containing at least one of the organic acids selected from oxalic acid, malonic acid, citric acid, maleic acid, tartaric acid, sulfosalicylic acid, sulfophthalic acid, etc., or by using a mixture electrolyte of inorganic acid with the organic acid.

The aluminum workpiece with an oxide film formed thereon is then dipped in a coating bath floating a coating material in a multilinear or multiannular pattern to deposit thereon a coating in a wood-grain pattern. In case where the coating material floats in the multilinear pattern there is formed a pattern of a straight grain of wood, while in case of the multiannular pattern or pattern of water rings there is formed a cross-grain pattern. After drying or directly, the workpiece is then subjected to a finish coating in a spray or dip coating process and to drying and baking. For mass production, it is preferred to carry out and finish coating after the pattern coating in a dip coating process like the pattern coating.

A fine wood-grain pattern can be applied to the whole surface of the workpiece by processing it in lengthwise hung-down state even if it is of a complicated shape, because the application of the pattern coating is carried out in a dip coating process.

Example 1: An aluminum extruded sheet, A-6063S, 20 cm long x 7 cm wide which had previously been degreased, etched and desmutted in the usual ways was subjected to an anodic oxidation in a 17.5 w/v % sulfuric acid electrolyte, washed with water and dried. A black modified acrylic lacquer enamel (acrylic resin/nitrocellulose) diluted with a thinner to an IHS cup consistency of 11 seconds was poured at five points on to the surface of water slowly flowing in one direction to form on the surface five thin streaks of the enamel extending in the direction of the flow of water. The flow of water and supply of enamel were stopped just before the arrival of the front ends of the lines of coating at the overflow end of the bath, and the aluminum sheet was slowly dipped in the bath to deposit thereon a patterned coating. The sheet was then drawn up, dried in air, dip-coated with an acrylic clear lacquer and baked at 180°C for 30 minutes to obtain a black pattern of wood grain on a silvery ground of the anodic oxide film.

Example 2: After anodic oxidation and washing with water in the same manner as in Example 1, the aluminum sheet was electrolytically colored to bronze by ac electrolysis under a voltage of 15 V for 2 min using a carbon counterelectrode in an electrolytic bath with the following composition: 30 g/ℓ nickel sulfate (hexahydrate); 15 g/ℓ magnesium sulfate (heptahydrate); 20 g/ℓ boric acid; 30 g/ℓ ammonium sulfate; and 0.5 g/ℓ sodium dithionite. The pH was 5.6 with a bath temperature of 20°C.

The sheet was then washed with water, dried and subjected to the pattern coating and clear lacquer coating in the same manner as in Example 1 to obtain a black wood-grain pattern on a bronze ground of the electrolytic color film.

Example 3: The same procedure as in Example 1 was repeated except that the aluminum sheet was, in place of the anodic oxidation in the sulfuric acid electrolytic bath, subjected to a dc electrolysis at a current density of 2 A/dm^2 for 30 minutes in a mixed electrolytic bath containing sulfosalicylic acid in a strength of 100 g/ℓ and sulfuric acid in a strength of 5 g/ℓ to electrolytically color the sheet to a light amber. Thus, there was formed a black wood grain pattern on a light amber ground of the electrolytically colored coating.

In this process of *H. Hirono, K. Nagata and N. Doguchi; U.S. Patent 4,210,695; July 1, 1980; assigned to Yoshida Kogyo KK, Japan* a colored pattern imitating the grain of wood is formed on the surface of an article formed of aluminum or its alloy by dipping the article in a coating bath floating a coating material in a multilinear pattern or multiannular pattern to form on the surface of the article a masking film in a pattern of the wood grain, subjecting the article to an oxide film application or etching and, after removal of the masking film, subjecting the article to an electrolytic coloring process.

The aluminum or its alloy workpiece is subjected to degreasing, washing, drying and other conventional pretreatments and, optionally, to etching, desmutting and the like special treatments and then dipped in a coating bath floating a coating material in a multilinear or multiannular pattern to deposit thereon a wood-grain pattern, as disclosed above in U.S. Patent 4,210,499.

The workpiece thus masked with a coating film in the pattern of a wood grain is, after washing and other optional processes, subjected to a surface-modifying process. The surface-modifying process aims to make a difference in surface property between the masked and unmasked areas in the course of the following anodic oxidation where there is formed a chemically oxidized film, an anodic oxide film or a colored anodic oxide film and a barrier-type oxide film. The workpiece can be subjected to electrolytic coloring after removal of the masking film with a solvent, sulfuric acid or an organic remover.

ADDITIONAL METAL SURFACE TREATMENT PROCESSES

COMPOSITIONS FOR TEMPORARY CORROSION RESISTANCE

Acid-Amine Salts and a Water-Dispersible Polymer

There are numerous instances in industry where an easily-corrodible material (e.g., iron, steel, or other metal) is subjected to a variety of corrosive environments between the time when it is manufactured and the time when it is actually used as a raw material in the manufacture of an intended product. For example, sheet steel or iron, after it leaves the smelter and before it is incorporated by the purchaser into an end product, undergoes a considerable period (e.g., 3 to 12 mo) of shipment, storage and handling.

During such shipment, storage and handling, the metal may be subjected to environments which are capable of quickly corroding the surface of such material (e.g., environments such as air, air containing high humidity, air containing corrosive salts such as sodium chloride found in marine or industrial environments, rain, surface condensation, etc.). With increasing lengths of time before use, or with increasing severity of the environment to which the material is subjected before use, the greater is the degree of corrosion.

Consequently, it is usually necessary to remove the undesirable corrosion which is present on the surface of the material before such material may be satisfactorily used by the purchaser.

The process of *F.P. Boerwinkle and T. Szauer; U.S. Patent 4,130,524; December 19, 1978; assigned to Northern Instruments Corporation* relates to compositions which are useful for coating surfaces to provide temporary protection against corrosion. The corrosion inhibiting composition comprises:

(a) An effective amount of a salt of a carboxylic acid and an organic amine, wherein the acid has 8 to 20 carbon atoms, and wherein the amine is selected from compounds of the formula: $R_1NR_2R_3$ where R_1 is alkyl or alkenyl of 11 to 20 carbons, R_2 and R_3 are $-H$, $-CH_3$ or $-CH_2CH_2OH$; and

$$R_1{-}\underset{}{\overset{R_4}{\overset{|}{N}}}{-}(CH_2)_n{-}\overset{R_3}{\overset{|}{N}}{-}R_2$$

where n is an integer of 2 to 4, R_4 is $-H$, $-CH_3$ or $-CH_2CH_2OH$; and

(b) An effective amount of a salt selected from:

(1) Salt of phosphoric acid and a second organic amine of the formula: $R_5NR_6R_7$, where R_5, R_6 and R_7 are selected from $-H$, alkyl and cycloalkyl, hydroxy-substituted alkyl and cycloalkyl, alkenyl and cycloalkenyl, and hydroxy-substituted alkenyl and cycloalkenyl, wherein the sum of the carbon atoms in R_5, R_6 and R_7 is from 6 to 12; and

(2) Salt of the second organic amine and an alkyl acid phosphate, wherein the alkyl group has from 1 to 8 carbons; and

(c) A water-dispersible polymer derived from the free-radical-initiated addition polymerization of the following monomers in the mol percentage given:

(1) An amine monomer of the formula:

$$CH_2{=}\overset{R_{10}}{\overset{|}{C}}{-}\overset{O}{\overset{\|}{C}}{-}X{-}R_{11}{-}\overset{R_{13}}{\overset{|}{N}}{-}R_{12}$$

where R_{10} is $-H$, $-CH_3$ or $-CH_2CH_3$, R_{11} is $-CH_2CH_2-$ or $-CH_2CH_2CH_2-$, R_{12} and R_{13} are $-H$ or alkyl of 1 to 4 carbons; X is $-O-$, $-NH-$ or $-S-$; present in an amount of about 40 to 100 mol %;

(2) An acidic monomer of the formula:

$$CH_2{-}\overset{R_{10}}{\overset{|}{C}}{-}\underset{O}{\underset{\|}{C}}{-}OH$$

present in an amount ranging from 0 to 50 mol mol %; and

(3)

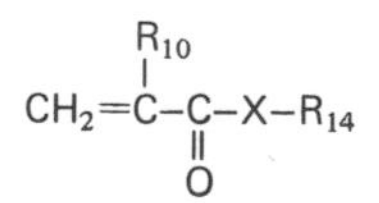

where R_{14} is hydrogen or alkyl of 1 to 4 carbons, present in an amount ranging from 0 to 20 mol %;

wherein the total mol percent of acidic monomer in the polymer is not greater than the mol percent of the amine monomer.

The compositions are water-dispersible and accordingly are preferably applied to substrates from an aqueous medium.

Example: A corrosion inhibiting composition is prepared by first preparing a water-dispersible ionic polymer, followed by adding to a water solution of such polymer appropriate amounts of carboxylic acid and amine such that the desired salt is formed in situ.

The water-dispersible polymer is prepared by the free-radical-initiated addition polymerization of the following monomers in the mol percentage stated: 50 mol % dimethylaminoethylmethacrylate, and 50 mol % methacrylic acid.

These monomers are dissolved in water in a suitable reaction vessel in an amount such that the total weight percentage of monomer in water is 24%. This solution is heated to approximately 70°C with stirring, after which 0.5% (based on the weight of monomer) of ammonium persulfate catalyst is added.

The reaction mixture exotherms to about 90°C in about 15 min, and the reaction mixture is maintained at 90°C for another 15 min, after which the polymer solution is cooled to room temperature. The polymer solution (23.5 to 24% solids) has a Brookfield viscosity in the range of 5,000 to 10,000 cp at room temperature, and a pH of 6.2.

A 12.5 ml sample of the above polymer solution is diluted with 87.5 ml of water to provide a water solution having 3 wt % polymer present therein. To this resulting polymer solution is added 6 g of cyclohexylamine butyl acid phosphate salt. Then 0.75 g oleyl amine is added, with stirring, after which 0.75 g oleic acid is added, with stirring. To improve the dispersion of the acid-amine salt in the polymer solution, the mixture is subjected to mixing, for 3 to 5 min, by a nonaerating stirrer (Kraft Apparatus, Inc., Model S-30). A slightly milky dispersion is obtained having a pH of about 6.4.

The product composition is applied to clean mild steel (number 1010) test panels (4" x 6") by spraying so as to obtain a continuous coating on such panels. The thus coated panels are immediately suspended vertically, and allowed to dry at room temperature for at least 4 hr.

The coated test panels, and clean untreated control panels, are then subjected to condensing humidity at 100°F according to the procedure of ASTM D-1735-62. The degree of corrosion is reported according to the Standard Method of Evaluating Degree of Rusting on Painted Steel Surfaces, ASTM D-610.68.

The untreated control test panels exhibited Grade 8 corrosion (0.1% of surface corroded) after 3 hr, and Grade 1 corrosion (greater than 50% of surface corroded) after 24 hr.

The test panels coated with the corrosion inhibiting composition of this Example, exhibited no corrosion at all (Grade 10) even after 1 wk.

Additional test panels of clean mild steel (number 1010), 2" x 4", are also coated with the composition of this Example by spraying and are hung vertically for 4 hr at room temperature. The coated test panels, and clean untreated control panels, are then partially immersed in 100 ml of 3.5% sodium chloride aqueous solution in a 250 ml beaker at room temperature. Visible corrosion of the untreated control panels is observed within 15 min, whereas no visible corrosion of the coated test panels is observed even after 10 hr. This demonstrates that the compositions of this process are extremely effective in imparting resistance to chloride corrosion of mild steel under atmospheric conditions.

In this process, *F.P. Boerwinkle; U.S. Patent 4,131,583; December 26, 1978; assigned to Northern Instruments Corporation* describes additional compositions which are useful for coating surfaces to provide temporary protection against corrosion. The corrosion inhibiting composition comprises:

(a) An effective amount of a salt of a carboxylic acid and an organic amine, wherein the acid has 11 to 20 carbon atoms, and wherein the amine is selected from compounds described above in U.S. Patent 4,130,524; and

(b) A water-dispersible polymer derived from the free-radical-initiated addition polymerization of the monomers in the mol percentage also given above in U.S. Patent 4,130,524.

Example: A corrosion inhibiting composition is prepared by first preparing a water-dispersible ionic polymer, followed by adding to a water solution of such polymer appropriate amounts of carboxylic acid and amine such that the desired salt is formed in situ.

The water-dispersible polymer is prepared by the free-radical-initiated addition polymerization as described above in the Example of U.S. Patent 4,130,524.

A 12.5 ml sample of the above polymer solution is diluted with 87.5 ml of water to provide a water solution having 3 wt % of polymer present therein. To this resulting polymer solution is added 0.5 g oleyl amine, with stirring, after which 0.5 g oleic acid is added, with stirring. To improve the dispersion of the acid-amine salt in the polymer solution, the mixture is subjected to mixing, for 3 to 5 min, by a nonaerating stirrer (Kraft Apparatus, Inc., Model S-30). A slightly milky dispersion is obtained having a pH of about 6.4.

A product composition is applied and evaluated as described above in the Example of U.S. Patent 4,130,524.

The untreated control test panels exhibited Grade 8 corrosion (0.1% of surface corroded) after 3 hr, and Grade 1 corrosion (greater than 50% of surface corroded) after 24 hr.

The test panels coated with the corrosion inhibiting composition of this Example exhibited no corrosion at all (Grade 10) even after one week.

COMPOSITE COATINGS

Isocyanate Adduct Incorporated in Topcoating

In composite coating techniques, it has been known to dry an undercoating, which needs curing, and then apply a topcoating. A single curing step then follows to cure both the undercoating and the topcoating as has been disclosed, for example, in British Patent 845,259.

After the precoat application, then the topcoat application, and the final curing, the resulting coated article must usually exhibit a wide variety of desirable characteristics.

T. Sato; U.S. Patent 4,098,620; July 4, 1978; assigned to Diamond Shamrock Corporation found that a final curing operation will not only fully cure the undercoating but will also, at the same time, cure the topcoating while providing a composite coating having enhanced characteristics. This final cure operation for both the undercoating and the topcoating is referred to herein for convenience as "monobaking." By incorporating isocyanate adduct into the applied topcoating, this monobaking can be effected without variation from the procedure with coatings not containing such adduct.

For the resulting composite coating, the sought after enhancement of resistance to chemical attack, such as acid resistance, as well as resistance to solvent attack is achieved. Other expected coating characteristics are not diminished. Rather, shear adhesion can be enhanced. Also, the storage life of the topcoat composition is unaltered. Unexpectedly, corrosion resistance for the coated metal substrate such as exhibited in salt spray testing is most desirably augmented.

The method first comprises establishing on the surface of the substrate an undercoating containing from 50 to 5,000 mg/ft^2 of coated substrate of pulverulent metal in intimate mixture with a substantially resin-free hexavalent-chromium-containing coating composition.

Such composition contains a hexavalent-chromium-providing substance and reducing agent therefor in liquid medium. The composition is present in an amount sufficient to provide the coating with from 5 to 500 mg/ft^2 of coated substrate of chromium.

The method next comprises drying the resulting undercoated substrate at a temperature and for a period of time sufficient to vaporize volatile substituents from the coating composition, but insufficient to provide a water-resistant undercoating having substantial orientation of the hexavalent chromium towards reduction. The method next comprises applying on the dry undercoating a topcoat composition comprising hydroxyl-containing resin, particulate, electrically conductive pigment and less than 5%, basis total topcoat composition weight, of isocyanate adduct capable of unblocking at a temperature above 350°F.

The method lastly comprises heating the resulting coated article to a temperature above 350°F and for a period of time sufficient to: (a) orient at least a portion of the undercoating hexavalent chromium toward reduction; commensurate with (b) unblocking the isocyanate adduct and providing crosslinking reaction of resulting isocyanate with topcoat resin.

Preparation of Test Panels: Unless otherwise specifically described, test panels are typically 4" x 8" cold rolled, low carbon steel panels. They are prepared for coating by first scrubbing with a cleaning pad which is a porous, fibrous pad of synthetic fiber impregnated with an abrasive. Thereafter, the scrubbed panels are immersed in a cleaning solution typically containing chlorinated hydrocarbon and maintained at about 180°F, or containing 1 to 5 oz/gal of water, of a mixture of 25 wt % tripotassium phosphate and 75 wt % potassium hydroxide. This alkaline bath is maintained at a temperature of about 150° to 180°F. Following the cleaning, the panels are rinsed with warm water and preferably dried.

Solvent Resistance Test: Painted panels are immersed for 16 hr in organic solvent. Panels are removed from the solvent and are immediately manually rubbed across the panel about twenty strokes with a paper tissue while using moderate pressure. Rubbed panels are then visually inspected to see if the undercoating has been exposed during the rubbing, with the film being regarded as soluble if the undercoating has been exposed. Solvents used include benzene, toluene and cellulose acetate.

Corrosion Resistance Test (ASTM B-117): Corrosion resistance of coated parts is measured by means of the standard salt spray (fog) test for paints and varnishes ASTM B-117.

Example: There is formulated, with blending, an undercoating composition containing 20 g/ℓ of chromic acid, 3.3 g/ℓ of succinic acid, 1.7 g/ℓ of succinimide, 1.5 g/ℓ of xanthan gum hydrophilic colloid, which is a heteropolysaccharide prepared from the bacteria species *Xanthomonas campestris* and has a molecular weight in excess of 200,000. Additionally, the composition contains 1 ml of formalin, 7 g/ℓ of zinc oxide, 120 g/ℓ of zinc dust having an average particle size of about 5 μ and having all particles finer than 16 μ, and 1 drop per liter of a wetter which is a nonionic, modified polyethoxide adduct having a viscosity in centipoises at 25°C of 180 and a density at 25°C of 8.7 lb/gal. After mixing all of these constituents, this undercoating composition is then ready for coating test panels.

Panels, prepared as described hereinabove, are dip coated in the undercoating composition. They are then removed from the composition and excess composition is drained from the panels. Some panels are then baked for 1.5 min, and are thereby the "monobake" panels, and some for 4 min, which are "dibake" panels, all in an oven at an oven temperature of 500°F.

Panels are then primer topcoated. The primer used is a commercially available primer which is a zinc-rich weldable primer having a weight per gallon of about 15.2 lb, a solids volume of about 29%, and containing about 64 wt % solids content. The binder component is prepared from a high molecular weight epoxy resin. The primer has a typical viscosity of about 80 sec as measured on a Number 4 Ford cup.

Prior to use, there is blended, with vigorous agitation, into one batch of the primer 0.5 wt %, basis total batch weight, of a phenol blocked polyisocyanate adduct. The adduct is a yellowish solid having a weight per gallon of 10.6 lb, an average equivalent weight of 336 and a percentage of available NCO of 11.5 to 13.5. Each respective batch of primer is applied to undercoated panels by drawing the primer down over the panel with a Number 18 draw bar to provide a smooth, uniform primer coat, generally of about 0.5 mil thickness. Resulting coated panels are cured for either 2½, 3 or 4½ min in an oven at 500°F.

In the manner as described hereinabove, panels that have been subjected to the abovedescribed corrosion resistance (salt spray) test are quantitatively evaluated for corrosion by visual inspection, comparing panels with one another and employing the above-discussed rating system. From the resulting ratings, the results for percentage changes are determined by straightforward calculation. Panels subjected to the above-discussed solvent resistance test gave the following results.

Undercoating cure time–monobake, 1.5 min; dibake, 4 min. Solvent resistance–no adduct, soluble; with adduct, insoluble. Corrosion resistance improvement–no adduct, at 330 test hours, 100%, and at 520 test hours, 100%; with adduct, at 330 test hours, 120% and at 520 test hours, 130%.

Prepaint and Topcoating Compositions Formulated with Multivalent Metal

V.V. Germano; U.S. Patent 4,020,220; April 26, 1977; assigned to Diamond Shamrock Corporation found that corrosion resistance of coating composites can be desirably enhanced when both the prepaint and the topcoating compositions are formulated with multivalent metal and careful control is exercised over cure conditions.

Additionally, water sensitivity of the base coating can be sufficiently suppressed to permit not only topcoating with compositions that are water-based, but also to permit water quenching of base coatings after heat curing.

A portion of the coating composite comprises an undercoating and a subsequent coating, each established from heat curable compositions that before curing contain, in liquid medium, a metal in nonelemental form, which metal can exhibit multivalency and is susceptible to valency reduction to a lower valence state during curing of applied composition. Moreover, such metal is at least partially present in the composition in a higher valency state and the compositions each further contain reducing agent for the metal.

The process provides extended substrate corrosion resistance protection by first establishing the undercoating in the composite, for providing substrate protection, but in non-water-resistant condition, and then precuring the established undercoating, at elevated temperature, to a dry and water-resistant coating, with the conditions of the elevated temperature precuring being selected to orient the undercoating towards containing a minimum amount of the metal in a higher valency state.

Next, the process provides for establishing the subsequent coating of the composite for providing substrate protection, with the subsequent coating being established in non-water-resistant condition, and finally, curing the subsequent coating through the conditions of the precuring, thereby initially orienting the subsequent coating towards minimization of higher valency state metal, and with the elevated temperature conditions by continuation thereof, then orientating this subsequent coating away from the minimization of higher valency state metal.

Of particular interest are the bonding coatings, i.e., coatings from compositions containing hexavalent-chromium-providing substance and a reducing agent therefor.

Although multivalent metals other than chromium can be present in the cured undercoating, such as for example, molybdenum and tungsten, chromium is selected as representative for determination of the appropriate cure conditions. In general, the precure conditions for chromium-containing undercoatings are cure temperatures below 550°F air temperature, and at such temperature, for times of less than about 10 min. However, lower temperatures such as 450° to 500°F, with commensurately longer cure times, such as up to 25 min or more, can be typically used.

It is also contemplated that the pulverulent-metal-containing base coating also form the topcoating. In general such topcoating may be a second application without variation, of the base coating. However, the undercoating may contain multivalent metal differing from such metal of the topcoat.

Again selecting chromium as representative for determination of the appropriate cure conditions, such cure conditions for the subsequent coatings is a cure temperature above 550°F air temperature or more for a time of greater than about 10 min. Preferably, for greater efficiency, curing proceeds at 600° to 700°F air temperature, with cure times of 10 to 20 min.

Example: Test panels, 4" x 8", that are all cold rolled, low carbon steel panels are used. These panels are prepared for coating by first scrubbing with a cleaning pad which is a porous, fibrous pad of synthetic fiber impregnated with an abrasive. Thereafter, the scrubbed panels are immersed in a cleaning solution typically containing 1 to 5 oz/gal of water, of a mixture of 25 wt % tripotassium phosphate and 75 wt % potassium hydroxide. This alkaline bath is maintained at a temperature of about 150° to 180°F. Following the cleaning, the panels are rinsed with warm water and preferably dried.

A test composition is prepared from 200 ml dipropylene glycol, 4 ml of wetter which is a nonionic, modified polyethoxy adduct having a viscosity in centipoises at 25°C of 180 and a density at 25°C of 8.7 lb/gal, 350 g of zinc flake having particle thickness of about 0.1 to 0.2 μ and a longest dimension of discrete particles of about 15 μ, 700 ml of deionized water, 50 g of chromic acid and 2 g of hydroxyethyl cellulose thickener. The thickener is a cream to white colored powder having a specific gravity of 1.38 to 1.40 at 20/20°C, an apparent density of 22 to 38 lb/ft^3, and all particles pass through 80 U.S. mesh.

Panels are coated by dipping into the coating composition, removing the panels and draining excess composition therefrom. This draining is then immediately followed by baking. Some panels thus coated have a high coating weight of 1,400 mg/ft^2 and others, by diluting the bath with distilled water before dipping, have a low coating weight of 700 mg/ft^2. Baking proceeds in a convection oven at an air temperature of about 450°F for a time of 10 min.

Subsequently, some of the low coating weight panels are redipped in the water diluted bath, so that they will achieve a final, two-coat weight of 1,400 mg/ft^2. Coating weights are determined by weighing the panel before coating, and then reweighing the coated panel. After the second coat, the two-coat panels are baked at an air temperature of 600°F for 15 min.

Some of the coated panels are then subjected to testing for leachable hexavalent chromium. The test method involved is a standard iodimetric titration. First, a test panel is immersed in 100 ml of a 2% ammonium hydroxide solution for 15 min. The panel is removed and the solution is titrated. In this titration, and in brief, the solution is acidified with concentrated hydrochloric acid to insure that the pH of the sample is less than 7. To the acidified sample there is then added a KI/starch ingredient.

If the solution turns purple, the presence of leachable chromium is thus indicated. A purple sample can then be titrated with standardized thiosulfate to the purple/white color change. This titration measures the free iodine of the sample which is quantitatively associated with the hexavalent chromium. By this test, both a low coating weight panel (one coat) and a one coat, but high coating weight panel, both cured at 450°F, show no leachable chromium. However, a two-coat panel indicates a color change, and thus the presence of leachable hexavalent chromium in the coating.

Selected heavy coating weight one coat panels, as well as two-coat panels, are then subjected to a corrosion resistance test by means of the standard salt spray (fog) test for paint and varnishes as described in ASTM B-117-64. In this test, panels are placed in a chamber held at constant temperature where they are exposed to a fine spray (fog) of a 5% salt solution for a period of time until first red rust is noted on the panel. By this testing, the two-coat panel is observed to proceed in the test more than one thousand hours, which is more than four times longer than the one coat panel, although each have comparable coating weights.

Phosphatized Base Coating and Water-Reducible Paint Topcoat

The work of *E.A. Rowe, Jr. and W.H. Cawley; U.S. Patent 4,143,205; March 6, 1979; assigned to Diamond Shamrock Corporation* is directed to a coated metal substrate having on the surface thereof an adherent, corrosion-resistant and water-insoluble composite coating. The coating comprises a water-insoluble phosphatized base coating on the metal obtained by contacting the substrate with an organic solvent and water-containing phosphatizing composition, containing water in minor amount, and on the base coating a paint topcoating from a paint topcoat composition.

The base coating is obtained by contacting the metal substrate with organic phosphatizing composition that comprises organic solvent providing liquid phase homogeneity with an organic solubilizing liquid while being a nonsolvent for a phosphatizing proportion of phosphoric acid in the composition; the organic solvent is unreactive with phosphoric acid in the composition. The phosphatizing composition further comprises a solubilizing liquid capable of solubilizing phosphoric acid in the composition while retaining liquid phase composition homogeneity, and with the solubilizing liquid being unreactive with phosphoric acid in the composition.

Further, the composition comprises a phosphatizing proportion of phosphoric acid, and water in an amount exceeding the proportion of phosphoric acid while being sufficient for the composition to provide a phosphatized coating of substantial water insolubility on a metal substrate in phosphatizing contact with such composition and while retaining liquid phase homogeneity.

The water-insolube base coating on a ferruginous substrate will be complex phosphatized coating of the iron phosphate type, and contain, in addition to trace elements, the elements iron, phosphorus, and oxygen plus carbon and nitrogen, and have a coating surface ratio of oxygen atoms to phosphorus atoms of at least 4:1, and a coating surface ratio of carbon atoms to phosphorus atoms of greater than 1.5:1.

Preferably, for best coating characteristics including augmented topcoat adhesion and corrosion protection, the base coating will be present in an amount between 20 and 100 mg/ft^2.

Preparation of Test Panels: Bare steel test panels, typically 6" x 4" unless otherwise specified, and all being cold rolled, low carbon steel panels are typically prepared for phosphatizing by degreasing for 15 sec in a commercial, methylene chloride degreasing solution maintained at about 104°F. Panels are removed from the solution, permitted to dry in the vapor above the solution, and are thereafter ready for phosphatizing.

Phosphatizing of Test Panels and Coating Weight: In the examples, cleaned and degreased steel panels are phosphatized by typically immersing the panels into hot phosphatizing solution maintained at its boiling point, for from 1 to 3 min each. Panels removed from the solution pass through the vapor zone above the phosphatizing solution until liquid drains from the panel; dry panels are then removed from the vapor zone.

Unless otherwise specified, the phosphatized coating weight for selected panels, expressed as weight per unit of surface area, is determined by first weighing the coated panel and then stripping the coating by immersing the coated panel in an aqueous solution of 5% chromic acid which is heated to 160° to 180°F during immersion. After panel immersion in the chromic acid solution for 5 min, the stripped panel is removed, rinsed first with water, then acetone, and air dried. Upon reweighing, coating weight determinations are readily calculated. Coating weights are expressed in milligrams per square foot. The tests used were:

Mandrel Test Bending ASTM D-522,
Reverse Impact ASTM D-2794-69,
Cross-Hatch Test, and
Relative Humidity Test ASTM D-1748.

Example: To 76.3 parts of a stabilized, commercially available technical grade methylene chloride there is added, with vigorous agitation, 16.3 parts methanol, 0.43 part orthophosphoric acid, 3.0 parts N,N-dimethylformamide, 3.9 parts water and 0.07 part dinitrotoluene. These blended ingredients are thereafter brought to reflux and the resulting solution is then used to phosphatize panels in the manner described hereinabove. For test purposes, panels having a 30 mg/ft^2 phosphatized coating weight are used.

The resulting phosphatized panels, as well as some additional of the abovedescribed bare steel test panels, used for comparative purposes, are all coated with a standard metal coating primer. More particularly, the primer is a gray zinc chromate primer number 63-12519 (Du Pont). The phosphatized panels are not thereafter coated. However, the bare steel panels with the zinc chromate primer are further topcoated with a solvent-based acrylic enamel. This is an off-white enamel number E525 (Du Pont).

A set of two resulting panels are then tested in the abovementioned cross-hatch test, which uses seventeen scribe lines per inch; another set of two panels are tested in the conical mandrel test. Results of such testing are reported in the following table.

Results are also reported in the following table for the reverse impact test at 40 inch-pounds. These results for reverse impact, as well as for the cross-hatch test, are reported numerically using the following rating system.

(10) Complete retention of film, exceptionally good for the test used;
(8) Some initial coating degradation;
(6) Moderate loss of film integrity;
(4) Significant film loss, inacceptable degradation of film integrity;
(2) Some coating retention only; and
(0) Complete film loss.

Results are also reported for panels subjected to the abovementioned relative humidity test. In this test, panels are subjected to 24 hr exposure.

Panel	Cross-Hatch	Conical Mandrel (mm)	Reverse Impact	Relative Humidity (blisters)
Zinc chromate primed	10	48	5	#6 medium-dense
Phosphatized	10	33	7	No blisters

The results show the highly desirable topcoat adhesion characteristic for the solvent phosphatized panels. This is for the phosphatized panels having a coating that, although it is water-insoluble, will also provide excellent topcoat adhesion for solvent-reduced topcoats. The results are particularly desirable since the total paint film thickness for the phosphatized test panels is measured at 0.7 mil, versus a total film thickness for the zinc chromate primed panels of 1.75 mils. Such measurements are determined by subjecting the panels to a Permascope, type ES-4 (from Twin City Testing Corp.), which operates by magnetic permeability.

Topcoating of Chromium Compound for Phosphated Nuts and Bolts

A. Teramae, K. Yamada and H. Kawasaki; U.S. Patent 4,074,011; February 14, 1978; assigned to Nippon Steel Corporation and Kobe Steel Ltd., both of Japan describe a two-step coating technique in which after having been cleaned by shot blast or acid-washing, a bolt, nut and washer are first treated with a phosphating solution containing phosphate of metals such as of iron, manganese, zinc and the like to form an adherent phosphate coating as a base coat, and is then coated thereon with a topcoat comprising essentially a chromium compound contained in a vehicle solution and further containing one or more compounds selected from metal salts of higher fatty acid, polyolefin compounds and ethylene tetrafluoride polymers either dispersed, or dissolved, or swelled in the solvent of the vehicle solution.

The phosphate coating which is suitable for the process should have a thickness corresponding to a coating weight of 0.1 to 50 g/m^2, and is preferably of the so-called thick film type with a coating weight of 5 to 30 g/m^2.

The content of chromium compound in the topcoating composition should be within the range of from 0.5 to 25% of chromium based on the dried topcoating weight.

In view of the performance of bolts, nuts and washers, it is undesirable to use slow drying vehicles, such as, oil-paint. Among the quick drying vehicles which are effective to serve the purposes of this process are those of vinyl polymers, acrylic resin, epoxy resin and melamine-alkyd resin. These polymers and resins are used in the form of solutions or dispersions in water or organic solvents, for example, alcohol solutions of polyvinyl butyral which is conventionally used as an etching primer, aqueous acrylic resin emulsions, epoxy primers which are used to provide a base for enamel, and aqueous melamine-alkyd resin solutions.

As a lubricity-imparting ingredient, use may be made of a metal salt of a higher fatty acid, such as, sodium stearate, calcium stearate, zinc oleate and magnesium linolenate. In order to ensure the sufficient lubricating action of the composite coating, the higher fatty acid metal salt employed must be uniformly distributed in the vehicle of the resultant topcoating. In other words, it must be either dissolved in, or swelled with the solvent of the vehicle solution, or otherwise no satisfactory lubricious properties can be imparted to the resultant composite coating. The content of higher fatty acid metal salt is preferably within the range of from 1 to 20 wt % based on the topcoating weight.

Another example of the lubricity-imparting ingredient is polyolefin compounds, namely, polymers of ethylenic hydrocarbons, such as polyethylene, isotactic polypropylene and polybutene. The polyolefin compound should have a molecular weight ranging from 1,000 to 10,000.

The content of the polyolefin compound should be within the range of 1 to 50 wt % based on the topcoating weight, and preferably from 5 to 20 wt %.

The topcoating weight suitable for the purpose is preferably within the range of 0.1 to 50 g/m^2. In terms of thickness, the coating weight of 1 g/m^2 corresponds to a thickness of about 0.67 to 1 μ, because the specific ratio of the topcoating is found to be about 1.0 to 1.5 μ.

The improvement of the lubricating property of the bolt and nut coated in accordance with the process is intended to reach a level of less than 0.150, and preferably less than 0.100 in terms of torque coefficient while still minimizing the range of distribution of the torque coefficients, thereby facilitating the fastening operation of each of the bolt and nut assemblies by application of an almost constant rating of torque force.

Example: After having been cleaned, a number of bolts made of high tensile strength steel (F11T) and having a size (nominal size of M30; length under head of 15 mm) in combination with corresponding nuts and washers were treated with a phosphating solution containing phosphate and zinc phosphates to form an undercoating of a weight of 10 g/m^2. Next, the phosphated articles were coated with the following coating composition to form a topcoat of a weight of 20 g/m^2.

Topcoating Composition

	Weight (g/m^2)
Ammonium chromate (chromium)	1.00
Sodium stearate	1.25
Polyethylene tetrafluoride particles of diameters less than 1 μ	1.00
Polyethylene particles of diameters less than 0.5 μ	1.25
Acrylic resin aqueous emulsion vehicle	

The results of various tests are as follows. There was no formation of rust after exposure to the outdoor humidity for 6 mo. The lubricity torque coefficient was 0.080 (means of $n = 100$) with a standard deviation of 0.008. In paintability tests, with epoxy type film at 30 μ, chlorinated rubber type at 30 μ, phthalic acid type at 50 μ, and tar-epoxy type at 70 μ, all peeled off slightly.

ORGANIC COATINGS

Organic Film Forming Resin, Phosphoric Acid, m-Nitrobenzoic Acid

"Autophoresis" process for coating metal surfaces have been described which comprise contacting the metal surface with an acidic aqueous resin dispersion which is such that the acid etches the metal surface to generate metal ions and as a result the dispersion becomes destabilized adjacent the metal surface to deposit thereover a substantially uniform coating comprising the resin, and probably also the metal ions. The process has the particular advantage that it will give substantially uniform thickness irrespective of the shape of the article, whereas conventional techniques of applying a coating, such as brushing, give variable thickness.

The autophoretic aqueous acidic coating composition of *D.B. Freeman and J.L. Prosser; U.S. Patent 4,199,488; April 22, 1980; assigned to Oxy Metal Industries Corporation* is free of fluoride and hydrogen peroxide and contains an organic film forming resin, phosphate and a nitrobenzoic acid or derivative thereof. The composition imparts corrosion resistance when contacted with a metal surface.

The preferred nitrobenzoic acid is m-nitrobenzoic acid, in the free acid form, in amounts of from preferably 0.25 to 3 g/ℓ with best results generally being obtained at about 1.5 g/ℓ.

The amount of phosphoric acid is preferably from 0.01 to 0.2 N with best results being obtained at about 0.02 to 0.05 N. Thus, preferably 0.3 to 3 g/ℓ phosphoric acid is used, generally about 1 g/ℓ. The amount of resin is preferably from 20 to 200 g/ℓ with best results generally being obtained at from 40 to 120 g/ℓ. The aqueous resin dispersion will contain dispersing agents in conventional manner which are generally nonionic. The resin should be a film forming resin. Preferably it is an acrylic resin.

Contact between the aqueous acidic resin dispersion and the surface is best achieved by immersion for from 15 sec to 2 min.

Example: A composition was made by mixing 50 g of an acrylic resin dispersion of "Primal" AC 64, 7.5 ml of 10% phosphoric acid solution, 1.5 g m-nitrobenzoic acid and sufficient water to make 1 ℓ. A clean steel panel was immersed in this solution for 5 min and was then removed, allowed to drain for 90 sec and baked at 140°C for 10 min. This steel panel was then subjected to a 24 hr salt spray test in accordance with ASTM-B114. 5% rusting was observed.

Comparative Example: The nitrobenzoic acid was omitted and 2.5 g of 30% hydrogen peroxide was added in its place. When subjected to the same salt spray test this gave 100% rusting.

Incorporating Bis(Chloroalkyl) Vinyl Phosphonates in PTFE Coatings

It is conventional to provide razor blade cutting edges with a coating of certain polyfluorocarbons, particularly polytetrafluoroethylene (PTFE). However, when PTFE is applied by conventional techniques to substrates other than steel and platinum/chromium alloys inadequate adhesion is obtained. Specifically, there are a number of materials, in particular alumina, silica (quartz), tungsten, titanium and tantalum, which have properties which would make them very suitable for use as razor blade cutting edges or as coatings on steel razor blade cutting edges, but which cannot effectively be used for this purpose because it is not possible, using available techniques, to form adherent polyfluorocarbon coatings thereon and such polymer coatings are as essential in obtaining the desirable overall combination of shaving properties as the nature of the underlying metallic structure.

K.S. Downing and R. Johnston; U.S. Patent 4,102,046; July 25, 1978; assigned to The Gillette Company found that satisfactory adhesion of polyfluorocarbon coatings to substrates of alumina, silica (quartz), tungsten, titanium or tantalum can be obtained by applying the polyfluorocarbon in the usual way together with certain bis(chloroalkyl) vinyl phosphonates and then sintering the polyfluorocarbon, again in the usual way.

The process comprises applying a dilute dispersion of the polyfluorocarbon, which dispersion also contains a bis(chloroalkyl) vinyl phosphonate of the formula:

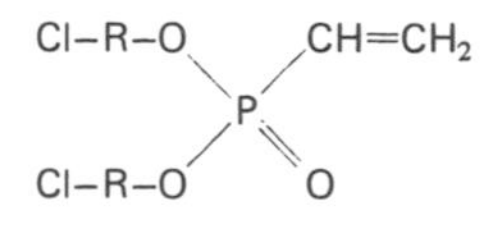

in which each R is an alkylene group with 1 to 4 carbon atoms, to the substrate and drying and sintering the coating formed.

The preferred bis(chloroalkyl) vinyl phosphonate is bis(chloroethyl) vinyl phosphonate (Pfaltz and Bauer, Inc.), which is a known compound. The amount of bis(chloroalkyl) vinyl phosphonate used is preferably 10 to 100%, based on the weight of polyfluorocarbon in the dispersion.

Examples 1 through 5: Stainless steel razor blades having sputtered coatings of the following materials on their cutting edges: alumina (Example 1), silica (Example 2), tungsten (Example 3), titanium (Example 4), and tantalum (Example 5), were provided with PTFE coatings on their cutting edges as follows.

A stack of the blades was heated in air to 100°C and their cutting edges were sprayed with a dispersion of 1% w/v of PTFE and 1% w/v of bis(chloroethyl) vinyl phosphonate in a mixture of 90% by volume of t-butanol and 10% by volume of 1,1,2-trifluoro-1,2,2-trichloroethane (Freon 113). The amount of dispersion applied to the cutting edges was such that the final coating thickness was 2000 to 5000 Å. After drying, the coatings were sintered in a nonoxidizing atmosphere at a temperature of 350°C for 15 min.

The PTFE-coated blades were subjected to cutting evaluation. For this purpose, samples of the new, freshly coated blades were used to cut through a standard thickness of wet wool felt and the cutting force (L_5) required to cut through the felt was measured (in pounds). The same blades were then used to cut 100 times through the same standard thickness of wet wool felt and the cutting force ($\overline{X}_{100}$) required for the 100th cut was measured (in pounds). The difference, $\overline{X}_{100} - L_5$ is termed the Δ value; the lower the Δ value, the better the adhesion of the PTFE coating. The results obtained are as follows.

Example	Substrate	L_5 (lb)	$\overline{X}_{100}$ (lb)	Δ (lb)
1	Alumina	1.15	1.43	0.28
2	Silica	1.26	1.53	0.27
3	Tungsten	1.15	1.53	0.38
4	Titanium	1.12	1.51	0.36
5	Tantalum	1.20	1.52	0.32

The Δ values shown in the right hand column of this table are entirely satisfactory for commercially acceptable razor blades and are comparable with those obtained by the conventional coating of PTFE on stainless steel razor blades (not having any metallic coating on the cutting edges).

When the abovedescribed PTFE coating precedure was repeated, but omitting the bis(chloroethyl) vinyl phosphonate from the coating dispersion, on stainless steel blades having the same coatings as those of Examples 1 through 5 on the cutting edges, the $\overline{X}_{100}$ cutting force was, in each case, too high to measure (the maximum cutting load which the equipment is designed to accept is 5 lb); that is to say, the cutting edges did not survive the 100 cuts.

Synergistic Mixture of Anodic and Cathodic Organic Inhibitors

Included among the several forms of corrosion which can cause the failure of metallic equipment exposed to oxygen-bearing water are general and pitting corrosion. General corrosion proceeds over the entire metal surface. Pitting corrosion occurs at small discrete locations on the metal surface and penetrates the metal 20 to 100 times faster than general corrosion. This leads to much earlier failure of metallic equipment than if just general corrosion were present.

Although organic inhibitors have been widely used to inhibit general corrosion, with the hope that they would also reduce pitting corrosion, it is conventionally believed that pitting corrosion could not be effectively inhibited in highly oxygenated systems without employing heavy metals such as Zn or Cr.

W.S. Tait; U.S. Patent 4,240,925; December 23, 1980; assigned to Petrolite Corporation found that certain synergistic mixtures of organic inhibitors effectively inhibit pitting corrosion, particularly where the mixture of organic inhibition is a blend of both anodic and cathodic organic inhibitors.

For this process, an anodic inhibitor was distinguished from a cathodic inhibitor by noting its effect on the height of the prospective nonlinear anodic polarization curve with respect to nonlinear curves obtained from an identical uninhibited system. Thus, a cathodic inhibitor lowers the height of the nonlinear cathodic polarization curve and an anodic inhibitor the nonlinear anodic polarization curve. Inhibitors which lower nonlinear polarization curves will not usually lower the height of linear polarization curves used for measuring general corrosion. Thus, linear polarization cannot be usually used to measure pitting corrosion.

These nonlinear polarization curves are obtained from potentiodynamic scanning.

Example 1: Examples of organic anodic and cathodic inhibitors suitable as pitting corrosion inhibitors are presented below.

Inhibitor	% Pitting Inhibition
Cathodic	
Phosphorylated pentaerythritol	86
Phosphorylated (trimethoxysilylpropyl)-ethylenediamine	98
Phosphonylated diethylenetriamine	89
Anodic	
2-Hydroxy-1,2,3-propanetricarboxylic acid	82
3,4,5-Trihydroxybenzoic acid	50
2-Phosphonobutane-1,2,4-tricarboxylic acid	75
2-Mercaptobenzothiazole	8
5,5'-Indigosulfonic acid	32
Maleic acid	33
4-Hydroxy-3-methoxybenzaldehyde	21

Example 2: This example illustrates how blending an anodic and cathodic inhibitor can produce a synergistic effect on pitting corrosion.

Inhibitor	% Pitting Inhibition	% General Corrosion Inhibition
Anodic–4-hydroxy-3-methoxybenzaldehyde	21	30
Cathodic–phosphorylated pentaerythritol	86	81
Blend of the above (1:1 by weight)	95	81

In addition, where phosphorylated pentaerythritol with a percent pitting inhibition of 86% is blended with an equal weight of 3,4,5-trihydroxybenzoic acid, with a percent pitting inhibition of 50%, the resulting blend has a percent pitting inhibition of 86%, thus upgrading 3,4,5-trihydroxybenzoic acid.

The synergistic blends of cathodic and anodic corrosion inhibitors of this process can also be employed in conjunction with other materials whether they are organic or inorganic. For example, they can be employed with inorganic phosphates, polyphosphates, etc., such as sodium hexametaphosphate.

It was also found that alkoxysilyl alkyl alkylene polyamines such as (trimethoxysilylpropyl)ethylene diamine are superior pitting corrosion inhibitors even when employed alone, yielding about 98% pitting corrosion inhibition. Although such compositions are known as corrosion inhibitors as illustrated in U.S. Patents 3,716,569 and 3,816,184, their superiority as pitting corrosion inhibitors has now been demonstrated. In addition such silyl compositions are useful in preparing synergistic blends with anodic corrosion inhibitors which upgrade the blend.

Preventing Formation of Pinholes in Cured Organic Autodeposited Coatings

Although autodepositing compositions are capable of forming organic coatings of excellent quality, it has been observed that under certain conditions where autodepositing compositions of the type which utilize a ferric-containing compound are employed, coatings having defects are formed. Such defects appear in the cured coatings as pinholes which are present also in the wet uncured coatings. Generally, pinholes are defects which appear as round holes in the organic film, approximately 1 to 2 mm in diameter extending to the metal surface. The term pinholes as used herein is intended to include other versions of the same defect which may appear as craters or blisters.

For use in some applications, such defects may be of little or no consequence. However, there are applications where coatings containing such defects would not be tolerable or at least, undesirable. Such defects generally lower the corrosion resistant properties of the coatings. In applications where such properties are important, such coatings would be unsuitable. They would be unsuitable also in applications in which it is desired that the coating appearance comprises a smooth unbroken film.

In the process of *T.J.C. Smith; U.S. Patent 4,199,624; April 22, 1980; assigned to Amchem Products, Inc.* a substrate capable of receiving an autodeposited coating is pretreated with an acid solution prior to subjecting the substrate to an autodepositing composition. Acid pretreating the substrate in applications in which an autodepositing composition tends to form coatings having pinholes has been found to be effective in deterring or preventing the formation of pinholes.

Acid solutions used in the pinhole preventing pretreatment will have at least 0.1 vol % of acid, preferably 2 to 5 vol %, and the temperature of the solution is desirably within the range of 75° to 200°F, preferably 100° to 150°F and the contact time between the solution and the metal surface is 30 sec to 3 min.

Phosphoric acid is particularly preferred in this process. When using phosphoric acid, it is recommended that the treating conditions include the use of an aqueous solution comprising 1 to 5 vol % of phosphoric acid at a temperature of 125° to 160°F for a contact time of 1 to 2 min. Generally, the acid solution used in pretreating worked or roughened metal surfaces will have a pH of between 2 and 3, and preferably about 2.

An optional ingredient that can be used in the acidic solution is a surfactant. The surfactant can be used in relatively small amounts, for example, 0.001 to 1 vol % of the solution.

Examples 1, 2 and C-1 through C-8: The examples show the acid pretreatment of metal articles according to the process and the application thereto of autodeposited coatings. For comparative purposes, this group of examples also shows the pretreatment of metal articles with alkaline cleaning materials.

Each of the articles used in this group of examples was made from hot rolled steel sheet that was dished by stamping. The articles contained gall marks along the junction of the dished and lip portions on the convex side of the articles. The articles were soiled with forming oil and shop dirt, but they were not rusted. Soils of this type would generally be removed by cleaning the articles with alkaline cleaning solutions.

Acid pretreating solutions were prepared from the following composition A.

Composition A

Ingredient	Weight Percent
Phosphoric acid (75 wt %)	33.5
Butyl ether of ethylene glycol (butyl Cellosolve)	7.9
Nonyl phenol ethoxylate (Triton N-100)	0.5
Sodium alkyl benzene sulfonate (Ultrawet 45DS)	0.3
Oleyl alcohol (technical grade)	0.1
Water	57.7

The pretreating solutions and the application conditions used are identified in the table below.

Ex. No.	Pretreating Material	Amount of Material in Aqueous Pretreating Solution	Application Conditions
1	Composition A	1-2 parts by vol (water)	3 min immersion, 110°F
2	Ex. 1 pretreating solution + 2.5% by vol of 21 wt % HF		3 min immersion, 110°F
C-1	Strongly alkaline spray cleaner*	2 oz/gal	3 min spray, 160°F
C-2	Ex. C-1 pretreating solution + 1% by vol of aqueous solution of 38 wt % NH_3		3 min spray, 160°F
C-3	Chlorinated alkaline cleaner**	2 oz/gal	3 min spray, 160°F
C-4	Oxygenated alkaline cleaner***	2 oz/gal	3 min spray, 160°F
C-5	Strongly alkaline immersion cleaner†	6 oz/gal	5 min immersion, 190°F
C-6	Ex. C-5 pretreating solution + 1% by wt of Glucoquest AC††		3 min immersion, 160°F
C-7	Ex. C-5 pretreating solution + 1% by vol Hampenol OH-1†††		3 min immersion, 160°F
C-8	Caustic gluconate type deruster (Ridoline 32)	1 lb/gal	immersion at 190°F (time not recorded)

*Ridoline 5290, a powder mixture of caustic soda, soda ash, sodium orthosilicate and surfactant (phosphate-free).

**Powder cleaner RL 69-364 (Amchem Products, Inc.).

***Powder cleaner containing sodium perborate, RL 69-390 (Amchem Products, Inc.).

†Ridoline 27, a powder mixture of sodium tripolyphosphate, caustic soda, sodium gluconate and wetting agent.

††Sodium alpha-glucoheptonate

†††Ethylenediaminetetraacetic acid type chelate for iron.

The articles treated with the acid pretreating solutions of Examples 1 and 2 above were first cleaned with the alkaline cleaning solution of Example C-5 above by immersing the articles in the cleaning solution for 5 min at a temperature of 160°F. The cleaned articles were then rinsed with tap water and thereafter subjected to the solutions of Examples 1 and 2.

After subjecting the articles to the pretreating solutions identified in the above table, they were rinsed by immersion for about 30 sec in agitated tap water and then further rinsed with an aerated 5 sec spray of deionized water. Thereafter, the articles were coated by immersing them for about 90 sec in an 800 gal bath of a stirred autodepositing composition comprising, on a 1 ℓ basis: 180 g of latex containing about 54% solids, 3 g of ferric fluoride, 2.1 g hydrofluoric acid, 5 g black pigment dispersion, and water to make 1 ℓ.

The resin of the latex used in the above composition comprised about 62% styrene, 30% butadiene, 5% vinylidene chloride and 3% methacrylic acid. A film formed from the resin is soluble in refluxing chlorobenzene to the extent of about 13%. That the resin is crosslinked is indicated by its insolubility in Soxhlet extraction with chlorobenzene. The water-soluble content of the latex is about 2% based on the weight of dried resin, with the water-soluble content comprising about 10% sodium phosphate, about 13% sodium oleoyl isopropanolamide sulfosuccinate and about 75% sodium dodecylbenzene sulfonate, the first mentioned ingredient being a buffering agent used in preparing the latex, and the last two mentioned ingredients being emulsifiers.

The pH of the latex was about 7.8 and the surface tension thereof about 45 to 50 dynes per centimeter. The average particle size of the resin was about 2000 Å.

The black pigment dispersion used in the above composition is an aqueous dispersion having a total solids content of about 36%. Carbon black comprises about 30% of the dispersion. It has a pH of about 10 to 11.5 and a specific gravity of about 1.17. The dispersion contains a nonionic dispersing agent (Aquablak 115).

After the resinous coated articles were withdrawn from the autodepositing composition, they were air-dried for about 1 min, immersed in tap water for about 30 sec, and thereafter immersed in an aqueous chromium-containing solution for about 30 sec. The chromium-containing solution was an aqueous solution containing 0.1 vol % of an aqueous solution of a surfactant, namely an aqueous solution of octyl phenol ethoxylate (Triton X102), and 3 vol % of an aqueous chromium-containing concentrate containing about 8.6 wt % CrO_3, about 8.6 wt % formaldehyde-reduced CrO_3 and about 3.9 wt % phosphoric acid.

After rinsing with the chromium solution, the articles were placed in a forced air draft oven having a temperature of about 275°F for about 20 min to fuse the resin coating.

Upon withdrawal from the oven, examination of the coated articles showed that the coatings of the articles pretreated with the solution of Examples C-1 through C-8 contained hundreds of pinholes which extended through the coating to the underlying metal surface. The pinholes were located in those portions of the coatings that overlay the gall marks on the articles. On the other hand, the coatings of the articles that were treated in accordance with the process (pretreating solutions of Examples 1 and 2), contained but a few pinholes which were located in some of the portions of the coatings that overlay some of the gall marks.

Thus, pretreatment of the articles according to the process resulted in a vast improvement in deterring pinhole formation.

Use of Cr(VI), Reduced Cr Rinse for Surface Coated with Polymeric Resin

The process of *W.S. Hall and L. Steinbrecher; U.S. Patent 4,030,945; June 21, 1977; assigned to Amchem Products, Inc.* relates to a metallic surface which is coated with a polymeric resinous coating by immersing the surface in an acidic aqueous coating composition comprising polymeric resinous coating-forming material and an oxidizing agent. The corrosion resistance and/or the surface appearance of the resinous coating is modified by rinsing the coating with an aqueous rinse solution containing hexavalent chromium or an aqueous rinse solution containing hexavalent chromium and reduced forms of chromium.

Examples 1 through 6: Cold rolled steel panels, 4" x 12", were cleaned with an alkaline cleaner, rinsed with water, and immersed for 5 min in an aqueous coating composition containing: 100 g/ℓ of styrene-butadiene copolymer (Pliolite 491 latex, Goodyear Tire and Rubber Co.); 2.1 g/ℓ of HF; 2.3 g/ℓ of H_2O_2; and 5 g/ℓ of carbon black pigment.

Upon being withdrawn from the coating composition, it was observed that resinous coatings were formed on the panels. The coatings were adherent and had thicknesses of about 1 mil.

The adherent resinous coatings were rinsed with running tap water and either rinsed with an aqueous Cr rinse solution or not rinsed as indicated in the table. Thereafter, the resinous coated panels were placed in an oven having a temperature of 220°C and baked for 10 min. After being withdrawn from the oven, the appearances of the resinous coatings were observed and the panels were subjected to a salt spray test (ASTM B-117), as indicated, for the purpose of evaluating the corrosion resistance of the coatings.

The results of the salt spray tests are reported as the distance of coating failure measured from the scribe. In those examples in which the failure was not uniform, the maximum and minimum distances of failure are given.

Each of the Cr^{6+}/reduced Cr rinse solutions that were used in the examples was prepared according to the method described in U.S. Patent 3,063,877. Generally, this involved the preparation of a Cr^{6+}/reduced Cr aqueous concentrate by reacting appropriate amounts of formaldehyde and CrO_3 in aqueous solution at elevated temperature to give the desired Cr^{6+}/reduced Cr molar ratio and then diluting the aqueous concentrate with additional water to give the desired rinse solution.

It is noted that in these examples, where the Cr^{6+}/reduced Cr molar ratio was 1, the composition also contained 3 wt % H_3PO_4 to prevent gelling of the concentrate.

Shown below are the results obtained when aqueous Cr rinse solutions containing varying amounts of total Cr are used. Each of the rinse solutions had a molar ratio of Cr^{6+}/reduced Cr of 1, but the total Cr content of the rinse solutions was varied as indicated.

The resinous coated panels were immersed in the rinse solutions, which were maintained at a temperature of about 25°C for 30 sec. After the rinsed coatings were baked as indicated hereinabove, they were subjected to the salt spray test for 144 hr.

The test results and the appearance of the coatings are set forth in the following table.

Ex. No.	Total Cr (g/ℓ)	Scribe Failure (inch)	Appearance of Coating
1	not rinsed	5/16-5/32	glossy, textured
2	0.52	3/16-5/32	very slightly matted
3	1.09	1/8	slightly matted
4	2.18	3/32-0	more matted than Ex. 3
5	4.37	0	more matted than Ex. 4
6	8.84	0	more matted than Ex. 5

Example 7: *(A)* – A steel panel was immersed in an aqueous coating composition containing 50 g/ℓ of acrylic resin (Catalin A-1316 latex), 2 g/ℓ of fluoride and 1.5 g/ℓ of hydrogen peroxide for 3 min. Upon withdrawal from the composition, the coated panel was rinsed with an aqueous solution containing hexavalent chromium and reduced chromium which was prepared according to the method described in U.S. Patent 3,063,877. The concentrations of the hexavalent and the reduced chromium were each about 0.5 g/ℓ (expressed as CrO_3).

The coated panel was treated for 30 sec with the rinse composition which had a temperature of 130°F. After the rinsing step was completed, the coated panel was baked to complete fusion of the coating.

(B) – A steel panel was treated in the same manner as set forth in (A) above, except that the rinse composition used in this example contained polyacrylic acids in addition to hexavalent chromium and reduced chromium. This type of composition is disclosed in U.S. Patent 3,185,596. The rinse composition contained 10 g/ℓ of Cr^{6+} and a like amount of reduced chromium (expressed as CrO_3) and 4.1 g/ℓ of polyacrylic acids (Acrysol-A-1, a solution of water-soluble polyacrylic acids).

After each of the panels of (A) and (B) above was rinsed, they were subjected to a salt spray test (ASTM B-117-61). Another steel panel, which was coated and baked in the same way as (A) and (B), but which was not rinsed, was subjected also to the salt spray test. The results of the salt spray tests showed that the rinsed panels had much better corrosion resistance properties than the unrinsed panel, with the panel of (A) out-performing slightly the panel of (B).

COPPER AND COPPER ALLOYS

Sealing Formulation for Bright Finish

The process of *J.M. Tucker; U.S. Patent 4,070,193; January 24, 1978; assigned to Kaddis Mfg. Co.* relates to an aqueous sealing solution for sealing copper and its alloys against corrosion following dipping of the copper or its alloys in a conventional brightening solution.

It is preferred that the copper metal pieces be first thoroughly cleaned to remove all foreign matter from the surfaces thereof. Following cleaning the metal pieces are dipped in a standard acidic brightening solution, such as "Super B" (Ashland Chemical Company), to remove surface metal oxides and other elements which dull the surface of the metal parts and otherwise tarnish or discolor the metal surface. After bright dipping the metal pieces should be thoroughly rinsed in cold water (temperatue of 60° to 90°F) and prepared for sealing.

It has been found that superior results are achieved by dipping the brightened article in an aqueous solution of sodium bichromate, chromic acid and sulfuric acid. The amount of sodium bichromate present may vary between about 1 and 3 lb; the amount of chromic acid may vary between ¼ and 1 oz; and the amount of sulfuric acid may vary between 1 and 25 ml, for each gallon of water in the mixture.

The previously cleaned and brightened copper articles are dipped in this aqueous solution for a period of time of from 30 sec to 5 min. Longer periods of immersion in excess of 2 min may be utilized without harmful results but such longer periods appear to serve no useful purpose.

After dipping in the solution, it is critical that the article be first rinsed in cold water at a temperature of from 60° to 90°F for 1 min and then rinsed in hot water for about 1 min. It has been found that the temperature of the hot water should range between 140° and 175°F. It should be understood that these two rinse steps must be accomplished by dipping the article first in cold water and then hot water.

Copper metal articles treated according to the process have been subjected to salt spray tests for as long as 100 hr without corrosion.

Example: 6 brass metal parts are first cleaned and then dipped in a standard brightening solution and then dipped for 1 min in an aqueous solution of the following: 1½ lb sodium bichromate, ½ oz chromic acid, 2 ml sulfuric acid, and 2 gal water.

The articles are then dipped for 1 min in a cold water rinse at tap water temperature. The articles are then dipped for 1 min in a hot water rinse at a temperature of 160°F. The articles are then dried.

The articles are then suspended in a salt spray chamber by means of nylon cord. The articles are subjected to salt spray exposure in this chamber according to ASTM-B-117-73. The temperature and salt exposure conditions are maintained at 95°±3°F with a 5% (neutral pH) salt spray solution. These conditions are maintained for a period of 100 hr. The articles are removed from the chamber at the completion of 100 hr exposure, and rinsed with warm tap water at a temperature of 90°F. None of the 6 brass metal articles had any visible evidence of corrosive attack.

Coating for Excellent Solderability and Tarnish Resistance

E.J. Caule; U.S. Patent 4,264,379; April 28, 1981; assigned to Olin Corporation provides a process resulting in a coated copper or copper alloy sheet, wherein the coated sheet displays high values of solderability, while other desirable properties such as substantial resistance to tarnish are also accomplished.

This is achieved by applying to a copper or copper alloy sheet or foil a solution of an organophosphonic acid, or salt thereof, for 4 to 20 or more seconds at room temperature up to about 100°C, draining off excess solution, rinsing, and drying, the treatment being combined with or preceded by oxidation of the sheet surface.

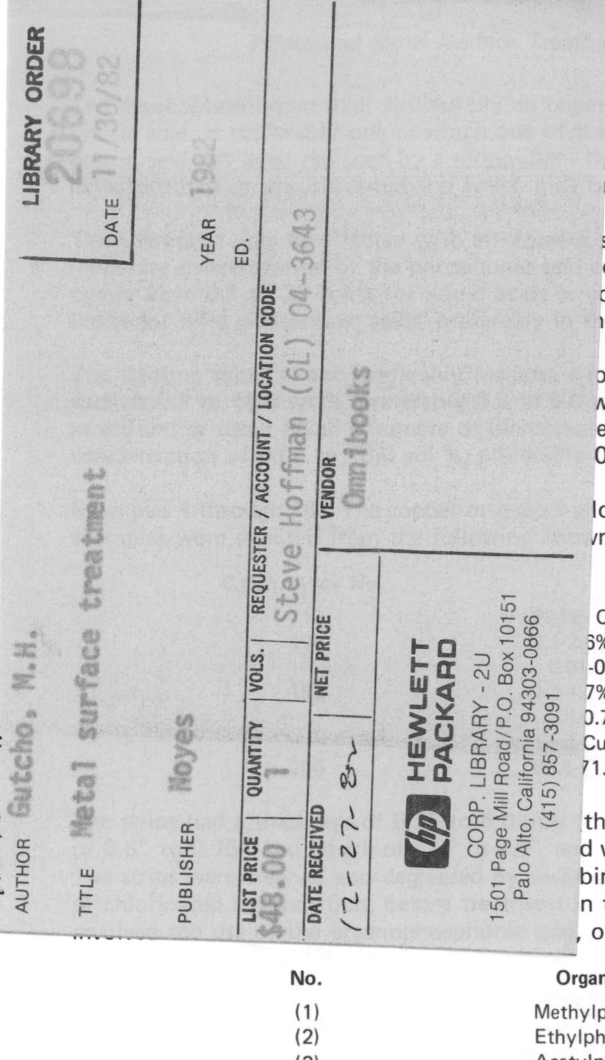

...ic substitution product of phos- ... three hydroxyl groups of phos- ...ydrocarbon radical, which may ...e saturated or unsaturated.

...solution containing a low to ...omponent or components, pref- ...orresponding weight percent ...e range of 0.1 to 40 wt %.

...ow to moderate concentration, ...wt % of oxidizing agent, such ...e, or nitric acid (100%) at a ...0.05 to 2.0 vol % HNO_3.

...loy strips treated in the following ...n compositions:

Composition

... Cu (min), 0.04% O
...% Fe, 0.05-0.2% Zn,
...-0.04% P, bal Cu
...7% Fe, 0.6-1.0% Co,
...0.7% Sn, 0.08-0.12% P,
... Cu
...71.5% Cu, bal Zn

...thousandths of an inch), a width ...d were in the annealed condition. ...bing with reagent grade benzene ...n the following examples which ..., or salt thereof, listed below.

No.	Organophosphonic Acid
(1)	Methylphosphonic acid
(2)	Ethylphosphonic acid
(3)	Acetylphosphonic acid
(4)	Propionylphosphonic acid
(5)	Hydroxyethylidene diphosphonic acid*
(6)	Ethylenediamine tetra(methylenephosphonic acid)*
(7)	Hexamethylenediamine tetra(methylenephosphonic acid)*

*Dequest 2010, 2041, 2051 respectively.

Reagents and conditions for the specific examples are listed in the following table.

Ex. No.	Organophosphonic Acid No.	Organophosphonic Percent	C.D.A. Alloy	Oxidizing Agent*	Oxidizing (%)	Temperature (°C)	Time (sec)
1	1	18**	110	A	3**	25	15
2	2	18**	110	A	3**	25	15
3	3	18**	110	A	3**	25	15
4	4	18**	110	A	3**	60	10

(continued)

Ex. No.	Organophosphonic Acid No.	Percent	C.D.A. Alloy	Oxidizing Agent*	(%)	Temperature (°C)	Time (sec)
5	1	1**	110	B	0.1***	60	15
6	5	10***	110	A	3**	25	20
7	5	10***	110	A	0.3**	25	15
8	7	5**	110	A	0.3**	25	15
9	6	5**	110	A	0.3**	25	15
10	5	1***	110	B	0.1***	100	15
11	5	1***	194	B	0.1***	100	15
12	5	1***	195	B	0.1***	100	15
13	5	1***	260	B	0.1***	100	15
14	6	1**	110	B	0.5***	100	15
15	6	1**	194	B	0.5***	100	15
16	6	1**	195	B	0.5***	100	15
17	7	1***	110	B	3***	100	15
18	7	1***	110	B	1***	100	15

*A is $Na_2Cr_2O_7$ and B is HNO_3.
**By weight.
***By volume.

The treatments were carried out by preparing aqueous solutions of the indicated compositions and maintaining at the stated temperature, and partially immersing clean strips of copper or copper alloy, about 0.5" in width, to a depth of about 1.5" for the stated time period. After immersion, each strip was rinsed in cold water and dried.

The coated strips were tested for resistance to tarnish by holding each sample in the vapor emerging from a freshly shaken bottle containing a 10 to 20 wt % aqueous solution of ammonium sulfide for 10 to 15 sec, whereupon untreated or incompletely treated areas displayed mottled surface colors of metal sulfide much duller and darker than the uniform bright metallic color of the well-coated treated areas.

Each of the above specific examples yielded strips displaying perfect resistance to tarnish over the entire portion which had been immersed in the treating solution, as such portions after exposure to hydrogen sulfide vapor, retained the lustrous bright metallic color of the initial strip. This was in sharp contrast to the variegated murky and dark color shades shown by the unimmersed strip portions.

Such resistance to tarnish was found to be retained even after lengthy storage, as for 500 hr, in a laboratory cabinet.

The surface film produced on the metal strips by the immersion treatment is transparent and invisible to the naked eye, but its presence as a coating which is substantially free of pores is established through the improvements effected thereby in a number of properties, as illustrated by the vastly increased resistance to tarnish and the restoration of solderability to a desired extent.

Solderability tests were carried out on treated strips resulting from the above examples, and yielded the highest test rating without exception, even after lengthy storage of the treated strips in a laboratory cabinet, as for 500 hr. The solderability rating was similar to that observed in testing freshly cleaned initial metal strips which had not been coated.

TITANIUM AND TITANIUM ALLOYS

Phosphate Doped Oxide Coating

The advent of high altitude and high speed aircraft and missiles has created a need for materials which exhibit high strength and high resistance to oxidative and corrosive degradation at elevated temperatures. The need for such materials becomes even more acute when one considers the great strains and stresses produced in structural elements during operation within a high speed, high altitude environment.

Titanium and titanium alloys are well known for their corrosive resistance and find their main application in the aircraft industry, where high strength, coupled with light weight, is a design criterion. These materials, however, do not possess sufficient stress corrosion resistance to meet the demands of aircraft and missiles.

In accordance with the process of *B.A. Manty; U.S. Patent 4,026,734; May 31, 1977; assigned to U.S. Secretary of the Air Force*, stress corrosion resistance can be imparted to titanium and titanium alloy articles by a coating process which comprises immersing the titanium article in an aqueous solution of ammonium phosphate in a vacuum chamber, pulling a vacuum over the article of the solution, releasing the vacuum, removing the article and draining the excess solution therefrom, placing the coated article in an oven to heat it at 800°F, and then removing the article from the oven.

The resulting coated titanium article possesses a high degree of resistance to stress corrosion because of the formation of a phosphate doped oxide coating on the surface of the article.

Example: A titanium article of conventional structure is immersed into a 5 wt % aqueous solution of ammonium phosphate positioned within a vacuum chamber. A vacuum of from 28" to 30" of mercury is pulled over the system. The vacuum is released while the titanium article remains in the phosphate solution. The article is then removed and excess solution is allowed to drain off. The article is then placed in an oven and heated to a temperature of about 800°F for a period of about 20 hr. The article is then removed from the oven and allowed to cool to room temperature.

The above procedure provides a phosphate doped oxide coated substrate with useful engineering properties. The modified titanium substrate is highly resistant to salt penetration and stress corrosion. The coating can be applied to any structure composed of titanium or titanium based alloys.

Surface Treatment to Improve Subsequent Bonding

The titanium alloy bonding difficulties encountered in industry, especially the aerospace industry, stem from the surface preparation of such metals.

In accordance with the method of *R. Villain; U.S. Patent 4,075,040; February 21, 1978; assigned to Societe Nationale Industrielle Aerospatiale, France*, an initial operation involves a degreasing (followed preferably by rinsing in water for 1 to 2 min and subsequent possible drying), and then a fluonitric scouring and an ordinary rinsing in demineralized water for 1 or 2 min.

This initial operation is intended primarily to provide a surface substantially free of foreign substances. It is followed by a second operation consisting of an immersion in a bath of sodium bifluoride containing about 12 g/ℓ of NaF-HF, for a period of 1 to 3 min, followed by an ordinary rinsing.

The third operation, which consists in rinsing for a few minutes in demineralized water at a temperature of at least 50°C, leads through hydrolysis to the elimination of the undesirable elements produced during the second operation.

Checking for proper elimination of these elements is effected by examining the surface of the titanium with a microscope. This examination, performed with a JEOC-JS MU3, 5 to 50 keV scanning electron microscope with a magnification of 3,000X, can be carried out on any surface containing residual elements following immersion in the fluoride-charged bath.

By the final stage in the second operation, the surface examination will already reveal the weight distribution of the fluoride possibly associated with other residual elements.

A comparison between the photographs taken after the second and third operations respectively will indeed provide the required verification means for revealing whether or not the third operation has been applied, and correctly so.

Example 1: Two titanium alloy plates of the TA6V kind, the surfaces of which have been prepared in accordance with this process, can be bonded together over an area of 3 cm^2 by means of a polyimide glue and, after an ageing period of approximately 2,000 hr at 250°C, possess a shear strength of about 500 kg at 250°C.

Example 2: In the case of a metal such as TA6V4, it is possible to effect an alkaline degreasing, by means of washing agents consisting of mixtures of phosphates and alkaline carbonates, using tensioactive substances. This washing is followed by a rinsing in ordinary cold running water, which is adequate. The next step is to effect a scouring of the surface of the treated metal, with degreasing and rinsing in a fluonitric bath, for example, such bath containing 50 vol % of nitric acid at 36°Bé, i.e., density of 1.38 g/cc, and 2 vol % of hydrofluoric acid having a concentration of 40 vol %, the remainder being water.

The temperature of such scouring-degreasing bath is maintained between ambient temperature and +60°C. Choice of the temperature will depend on the time lapse desired before the required result is obtained.

Such scouring-degreasing is followed by abundant rinsing in ordinary running water. This rinsing may be carried out in a possibly agitated bath, or by projection or sprinkling.

Example 3: *Evaluation* – The results obtained with an adhesive substance and a shear strength test on standard test specimens, by comparison with a surface preparation by sanding in the case of TA6V4, have been as follows: sanding at ambient temperature, failure under a load of 4,200 Newtons; and abovedescribed treatment at ambient temperature, failure under a load of 8,100 to 8,400 Newtons, the adhesive substance employed being that produced by the Rhone-Poulenc

Company under the designation IPA 380. At a temperature of 250°C, the results would be 3,900 to 4,200 Newtons and 7,200 Newtons respectively.

With a different adhesive substance, on TA6V4 metal treated according to this method and treated by blue anodic oxidation respectively, the figures would be in the region of 15,000 and 6,000 Newtons respectively.

In the case of adhesive substances suitable for use as painting primers on similar metals with a very high titanium content, treatment by the known phosphate-fluoride method results in a perpendicular rip-off force after damp ageing under heat of 156 Newtons, whereas the subject preparation treatment of this process makes it possible to obtain, under similar conditions, a force of 215 Newtons, assuming equivalent surfaces.

Forming Friction Surface Composed of Titanium Oxide

The object of the process of *A. Gaucher and B. Zabinski; U.S. Patent 4,263,060; April 21, 1981; assigned to Centre Stephanois de Recherches Mecanique Hydromecanique et Frottement, France* is to produce a friction surface on a titanium or titanium alloy part, or a part coated with titanium or titanium alloy, by forming a surface layer of a number of titanium oxides, in gradually decreasing ratios of oxygen to titanium, having only a very thin outermost surface of TiO_2, and oxides such as $Ti_{10}O_{19}$, Ti_8O_{15}, Ti_7O_{13}, Ti_4O_7, Ti_3O_4, Ti_2O, TiO between the TiO_2 outer surface and the Ti part material, by precisely controlling the amount of oxygen available in a low pressure gaseous atmosphere and the treatment temperature and time.

The method of treating a part which contains titanium in its external surface, includes the steps of removing at least a portion of a natural layer of oxide which covers the part, and placing the part within a fluid-tight enclosure. The enclosure is then evacuated until the pressure therein ranges from about 1 to 10^{-8} Torr. Then the enclosure is isolated from the evacuating means. Oxygen is then introduced into the evacuated fluid-tight enclosure in an amount ranging from about 10^{-3} to about 2.55 mg per each square centimeter of total external surface area of the part or parts within the evacuated fluid-tight enclosure. The enclosure is heated to a temperature from about 450° to 880°C.

Figure 9.1 is a graph showing the law of variation of the thickness ϵ of the layer, of the friction coefficient f, and of the total hardness δ, as a function of the amount of oxygen introduced per unit area, the curves ϵ, f and δ merging in a single curve C.

A test piece made of titanium and having an area S was heated under conditions of temperature, time, pressure, ambience, and the like, which will be made precise hereinafter, in a fluid-tight furnace into which a weight Q of oxygen was introduced. A layer formed on the surface of the test piece consisted mainly of titanium oxides.

The experiment was repeated a number of times, while varying the ratio Q/S of the weight of oxygen introduced to the part or parts treated.

For all the parts treated in this way, the total hardness δ, the thickness ϵ of the layers formed within the time unit, and the friction coefficient f of the test piece thus treated when the latter rubs on steel, were carefully measured for each experiment.

If the ratio Q/S is plotted as abscissa and the values δ, ϵ, and f are plotted as ordinate on a system of axes suitably selected, it will be seen that the three curves obtained merge in single curve C, the peculiarity of which lies in the fact that it presents a very high peak and a very narrow base. This scientific phenomenon leads very naturally to the oxidation treatment of titanium according to this process. It is a question of replacing, at least partially, the natural oxide layer TiO_2, which coats any titanium part, by the oxide layer obtained when the oxidation takes place according to the peak of the curve **C** of Figure 9.1.

Figure 9.1: Graph Showing Characteristics of Layers of Titanium Oxides

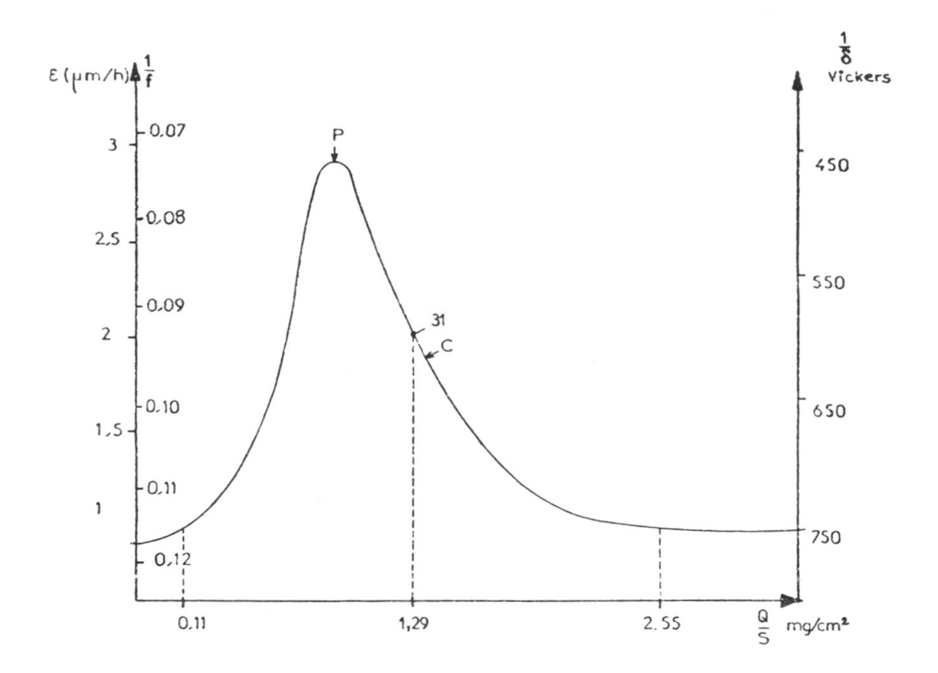

Source: U.S. Patent 4,263,060

Example: *Faville type tests* – This type of test is carried out on a Faville-Levally apparatus, in which a cylinder of a diameter of 6.5 mm and a height of 40 mm, the surface S of which is equal to 9.35 cm^2, is rotated between two jaws cut as Vs having 90° angles. A load, which increases linearly as a function of time, is applied on the jaws.

The test is carried out in the ambient air. A test part made of titanium or titanium alloy and conventionally oxided, for instance, in a bath of sulfuric acid at environmental temperature with a current density of 0.3 A/cm^2 for 20 min, seizes practically instantaneously between the steel jaws.

In contrast, a test part treated by the method, that is, having first been subjected to a sanding operation in order to remove the natural layer of oxide to a depth of 2.5 μ, and then heated to 650°C for 8 hr in a furnace containing 12 mg of oxygen, so as to treat with a ratio Q/S equal to 1.29 (point **31**), the auxiliary gas being argon, can rub under a load of 600 daN without any seizing occurring. In this case the coefficient of friction throughout the test remains lower than 0.12.

Chromium-Molybdenum Coated Titanium Alloy

In accordance with the process of *B.A. Manty, H.R. Liss and H. Lada; U.S. Patent 4,236,940; December 2, 1980; assigned to United Technologies Corp.* an optimum surface conditioning treatment for consistently satisfying the coating adherence requirement involves the formation of a chromium conversion coating on a cleaned titanium substrate prior to electrodeposition. The preferred method involves wet abrasive blasting of the substrate surface with a slurry of metal oxide, (such as novaculite, Novasite 200) at a sufficient pressure, such as 0.4 to 0.7 MN/m^2 (50 to 100 psig) air pressure, to remove foreign matter and oxide from the surface. After vapor blasting, the substrate is rinsed in clean water and preferably is kept wet prior to the chromium conversion treatment.

The chromium conversion treatment is satisfactorily effected with an acid chromate solution comprising 1 to 3 vol % hydrofluoric acid, 7.5 to 15 g/ℓ (1.0 to 2.0 oz/gal) sodium chromate (Na_2CrO_4), balance water at ambient temperature. A convenient method for providing contact between the substrate surface and acid chromate solution is by simple immersion of the substrate therein, for example, immersion times of 5 to 15 seconds being satisfactory. Of course, the coating will be very thin, typically less than about 0.0025 mm (0.0001 inch).

The acid chromate treated substrate is then rinsed in water to remove excess solution and preferably is kept wet prior to electroplating. Electroplating of the chromium molybdenum alloy coating can be effected conveniently in a conventional, self-regulating chromium plating bath containing a source of dissolved molybdenum such as molybdic acid, sodium molybdate and ammonium molybdate. In a bath comprising 263 to 338 g/ℓ (35 to 45 oz/gal) chromic acid (CrO_3), 75 g/ℓ (10 oz/gal) ammonium molybdate, balance water, a current density of 24 to 47 ASD (1.5 to 3 ASI) is satisfactory.

The current is maintained until the desired thickness of chromium molybdenum coating is achieved. For most applications, an electroplate of about 0.013 to 0.076 mm (0.0005 to 0.003 inch) is satisfactory. Upon removal from the electroplating bath, as a result of the pretreatment, it is found that the electroplated substrate can be readily handled without a loss of coating adhesion.

The electroplated substrate is next heat treated to obtain a diffusion bond between the substrate, the conversion coating, and the chromium molybdenum electroplate. It is not an object to completely interdiffuse the coating with the substrate. That is, the surface desired is a chromium molybdenum alloy, not an alloy also containing titanium.

The temperature of heat treatment should be in about the 700° to 820°C (1300° to 1500°F) range. This was established through wear resistance tests. Preferably the heat treatment of the coated substrate is at about 760°C (1400°F) for about 3 hr.

By way of example, the alloy which comprises 6% Al, 2% Sn, 4% Zr, 6% Mo, balance Ti (all by wt) has a preferred solution heat treatment of 845° to 915°C (1550° to 1675°F)/1hr followed by a precipitation treatment of 595°C (1100°F)/4 to 8 hr.

Acid Resistant Oxide Layers for Titanium Alloy Containing Aluminum

The work of *C.-C. Wang; U.S. Patent 4,266,987; May 12, 1981; assigned to Kennecott Copper Corp.* relates to a method for providing an acid-resistant oxide layer on an alloy comprising the steps of preliminarily cleaning the surface of the alloy and then exposing the clean surface to an environment specifically selected to produce an acid resistant oxide by selective oxidation of one of the alloy's constituents while providing a reducing environment for its other constituents.

Example: A titanium alloy containing 8% Al, 1% Mo, and 1% V, after cleaning, is placed in an evacuated chamber, with TiO_2 powder, and heated to 1200°C. A thin layer of pure Al_2O_3 results. Once the aluminum oxide layer is formed, the titanium, molybdenum, and vanadium base metals are protected and a thicker layer of Al_2O_3 may be grown by subjecting the alloy to air or oxygen at a temperature suitable to attain a given reaction rate such as 1200°C.

MISCELLANEOUS TREATMENTS

Treatment Liquid Containing Oxytitanic or Peroxytitanic Ions

H. Kogure; U.S. Patent 4,130,431; December 19, 1978; assigned to Kansai Paint Co., Ltd., Japan has shown that when a metal surface treatment liquid containing oxytitanic ion and/or peroxytitanic ion is applied to a metal substrate, titanic acid resulting from the reduction of the oxytitanic ion or peroxytitanic ion reacts with the metal substrate to form a composite oxide film thereon. The composite oxide film thus obtained is superior in water resisting property, corrosion resisting property, mechanical strength and adhesive property to coated film, etc. The metal surface treatment liquid has a pH of less than 6 and contains oxytitanic ion TiO^{2+} and/or peroxytitanic ion TiO_2^{2+}.

In addition, a rust preventive paint containing oxytitanic ion or peroxytitanic ion has excellent water resistance, corrosion resistance, mechanical strength, adhesiveness to finishing paint, etc.

Although it is possible to oxidize titanium compound or metal titanium dissolved in the acid solution by standing the dissolved material in the air or by blowing air or oxygen therein, it is advisable to add peroxides in order to perform the oxidation reaction quickly. Hydrogen peroxide is especially preferable.

In the abovementioned reaction, oxytitanic ion can be obtained by merely dissolving in acid barium titanate, iron titanate, calcium titanate, magnesium titanate, aluminum titanate, lead titanate and titanium sulfate as the titanium compound but further oxidation is necessary for obtaining peroxytitanic ion. Also, it is possible to obtain a solution of oxytitanic ion and/or peroxytitanic ion without precipitation by gradually adding the alkyl titanate into the mixture solution having a temperature of below 10°C of the acids and the peroxides (preferably, hydrogen peroxide), in the case of using the alkyl titanate. In this case, when the alkyl titanate is not added gradually or the reaction temperature rises over

10°C, precipitation of a great quantity of sparingly water-soluble titanium dioxide is caused so that a desirable oxytitanic ion and/or peroxytitanic ion cannot be obtained.

Comparing oxytitanic ion with peroxytitanic ion in this process, the peroxytitanic ion is preferable as the peroxytitanic ion has higher activity than the former to metal substrate.

In these manufacturing methods, colorless oxytitanic ion is easily formed in the pH range of less than 6 and peroxytitanic ion is also likely to be formed within the pH range of less than 6 and is orange-colored (red brown in case of high concentration). Also, the pH of the metal surface treatment liquid should be in the range of less than 6, preferably 0 to 3.5.

In the case of the metal surface treatment liquid, each oxytitanic ion or peroxytitanic ion may be independently contained or both may be mixed. The concentration of these ions is preferably within the range of 0.0025 to 2.5 ion mols in 1,000 g of the surface treatment liquid.

Example 1: 12 parts of magnesium titanate were added into 63 parts of 75% aqueous solution of sulfuric acid and dissolved at 80°C. After the obtained solution was cooled to the room temperature, the solution was diluted with 800 parts of distilled water and 125 parts of isopropyl alcohol. Thus a metal surface treatment liquid having pH of 0 and concentration of 0.1 mol oxytitanic ion per 1,000 g of the liquid was obtained.

Example 2: 35 parts of 35% hydrogen peroxide were added to 3 parts of 35% aqueous solution of hydrochloric acid and the solution was kept below 10°C. 50 parts of tetrabutyl titanate were dropped into the solution in the course of 30 minutes with stirring and 500 parts of water of 5°C was dropped over a period of 10 minutes. Then 50 parts of tetrabutyl titanate were added dropwise in the course of 60 minutes, and then 360 parts of ethyl alcohol of 5°C were added and the mixture obtained was stirred for 20 minutes. Thus a red brown metal surface treatment liquid having a pH of 2 and a concentration of 0.295 mol peroxytitanic ion per 1,000 g of the liquid was obtained.

Example 3: 12.0 parts of magnesium titanate were added to 60 parts of 70% sulfuric acid. The mixture was heated at 80°C for 2 hr to obtain a sulfuric acid solution in which magnesium titanate was completely dissolved. After the solution was cooled to room temperature, 200 parts of ethylene glycol monoethyl ether, 700 parts of distilled water and 28 parts of titanium dioxide were added to the solution and mixed. Thus a metal surface treatment liquid having pH of 0.1 and concentration of 0.1 mol peroxytitanic ion per 1,000 g of the liquid was obtained.

Comparative Example 1: 50 parts of zinc peroxide was added to 80 parts of phosphoric acid solution adjusted to pH 6.5, and furthermore 50 parts of titanium dioxide was added thereto with stirring at room temperature. Reaction proceeded for 16 hr but titanium dioxide was not dissolved into the solution and metal surface treatment liquid containing oxytitanic ion was not obtained.

Comparative Tests: The metal surface treatment liquids and the rust preventive paints in the foregoing examples were subjected to various tests, the results of which are shown in the following table. With respect to Examples 1 through 3 and Comparative Example 1, the metal surface treatment liquids were applied to various metal substrates by dipping. Then, the coated metal substrates were dried by hot air of 50°C and washed with water to form a composite oxide film thereon.

Ex. No.	Substrate	Dipping Time	Surface Condition of Film	Salt Spray Test	Corrosion Electric Current Density ($\mu A/cm^2$)	Pencil Hardness
1	Steel	1 min	Yellowish brown	Good	12.0	9H
	Aluminum	1 min	Silver white	Good	<10.0	9H
2	Steel	2 min	Brown	Good	10.5	9H
	Aluminum	2 min	Silver white	Good	<10.0	9H
3	Galvanized steel	50 sec	Blackish gray	Good	25.0	9H
	Aluminum	1 min	Silver white	Good	<10.0	9H
1*	Steel	10 min	No film formation	Rusty	145.0	7H

*Comparative.

Example 4: *Rust preventative paint containing oxytitanate and peroxytitanate* – 9 parts of magnesium titanate were dissolved in a mixed solution consisting of 0.3 part of 10% aqueous phosphoric acid solution and 5.2 parts of 20% aqueous hydroperoxide solution with stirring for 3 hr at room temperature. Thus, a solution containing oxytitanic ion and peroxytitanic ion (less than 5.172 mols ion per 1,000 g of the solution) and magnesium titanate dispersed therein was obtained.

The following components were premixed and the mixture was dispersed by ball mill for 12 hr to obtain a base paint.

	Parts
Epikote #1007	10.0
Methyl isobutyl ketone	20.0
Ethylene glycol monoethyl ether	23.0
Silica powder	3.0
Mica powder	2.0
Red iron oxide	1.0
Toluene	23.0

Then, 82.0 parts of the base paint were mixed with 14.5 parts of the above solution to prepare a rust preventive paint of the process. The pH of the paint obtained was 2.6. The concentration of oxytitanic ion and peroxytitanic ion in the paint was 0.777 mol ion per 1,000 g of the paint.

Hydrophilization of Metal Surface

H.-J. Schlinsog; U.S. Patent 4,292,095; September 29, 1981; assigned to The Continental Group, Inc. has found that plates, especially aluminum plates, tin-coated iron plates, chrome-plated iron plates and iron plates per se may be subjected to mechanical forming processes, especially to deep-drawing or wall-ironing, without the use of lubricants hitherto considered indispensable, if the plate surface is hydrophilized, i.e., made hydrophilic.

Preferably, such hydrophilization is accomplished by the generation of a hydroxide of the metal involved or a hydroxide-containing compound of the metal involved on the surfaces of such metals and/or metal oxides, especially a hydroxide of the lowest valence state of the metal which is generated on the surfaces.

Metal surfaces hydrophilized in accordance with this process show technologically extremely interesting, unexpected properties. In particular, apart from mechanical forming processes like deep-drawing or wall-ironing without lubricants hitherto considered indispensable, much more efficient and thus more economical coatings can be achieved.

Example 1: *Mechanical* – A sheet of aluminum of DIN A4 dimensions, 0.3 mm thick, composition: silicon 0.30, iron 0.70, copper 0.25, manganese 1.0 to 1.5, magnesium 0.3 to 1.3, zinc 0.25, balance aluminum (% by wt), is hydrophilized by reciprocating a paper fleece across the surface of the sheet five times, with an average pressure of 1 kg/cm^2 being exerted. As a rule, this frictional movement is performed as many times as are required to make a residue from the aluminum surface visible on the paper fleece used for rubbing (black discoloration of the paper fleece). Evidence of the change of the hitherto hydrophobic aluminum surface to a hydrophilic surface due to this treatment is obtained as follows.

Prior to the mechanical hydrophilization, the completely degreased aluminum surface is hydrophobic, which can easily be demonstrated by water being poured on the vertical aluminum plate and running down in small droplets, or by adsorption of conventional offset printing inks on the aluminum surface, i.e., a hydrophobic reaction.

Aluminum plate treated by the hydrophilization process described can be identified as hydrophilic by pouring water on a vertical plate which causes complete wetting of the aluminum surface, which has been mechanically treated as described above, and remains there for about 60 seconds, after which time the water evaporates gradually from top to bottom, and the plate surface no longer adsorbs offset printing ink.

Example 2: *Chemical* – A sheet of aluminum of the type specified in Example 1 is chemically hydrophilized by immersion in a 1 N sodium hydroxide solution for 30 minutes, the sodium hydroxide solution having a temperature of 60° to 80°C. The aluminum sheet is then removed from the sodium hydroxide solution and rinsed with distilled water until the rinsing water no longer shows alkalinity. Then the hydrophilization test described in Example 1 will be performed by observing the speed of the water running down the vertical sheet. The tests will show that the degree of hydrophilization achieved by the chemical treatment described in this example is equal to that for the mechanical hydrophilization described in Example 1.

Example 3: *Electrochemical* – A sheet of aluminum of the type specified in Example 1 is immersed in an electrolyte consisting of 0.5% sodium hydroxide solution at room temperature (25°C). Anodic current of 70 A/m^2 is applied (related to the surface area of the aluminum). After not more than 2 seconds the entire aluminum sheet will be of equal hydrophilic nature as the sheets treated in accordance with Examples 1 and 2. Also in this case the sheet is rinsed with distilled water until the draining distilled water is free from alkali. The method of determining hydrophilicity is the same as described in the foregoing examples.

Example 4: *Thermal* – An aluminum sheet of the type described in Example 1 is placed in an electric oven and heated to a temperature of 200°C for about 6 minutes. The sheet is then removed from the electric oven and cooled to room temperature in standard laboratory atmosphere. Then the hydrophilicity test described in Example 1 was carried out. The test results shows that the sheet exposed to such thermal treatment has the same degree of hydrophilicity as the sheets described in Examples 1 through 3. In this particular case, an even longer hydrophilic condition is accomplished. It lasts for at least 36 hours.

Example 5: An aluminum sheet hydrophilized as per Example 1 is preserved immediately upon completion of the hydrophilization treatment by applying tetraethylene glycol, for instance, by spraying; alternatively, the preservation effect can be achieved by passing the hydrophilized metal sheet through a tetraethylene glycol bath immediately after hydrophilization.

Decorative Oxidation Process on Artwork

Artwork and decorative articles employing the natural earth colors, including oranges, browns, yellows and reds are very popular. Such artwork includes both abstract art and simulated primitive art. Such art attempts to use natural materials in order to obtain the natural earth colors.

According to *D.J. Greenspan; U.S. Patent 4,247,589; January 27, 1981*, decorative articles and methods of producing them are provided in which a layer or layers of an oxidizable metal are selectively oxidized to produce color changes in the layer(s), and the oxidizing process is then selectively terminated to set the colors in the layer(s) with the desired decorative effect. The resulting decorative article exhibits predetermined color patterns characteristic of various stages of the oxidation of the metal. The metal is preferably in the form of finely divided particles of elemental metals, metal salts, compounds, alloys and metals with combined impurities, preferably multivalent metals such as iron, copper and brass.

The metal particles may be adhesively or cohesively secured to a substrate, which preferably has a porous surface. The oxidation of the metal includes wetting the metal with a corrosive liquid, such as water, acid, base or electrolyte solutions, or humidity, in the presence of an oxygen containing atmosphere. The rate and extent of oxidation may be controlled by at least partially covering the metal with occlusive dressings (i.e., absorbent dressings), or chemically treating the metal with catalysts.

The oxidation step may be terminated simply by drying the oxidized metal and coating the oxidized metal with an at least partially transparent sealant to prevent further oxidation. In a particularly preferred variation, metal particles, such as iron powder, are sprinkled in thin layers on a relatively smooth substrate of plaster of Paris which may be reinforced with a fabric matrix. During oxidation of the metal particles, portions of the surface may be covered with wet paper towels (occlusive dressings). The wet paper towels keep the contacted portions wet and speed the oxidation in those portions of the article.

A preferred base comprises a sheet or slab of polymeric foam, preferably a closed cell polystyrene foam.

The base is preferably covered with a porous matrix which serves as a substrate for the oxidizable metal particles used to form the decorative effect. The porous matrix may advantageously comprise plaster of Paris, which has the characteristic of providing a smooth but porous surface to which the oxidizable metal may cohere.

The base with porous matrix is next provided with a layer of oxidizable metal for forming the decorative surface of the article. The layer of oxidizable metal may comprise any of a large number of different forms of metal, such as powder, shavings, turnings, wire, fragments, steel wool, screening, sintered particles, etc. However, it is preferred that the layer of oxidizable metal be relatively thin so that it will adhere or cohere rather closely to the substrate and provide optimum light reflection to enhance color contrasts. Layer(s) 10 mils or less are preferred.

A particularly suitable oxidizable metal for use in this process is commercially available iron powder which comprises 98% iron and trace amounts of various impurities. Whereas pure iron corrodes slowly, it is believed that impurities in the iron accelerate the corrosion process. Similarly, since oxidation or rusting occurs slowly at room temperature, an electrolyte such as sodium chloride or mineral acids such as hydrochloric or sulfuric acid may be used to accelerate the corrosion. Dust particles and other pollutants in the atmosphere also greatly enhance the oxidation process. As the rusting or oxidation of the iron powder proceeds, oxide forms on the surface of the particles resulting in bridging from one particle to the next. In this manner a lattice work is established which enhances the strength of the metal layer as well as the cohesion of the metal layer to the substrate.

Preferably this oxidation and bridging is allowed to proceed for 10 or 20 min before applying any occlusive dressing over the iron. By overlapping the occlusive dressings or having the dressings not touch in certain areas, the oxidation process is caused to proceed at different rates. In addition, the occlusive dressings affect the texture of the surface of the metal layer. Thus, the weight of the water applied on top of the occlusive dressings has a tendency to mat down the underlying metal surface making it smoother in texture than portions which are not contacted by occlusive dressings.

Where the occlusive dressing is matted flat against the iron particles, the rust assumes a lighter color and also a smoother, nontextured finish. By overlapping occlusive dressings, one can establish a rhythmic harmony of color tone and variations in the oxidation color spectrum including reds, oranges and yellows. Black, green and dark brown colors, which are established early in the rusting process, can also be achieved.

Wet Impact Plating

One problem associated with metal surface treating processes is the pollution which is caused by the effluent streams containing unused particles of plating metals, acid and other chemicals suspended or dissolved in the process liquids which are discarded. Another problem heretofore associated with impact plating processes has been the great volume of fresh water required to conduct the required cleanings, rinsings, etc.

L. Coch; U.S. Patent 4,062,990; December 13, 1977; assigned to Walders Kohinoor, Inc. is concerned with providing a cyclic process of the surface treatment of metal parts in aqueous media whereby the discharge of pollutant-bearing effluents is substantially reduced or eliminated.

The process is suitable for use with either a continuous wet impact plating process, in which the cleaning step, the preplating step, and the principal plating step are done in sequence with no intermediate rinsing; or with an interrupted wet impact plating process in which rinsing occurs between any of the steps.

The process is particularly advantageously applicable as an interrupted one, in which the impacting media are cleaned after each plating step. As more fully described in U.S. Patent 3,690,935, this known process has characteristically comprised:

(a) Admixing the steel parts to be plated with impacting media and an aqueous acidic cleaner such as sulfuric acid (which again usually contains a pickling inhibitor such as dibutylthiourea) in an agitatable container;

(b) Pickling the parts by agitating the container for several minutes;

(c) Adding an aqueous solution of a water-soluble copper salt and inorganic acid, e.g., sulfuric acid, such that copper metal plates out onto the metal articles with continued agitation;

(d) Draining the solution from the agitatable container as a nonreusable effluent after several minutes of additional agitation;

(e) Thoroughly rinsing the impacting media and metal parts and draining off the rinse water as a nonreusable effluent;

(f) Adding the zinc or cadmium and conventional plating accelerator or promoter chemical, e.g., stannous chloride or sulfate;

(g) Agitating the container and contents for a suitable additional period, e.g., 30 to 90 minutes, until the parts are plated;

(h) Opening the container at this point and draining off the liquids as nonreusable effluent;

(i) Separating the parts from the impacting media;

(j) Rinsing the parts several times, with each batch of rinse water being allowed to flow into a discharge basin and discarded at least to a major extent; and

(k) Loading the media back into the container in preparation for the next load to be plated.

In applying the process to the preferred interrupted impact plating process, the cleaning-coppering effluent is not discarded as described in step (d) above, but is caught in a suitable secondary container as the agitatable container is emptied and the effluent is pumped into its own suitable storage tank. This storage tank is constructed in such a manner that particulate material, such as any particles of zinc or cadmium which may be in suspension in the effluent, may precipitate spontaneously to the bottom of the tank and clear liquid may be drawn off from a decanting spigot located a suitable distance above the bottom of the tank. The tank is also constructed so that it has means for removal of the solid particles which accumulate by precipitation at the bottom of the tank.

The plating effluents as described in step (h) and the rinsing effluents described in steps (e) and (j) are each individually caught in a secondary container and each pumped into its separate storage tank of the design stated above, thus keeping each of the plating and rinsing effluents segregated from each other.

The cleaning-coppering effluent which is drained from one cycle and caught and stored is recycled by being reintroduced into the agitatable container and used as the pickling solution in step (a) in a later cycle. Some new acidic cleaner or pickling solution must be periodically added to the cleaning solution to make up dragout losses and to insure proper cleaning. The process thus reduces by many orders of magnitude the volume of fluid to be subjected to rectification procedures. By allowing virtually perpetual recycling of all process fluids, it completely avoids the necessity for conventional antipollution procedures because no part of such fluids ever needs to be run off into the sewer. Nevertheless, excellent plating quality is obtained.

Another known variation of the interrupted wet process involves the use of a perforate agitatable container. This comprises:

(1) Placing the parts to be plated in an agitatable perforate container;

(2) Lowering the entire container into an acid bath;

(3) Agitating the container;

(4) Removing the container from the acid bath;

(5) Lowering the container into a rinse bath and agitating;

(6) Removing the container from the rinse;

(7) Lowering the container into a bath of a water-soluble copper salt and an inorganic acid;

(8) Agitating the container for several minutes;

(9) Removing the container from the copper bath;

(10) Lowering the container into the rinse bath and agitating;

(11) Removing the metal parts from the perforate container;

(12) Placing the metal parts, impacting media, promoter chemical, zinc or cadmium, and water into an imperforate agitatable plating container;

(13) Agitating the container for a suitable period until the parts are plated;

(14) Opening the container and draining off the liquid as nonreusable effluent;

(15) Separating the parts from the media;

(16) Rinsing the parts several times, with the rinse water flowing into the discharge basin as effluent; and

(17) Reloading the media into the container for the next cycle.

After several cycles, an acid cleaning step (18) is generally required for the media, the cleaning being done by placing the media in an agitatable container along with a strong acid cleaning solution and agitating the mass for a suitable period. In prior art, the acid cleaning solution was drained off as nonreusable effluent after cleaning.

In applying the process to this version of the interrupted process, the plating effluent drained off in step (14), the rinsing effluent drained off in step (6), and the media cleaning effluent drained off in step (18) are caught in secondary containers and pumped into suitable storage tanks, all constructed in such a manner that particulate matter such as metallic particles of zinc or cadmium which are in suspension in the effluent may precipitate out spontaneously to the bottom of the container and clear liquid may be drawn off from a decanting spigot located a suitable distance from the floor of the container. The containers are also constructed such that they have a method of removal and reuse for the accumulation of solid particles which accumulate by precipitation at the bottom of the containers.

Plating and rinsing effluents are rectified by reversing the pH to about 8.5 by the addition of a base such as NaOH, which will precipitate the dissolved metals as an insoluble precipitate, and then separating the precipitate from the solution. After rectification this solution is returned to a pH of 6 or less by the addition of the conventional inhibited acid solution, e.g., sulfuric acid, and may then again be used as the liquid component of the plating solution for many more cycles, until the rectification process is repeated.

The acid stripping effluents are recycled in the interrupted wet process.

Sulfide Stain Resistant Finish for Tinplate

Untreated tinplate develops a dark irregular stain when in contact with a range of sulfur-bearing natural products under the conditions used in food processing. Sulfur-bearing proteins are present in many foods which are preserved in cans, and after processing the interior surface of the can is often discolored to a marked degree due to the formation of metal sulfides, while the food in contact with such discolored areas may itself become stained. The stain appears to be harmless, but is objectionable and should be avoided.

Protective coatings utilizing chromic acid have been used. In view of the toxic nature of chromium(VI) compounds, it would be preferable to use an alternative, nontoxic material to produce a stain resistant finish on tinplate.

According to *P.J. Heyes; U.S. Patent 4,294,627; October 13, 1981; assigned to Metal Box Ltd., England* a metal surface, particularly a tinplate surface of a can for canned food, is treated by contact with a solution containing a zirconium compound, particularly ammonium zirconium carbonate or zirconium acetate. An inorganic salt may also be present in the solution. The concentration of the zirconium compound, calculated as ZrO_2, is between 0.1 and 10% w/w. The surface is thereafter heated to between 20° and 300°C until it is dry. The surface may be cleaned prior to contact with the solution. In this way a coating is provided on the surface which improves the stain resistance of the surface.

Example 1: To establish the conditions under which a sulfide stain resistant finish can be produced on drawn tinplate using ammonium zirconium carbonate (AZC), the following experiments were made.

Unwashed, drawn and wall-ironed (DWI) can sections were solvent cleaned by immersion in butyl Cellosolve, followed by washing in hot (>90°C) 25% Decon 90 solution and a distilled water rinse. The sections were then immersed in stabilized AZC solutions (0.002 to 20% w/w ZrO_2) for a few seconds. The specimens were dried in an oven at 100°C. Similarly cleaned sections were briefly immersed in a dilute AZC solution (0.5% w/w ZrO_2) and dried at temperatures ranging from 20° to 300°C.

The samples were subjected to the sulfide staining test in which they were placed in a dried pea and brine staining medium and heated to 121°C for 1 hour in a pressure cooker. They were then assessed visually. Some samples were examined in a scanning electron microscope using an x-ray analyzer.

Concentration of AZC		Oven	
As Bacote 20	As ZrO_2	Temperature	Sulfide
(%)		(°C)	Stain*
100	20	100	3-4
50	10	100	3-4
10	2	100	1
5	1	100	1
2.5	0.5	100	1
1	0.2	100	2-3
0.5	0.1	100	3-4
0.1	0.02	100	4-5
0.01	0.002	100	4-5
2.5	0.5	20	2
2.5	0.5	50	1
2.5	0.5	150	1-2
2.5	0.5	200	1-2
2.5	0.5	250	1-2
2.5	0.5	300	1-2
Untreated DWI samples	—	—	5

*1 is none, 2 is slight, 3 is borderline acceptability, and 5 is severe (as received samples).

It is thus apparent that adequate sulfide stain resistance can be achieved by briefly immersing clean tinplate in an AZC solution of ZrO_2 content between 0.1 and 10% and drying at temperatures between 20° and 300°C. The ZrO_2 content is preferably between 0.2 and 2% w/w.

The results indicate that a protective film was produced on the clean surface when a dilute AZC solution was dried on the drawn tinplate. The film once formed was not destroyed by rinsing the can in water and redrying. The stain resistance did not arise either when undiluted AZC was dried or when the dilute AZC was not dried before staining. Of the specimens examined in the scanning electron microscope after staining, only those that did not stain had detectable zirconium on the surface.

Example 2: Conventionally, tinplate when received from the manufacturers has already been subjected to a passivation treatment, and in the previous example such passivated tinplate was used, the drawing processes described destroying the effectiveness of any coatings on the tinplate.

INVENTOR INDEX

U.S. PATENT NUMBER INDEX

Copies of U.S. patents are easily obtained
from the U.S. Patent Office at 50¢ a copy.

COMPANY INDEX

The company names listed below are given exactly as they appear in the patents, despite name changes, mergers and acquisitions which have, at times, resulted in the revision of a company name.

NOTICE